PEDOE'S HIGHER MATHEMATICS

VOLUME II

HIGHER TECHNICIAN SERIES

General Editor

M. G. Page

B.Sc.(Hons.)(Eng.), C.Eng., F.I.Mech.E., F.I.Prod.E., M.B.I.M., F.S.S.

Head of Department of Production Engineering
The Polytechnic, Wolverhampton

PEDOE'S HIGHER
MATHEMATICS

VOLUME TWO

J. PEDOE

M.A. (Cantab.), B.Sc. (Lond.), F.S.S.
Formerly Senior Lecturer, the College of Technology, Leicester

HODDER AND STOUGHTON
LONDON SYDNEY AUCKLAND TORONTO

" I made myself familiar with the elements of
mathematics together with the principles of differential and integral calculus. In doing so I had
the good fortune of hitting on books which were
not too particular in their logical rigour but which
made up for this by permitting the main thoughts
to stand out clearly."

A. Einstein, *Autobiographical Notes.*

First printed 1954
Second impression (with corrections) 1957
Third impression (with corrections) 1961
Fourth impression 1965
Fifth impression 1967
Second edition 1971
Third impression 1977
Fourth impression 1978
Fifth impression published as Pedoe's Higher Mathematics
Volume 2 1985

Second edition copyright © 1970
J. Pedoe

ISBN 0 340 06715 2

*Printed in Great Britain for
Hodder and Stoughton Educational,
a division of Hodder and Stoughton Ltd,
Mill Road, Dunton Green, Sevenoaks, Kent
by Page Bros (Norwich) Ltd*

GENERAL EDITOR'S FOREWORD

THE rate of technological change is continually increasing, and consequently the pattern of courses available in further and higher education tends to have a similarly increasing rate of change. This could lead to the cessation of certain well-known courses concurrently with the introduction of entirely new courses. In other cases, new courses may be only a re-arrangement of existing syllabus material, but the approach to the individual topics alters due to different learning techniques and/or the increasing attention being paid to industrial training.

It is becoming evident that the provision of textbooks for distinct stages of particular courses is not necessarily an ideal answer to the demands of education. Whilst in the engineering industry the distinctions between operatives, craftsmen, technicians and technologists can on certain occasions be clearly delineated, the dividing lines tend to become blurred when considering the provision of the raw material of learning. In other technologies the primary divisions themselves are by no means realistic.

The Higher Technician Series has been introduced to provide a suitable selection of textbooks for the entire area of further and higher education. Some may be particularly suitable for certain well established courses, but others may satisfy demands which arise from quite different courses of differing educational levels. It will be by no means unusual in the future for a particular textbook to be written to meet concurrently the requirements of both technological and technician studies.

This viewpoint is reflected in the titles occurring in the series. The lively atmosphere which now exists in further and higher education provides a remarkable challenge to publishers. This series, it is hoped, will competently meet that challenge.

M. G. PAGE

PREFACE

THIS book is a continuation of Volume I. The two together should cover the courses of most Higher National (H1 and H2) courses. They will also be found to provide an adequate course for such an examination as B.Sc.(Eng.) London (revised regulations), Parts I and II. They do not cover the whole course, but a student who has worked through them will find himself well prepared for a high proportion of the questions which appear in the examination papers.

Attention is drawn to the chapter on Laplace Transforms, which provides an alternative to the D (operator) method for solving linear differential equations with constant coefficients. After some nine years of teaching both methods to " Honours " engineers I can assure the student that this alternative method is well worth knowing. [Those too busy will at least, I hope, refer to 10.12 (i.e., section 12 of chapter 10) where it is shown that the H and δ functions can be used to advantage in the solution of problems on bending beams.] I have used the definition

$$\int_0^\infty e^{-st}F(t)dt \quad \text{and not} \quad p\int_0^\infty e^{-pt}F(t)dt$$

since the former provides an easier algebra. The parameter s is used and not p to remind the student that the results are not those associated with the name of Heaviside.

I am very grateful to the many busy experts who, through the publishers, provided me with detailed criticisms and suggestions for this volume. I am particularly indebted to Mr. F. Bowman, formerly Head of the Department of Mathematics, College of Technology, Manchester, Dr. E. T. Davies, Professor of Mathematics, University of Southampton, Dr. M. R. Gavin, M.B.E., Professor of Applied Electricity, University College of North Wales, Bangor, Mr. A. B. Greene, Senior Lecturer in the Department of Automobile Engineering, Loughborough College of Technology, Mr. C. G. Paradine, Senior Lecturer in the Department of Mathematics, Battersea Polytechnic, Mr. W. G. L. Sutton, Vice-Principal, Leeds College of Technology, and Mr. E. O. Taylor, Senior Lecturer in the Department of Electrical Engineering, Heriot-Watt College. They will, I trust, find that I have paid a good deal of attention to their suggestions, and if not all have been adopted, at least I have tried to make a maximum of the overall satisfaction they will feel subject to the (non-Lagrangian) condition of retaining the original purpose of the book. A colleague, Mr. D. J. Memory, B.Sc., read the proofs of the sections on electromagnetic waves and

Maxwell's equations. The two or three sentences interpolated in amplification of the text are due to him, and I am very grateful to him for his help.

J. PEDOE

Plymouth,
Leicester

PREFACE TO THE FOURTH EDITION

WHEN this textbook was first published in 1954, its title and content reflected the patterns of education of that time. Since that date the structure of education has undergone considerable changes, one change in particular being the entire concept of National Certificate courses.

The changing pattern of courses means that its title somewhat belies this book's immediate and future usage. The National Certificate Courses are now recognised as Technician Courses in their own right, and while this text book will be eminently suitable for those courses as long as they exist, it will also provide a high proportion of the subject matter for a considerable number of other courses, particularly Diploma Courses in Engineering, C.E.I. courses, and courses for Technological Degrees.

Thus at the present time, and in the future if the recommendations of the Haslegrave Report are implemented, this textbook will be used more frequently for courses other than National Certificates. The publishers, however, see no reason to change its title; the book has served education magnificently, and its enviable reputation is well deserved.

In this new edition the text has been revised to introduce metric units, with SI units as a first order of preference. The other revisions are of less significance, and it is a tribute to the original writings of the late Mr. Pedoe that the presentation remains virtually unchanged.

M. G. PAGE

CONTENTS

ix

LIST OF FORMULAE

(continued from Vol. I)

Euler's Formulae

$$\sin x = \frac{1}{2j}(e^{jx} - e^{-jx}). \qquad \cos x = \tfrac{1}{2}(e^{jx} + e^{-jx}).$$

$$\text{sh } jx = j \sin x. \qquad \sin jx = j \text{ sh } x.$$

$$\text{ch } jx = \cos x. \qquad \cos jx = \text{ch } x.$$

Fourier Series

$$f(x) = \tfrac{1}{2}a_0 + \sum_{n=1}^{\infty}(a_n \cos nx + b_n \sin nx) \quad [\text{period } 2\pi]$$

where for the range $-\pi$ to $+\pi$,

$$a_n = \frac{1}{\pi}\int_{-\pi}^{+\pi} f(x)\cos nx \, dx \quad (n = 0, 1, \ldots);$$

$$b_n = \frac{1}{\pi}\int_{-\pi}^{+\pi} f(x)\sin nx \, dx \quad (n = 1, 2, \ldots);$$

Cosine series if $f(x) = f(-x)$; Sine series if $f(x) = -f(-x)$.

Leibnitz's Theorem

$$D^n(uv) = uv_n + {}_nC_1 u_1 v_{n-1} + \ldots {}_nC_r u_r v_{n-r} + \ldots u_n v.$$

Taylor's Series

$$f(x + h) = f(x) + hf'(x) + \ldots + \frac{h^r}{r!}f^r(x) + \ldots$$

Integrals

$$\int \frac{f'(x)}{f(x)}\, dx = \log f(x) + C, \quad \int \frac{f'(x)}{\sqrt{[f(x)]}}\, dx = 2\sqrt{[f(x)]} + C.$$

$$\int \sqrt{(a^2 - x^2)}dx = \tfrac{1}{2}x\sqrt{(a^2 - x^2)} + \tfrac{1}{2}a^2 \sin^{-1}\frac{x}{a} + C.$$

$$\int \sqrt{(x^2 - a^2)}dx = \tfrac{1}{2}x\sqrt{(x^2 - a^2)} - \tfrac{1}{2}a^2 \log\{x + \sqrt{(x^2 - a^2)}\} + C.$$

$$\int \sqrt{(x^2 + a^2)}dx = \tfrac{1}{2}x\sqrt{(x^2 + a^2)} + \tfrac{1}{2}a^2 \log\{x + \sqrt{(x^2 + a^2)}\} + C.$$

$$\int \sec^3 x \, dx = \tfrac{1}{2}\sec x \tan x + \tfrac{1}{2}\log(\sec x + \tan x) + C.$$

$$\int_0^{\frac{1}{2}\pi} \sin^n x \, dx = \int_0^{\frac{1}{2}\pi} \cos^n x \, dx \quad (n \text{ a positive integer})$$

$$= \frac{1.\,3.\ \ldots\ (n-3)(n-1)}{2.\,4.\ \ldots\ \ (n-2)n} \cdot \frac{\pi}{2} \quad (\text{if } n \text{ is even})$$

$$= \frac{2.\,4.\ \ldots\ (n-3)(n-1)}{3.\,5.\ \ldots\ \ (n-2)n} \quad (\text{if } n \text{ is odd}).$$

(See 2.4 for general formula).

The Catenary

$$s = c \tan \theta \qquad\qquad T = wy$$

$$s = c \operatorname{sh}\left(\frac{x}{c}\right) \qquad\qquad x = c \log(\sec\theta + \tan\theta)$$

$$y = c \operatorname{ch}\left(\frac{x}{c}\right) \qquad\qquad y = c \sec\theta$$

$$y^2 = s^2 + c^2.$$

Statistics

$$\sigma = \sqrt{\left\{\frac{1}{n}\Sigma(x_r - \bar{x})^2\right\}} = \sqrt{\frac{1}{n}\left[\Sigma x_r^2 - \frac{1}{n}(\Sigma x_r)^2\right]}$$

$$\sigma^2 = S^2 - (\bar{x} - A)^2.$$

If $y = ax + b$, for n pairs (x_r, y_r) the " normal " equations are:

$$a\Sigma x_r^2 + b\Sigma x_r - \Sigma x_r y_r = 0$$

$$a\Sigma x_r + nb - \Sigma y_r = 0$$

$$r = \frac{\Sigma(x_r - \bar{x})(y_r - \bar{y})}{n\sigma_x\sigma_y} = \frac{\Sigma x_r y_r - n\bar{x}\bar{y}}{\sqrt{(\Sigma x_r^2 - n\bar{x}^2)(\Sigma y_r^2 - n\bar{y}^2)}}.$$

TABLE OF SOME LAPLACE TRANSFORMS

	Function. $F(t).$	Transform. $\displaystyle\int_0^\infty e^{-st}F(t)dt.$
1.	$1.$	$1/s.$
2.	$e^{at}.$	$1/(s-a).$
3.	$\sin at.$	$a/(s^2+a^2).$
4.	$\cos at.$	$s/(s^2+a^2).$
5.	$t.$	$1/s^2.$
6.	t^n (n a $+$ ve integer).	$n!/s^{n+1}.$
7.	$\sinh at.$	$a/(s^2-a^2).$
8.	$\cosh at.$	$s/(s^2-a^2).$
9.	$t\sin at.$	$2as/(s^2+a^2)^2.$
9(a).	$t\cos at.$	$(s^2-a^2)/(s^2+a^2)^2.$
10.	$\sin at - at\cos at.$	$2a^3/(s^2+a^2)^2.$
11.	$e^{-at}t^n.$	$n!/(s+a)^{n+1}.$
12.	$e^{-bt}\cos at.$	$(s+b)/[(s+b)^2+a^2].$
13.	$e^{-bt}\sin at.$	$a/[(s+b)^2+a^2].$
14.	$e^{-bt}\cosh at.$	$(s+b)/[(s+b)^2-a^2].$
15.	$e^{-bt}\sinh at.$	$a/[(s+b)^2-a^2].$
16.	$H(t),\ H(t-a).$	$1/s,\ e^{-as}/s.$
17.	$\delta(t),\ \delta(t-a).$	$1,\ e^{-as}.$

Some Theorems used in Laplace Transforms.

1. If $f(s) = L\{F(t)\}$ then $f(s+a) = L\{e^{-at}F(t)\}.$

2. $L\left\{\dfrac{dx}{dt}\right\} = s\bar{x} - x_0$

$L\left\{\dfrac{d^2x}{dt^2}\right\} = s^2\bar{x} - sx_0 - x_1$

$L\left\{\dfrac{d^3x}{dt^3}\right\} = s^3\bar{x} - s^2x_0 - sx_1 - x_2$

and generally

$L\left\{\dfrac{d^nx}{dt^n}\right\} = s^n\bar{x} - s^{n-1}x_0 - s^{n-2}x_1 \ldots - sx_{n-2} - x_{n-1}$

where $\bar{x}$ is the transform of x, and $x_r = \left(\dfrac{d^rx}{dt^r}\right)_{t=0}.$

3. $L\left\{\displaystyle\int_0^t F(t)dt\right\} = \dfrac{1}{s}f(s).$

4. $\int_0^x H(X - a)dX = (x - a)H(x - a).$

$$\int_0^x (X - a)^n H(X - a)dX = \frac{(x - a)^{n+1}}{n + 1} H(x - a).$$

5. $\int_0^\infty \delta(x)dx = 1.$

$$\int_0^x \delta(X - a)dX = H(x - a).$$

CHAPTER 1

INDETERMINATE FORMS AND LIMITS. CURVATURE

"Having to evaluate an expression of the form 0/0 is the penalty you have to pay for over-idealization and is an indication that the treatment of the problem is at variance with nature."

Professor Sir C. Inglis, *Applied Mechanics for Engineers.*

1.1. The Expansion Method

In Vol. I (26.13) we evaluated expressions of the form 0/0. This will now be dealt with further.

Example 1. Obtain the expansion of tan θ in powers of θ as far as the term in θ^5.

A column of length l has a vertical load P and a horizontal load F at the top. The transverse deflection is

$$\delta = \frac{Fl}{P}\left[\frac{\tan ml}{ml} - 1\right],$$

where $m^2 = P/EI$. Show that as $P \longrightarrow 0$, $\delta \longrightarrow Fl^3/3EI$ and that a small value of P leads to an increase of this by about 40 Pl^2/EI per cent.   [L.U.]

We may consider that we have found, as will be done later, the deflection at the end of a vertical cantilever column subjected at the free end to a horizontal load F and a vertical load P. As a check on the answer we could put $P = 0$ in it and expect to obtain the usual formula $Fl^3/3EI$ for the deflection at the end of a light cantilever subject only to the load F. We obtain, however,

$$\frac{Fl}{0}\left[\frac{0}{0} - 1\right]$$

and must "evaluate" this indeterminate form.

Since $\tan \theta = \dfrac{\sin \theta}{\cos \theta}$

$$= \frac{\theta - \dfrac{1}{3!}\theta^3 + \dfrac{1}{5!}\theta^5 - \cdots}{1 - \dfrac{1}{2!}\theta^2 + \dfrac{1}{4!}\theta^4 - \cdots}$$

$$= \theta + \tfrac{1}{3}\theta^3 + \tfrac{2}{15}\theta^5 + \cdots \quad \text{(by long division)}$$

Using this expression with $ml = \theta$,

$$\delta = \frac{Fl}{P}\left[\frac{ml + \tfrac{1}{3}m^3l^3 + \tfrac{2}{15}m^5l^5 + \cdots}{ml} - 1\right]$$

$$= \frac{Fl}{P}\left[\tfrac{1}{3}m^2l^2 + \tfrac{2}{15}m^4l^4 + \cdots\right]$$

$$= \frac{Fl}{P}\left[\frac{1}{3}l^2\frac{P}{EI} + \frac{2}{15}l^4\frac{P^2}{E^2I^2} + \cdots\right] \quad (m^2 = P/EI)$$

$$= Fl\left[\frac{1}{3}\frac{l^2}{EI} + \frac{2}{15}\frac{l^4 P}{E^2I^2} + \text{further positive powers of P}\right].$$

1

We may now put P = 0 and obtain

$$\delta = \frac{Fl^3}{3EI} \quad \cdot \quad \cdot \quad \cdot \quad \cdot \quad \cdot \quad \cdot \quad \cdot \quad \cdot \quad (1)$$

If P is so small that powers of P above the first may be neglected, then the increase in the deflection is

$$\frac{2}{15} \frac{FPl^5}{E^2I^2}$$

which is $40Pl^2/EI$ per cent. of the deflection in (1).

1.2. l'Hopital's Rule

The expansion method will solve most problems on indeterminate forms, but an easier method is available in certain cases. Suppose

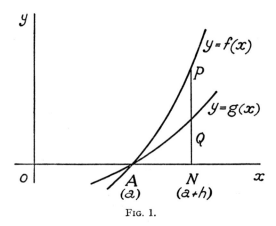

Fig. 1.

we wish to evaluate the ratio $f(x)/g(x)$ at the point $x = a$, but find $f(a) = 0 = g(a)$ so that the ratio becomes $0/0$. If the figure represents the curves $y = f(x)$ and $y = g(x)$, both of which pass through the point $A(a, 0)$ since $f(a) = 0 = g(a)$, then at a near point $N(a + h, 0)$

$$\frac{f(a + h)}{g(a + h)} = \frac{PN}{QN} = \frac{PN}{AN} \bigg/ \frac{QN}{AN} = \frac{\tan PAN}{\tan QAN}.$$

As N approaches A, tan PAN approaches $f'(a)$ and tan QAN approaches $g'(a)$. Therefore

$$\operatorname*{Lt.}_{x \to a} \frac{f(x)}{g(x)} = \operatorname*{Lt.}_{h \to 0} \frac{f(a + h)}{g(a + h)} = \frac{f'(a)}{g'(a)}$$

(provided that $f'(a)$ and $g'(a)$ are not both zero, in which case the value of the limit is still indeterminate).

Example 2. Find $\operatorname*{Lt.}_{x \to 1} \dfrac{x^3 - x^2 + 3x - 3}{x^2 + x - 2}.$

Since this is of the form $0/0$ for $x = 1$, we differentiate numerator and denominator to obtain

$$\frac{3x^2 - 2x + 3}{2x + 1}.$$

For $x = 1$ this gives $4/3$, which is the value of the limit. In this case the result is easily checked, since

$$x^3 - x^2 + 3x - 3 = (x - 1)(x^2 + 3),$$
$$x^2 + x - 2 = (x - 1)(x + 2).$$

1.3. It may happen that $f'(a) = 0 = g'(a)$, *i.e.*, the curves both touch the x axis at $x = a$. In this case assume that the figure above represents not y but $\dfrac{dy}{dx}$ drawn against x. The above proof now gives

$$f''(a)/g''(a)$$

as the value of the limit, provided again that both are not zero. Proceeding in this way, we find

$$\operatorname*{Lt.}_{x \to a} \frac{f(x)}{g(x)} = \frac{f^n(a)}{g^n(a)},$$

where n is the order of the lowest derivative obtained for which at least one of $f^n(a)$ or $g^n(a)$ is not zero.

It can be shown that an exactly similar process will evaluate the form ∞/∞.

Example 3. Find $\operatorname*{Lt.}_{x \to 0} \dfrac{x - \sin x}{x - \tan x}$.

(1) For $x = 0$ this is of the form $0/0$.
(2) Differentiate each expression in the ratio to obtain $(1 - \cos x)/(1 - \sec^2 x)$. This is again of the form $0/0$ for $x = 0$.
(3) Repeat the differentiation and obtain $\sin x/- 2 \sec^2 x \tan x$. This is again of the form $0/0$.
(4) Repeat the differentiation and obtain $\cos x/- 2(\sec^4 x + 2 \sec^2 x \tan^2 x)$. For $x = 0$ this gives the value $- 1/2$.

We therefore have

$$\operatorname*{Lt.}_{x \to 0} \frac{x - \sin x}{x - \tan x} = - \tfrac{1}{2}.$$

At stage (3) we could have changed the ratio to read $- \cos^3 x/2$ and obtained the value $- \tfrac{1}{2}$ for $x = 0$ without needing stage (4).

Note : The ratio evaluated by this method must be indeterminate, or the wrong result will be obtained. Thus

$$\operatorname*{Lt.}_{x \to 1} \frac{x^3 + 3x + 2}{x^2 - x - 2} = \frac{1 + 3 + 2}{1 - 1 - 2} = - 3$$

and is not of the indeterminate form. If this is not noticed and l'Hopital's Rule is used the value

$$\left[\frac{3x^2 + 3}{2x - 1}\right]_{x=1} = 6$$

results, and this is incorrect.

1.4. Other Indeterminate Forms

Other forms often occur, and each must be treated on its merits. In many cases a simple substitution or change such as considering the logarithm of the expression will reduce the form to the above type.

Example 4. Show that the area for x positive between the line $y = 1$ and the curve $y = \tanh x$ is $\log_e 2$. [L.U.]

The area is

$$\int_0^\infty (1 - \operatorname{th} x)dx = \left[x - \log \operatorname{ch} x \right]_0^\infty$$

We appear to have the form $\infty - \infty$, but

$$x - \log \operatorname{ch} x = \log e^x - \log \operatorname{ch} x$$
$$= \log (2e^x)/(e^x + e^{-x})$$
$$= \log \{2/(1 + e^{-2x})\}$$

therefore

$$\operatorname*{Lt.}_{x \to \infty} (x - \log \operatorname{ch} x) = \operatorname*{Lt.}_{x \to \infty} \log \{2/(1 + e^{-2x})\}$$
$$= \underline{\log_e 2.}$$

EXERCISE 1

Evaluate:

1. $\operatorname*{Lt.}_{x \to 0} \dfrac{\sin^2 x}{x}$.

2. $\operatorname*{Lt.}_{x \to 0} \dfrac{\sin^2 x}{x^2}$.

3. $\operatorname*{Lt.}_{x \to 0} \dfrac{\log (1 + x)}{x}$.

4. $\operatorname*{Lt.}_{x \to \frac{1}{2}\pi} (\sec x - \tan x)$.

5. $\operatorname*{Lt.}_{x \to \infty} \dfrac{x^2 - 3x + 2}{2x^2 + x + 100}$. $\left(\text{put } x = \dfrac{1}{t}\right)$.

6. $\operatorname*{Lt.}_{x \to 0} \dfrac{\sec x - 1}{x \sin x}$.

7. $\operatorname*{Lt.}_{x \to \frac{1}{2}} \dfrac{\log 2x}{2x - 1}$.

8. $\operatorname*{Lt.}_{x \to 0} \dfrac{1 - 2\sin^2 x - \cos^3 x}{5x^2}$.

9. $\operatorname*{Lt.}_{P \to 0} \dfrac{W}{2P} \left(\dfrac{\sin mx}{m \cos \frac{1}{2}ml} - x \right)$, where $m^2 = P/\mathrm{EI}$.

10. $\operatorname*{Lt.}_{p \to n} \dfrac{\cos pt - \cos nt}{n^2 - p^2}$

11. Show that:

(a) $\operatorname*{Lt.}_{h \to 0} \left[\dfrac{\sin (x + h) + \sin (x - h) - 2\sin x}{h^2} \right] = -\sin x$.

(b) $\operatorname*{Lt.}_{x \to \infty} \left[\log (1 + ax) - 2\log x + \log (a + x) \right] = \log a$.

12. Obtain in any manner the series for $\sec \theta$ as far as θ^4.

If a beam of length l is subject to a uniformly distributed load of intensity w and to an end thrust P, the central deflection is given by

$$\frac{w\mathrm{EI}}{\mathrm{P}^2} \left[\sec \frac{ml}{2} - 1 \right] - \frac{wl^2}{8\mathrm{P}},$$

where $m^2 = \mathrm{P/EI}$. Show that as $\mathrm{P} \to 0$ this tends to the limit $5wl^4/384\mathrm{EI}$.

[L.U.]

13. Evaluate: (a) $\operatorname*{Lt.}_{h \to 0} \dfrac{f(c + 2h) - f(c - 2h)}{h}$;

(b) $\operatorname*{Lt.}_{h \to 0} \dfrac{f(x + 2h) - 2f(x + h) + f(x)}{h^2}$.

14. Find the limit of $x\{\sqrt{(x^2 + a^2)} - x\}$ as $x \to \infty$.

15. Find the limits as $x \to 0$ of

$$\frac{1}{x^3}\left(\operatorname{cosec} x - \frac{1}{x} - \frac{x}{6}\right), \quad \frac{1}{x^3}\left(\cot x - \frac{1}{x} + \frac{x}{3}\right).$$

16. Show :

(a) $\operatorname*{Lt.}_{x \to 0} \dfrac{\sin x \cdot \sin^{-1}x - x^2}{x^6} = \dfrac{1}{18}$;

(b) $\operatorname*{Lt.}_{x \to 0} \dfrac{\tan x \cdot \tan^{-1}x - x^2}{x^6} = \dfrac{2}{9}$;

(c) $\operatorname*{Lt.}_{x \to 0} \dfrac{d}{dx}\left(\dfrac{1}{\sin x} - \dfrac{1}{x}\right) = \dfrac{1}{6}$.

17. The permeance of a trapezoidal slot is given by

$$\frac{h}{(w_1 + w_2)^2}\left[w_1 - \frac{w_1 - w_2}{2} - \frac{w_1{}^3}{(w_1 - w_2)^2} - \frac{w_1{}^2}{2(w_1 - w_2)} - \frac{w_1{}^4}{(w_1 - w_2)^3}\log\frac{w_2}{w_1}\right]$$

where w_1, w_2 are the slot widths and h is the slot depth.

If we let $w_1 = w_2 = w$ show that the above expression is consistent with the value $h/3w$ for the permeance of a rectangular slot.

18 Show that

$$x = \frac{cn}{n^2 - p^2}(n \sin pt - p \sin nt)$$

satisfies the differential equation

$$\frac{d^2x}{dt^2} + n^2x = n^2c \sin pt.$$

Find the value of x when $p \to n$, i.e., when the period of the applied force equals the natural period of the spring.

19. The magnetic field strength at the centre O of a multilayer solenoid of T turns is given by

$$H_0 = \frac{2\pi IT}{10(R - r)} \log \left(\frac{R + \sqrt{(R^2 + \frac{1}{4}L^2)}}{r + \sqrt{(r^2 + \frac{1}{4}L^2)}}\right).$$

Show that if $r \to R$, T, L, I being constant then

$$H_0 = \frac{4\pi IT}{10L}\left(1 + \frac{4R^2}{L^2}\right)^{-\frac{1}{2}}$$

and that provided $L > 30R$ then the usual formula for a " long " solenoid

$$H_0 = \frac{4\pi IT}{10L}$$

is true to within $\frac{1}{4}$ per cent.

20. A flat plate spring of uniform thickness t is fixed at one end (horizontally) and loaded at the free end with a weight W   The deflection at the free end is given as

$$\frac{12Wl^3}{Ebt^3}\left\{\frac{b}{2c} - \frac{b^2}{c^2} + \frac{b^3}{c^3}\log\left(\frac{b + c}{b}\right)\right\}$$

where b is the width of the free end and $b + c$ that of the fixed end. Show that as $c \to 0$ the deflection becomes that for a cantilever of constant width $4Wl^3/Ebt^3$.

21. The deflection at the end of a cantilever subject to a compressive force P at the free end is given as

$$\delta = \frac{wEI}{P^2}\left[1 - \frac{\theta^2}{2} - \sec \theta + \theta \tan \theta\right]$$

where $\theta = l\sqrt{(P/EI)}$. Show that as $P \longrightarrow 0$ the deflection assumes the usual value $wl^4/8EI$ for a cantilever sagging under its own (uniform) weight.

22. The polar moment of inertia of a regular polygon of n sides each of length a and area A is

$$\frac{Aa^2}{24}\left[\frac{2 + \cos (2\pi/n)}{\sin^2 (\pi/n)}\right].$$

Show that when $n \longrightarrow \infty$ the formula becomes $\frac{1}{2}Ar^2$, the polar moment of inertia for a circular area A of radius r.

23. Find the area bounded by the curve $y = -\log (1 - x^2)$ the x axis and the ordinate $x = x_1$ ($0 < x_1 < 1$). Show that as $x_1 \longrightarrow 1$ this area tends to the limit $2 - 2 \log 2$. [L.U.]

1.5. Radius of Curvature

In Vol. I (23) we obtained the formula

$$\rho = (1 + y_1^2)^{3/2}/y_2 \quad \cdot \quad \cdot \quad \cdot \quad \cdot \quad \cdot \quad (1)$$

for the radius of curvature at any point (x, y) on a curve and evaluated it for curves given by the equation $y = f(x)$. The method can easily be extended to curves given by parametric equations.

Example 5. Find the radius of curvature at the point θ on the curve $x = a \cos^3 \theta$, $y = a \sin^3 \theta$, and show that its maximum value is $3a/2$. [L.U.]

$$x = a \cos^3 \theta \qquad\qquad y = a \sin^3 \theta$$

$$\frac{dx}{d\theta} = -3a \cos^2 \theta \sin \theta \qquad\qquad \frac{dy}{d\theta} = 3a \sin^2 \theta \cos \theta$$

therefore

$$\frac{dy}{dx} = \frac{dy}{d\theta} \Big/ \frac{dx}{d\theta} = \frac{3a \sin^2 \theta \cos \theta}{-3a \cos^2 \theta \sin \theta}$$

$$= -\tan \theta$$

also

$$\frac{d^2y}{dx^2} = \frac{d}{dx}\left(\frac{dy}{dx}\right) = \frac{d}{d\theta}\left(\frac{dy}{dx}\right) \cdot \frac{d\theta}{dx}$$

$$= \frac{d}{d\theta}(-\tan \theta) \cdot \left(\frac{-1}{3a \cos^2 \theta \sin \theta}\right)$$

$$= 1/3a \cos^4 \theta \sin \theta.$$

By (1)

$$\rho = (1 + \tan^2 \theta)^{3/2} \cdot 3a \cos^4 \theta \sin \theta$$

$$= 3a \cos \theta \sin \theta = \frac{3a}{2} \sin 2\theta.$$

The maximum value of ρ is then $\frac{3a}{2}$, since the maximum value of $\sin 2\theta$ is 1.

$\left(\text{When using this method it is essential to simplify the value of } \dfrac{dy}{dx} \text{ in terms of } \theta\right.$
as much as possible before finding $\left. \dfrac{d^2y}{dx^2}.\right)$

1.6. Centre of Curvature

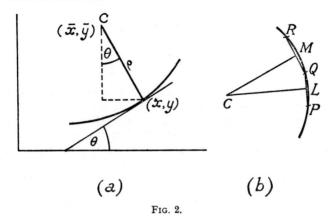

FIG. 2.

The centre of curvature at any point (x, y) of a curve clearly lies on the normal at this point to the curve and on the concave side of it. If $(\bar{x}, \bar{y})$ are the co-ordinates of the point, the figure shows that

$$\bar{x} = x - \rho \sin \theta, \quad \bar{y} = y + \rho \cos \theta$$

where we must give ρ the sign obtained from the formula

$$\rho = (1 + y_1^2)^{3/2}/y_2$$

i.e., the sign of y_2 for if y_2 is negative, the centre will not lie as above, but on the other side of the tangent. Using the values for $\sin \theta$, $\cos \theta$ [Vol. I (22.5)]

$$\sin \theta = \frac{dy}{ds} \qquad\qquad \cos \theta = \frac{dx}{ds}$$

$$= \frac{y_1}{\sqrt{[1 + y_1^2]}}, \qquad\qquad = \frac{1}{\sqrt{[1 + y_1^2]}},$$

and for ρ, the values of $\bar{x}$, $\bar{y}$ can be evaluated in a given case.

Example 6. Find the centre of curvature at the origin on the curve $y = x - 2x^2 + x^3$.

$$\frac{dy}{dx} = 1 - 4x + 3x^2 = 1 \text{ at } (0, 0)$$

$$\frac{d^2y}{dx^2} = -4 + 6x = -4 \text{ at } (0, 0)$$

therefore

$$\rho = (1 + 1)^{3/2}/(-4) = -2^{3/2}/4$$

and

$$\sin \theta = \frac{1}{\sqrt{(1 + 1)}}, \quad \cos \theta = \frac{1}{\sqrt{(1 + 1)}},$$

$$= \frac{1}{\sqrt{2}}, \qquad\qquad = \frac{1}{\sqrt{2}},$$

therefore
$$\bar{x} = x - \rho \sin \theta$$
$$= + \frac{2^{3/2}}{4} \cdot \frac{1}{\sqrt{2}} = \frac{1}{2}$$
$$\bar{y} = y + \rho \cos \theta$$
$$= - \frac{2^{3/2}}{4} \cdot \frac{1}{\sqrt{2}} = -\frac{1}{2}.$$

1.7. An easier way is as follows. The normals at two near points will intersect at a point which in the limit is the centre of curvature. This is clear from the figure Fig. 2(b), where LC is the perpendicular bisector of PQ and MC of QR, so that C is the centre of the circle through P, Q and R. As P tends to Q and R tends to Q, PQ, RQ become tangents, CL, CM become normals, and C, the centre of the circle which in the limit is the circle of curvature at Q, becomes the limiting point of intersection of consecutive normals.

If working with the normals in terms of a parameter t

$$N(x, y, t) = 0, \qquad N(x, y, t + \Delta t) = 0$$

are two consecutive normals, then for their point of intersection

$$N(x, y, t + \Delta t) - N(x, y, t) = 0,$$

and dividing by Δt and proceeding to the limit

$$\frac{\partial N}{\partial t} = 0,$$

also holds for their point of intersection. We have then two equations

$$N = 0, \qquad \frac{\partial N}{\partial t} = 0$$

from which to find $\bar{x}$, $\bar{y}$ the co-ordinates of the centre of curvature in terms of a parameter t.

Example 7. Find the centre of curvature at any point α on the ellipse $x = a \cos \alpha$, $y = b \sin \alpha$.

Since
$$\frac{dy}{dx} = \frac{b \cos \alpha}{- a \sin \alpha}$$

the equation to the normal is

$$y - b \sin \alpha = \frac{a \sin \alpha}{b \cos \alpha} (x - a \cos \alpha)$$

which simplifies to

$$ax \sec \alpha - by \operatorname{cosec} \alpha = a^2 - b^2 \quad . \quad . \quad . \quad . \quad (1)$$

Differentiating partially with respect to α

$$ax \sec \alpha \tan \alpha + by \operatorname{cosec} \alpha \cot \alpha = 0 . \quad . \quad . \quad . \quad (2)$$

From (1) and (2)

$$x = \frac{(a^2 - b^2)}{a} \cos^3 \alpha, \qquad y = - \frac{(a^2 - b^2)}{b} \sin^3 \alpha,$$

giving the co-ordinates of the centre of curvature. We can often use a "trick" method in (1). Rewriting it as

$$ax - by \frac{\cos \alpha}{\sin \alpha} = (a^2 - b^2) \cos \alpha,$$

in partial differentiation with respect to α the term ax becomes zero, and we obtain immediately the value of y as

$$- by \left(- \operatorname{cosec}^2 \alpha \right) = - (a^2 - b^2) \sin \alpha$$

or

$$y = - \left(\frac{a^2 - b^2}{b} \right) \sin^3 \alpha.$$

Similarly, rewriting (1) as

$$ax \frac{\sin \alpha}{\cos \alpha} - by = (a^2 - b^2) \sin \alpha,$$

partial differentiation this time gives

$$ax \sec^2 \alpha = (a^2 - b^2) \cos \alpha$$

or

$$x = \left(\frac{a^2 - b^2}{a} \right) \cos^3 \alpha.$$

1.8. The Evolute

As a point describes a curve, the corresponding centre of curvature will describe a curve called the *evolute* of the original curve. Thus the evolute of the ellipse is obtained by eliminating α from the equations

$$x = \left(\frac{a^2 - b^2}{a} \right) \cos^3 \alpha, \qquad y = - \left(\frac{a^2 - b^2}{b} \right) \sin^3 \alpha$$

and is

$$(ax)^{2/3} + (by)^{2/3} = (a^2 - b^2)^{2/3} (\cos^2 \alpha + \sin^2 \alpha)$$
$$= (a^2 - b^2)^{2/3}.$$

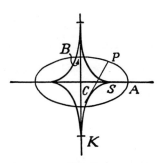

Fig. 3.

The diagram shows the ellipse and its evolute. As the point P describes the part AB of the curve, the centre of curvature C describes the part SK of the evolute. [From the method of obtaining the evolute it will be seen that the evolute can also be regarded as the envelope of normals (see 12).]

EXERCISE 2

1. Find the radius of curvature at any point of the curve $y^2 = 20x^5$. Show that the radius of curvature is a minimum at $x = 1/10$ and find the value. [L.U.]

2. If $x = a(1 - \cos \theta)$ $y = a \sin \theta$, find $\dfrac{dy}{dx}, \dfrac{d^2y}{dx^2}$. Deduce the value of ρ at the point θ. [L.U.]

3. Find the radius of curvature ρ at any point P of the parabola $y^2 = 4ax$ and prove that ρ^2 is proportional to SP^3 where S is the focus. [L.U.]

4. Prove that the radius of curvature at the point θ on the curve

(i) $x = 3a \cos \theta - a \cos 3\theta$, $y = 3a \sin \theta - a \sin 3\theta$ is $3a \sin \theta$.

(ii) $x = a \sin 2\theta(1 + \cos 2\theta)$, $y = a \cos 2\theta(1 - \cos 2\theta)$ is $4a \cos 3\theta$. [L.U.]

5. Find the radius of curvature in terms of θ at the point of an ellipse whose co-ordinates are $(a \cos \theta, b \sin \theta)$.

Show that the co-ordinates of the centre of curvature are $(b\lambda \cos^3 \theta, - a\lambda \sin^3 \theta)$, where $\lambda = (a^2 - b^2)/ab$ and deduce the locus of the centre of curvature. [L.U.]

6. For the curve $x = a \cos^3 \theta$, $y = a \sin^3 \theta$, prove that the length of the perpendicular from the origin on to the tangent at any point θ is $a \sin \theta \cos \theta$ and that $\rho = 3a \sin \theta \cos \theta$. [L.U.]

7. Find the radius of curvature at a point on the curve $y = a \operatorname{ch} (x/a)$.

If the normal at a point P of the curve meets the x axis in G show that PG is equal in length to the radius of curvature at P. [L.U.]

8. If (X, Y) is the centre of curvature at the point (x, y) of a plane curve, show that

$$X = x - y_1 \frac{(1 + y_1^2)}{y_2}, \qquad Y = y + \frac{(1 + y_1^2)}{y_2}.$$

Find the equation of the locus of the centre of curvature of $y^2 = 4ax$. [L.U.]

9. If $x = a \sin 2\theta$, $y = a \cos \theta$ prove that

$$\frac{d^2y}{dx^2} = - \frac{\cos \theta(1 + 2 \sin^2 \theta)}{4a \cos^3 2\theta}.$$

Sketch the curve given by the above values of x and y and find its radius of curvature at the point $(0, a)$. [L.U.]

10. Find the equations of the tangent and normal to the curve $xy = c^2$ at the point P, whose co-ordinates are $(cp, c/p)$.

If the normal at P cuts the curve again in the point $Q(cq, c/q)$ show that $q = - 1/p^3$, and prove that the radius of curvature at P is equal to $\frac{1}{2}$PQ. [L.U.]

11. Find the radius of curvature and the co-ordinates of the centre of curvature at the point $(\frac{1}{2}, \frac{3}{8})$ on the curve $y = x^3 + 2x^2 - 2x + 1$. [L.U.]

12. Prove that for the curve $axy = x^3 - 2a^3$, the ordinate y has a minimum value at the point $(- a, 3a)$ and find the equation of the circle of curvature at this point.

13. P is the point for which $\theta = \frac{1}{4}\pi$ on the curve $x = \sin^3 \theta$, $y = \cos^3 \theta$ and C is the centre of curvature at P. Show that the equation of the line CP is $y\sqrt{3} = x$ and find the length of CP.

14. Show that at any point on the curve $y = x/\sqrt{(x - 1)}$ the radius of curvature is given by

$$\rho = \frac{(4x^3 - 11x^2 + 8x)^{3/2}}{2(4 - x)(x - 1)^2}$$

15. The point P is given by
$$x = 2\cos 2t - \cos 4t, \qquad y = 2\sin 2t + \sin 4t.$$
Prove that the normal at P to the locus of P is
$$x\sin t + y\cos t = 3\sin 3t.$$
If (x_1, y_1) is the centre of curvature corresponding to P, and if C is $(\tfrac{1}{3}x_1, \tfrac{1}{3}y_1)$ and Q is $(-\tfrac{1}{3}x_1, -\tfrac{1}{3}y_1)$ prove that PQ = 2CP.

16. Writing $\dot{x}$ for $dx/d\theta$, $\ddot{x}$ for $d^2x/d\theta^2$, show that
$$\frac{d^2y}{dx^2} = \frac{d}{dx}\left(\frac{\dot{y}}{\dot{x}}\right) = \frac{d}{d\theta}\left(\frac{\dot{y}}{\dot{x}}\right)\frac{1}{\dot{x}}$$
$$= (\dot{x}\ddot{y} - \dot{y}\ddot{x})/\dot{x}^3.$$

Hence obtain the formula

$$\rho = \pm \frac{(\dot{x}^2 + \dot{y}^2)^{3/2}}{\dot{x}\ddot{y} - \ddot{x}\dot{y}}$$

when x, y are given in terms of a parameter θ. Apply this to the curve $x = a\cos^3\theta$, $y = a\sin^3\theta$ to obtain $\rho = 3a\cos\theta\sin\theta$.

17. The parametric equations of a curve are $x = 3t^2$, $y = 1 + 2t^3$. Find the equation of the tangent at the point P whose parameter is t and prove that the length of the radius of curvature at P is $6t(1 + t^2)^{3/2}$.   [L.U.]

18. The co-ordinates of a point P on an ellipse are $(2\cos t, \sin t)$, where t is a variable parameter. Prove that the normal at P is
$$\frac{2x}{\cos t} - \frac{y}{\sin t} = 3.$$
The normal at P meets the x axis at Q and the y axis at R. A point S is taken on QP produced such that QS = RQ. Prove that the locus of S is a circle whose equation is $x^2 + y^2 = 9$.   [L.U.]

19. A circle of radius a rolls on the outside of a fixed circle of radius $3a$ whose centre is A. Show by a diagram the directions of the tangent and normal to the locus of a point P on the rolling circle. If p is the length of the perpendicular from A to the tangent at P and r is the length of AP, show that $r^2 = 9a^2 + 16p^2/25$ and that the radius of curvature at P is $16p/25$.

$$\left[\text{It can be shown that } \rho = r\frac{dr}{dp}\cdot\right]$$ [L.U.]

20. The mid point of a rod AB of length $2a$ describes a fixed circle whose diameter is a; the rod passes through a fixed point O on the circumference. If OQ, perpendicular to the rod, meets the circle again in Q show that Q is the instantaneous centre for the motion of the rod and that the radius of curvature of the locus of A is $\tfrac{3}{2}$AQ.   [L.U.]

21. Show that the line $tx + y - at^3 - 2at = 0$ is normal to the parabola $y^2 = 4ax$ at the point $(at^2, 2at)$. Show also that it meets the curve again at $(aT^2, 2aT)$ where $T = -t - 2/t$. Hence find the equation of the locus of middle points of normal chords of the parabola.   [L.U.]

22. Find the equation of the normal at the point T $(at^2, 2at)$ of the parabola $y^2 = 4ax$.

A segment TP of constant length l is measured off in the outward direction along the normal. Find in terms of the parameter t the equations of the locus of P.   [L.U.]

CHAPTER 2

REDUCTION FORMULAE. FURTHER INTEGRATION

" It does not trouble the mathematician that he has to deal with unknown
things.   At the outset in algebra he handles unknown quantities x and y.
His quantities are unknown, but he subjects them to known operations—
addition, multiplication, etc.   Recalling Bertrand Russell's famous definition, the mathematician never knows what he is talking about nor whether
what he is saying is true; but we are tempted to add, at least he does know
what he is doing."

Sir Arthur Eddington, New Pathways in Science.

2.1. The General Method

The student wishing to find

$$\int x^n e^{-x} dx.$$

will note that since the integrand is the product of two different functions
of x, the method of integration by parts [Vol. I (18.9)] is suggested.
Using this :

$$\int x^n e^{-x} dx = x^n(-e^{-x}) - \int n x^{n-1}(-e^{-x})dx$$

$$= -x^n e^{-x} + n\int x^{n-1}e^{-x}dx . \quad \cdots \quad (1)$$

We have not evaluated the integral, but since the index of x has
been reduced by one, it is clear that if we keep on integrating by parts
finally the term in x in the integrand will be reduced to $x^0 = 1$ and
we will have a simple integral, that of e^{-x}, for the last stage.

Carrying the above integral one stage further, since

$$\int x^{n-1}e^{-x}dx = -x^{n-1}e^{-x} + (n-1)\int x^{n-2}e^{-x}dx,$$

therefore

$$\int x^n e^{-x}dx = -x^n e^{-x} - nx^{n-1}e^{-x} + n(n-1)\int x^{n-2}e^{-x}dx \quad (2)$$

An equation such as (1) above, in which an integral that cannot be
evaluated directly is connected with an integral containing one of the
functions raised to a lower power, is called a *reduction formula*.

It will be noticed that the form of the integrand $x^p e^{-x}$ (where p is n
to begin with and decreases at each stage) remains constant throughout

the successive stages. We may therefore shorten the work by denoting the integral by I_n, n being the index of x. Thus

$$I_n \equiv \int x^n e^{-x} dx \quad \text{and} \quad I_{n-2} \equiv \int x^{n-2} e^{-x} dx$$

Using this shorthand (1) becomes

$$I_n = -x^n e^{-x} + n I_{n-1}$$

whilst (2) becomes

$$I_n = -x^n e^{-x} - n x^{n-1} e^{-x} + n(n-1) I_{n-2}.$$

Example 1. If $I_{2n+1} = \int x^{2n+1} e^{-x^2} dx$ find a reduction formula and hence find I_5.

Using an x with the e^{-x^2} to make a perfect integral,

$$\int x^{2n} (x e^{-x^2}) dx = x^{2n} (-\tfrac{1}{2} e^{-x^2}) - \int 2n \cdot x^{2n-1} (-\tfrac{1}{2} e^{-x^2}) dx$$

$$\therefore \quad I_{2n+1} = -\tfrac{1}{2} x^{2n} e^{-x^2} + n I_{2n-1}$$

giving the reduction formula required.

When $n = 2$ this gives

$$I_5 = -\tfrac{1}{2} x^4 e^{-x^2} + 2 I_3 \quad \cdot \quad \cdot \quad \cdot \quad \cdot \quad \cdot \quad \cdot \quad \cdot \quad (3)$$

When $n = 1$:

$$I_3 = -\tfrac{1}{2} x^2 e^{-x^2} + I_1$$

but

$$I_1 = \int x^1 e^{-x^2} dx = \tfrac{1}{2} e^{-x^2}$$

$$\therefore \quad I_3 = -\tfrac{1}{2} x^2 e^{-x^2} - \tfrac{1}{2} e^{-x^2}.$$

Using (3),

$$I_5 = -\tfrac{1}{2} x^4 e^{-x^2} + 2[-\tfrac{1}{2} x^2 e^{-x^2} - \tfrac{1}{2} e^{-x^2}]$$

$$= \underline{-e^{-x^2}[\tfrac{1}{2} x^4 + x^2 + 1]}.$$

Example 2. If $I_n = \int_0^\pi x^n \cos x \, dx$, evaluate I_5 by using a reduction formula.

$$\int x^n \cos x \, dx = x^n \sin x - \int n x^{n-1} \sin x \, dx.$$

The integral on the right contains $\sin x$ instead of $\cos x$, so that this is not strictly a reduction formula.

Since

$$\int x^{n-1} \sin x \, dx = x^{n-1} (-\cos x) - \int (n-1) x^{n-2} (-\cos x) dx,$$

we find as a reduction formula

$$\int x^n \cos x \, dx = x^n \sin x + n x^{n-1} \cos x - n(n-1) \int x^{n-2} \cos x \, dx.$$

Inserting the limits

$$I_n = -n \pi^{n-1} - n(n-1) I_{n-2}.$$

When $n = 5$ this gives

$$I_5 = -5 \pi^4 - 5 \cdot 4 I_3,$$

$n = 3$:

$$I_3 = -3 \pi^2 - 3 \cdot 2 I_1$$

Clearly we will not eliminate the x by this process, if we continue with negative powers of x. However, the integral has been reduced to its simplest form, and now

$$I_1 = \int_0^\pi x \cos x dx = \Big[x(\sin x) \Big]_0^\pi - \int_0^\pi 1 . \sin x dx.$$

$$= -2$$

$$\therefore \quad I_5 = -5\pi^4 - 20[-3\pi^2 + 12]$$

$$= \underline{-5\pi^4 + 60\pi^2 - 240.}$$

EXERCISE 3

Prove the following:

1. If $V_n = \int x^n e^{ax} dx, \quad aV_n + nV_{n-1} = x^n e^{ax}.$

2. If $V_n = \int_0^\infty x^n e^{-x} dx, \quad V_n = nV_{n-1}$ where $n > 1$.

3. If $V_n = \int_0^{\frac{1}{2}\pi} x^n \cos x dx, \quad V_n = (\tfrac{1}{2}\pi)^n - n(n-1)V_{n-2}.$

4. If $V_{2n} = \int_0^1 \frac{x^{2n}}{1+x^2} dx, \quad V_{2n} = \frac{1}{2n-1} - V_{2n-2}$ where $n > \tfrac{1}{2}$.

5. If $V_n = \int (\log x)^n dx, \quad V_n + nV_{n-1} = x(\log x)^n.$ Hence find $\int (\log x)^3 dx.$

6. If $V_n = \int (x^2 + a^2)^n dx, \quad (2n+1)V_n = x(x^2 + a^2)^n + 2na^2 V_{n-1}.$

7. If $V_n = \int x^m (\log x)^n dx, \quad (m+1)V_n = x^{m+1}(\log x)^n - nV_{n-1}.$

8. By writing $\tan^n x$ as $\tan^{n-2} x(\sec^2 x - 1)$ show that if

$$V_n = \int_0^{\frac{1}{2}\pi} \tan^n x dx, \qquad V_n = \frac{1}{n-1} - V_{n-2}.$$

9. Prove that

$$\int \frac{x^n dx}{\sqrt{(a^2 + x^2)}} = \frac{x^{n-1}\sqrt{(a^2 + x^2)}}{n} - \frac{n-1}{n} a^2 \int \frac{x^{n-2}}{\sqrt{(a^2 + x^2)}} dx.$$

Hence evaluate

$$\int_0^2 \frac{x^5 dx}{\sqrt{(5 + x^2)}}. \qquad \text{[L.U.]}$$

10. If $\qquad\qquad I_n = \int t^n dt/(1 + t^2)$

show that

$$I_{n+2} = t^{n+1}/(n+1) - I_n.$$

Evaluate

$$\int_0^1 t^6 dt/(1 + t^2). \qquad \text{[L.U.]}$$

11. Obtain the reduction formula

$$\int x^n \sin x dx = nx^{n-1} \sin x - x^n \cos x - n(n-1) \int x^{n-2} \sin x dx.$$

Hence evaluate:

$$\text{(i)} \int_0^{\pi} x^5 \sin x dx; \qquad \text{(ii)} \int_0^{\frac{1}{2}\pi} x^4 \sin x dx. \qquad \text{[L.U.]}$$

12. If

$$I_n = \int_0^a \frac{x^n dx}{\sqrt{(3a^2 + x^2)}}$$

show that

$$I_n = \frac{2a^n}{n} - 3\frac{n-1}{n} a^2 I_{n-2}.$$

Evaluate I_7. [L.U.]

2.2. The Integrals $\int_0^{\pi/2} \cos^n x dx, \int_0^{\pi/2} \sin^n x dx$

Between the limits 0 and $\frac{1}{2}\pi$ the curve $y = \cos^n x$ is the reverse of $y = \sin^n x$. The two integrals above therefore represent the same area and are equal. The student should also put $x = \frac{\pi}{2} - z$ in one integral and convert it to the other.

We will deal with the first. Writing it as a product:

$$\int_0^{\pi/2} \cos^{n-1} x \cdot \cos x dx = \left[\cos^{n-1} x (\sin x) \right]_0^{\pi/2}$$

$$- \int_0^{\pi/2} (n-1) \cos^{n-2} x (- \sin x) \sin x dx$$

$$= (n-1) \int_0^{\pi/2} \cos^{n-2} x \sin^2 x dx$$

$$= (n-1) \int_0^{\pi/2} \cos^{n-2} x (1 - \cos^2 x) dx$$

$$= (n-1) \int_0^{\pi/2} \cos^{n-2} x dx - (n-1) \int_0^{\pi/2} \cos^n x dx.$$

If the original integral is denoted by I_n, this is

$$I_n = (n-1)I_{n-2} - (n-1)I_n$$

or

$$I_n = \frac{n-1}{n} I_{n-2} \quad . \quad . \quad . \quad . \quad . \quad . \quad (4)$$

giving a relation between two integrals, one with the power of $\cos x$ reduced by 2.

If $n = 4$, (4) gives

$$I_4 = \tfrac{3}{4} I_2.$$

To find I_2 put $n = 2$ in (4):

$$I_2 = \tfrac{1}{2}I_0$$

now

$$I_0 = \int_0^{\pi/2} 1\,dx = \tfrac{1}{2}\pi$$

$$\therefore \quad I_4 = \frac{3 \cdot 1}{4 \cdot 2} I_0 = \frac{3 \cdot 1}{4 \cdot 2} \times \frac{\pi}{2} = \frac{3\pi}{16}.$$

If $n = 5$, an odd number, (4) gives

$$I_5 = \tfrac{4}{5}I_3.$$

If $n = 3$:

$$I_3 = \tfrac{2}{3}I_1$$

now

$$I_1 = \int_0^{\frac{1}{2}\pi} \cos x\,dx = 1$$

therefore

$$I_5 = \frac{4 \cdot 2}{5 \cdot 3} \times 1 = \frac{8}{15}.$$

Returning to (4).

$$I_n = \frac{n-1}{n} I_{n-2}.$$

With $(n-2)$ for n this gives

$$I_{n-2} = \frac{n-3}{n-2} I_{n-4},$$

therefore

$$I_n = \frac{(n-1)(n-3)}{n(n-2)} I_{n-4},$$

and since similarly

$$I_{n-4} = \frac{(n-5)}{(n-4)} I_{n-6}$$

$$I_n = \frac{(n-1)(n-3)(n-5)}{n(n-2)(n-4)} I_{n-6}.$$

Continuing in this way we note that if n is even we end with $I_0 = \tfrac{1}{2}\pi$, whilst if n is odd we end with $I_1 = 1$.

The general formula is therefore

$$I_n = \frac{(n-1)(n-3)(n-5) \; \ldots \ldots 2}{n(n-2)(n-4) \; \ldots \ldots 3} \times 1 \quad \text{if } n \text{ is odd}$$

$$= \frac{(n-1)(n-3)(n-5) \; \ldots \ldots 1}{n(n-2)(n-4) \; \ldots \ldots 2} \times \frac{\pi}{2} \quad \text{if } n \text{ is even}.$$

Example 3. Write down the values of:

(a) $\int_0^{\frac{1}{2}\pi} \sin^6 x\,dx$; (b) $\int_0^{\frac{1}{2}\pi} \sin^3 \theta\,d\theta$; (c) $\int_{-\pi/2}^{+\pi/2} \cos^4 x\,dx$; (d) $\int_0^{\pi} \cos^7 \theta\,d\theta$.

Using I to denote the integral,

(a) $I = \dfrac{5 \cdot 3 \cdot 1}{6 \cdot 4 \cdot 2} \times \dfrac{\pi}{2} = \dfrac{5\pi}{32}.$

(b) $I = \dfrac{2}{3}.$

(c) A rough sketch of $\cos x$, and hence $\cos^4 x$, shows that the area between $-\dfrac{\pi}{2}$ and 0 is equal to that between 0 and $\dfrac{\pi}{2}$ and of the same sign·

$$\therefore \quad I = 2 \int_0^{\frac{1}{2}\pi} \cos^4 x \, dx = 2 \left(\dfrac{3 \cdot 1}{4 \cdot 2} \times \dfrac{\pi}{2} \right) = \dfrac{3\pi}{8}.$$

(d) The graph $y = \cos^7 \theta$ gives two areas equal but opposite in sign, a positive area 0 to $\frac{1}{2}\pi$ and an equal but negative area (*i.e.*, below the θ axis) from $\frac{1}{2}\pi$ to π. The value of the integral is therefore zero.

2.3. In the case of

$$\int_0^{\frac{1}{2}\pi} \sin^m x \cos x \, dx \quad \text{or} \quad \int_0^{\frac{1}{2}\pi} \cos^m x \sin x \, dx$$

the first integral can be written

$$\int_0^{\frac{1}{2}\pi} \sin^m x \, d(\sin x) = \left(\dfrac{\sin^{m+1} x}{m+1} \right)_0^{\pi/2}$$

$$= \dfrac{1}{m+1}$$

and similarly for the second.

2.4. For the general case of

$$I = \int_0^{\pi/2} \cos^n x \sin^m x \, dx$$

where m, n are *positive integers* greater than 1 it can be shown as above that when m and n are both even (positive integers)

$$I = \dfrac{(n-1)(n-3) \ldots \times (m-1)(m-3) \ldots}{(m+n)(m+n-2)(m+n-4) \ldots} \times \dfrac{\pi}{2}$$

but when either (or both) is an odd positive integer

$$I = \dfrac{(n-1)(n-3) \ldots \times (m-1)(m-3) \ldots}{(m+n)(m+n-2)(m+n-4) \ldots} \times 1$$

where in each formula each of the three products continues down to 2 or 1 according as its first term, *e.g.*, $m+n$ is even or odd.

Example 4.

$$\int_0^{\frac{1}{2}\pi} \sin^6 x \cos^5 x \, dx = \dfrac{5 \cdot 3 \cdot 1 \times 4 \cdot 2}{11 \cdot 9 \cdot 7 \cdot 5 \cdot 3 \cdot 1}$$

$$= \dfrac{8}{693}$$

$$\int_0^{\frac{1}{2}\pi} \cos^4 x \sin^4 x \, dx = \dfrac{3 \cdot 1 \times 3 \cdot 1}{8 \cdot 6 \cdot 4 \cdot 2} \times \dfrac{\pi}{2}$$

$$= \dfrac{3\pi}{256}.$$

EXERCISE 4

Write down the values of:

1. $\int_0^{\frac{1}{2}\pi} \sin^2 x\,dx.$

2. $\int_0^{\pi} \cos^2 \theta\,d\theta.$

3. $\int_{-\frac{1}{2}\pi}^{+\frac{1}{2}\pi} \cos^3 x\,dx.$

4. $\int_0^{\frac{1}{2}\pi} \sin^4 \theta\,d\theta.$

5. $\int_0^{\pi} \cos^5 x\,dx.$

6. $\int_0^{\pi} \cos^6 x\,dx.$

7. Using the substitution $\dfrac{\theta}{2} = x$, find:

(a) $\int_0^{\pi} \sin^3 \left(\dfrac{\theta}{2}\right) d\theta;$

(b) $\int_0^{\pi} \cos^4 \left(\dfrac{\theta}{2}\right) d\theta.$

8. Using the substitution $x = a \tan \theta$, find:

(a) $\int_0^{\infty} \dfrac{dx}{(a^2 + x^2)^{3/2}};$

(b) $\int_0^{\infty} \dfrac{x^2}{(a^2 + x^2)^2}\,dx.$

9. Using the substitution $x = \sin \theta$, find

$$\int_0^1 \dfrac{x^4 dx}{\sqrt{1 - x^2}}.$$

10. Find the volume generated when the curve $y = \cos^3 x$ between $x = 0$ and π revolves round the x axis.

11. Solve the above problem if the curve is $y = 1 - \cos^3 x$ and the limits are 0 and 2π.

12. Find the position of the centre of gravity of the area enclosed by the x axis and the curve $y = \sin^2 x$ between 0 and π.

13. Find the mean value between $x = 0$ and $x = \pi$ of:

(a) $\sin^3 x;$

(b) $\sin^4 x.$

14. Express the integrand in terms of the half angles and hence find

(a) $\int_0^{\pi} (1 - \cos \theta)^3 d\theta;$

(b) $\int_0^{\pi} \sin^2 \theta(1 + \cos \theta)^4 d\theta.$

15. Show that

$$\int_0^{\frac{1}{2}\pi} \sin^{2n} \theta\,d\theta = \dfrac{2n - 1}{2n} \int_0^{\frac{1}{2}\pi} \sin^{2n-2} \theta\,d\theta,$$

and deduce a formula for the first integral.

Express

$$\int_0^{\frac{1}{2}\pi} \dfrac{d\theta}{\sqrt{(1 - k^2 \sin^2 \theta)}}$$

as a series of ascending powers of k and evaluate correct to three significant figures when $k = 0.5$.   [L.U.]

16. Prove that

$$n \int \tan^n \theta \sec \theta\,d\theta = \tan^{n-1} \theta \sec \theta - (n - 1) \int \tan^{n-2} \theta \sec \theta\,d\theta.$$

Hence or otherwise evaluate

$$\int_0^{\frac{1}{4}\pi} \dfrac{\sin^3 \theta}{\cos^4 \theta}\,d\theta \quad \text{and} \quad \int_0^{\frac{1}{4}\pi} \dfrac{\sin^4 \theta}{\cos^5 \theta}\,d\theta.$$ [L.U.]

17. If $\qquad I_n = \int \dfrac{\sin{(2n-1)x}}{\sin x}\,dx, \quad J_n = \int \dfrac{\sin^2 nx}{\sin^2 x}\,dx,$

prove that apart from constants of integration

$$n(I_{n+1} - I_n) = \sin 2nx$$
$$J_{n+1} - J_n = I_{n+1}.$$

Evaluate I_3 and J_3. [L.U.]

Write down the value of:

18. $\displaystyle\int_0^{\frac12\pi} \sin^5 x \cos^4 x\,dx.$ **19.** (i) $\displaystyle\int_0^{\frac12\pi} \cos^4 x \sin^4 x\,dx;$ (ii) $\displaystyle\int_0^{\frac12\pi} \cos^5\theta\sin\theta\,d\theta.$

20. $\displaystyle\int_0^{\frac12\pi} \cos^7 x \tan^4 x\,dx.$ **21.** (i) $\displaystyle\int_0^{\frac12\pi} \cos^2\theta\sin^6\theta\,d\theta;$ (ii) $\displaystyle\int_0^{\frac12\pi} \sin^2\theta\cos\theta\,d\theta.$

22. By putting $\dfrac{\theta}{2} = x$ change to a standard form, and hence find:

(a) $\displaystyle\int_0^{\pi} \sin^4\left(\frac{\theta}{2}\right)\cos^3\left(\frac{\theta}{2}\right)d\theta;$ (b) $\displaystyle\int_0^{\pi} \sin^6\left(\frac{\theta}{2}\right)\cos^2\left(\frac{\theta}{2}\right)d\theta.$

23. By means of the substitution $x = a\sin\theta$, find

(a) $\displaystyle\int_0^{a} x^3(a^2 - x^2)^{3/2}\,dx;$ (b) $\displaystyle\int_0^{a} x^4(a^2 - x^2)^5\,dx.$

24. By means of the substitution $x = \sin^2\theta$ find:

(a) $\displaystyle\int_0^{1} x^{5/2}(1-x)^{3/2}\,dx;$ (b) $\displaystyle\int_0^{1} x\sqrt{(x - x^2)}\,dx.$

Find the value of:

25. $\displaystyle\int_0^{a} x^4(a^2 - x^2)^{\frac12}\,dx.$ **26.** $\displaystyle\int_0^{2a} x\sqrt{(2ax - x^2)}\,dx.$ **27.** $\displaystyle\int_0^{2a} x^2\sqrt{(2ax - x^2)}\,dx.$

28. Find the area of one loop of the curve $y^2 = x^4(4 - x^2)$.

29. Find the total area enclosed by the curve $y^2 = x^5(2 - x)$.

30. For the curve $x = a\sin^3\theta$, $y = a\cos^3\theta$, find: (1) The area enclosed by the curve in the first quadrant and the axes. (2) The centre of gravity of this area. (3) The volume generated when this area revolves round the x axis. (4) The length of the arc of the curve bounding this area. (5) The surface of revolution generated when this arc revolves round the x axis.

2.5. The Range a Multiple of $\frac12\pi$

In practice we often have to deal with a range that is a multiple of $\frac12\pi$, e.g., $-\pi$ to $+\pi$ or 0 to 2π. We will show that the area between the curve $y = \cos^m x \sin^n x$ and the x axis is the same in each interval of $\frac12\pi$, e.g., in the intervals $-\pi$ to $-\frac12\pi$, $-\frac12\pi$ to 0, 0 to $+\frac12\pi$, etc. As a particular case consider $y = \sin^3 x \cos^2 x$, where the areas A, B, C, D shown in Fig. 4 are all numerically equal but may differ in sign. Thus the area A is

$$\int_{-\pi}^{-\frac12\pi} \sin^3 x \cos^2 x\,dx.$$

If we put $x = -\pi + X$, this becomes

$$-\int_0^{\pi/2} \sin^3 X \cos^2 X dX$$

showing that A is equal in area to C but on the other side of the x axis. We therefore refer all such integrals to the value between

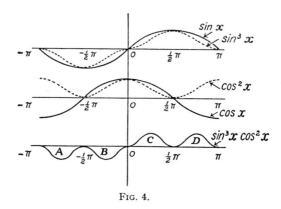

Fig. 4.

0 and $\frac{1}{2}\pi$, which is written down by the formula. In integrating across several such intervals we must note whether the curve is above or below the x axis. The sketch above shows that for

$$\int_{-\pi}^{+\pi} \sin^3 x \cos^2 x dx$$

there are two areas above and two below the x axis, and therefore the value of the integral is zero. But for

$$\int_{-\pi}^{+\pi} \sin^4 x \cos^2 x dx$$

the even indices show that the curve will be above the x axis. There are 4 intervals of $\frac{1}{2}\pi$ in the range, and therefore the value of the integral is

$$4\int_0^{\frac{1}{2}\pi} \sin^4 x \cos^2 x dx$$

which is

$$4\left(\frac{3 \cdot 1 \times 1}{6 \cdot 4 \cdot 2} \times \frac{\pi}{2}\right) = \frac{\pi}{8}.$$

It will be seen that there is no need to draw a figure, since all we require is the sign of the curve in each interval of $\frac{1}{2}\pi$.   Thus for

$$\int_{-\pi}^{+\pi} \sin^3 x \cos^2 x\,dx$$

we need only the signs as shown :

$-$	$-$	$+$	$+$	$\sin x$
$-$	$+$	$+$	$-$	$\cos x$

$$-\pi \qquad -\frac{\pi}{2} \qquad 0 \qquad \frac{\pi}{2} \qquad \pi$$

(a)	$+$	$+$	$+$	$+$	$\cos^2 x$
(b)	$-$	$-$	$+$	$+$	$\sin^3 x$
$(a) \times (b)$	$-$	$-$	$+$	$+$	$\sin^3 x \cos^2 x.$

Since $\sin x$ is negative between $-\dfrac{\pi}{2}$ and 0, so is $\sin^3 x$. $\cos^2 x$ is positive, and the product of these signs shows that $\sin^3 x \cos^2 x$ is negative in this range.   Our final result is two negatives and two positives, so that the value of the integral is zero.

Example 5.   The area of a circle of radius r revolves round an axis distant d from its centre so as to generate an anchor ring.   Find the moment of inertia of the solid ring about its axis.

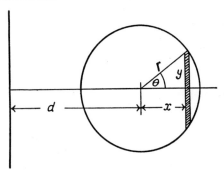

FIG. 5.

If M is the mass of the ring and ρ its density then by Pappus' Theorem [Vol. I (24.6)]

$$M = \rho(\pi r^2)(2\pi d)$$
$$= \rho 2\pi^2 r^2 d. \qquad \ldots \ldots \ldots \quad (5)$$

The element $2y\,dx$ revolves round the axis to generate a ring every part of which is $d + x$ from the axis.   The moment of inertia of this ring is therefore

$$[\rho(2y\,dx) \cdot 2\pi(x + d)](x + d)^2$$
$$\therefore \quad Mk^2 = 4\pi\rho\int_{-r}^{+r} y(x + d)^3 dx.$$

From the figure since $x = r \cos \theta$, $y = r \sin \theta$, the integral becomes

$$- 4\pi\rho r^2 \int_\pi^0 \sin^2 \theta (r \cos \theta + d)^3 d\theta$$

$$= 4\pi\rho r^2 \int_0^\pi (r^3 \sin^2 \theta \cos^3 \theta + 3dr^2 \sin^2 \theta \cos^2 \theta + 3d^2 r \sin^2 \theta \cos \theta + d^3 \sin^2 \theta) d\theta.$$

The integrals of the first and third terms give zero.

$$\therefore \quad Mk^2 = 4\pi\rho r^2 \int_0^\pi (3dr^2 \sin^2 \theta \cos^2 \theta + d^3 \sin^2 \theta) d\theta$$

$$= 4\pi\rho r^2 \left[3dr^2 \left(2\frac{1 \cdot 1}{4 \cdot 2} \times \frac{\pi}{2} \right) + d^3 \left(2\frac{1}{2} \times \frac{\pi}{2} \right) \right]$$

$$= M[\tfrac{3}{4}r^2 + d^2]$$

on using (5).

In the following exercises the student should wherever possible reduce the range of integration from the very beginning by noting the symmetry of the curve. Thus in finding the area of a cycloid [Vol. I (12.2)] the integral of ydx between the limits 0 and 2π could be used. But a knowledge of the symmetry of the curve shows that twice the same integral between the limits 0 and π is equivalent and leads more rapidly to the answer.

EXERCISE 5

Find the values of:

1. $\displaystyle\int_0^\pi \sin^7 \theta d\theta.$ 2. $\displaystyle\int_0^\pi \cos^7 x dx.$ 3. $\displaystyle\int_0^\pi \cos^8 x dx.$ 4. $\displaystyle\int_{-\frac{1}{2}\pi}^{+\frac{1}{2}\pi} \sin^5 x dx.$

5. $\displaystyle\int_0^\pi (l + m \cos x)^3 dx.$ 6. $\displaystyle\int_{-\frac{1}{2}\pi}^{+\frac{1}{2}\pi} \sin^2 x \cos^7 x dx.$

7. $\displaystyle\int_0^{2\pi} \sin^2 x \cos^5 x dx.$ 8. $\displaystyle\int_{\frac{1}{2}\pi}^{\frac{3\pi}{2}} \sin^3 x \cos^4 x dx.$

9. (a) With the substitution $x = a \tan \theta$ find

$$\int_{-\infty}^{+\infty} \frac{x^4 dx}{(a^2 + x^2)^3}.$$

(b) From graphical considerations write down the value of

$$\int_{-\infty}^{+\infty} \frac{x^5 dx}{(a^2 + x^2)^3}.$$

10. For the cycloid $x = a(\theta - \sin \theta)$, $y = a(1 - \cos \theta)$ show that the moment of inertia of the area under one arch about the x axis is $35\pi a^4/12$ and the square of the corresponding radius of gyration is $35a^2/36$.

11. In the above cycloid show that the square of the radius of gyration of the arc of the curve about the x axis is $32a^2/15$.

12. An anchor ring is generated by the revolution of a circle of radius r about an axis distant d from the centre.   Find the second moment of the surface of the ring about its axis.

13. For the hypocycloid $x = a \cos^3 \theta, y = a \sin^3 \theta$, show that the second moment and the radius of gyration of the area about the x axis are given by

$$AK^2 = \tfrac{21}{512}\pi a^4, \quad K^2 = \tfrac{7}{64}a^2.$$

14. Show that the total perimeter of an ellipse of semi axes a and b and eccentricity e is given by

$$s = 4a \int_0^{\frac{1}{2}\pi} \sqrt{(1 - e^2 \sin^2 \theta)}d\theta$$

$$= 2\pi a\{1 - \tfrac{1}{4}e^2 - \tfrac{3}{64}e^4 \ldots \}.$$

where $b^2 = a^2(1 - e^2)$.

Further Integration

2.6. Quadratic Expressions

In Vol. I we obtained the standard forms

$$\int \frac{dx}{\sqrt{(a^2 - x^2)}} = \sin^{-1}\left(\frac{x}{a}\right) + C$$

$$\int \frac{dx}{\sqrt{(x^2 + a^2)}} = \text{sh}^{-1}\left(\frac{x}{a}\right) + C, \quad \int \frac{dx}{\sqrt{(x^2 - a^2)}} = \text{ch}^{-1}\left(\frac{x}{a}\right) + C.$$

Any integral of the form

$$\int \frac{dx}{\sqrt{P}} \quad \cdot \quad \cdot \quad \cdot \quad \cdot \quad \cdot \quad \cdot \quad (1)$$

where $P = ax^2 + bx + c$ can be reduced to one of the above forms.

Example 6.   Evaluate the above integral (1) when P is:

    (a) $2 - 5x - 3x^2$;   (b) $8x^2 + 7x + 50$;   (c) $4x^2 + 2x - 9$.

(a) $2 - 5x - 3x^2 = 3\{\tfrac{2}{3} - \tfrac{5}{3}x - x^2\}$

$$= 3\{\tfrac{49}{36} - (x + \tfrac{5}{6})^2\}.$$

so that

$$\int \frac{dx}{\sqrt{(2 - 5x - 3x^2)}} = \frac{1}{\sqrt{3}} \int \frac{dx}{\sqrt{\{(\tfrac{7}{6})^2 - (x + \tfrac{5}{6})^2\}}}.$$

The substitution

$$x + \tfrac{5}{6} = z$$

reduces the integral to the standard form and its value is

$$\frac{1}{\sqrt{3}} \sin^{-1}\left(\frac{z}{7/6}\right) = \frac{1}{\sqrt{3}} \sin^{-1}\left(\frac{6x + 5}{7}\right) + C.$$

(b) Since $8x^2 + 7x + 50 = 8\{(x + \tfrac{7}{16})^2 + \tfrac{1551}{256}\}$

$$\int \frac{dx}{\sqrt{(8x^2 + 7x + 50)}} = \frac{1}{\sqrt{8}} \text{sh}^{-1}\left(\frac{16x + 7}{\sqrt{1551}}\right) + C.$$

(c) $4x^2 + 2x - 9 = 4\{(x + \tfrac{1}{4})^2 - \tfrac{37}{16}\}$

and

$$\int \frac{dx}{\sqrt{(4x^2 + 2x - 9)}} = \tfrac{1}{2} \text{ch}^{-1}\left(\frac{4x + 1}{\sqrt{37}}\right) + C.$$

We have also seen that

$$\int \frac{dx}{x^2 + a^2} = \frac{1}{a} \tan^{-1}\left(\frac{x}{a}\right)$$

$$\int \frac{dx}{x^2 - a^2} = \frac{1}{2a} \log \frac{x - a}{x + a} \qquad (x^2 > a^2)$$

$$\int \frac{dx}{a^2 - x^2} = \frac{1}{2a} \log \frac{a + x}{a - x} \qquad (x^2 < a^2)$$

and any integral of the form

$$\int \frac{dx}{P} \qquad \cdots \cdots \cdots \quad (2)$$

where $P = ax^2 + bx + c$ can be reduced to one of these forms.

Example 7. Find the integral (2) above when P is:

(a) $2x^2 + 2x + 5$; (b) $2x^2 + 12x - 8$; (c) $6x - 8 - x^2$.

(a) Since $\qquad 2x^2 + 2x + 5 = 2\{(x + \frac{1}{2})^2 + \frac{9}{4}\}$

$$\int \frac{dx}{2x^2 + 2x + 5} = \frac{1}{2} \int \frac{dx}{(x + \frac{1}{2})^2 + (\frac{3}{2})^2}.$$

The substitution

$$x + \tfrac{1}{2} = z$$

is hardly necessary to enable the student to see that this is the standard form slightly disguised, and its value is

$$\frac{1}{2} \cdot \frac{1}{\frac{3}{2}} \tan^{-1}\left(\frac{x + \frac{1}{2}}{\frac{3}{2}}\right) = \frac{1}{3} \tan^{-1}\left(\frac{2x + 1}{3}\right) + C.$$

(b) $\qquad 2x^2 + 12x - 8 = 2(x^2 + 6x - 4)$
$$\qquad\qquad\qquad = 2\{(x + 3)^2 - 13\}$$

so that

$$\int \frac{dx}{2x^2 + 12x - 8} = \frac{1}{2} \int \frac{dx}{(x + 3)^2 - 13}$$

$$= \frac{1}{2} \cdot \frac{1}{2\sqrt{13}} \log \left(\frac{x + 3 - \sqrt{13}}{x + 3 + \sqrt{13}}\right) + C.$$

(c) Since $\qquad 6x - 8 - x^2 = 1 - (x - 3)^2$

$$\int \frac{dx}{6x - 8 - x^2} = \int \frac{dx}{1 - (x - 3)^2}$$

$$= \frac{1}{2} \log \frac{1 + (x - 3)}{1 - (x - 3)}$$

$$= \frac{1}{2} \log \left(\frac{x - 2}{4 - x}\right) + C.$$

2.7. Variations of all the above occur with a first-degree factor in the numerator. We use up the x's in the numerator to obtain the differential of the quadratic in the denominator and thus separate it into two parts one of which integrates at sight since the numerator is the differential of the denominator and the other part with a constant denominator integrates as one of the forms dealt with above. Thus

$$\int \frac{3x + 2}{\sqrt{(x^2 + 2x + 3)}} \, dx$$

$$= \int \frac{\frac{3}{2}(2x + 2) - 1}{\sqrt{(x^2 + 2x + 3)}} \, dx$$

$$= \frac{3}{2} \int \frac{(2x + 2)dx}{\sqrt{(x^2 + 2x + 3)}} - \int \frac{dx}{\sqrt{\{(x + 1)^2 + 2\}}}$$

$$= 3(x^2 + 2x + 3)^{\frac{1}{2}} - \text{sh}^{-1}\left(\frac{x + 1}{\sqrt{2}}\right) + C.$$

2.8. Again the differentiation of

$$y = \sin^{-1}\left(\frac{1}{x}\right) \quad \left[= \text{cosec}^{-1}(x)\right]$$

gives

$$\frac{dy}{dx} = \frac{1}{\sqrt{\left(1 - \frac{1}{x^2}\right)}} \times -\frac{1}{x^2}$$

$$= -\frac{1}{x\sqrt{(x^2 - 1)}}$$

We then expect general expressions of this type to reduce to a standard form when the substitution corresponding to $z = \frac{1}{x}$ is made.

Example 8. Find $\quad I = \int \dfrac{dx}{(x - 2)\sqrt{(2x^2 - 6x + 5)}}.$

Put $\quad\quad\quad x - 2 = \dfrac{1}{y}$

logarithmic differentiation: $\dfrac{dx}{x - 2} = -\dfrac{dy}{y}$

also $\quad 2x^2 - 6x + 5 = 2\left(2 + \dfrac{1}{y}\right)^2 - 6\left(2 + \dfrac{1}{y}\right) + 5$

$$= (y^2 + 2y + 2)/y^2.$$

Then $\quad\quad\quad I = -\int \dfrac{dy}{\sqrt{(y^2 + 2y + 2)}}$

$$= -\int \frac{dy}{\sqrt{\{(y + 1)^2 + 1\}}}$$

$$= -\text{sh}^{-1}(y + 1)$$

$$= -\text{sh}^{-1}\left(\frac{x - 1}{x - 2}\right) + C.$$

2.9. In Vol. I (18.3) and (29.9) we obtained

$$\int \sqrt{(a^2 - x^2)}dx = \tfrac{1}{2}x\sqrt{(a^2 - x^2)} + \tfrac{1}{2}a^2 \sin^{-1}\left(\frac{x}{a}\right) + C$$

$$\int \sqrt{(a^2 + x^2)}dx = \tfrac{1}{2}x\sqrt{(a^2 + x^2)} + \tfrac{1}{2}a^2 \operatorname{sh}^{-1}\left(\frac{x}{a}\right) + C.$$

Similarly

$$\int \sqrt{(x^2 - a^2)}dx = \tfrac{1}{2}x\sqrt{(x^2 - a^2)} - \tfrac{1}{2}a^2 \operatorname{ch}^{-1}\left(\frac{x}{a}\right) + C.$$

Any integral of the form

$$\int \sqrt{(ax^2 + bx + c)}dx$$

can be reduced to one of these.

Example 9. Find $\int \sqrt{P}\,dx$ when P is (a) $2 - 4x - 7x^2$; (b) $x^2 + 2x + 3$.

(a)
$$2 - 4x - 7x^2 = 7\{\tfrac{2}{7} - \tfrac{4}{7}x - x^2\}$$
$$= 7\{\tfrac{18}{49} - (x + \tfrac{2}{7})^2\}.$$

$$\therefore \int \sqrt{(2 - 4x - 7x^2)}dx = \sqrt{7}\int \sqrt{\{\tfrac{18}{49} - (x + \tfrac{2}{7})^2\}}dx$$

with $a^2 = \tfrac{18}{49}$ and $x + \tfrac{2}{7}$ as the x in the first formula above this integral is

$$\sqrt{7}\left[\tfrac{1}{2}(x + \tfrac{2}{7})\sqrt{\{\tfrac{18}{49} - (x + \tfrac{2}{7})^2\}} + \tfrac{1}{2}\cdot\tfrac{18}{49}\sin^{-1}\left(\frac{x + \tfrac{2}{7}}{\tfrac{1}{7}\sqrt{18}}\right)\right] + C$$

$$= \frac{\sqrt{7}}{98}\left[(7x + 2)\sqrt{(14 - 28x - 49x^2)} + 18\sin^{-1}\left(\frac{7x + 2}{\sqrt{18}}\right)\right] + C.$$

(b) $x^2 + 2x + 3 = (x + 1)^2 + 1$.

By the second formula above

$$\int \sqrt{(x^2 + 2x + 3)}dx = \tfrac{1}{2}(x + 1)\sqrt{(x^2 + 2x + 3)} + \tfrac{1}{2}\operatorname{sh}^{-1}(x + 1) + C.$$

2.10. Integrals Involving Logarithms

Since if $y = \log_e x$, $x = e^y$ and we must take the exponential function as positive and single valued (note also its graph), it follows that x must be positive when we write

$$\int \frac{dx}{x} = \log x.$$

If x is negative, then the substitution $x = -z$ produces (with z positive)

$$\int \frac{dz}{z} = \log z$$
$$= \log(-x).$$

We can combine both cases into one formula, and also avoid the difficulties of combining the logarithms of negative numbers [Vol. 1 (18.6)] if we write

$$\int \frac{dx}{x} = \log |x|.$$

We also found, with $t = \tan \frac{1}{2}x$

$$\int \frac{dx}{\sin x} = \int \frac{dt}{t}$$

and can only proceed to the next step

$$= \log (\tan \tfrac{1}{2}x)$$

if $\tan \frac{1}{2}x > 0$. If $\tan \frac{1}{2}x < 0$ the substitution $x = -y$ will produce the result

$$\log (- \tan \tfrac{1}{2}x)$$

showing that the integral is the logarithm of the positive value of $\tan \frac{1}{2}x$, so that again

$$\int \frac{dx}{\sin x} = \log | \tan \tfrac{1}{2}x |.$$

Similarly,

$$\int \tan x\,dx = - \log |\cos x| = \log |\sec x|$$

$$\int \cot x\,dx = \log |\sin x|.$$

To show, for example, that

$$\int \sec x\,dx = \log | \tan (\tfrac{1}{2}x + \tfrac{1}{4}\pi) | = \log |\sec x + \tan x|$$

we may write

$$\int \sec x\,dx = \int \frac{\sec x(\tan x + \sec x)}{\sec x + \tan x}\,dx.$$

The numerator is now the differential of the denominator, and therefore if this is positive the integral is

$$\log (\sec x + \tan x).$$

If $\sec x + \tan x < 0$, let $x = \pi - y$, and we obtain

$$\log \{- (\sec x + \tan x)\}$$

so that the above formula gives both cases.

Finally, the two formulae previously quoted may be combined as

$$\int \frac{dx}{x^2 - a^2} = \frac{1}{2a} \log \left| \frac{x - a}{x + a} \right|.$$

Example 10. Find:

(a) $\int \operatorname{cosech} x\,dx$; (b) $\int_5^6 \frac{dx}{16 - x^2}$; (c) $\int \frac{2x^2 - x - 5}{2x^2 - 5x - 3}\,dx$.

(a)
$$\int \operatorname{cosech} x\,dx = \int \frac{2}{e^x - e^{-x}}\,dx$$

$$= \int \frac{2e^x}{(e^x)^2 - 1}\,dx$$

$$= \int \left(\frac{e^x}{e^x - 1} - \frac{e^x}{e^x + 1} \right)\,dx$$

$$= \log |e^x - 1| - \log (e^x + 1)$$

$$= \log \left| \frac{e^x - 1}{e^x + 1} \right| + C.$$

(When an expression is obviously positive we may, if we wish, omit the modulus in its logarithm.)

(b)
$$\int_5^6 \frac{dx}{16 - x^2} = -\int_5^6 \frac{dx}{x^2 - 16}$$

$$= -\frac{1}{2 \cdot 4} \log \left| \frac{x - 4}{x + 4} \right| \Big]_5^6$$

$$= -\tfrac{1}{8} \log \tfrac{9}{5}.$$

(c)
$$\frac{2x^2 - x - 5}{2x^2 - 5x - 3} = 1 + \frac{4}{7(2x + 1)} + \frac{10}{7(x - 3)}.$$

$$\therefore \quad I = x + \tfrac{2}{7} \log |2x + 1| + \tfrac{10}{7} \log |x - 3| + C.$$

2.11. Trigonometrical Functions

We have dealt with integrals of the form $\sin^m \theta$ and $\sin^m \theta \cos^n \theta$ when these are simply evaluated [Vol. I (18.8)].

For other circular functions we can proceed as

(i)
$$\int \tan^4 \theta\,d\theta = \int \tan^2 \theta(\sec^2 \theta - 1)\,d\theta$$

$$= \int \tan^2 \theta\,d(\tan \theta) - \int \tan^2 \theta\,d\theta$$

$$= \tfrac{1}{3} \tan^3 \theta - \int (\sec^2 \theta - 1)\,d\theta$$

$$= \tfrac{1}{3} \tan^3 \theta - \tan \theta + \theta + C.$$

(ii)
$$I = \int \operatorname{cosec}^3 \theta\,d\theta = \int \operatorname{cosec} \theta\,d(- \cot \theta)$$

$$= \operatorname{cosec} \theta(- \cot \theta) - \int (- \cot \theta)(- \operatorname{cosec} \theta \cot \theta)\,d\theta$$

$$= - \operatorname{cosec} \theta \cot \theta - \int \operatorname{cosec} \theta\,(\operatorname{cosec}^2 \theta - 1)\,d\theta$$

$$= - \operatorname{cosec} \theta \cot \theta - \int \operatorname{cosec}^3 \theta\,d\theta + \int \operatorname{cosec} \theta\,d\theta.$$

Therefore
$$I = -\tfrac{1}{2} \operatorname{cosec} \theta \cot \theta + \tfrac{1}{2} \log |\tan \tfrac{1}{2}\theta| + C.$$

Finally, to find an integral such as

$$I = \int \frac{\sin \theta + 7 \cos \theta}{\sin \theta + 2 \cos \theta}\,d\theta.$$

Let $\sin \theta + 7 \cos \theta \equiv A(\sin \theta + 2 \cos \theta) + B(\cos \theta - 2 \sin \theta)$, where the expression with A is the denominator and that with B is its differential coefficient. From this

$$A - 2B = 1$$
$$2A + B = 7$$

and

$$A = 3, \quad B = 1$$

therefore

$$I = \int \frac{3(\sin \theta + 2 \cos \theta)}{\sin \theta + 2 \cos \theta} \, d\theta + \int \frac{\cos \theta - 2 \sin \theta}{\sin \theta + 2 \cos \theta} \, d\theta$$

$$= \int 3 d\theta + \int d \log |\sin \theta + 2 \cos \theta|$$

$$= \underline{3\theta + \log |\sin \theta + 2 \cos \theta| + C.}$$

EXERCISE 6

Find the integrals of:

1. $\dfrac{1}{4x^2 + 4x + 5}$.

2. $\dfrac{1}{\sqrt{(x^2 - 4x - 21)}}$.

3. $\sqrt{x^2 - 4x - 21}$.

4. $\dfrac{1}{\sqrt{(15 + 4x - 4x^2)}}$.

5. $\sqrt{(9x^2 - 30x - 119)}$.

6. $\dfrac{1}{\sqrt{(x - 1)(2 - x)}}$.

7. $\dfrac{1}{x\sqrt{(3x^2 - 2x - 1)}}$.

8. $\dfrac{1}{(x + 1)\sqrt{(x^2 + 2x)}}$.

9. $\sqrt{(x^2 + 2x + 6)}$.

10. $x/\sqrt{(x^4 - 1)}$.

11. $x\sqrt{(x^4 - 1)}$.

12. $x^2/(x^2 + 1)$.

13. $\dfrac{x^2 + 9}{(x - 2)(x^2 + 2x + 5)}$.

14. $\dfrac{1}{(x^2 + 1)(x + 1)^2}$.

15. xe^{-x^2}.

16. $\dfrac{1}{x^2} \cos \dfrac{1}{x}$.

17. $\dfrac{x}{1 + x^4}$.

18. $\dfrac{\sec^2 \theta}{4 + \tan^2 \theta}$.

19. $\dfrac{\sin x}{(1 + \cos x)^2}$.

20. $x\sqrt{(1 + x^2)}$.

21. $\dfrac{1}{1 + 2 \cos \theta}$.

22. $\dfrac{1}{1 + 2 \operatorname{ch} x}$.

23. $\dfrac{1}{1 + \sqrt{x}}$.

24. $\dfrac{1}{\sqrt{(x + a)} - \sqrt{(x - a)}}$.

25. $\dfrac{x^2 - x + 1}{2x - 1}$.

26. $\sqrt{\{(x - 1)/(x + 2)\}}$.

Evaluate:

27. $\displaystyle\int_1^2 \frac{\log x}{(x + 1)^2} \, dx$.

28. $\displaystyle\int_0^\pi \cos^2 x(1 + \cos x)^3 dx$.

29. $\displaystyle\int_0^{\frac{1}{2}\pi} \sqrt{(16 \sin^4 \theta + 64 \sin^2 \theta \cos^2 \theta)} d\theta$.

30. $\displaystyle\int_0^{\frac{1}{2}\pi} \frac{dx}{2 + \cos x}$.

31. $\displaystyle\int_0^a (x + \sqrt{(a^2 - x^2)}) dx$.

32. $\displaystyle\int_b^a \sqrt{\frac{a - x}{x - b}} dx$.

33. $\displaystyle\int_0^{\frac{1}{2}\pi} \tan^4 \theta d\theta$.

34. $\displaystyle\int_0^1 \frac{dx}{(1+x^2)^2}.$

35. $\displaystyle\int_0^{\frac{1}{2}\pi} \frac{dx}{2\cos x + \sin x}.$

36. $\displaystyle\int_0^{\frac{1}{2}\pi} \frac{d\theta}{16\cos^2\theta + \sin^2\theta}.$

37. $\displaystyle\int_0^\infty e^{-2x}\operatorname{sh} x\,dx.$

38. Find A, B, C so that
$$4 + 5\sin x + \cos x = A + B(\cos x - \sin x) + C(1 + \sin x + \cos x).$$
Hence find
$$\int_0^{\frac{1}{2}\pi} \frac{4 + 5\sin x + \cos x}{1 + \sin x + \cos x}\,dx.$$

39. $\displaystyle\int_3^5 \frac{dx}{x(41 - x^2)\sqrt{(25 - x^2)}}.$ (Put $25 - x^2 = z^2$ and use partial fractions.)

40. Show that
$$\int_0^\infty \frac{dx}{1+x^3} = \int_0^\infty \frac{x\,dx}{1+x^3} = \frac{1}{2}\int_0^\infty \frac{dx}{1 - x + x^2}.$$
Hence evaluate the first integral.

41. Prove $\displaystyle\int_0^\pi \frac{d\theta}{4 - 3\sin\theta} = \frac{\pi}{\sqrt{7}} + \frac{2}{\sqrt{7}}\tan^{-1}\left(\frac{3}{\sqrt{7}}\right).$

The following examples are all taken from London University examination papers B.Sc.(Eng.) Part II.

42. Prove that
$$\int_0^n x^2(n - x)^p dx = \int_0^n x^p(n - x)^2 dx$$
and find the common value of the two integrals.

43. Evaluate:

  (i) $\displaystyle\int_0^1 \sin^{-1} x\,dx\,;$ (ii) $\displaystyle\int_2^4 \frac{(3x + 2)}{x^2 + 3x - 4}\,dx\,;$ (iii) $\displaystyle\int_0^{\pi/6} \frac{\sin^2 x}{\cos x}\,dx.$

44. Evaluate: (i) $\displaystyle\int_{\pi/6}^{\pi/4} \frac{dx}{\sin^4 x \cos^4 x}\,;$ (ii) $\displaystyle\int_{1/2}^{2/3} \frac{dx}{x(3x - 2x^2 - 1)^{1/2}}.$

45. Find the centroid of a sector of a circle when the surface density varies as the distance from the centre. [Assume that the radius is a and the angle of the sector is 2α.]

46. If $\displaystyle I_n = \int e^{ax}\cos^n x\,dx$, where n is a positive integer prove that
$$I_n = \frac{1}{a^2 + n^2}e^{ax}\cos^{n-1}x(a\cos x + n\sin x) + \frac{n(n-1)}{a^2 + n^2}I_{n-2}.$$
Hence evaluate
$$\int_0^{\frac{1}{2}\pi} e^x\cos^4 x\,dx.$$

47. Evaluate:

  (i) $\displaystyle\int \frac{(x - 1)dx}{(x + 1)(x^2 + 1)}\,;$ (ii) $\displaystyle\int_1^2 x^2\log x\,dx\,;$

                 (iii) $\displaystyle\int_0^{\pi/4} \sin^2\theta\cos^4\theta\,d\theta\,;$

(iv) The density of a sphere of radius a at distance r from the centre is $k \left(1 - \dfrac{2r}{3a}\right)$. Prove that the mean density is half the density at the centre.

48. Find the following indefinite integrals:

(i) $\displaystyle\int \tan^4 \theta d\theta$;   (ii) $\displaystyle\int \cos \theta d\theta / (5 + 4 \cos \theta)$;   (iii) $\displaystyle\int x^3 dx / (a^2 + x^2)^2$

and evaluate correct to three significant figures;

$$\text{(iv)} \int_{0\cdot 5}^{2} x^3 \log x dx.$$

49. Evaluate: (i) $\displaystyle\int_{0}^{2} (1 + x) \sqrt{\left(\dfrac{2 - x}{2 + x}\right)} dx$;   (ii) $\displaystyle\int_{0}^{1} \dfrac{dx}{(3 - x^2)\sqrt{(1 + x^2)}}$.

50. Find:

(i) $\displaystyle\int \dfrac{dx}{x^2 + 4x + 5}$;

(ii) $\displaystyle\int \dfrac{dx}{x^2 + 4x + 3}$;

(iii) $\displaystyle\int x^2 \sin x dx$;

(iv) $\displaystyle\int \dfrac{d\theta}{5 + 3 \cos \theta}$.

51. Prove that:

(i) $\displaystyle\int_{0}^{\frac{1}{2}\pi} \log \sin x dx = \int_{0}^{\frac{1}{2}\pi} \log \cos x dx = \tfrac{1}{2} \int_{0}^{\pi/2} \log \sin 2x dx - \dfrac{\pi}{4} \log 2$;

(ii) $\displaystyle\int_{0}^{\frac{1}{2}\pi} \log \sin 2x dx = \tfrac{1}{2} \int_{0}^{\pi} \log \sin x dx = \int_{0}^{\pi/2} \log \sin x dx.$

Deduce that
$$\int_{0}^{\frac{1}{2}\pi} \log \sin x dx = \dfrac{\pi}{2} \log \dfrac{1}{2}.$$

52. Find: (a) $\displaystyle\int \dfrac{\sin \theta d\theta}{\sqrt{(\sin^2 \alpha - \sin^2 \theta)}}$;   (b) $\displaystyle\int \dfrac{\cos \theta d\theta}{\sqrt{(\sin^2 \alpha - \sin^2 \theta)}}$.

Evaluate to 3 decimal places:

(c) $\displaystyle\int_{0}^{1} (1 - x)^3 \sqrt{x} dx$;   (d) $\displaystyle\int_{0}^{\frac{1}{4}\pi} \sec^3 \theta d\theta.$

53. Find:

(i) $\displaystyle\int \sqrt{\left(\dfrac{1 - x}{1 + x}\right)} dx$;   (ii) $\displaystyle\int \dfrac{dn}{\cosh n}$.

(iii) Evaluate correct to three significant figures
$$\int_{0}^{2} \dfrac{dx}{x^3 + 8}.$$

54. Prove that
$$\int_{0}^{2a} x^m (2ax - x^2)^{\frac{1}{2}} dx = \dfrac{a^{m+2} \pi (2m + 1)\,!}{2^m m\,! \,(m + 2)\,!}$$

where m is a positive integer.

Hence or otherwise find: (a) the area enclosed by the curve $a^2 y^2 = x^2 (2ax - x^2)$ and (b) the second moment of this area about the y axis.

55. Evaluate:

(i) $\displaystyle\int \frac{(x-1)dx}{\sqrt{(4+3x-x^2)}}$; (ii) $\displaystyle\int \frac{dx}{(x-1)\sqrt{(4+3x-x^2)}}$;

(iii) $\displaystyle\int \frac{2\cos x + 3\sin x}{3 + 5\cos x}\,dx.$

56. Prove that

$$\int_0^{\phi-e} \frac{\sin\phi\,d\theta}{\cos\theta - \cos\phi} = \log\frac{\sin\left(\phi - \dfrac{e}{2}\right)}{\sin\frac{1}{2}e}.$$

Hence or otherwise prove that

$$\lim_{e \to 0}\left[\int_0^{\phi-e}\frac{d\theta}{\cos\theta - \cos\phi} + \int_{\phi+e}^{\pi}\frac{d\theta}{\cos\theta - \cos\phi}\right] = 0.$$

57. Evaluate: (i) $\displaystyle\int_{-1}^{+1} t^4(1-t^2)^{3/2}dt$;

(ii) $\displaystyle\int_0^{\pi/2}\frac{dx}{(1 + a^2\cos^2 x + b^2\sin^2 x)}$;

(iii) $\displaystyle\int_0^1 \frac{dx}{(1+x^2)\sqrt{(2+x^2)}}.$

58. If $\tan\frac{1}{2}\theta = \sqrt{\{(1+k)/(1-k)\}}\,.\,\tan\frac{1}{2}\phi$ show that:

(i) $(1 + k\cos\theta)(1 - k\cos\phi) = 1 - k^2$;

(ii) $d\theta/d\phi = \sqrt{(1-k^2)}/(1 - k\cos\phi).$

Hence evaluate $\displaystyle\int_0^{\pi} d\theta/(1 + k\cos\theta)^3.$

59. Prove that

$$\int_0^a f(x)dx = \int_0^a f(a-x)dx.$$

Evaluate

$$\int_0^{\frac{1}{2}\pi} \frac{x\sin 2x}{1 + \cos^2 2x}\,dx$$

and show

$$\int_0^{\frac{1}{2}\pi} \frac{a\sin x + b\cos x}{\sin x + \cos x}\,dx = \frac{(a+b)\pi}{4}$$

60. By putting $t = \tan\frac{1}{2}x$ or otherwise prove that

$$\int_0^{\pi} \frac{dx}{1 - 2a\cos x + a^2} = \frac{\pi}{1 - a^2}.$$

Evaluate the above integral approximately by Simpson's Rule with three ordinates and show that Simpson's Rule will give the exact value if $3a^2 = 1$.

61. A square of side 6 mm rolls outside a fixed circle of radius 2 mm starting with the mid point of one side touching the circle at a point A. Show that when the point of contact has moved to B, the length of the arc of the path of the centre of the square is given by

$$s = \int_0^{\theta} \sqrt{(9 + 4\phi^2)}d\phi$$

where θ is the angle subtended at the centre of the circle by the arc AB, provided $\theta \leqslant 3/2$. Find this length correct to 3 significant figures when the arc AB is 2 mm long.

62. Evaluate: (i) $\displaystyle\int \frac{x\,dx}{x^2 + 2x - 4}$; (ii) $\displaystyle\int x^2 \sin^2 x\,dx$ and prove that

$$\int_0^{\frac{1}{2}\pi} \frac{dx}{a^2 \cos^2 x + b^2 \sin^2 x} = \frac{\pi}{2ab}.$$

63. Find: (i) $\displaystyle\int_2^5 \frac{2x^2 - 3x + 4}{(x-1)^2}\,dx$; (ii) $\displaystyle\int_0^\infty \frac{dx}{x^3 + 8}$.

64. (a) Put $e^x = z$ and show that

$$\int_0^\infty \operatorname{sech} x\,dx = 2\tan^{-1}(e^x)\Big]_0^\infty = \frac{\pi}{2}.$$

(b) Find $\displaystyle\int \frac{dx}{1 + 3e^x + 2e^{2x}}.$

65. Show that

$$\int_0^{\log 2} \frac{dx}{\operatorname{sh} x + 5\operatorname{ch} x} = \frac{1}{\sqrt{6}}\{\tan^{-1}\sqrt{6} - \tan^{-1}\tfrac{1}{2}\sqrt{6}\}.$$

CHAPTER 3

TAYLOR'S THEOREM. EXPANSION IN SERIES. APPROXIMATE INTEGRATION

"The age of chivalry is gone. That of . . . calculators has succeeded and the glory of Europe is extinguished for ever."

E. Burke (1792).

3.1. Taylor's Theorem

We have had Maclaurin's Theorem [Vol. I (26.3)]

$$f(h) = f(0) + hf'(0) + \frac{h^2}{2!} f''(0) + \dots \quad . \quad . \quad (1)$$

In Fig. 6(a) this may be taken as expressing the height $f(h)$ in terms of the values of the function at the origin. If we now transfer the

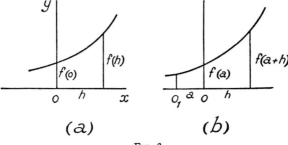

(a)   (b)

FIG. 6.

origin to 0_1, a distance a to the left, the height is now $f(a + h)$ expressed in terms of the values at a and putting $a + h$ for h and a for 0 in the above expansion gives the more general expansion

$$f(a + h) = f(a) + hf'(a) + \frac{h^2}{2!} f''(a) + \dots \quad . \quad (2)$$

in which, when required, there are two variables a and h.

Example 1. Expand $\sin (x + h)$ in powers of h and find the value of $\sin 46°$ correct to 5 decimals.

If $f(x) = \sin x, f'(x) = \cos x, f''(x) = -\sin x$, etc.

By (2), with x instead of a,

$$\sin (x + h) = \sin x + h \cos x + \frac{h^2}{2!} (- \sin x) + \frac{h^3}{3!} (- \cos x) + \dots$$

When $x = \frac{1}{4}\pi$ and $h = 1° = 0.017\ 453$,

$$\sin 46 = \frac{1}{\sqrt{2}} + (0.017\ 453)\,\frac{1}{\sqrt{2}} - \frac{(0.017\ 453)^2}{2}\left(\frac{1}{\sqrt{2}}\right)\ \cdots$$

$$= 0.707\ 106\ 8 + 0.012\ 341\ 4 - 0.000\ 107\ 7$$
$$= 0.719\ 340\ 5$$
$$= 0.719\ 34 \text{ to 5 decimal places.}$$

The value from 7-figure tables is $0.719\ 339\ 8$.

Example 2. Put $x^5 + 2$ into powers of $x + 1$ and hence find

$$\int \frac{x^5 + 2}{(x+1)^7}\, dx.$$

If $f(x) = x^5 + 2$, then with

$$x = -1 + (x + 1) = a + h$$

so that $a = -1$, $\ h = x + 1$

$$f(x) = f(a + h)$$

$$= f(-1) + (x+1)f'(-1) + \frac{(x+1)^2}{2!}f''(-1) + \frac{(x+1)^3}{3!}f^3(-1)$$

$$+ \frac{(x+1)^4}{4!}f^4(-1) + \frac{(x+1)^5}{5!}f^5(-1) + \ldots \qquad \cdot \ \cdot \ (1)$$

Since
$$\begin{array}{ll} f(x) = x^5 + 2 & f(-1) = 1 \\ f'(x) = 5x^4 & f'(-1) = 5 \\ f''(x) = 20x^3 & f''(-1) = -20 \\ f^3(x) = 60x^2 & f^3(-1) = 60 \\ f^4(x) = 120x & f^4(-1) = -120 \\ f^5(x) = 120 & f^5(-1) = 120 \\ f^n(x) = 0 \quad \text{for } n > 5. & \end{array}$$

By (1)
$$x^5 + 2 = 1 + 5(x+1) - \frac{20}{2!}(x+1)^2 +$$
$$\frac{60}{3!}(x+1)^3 - \frac{120}{4!}(x+1)^4 + \frac{120}{5!}(x+1)^5.$$

From this
$$\int \frac{x^5 + 2}{(x+1)^7}\,dx = \int \left[\frac{1}{(x+1)^7} + \frac{5}{(x+1)^6} - \frac{10}{(x+1)^5} + \right.$$
$$\left. \frac{10}{(x+1)^4} - \frac{5}{(x+1)^3} + \frac{1}{(x+1)^2}\right]dx$$

$$= -\frac{1}{6(x+1)^6} - \frac{1}{(x+1)^5} +$$
$$\frac{5}{2(x+1)^4} - \frac{10}{3(x+1)^3} + \frac{5}{2(x+1)^2} - \frac{1}{x+1} + C.$$

3.2. We have met with several methods of expanding a given function in powers of x, and must now consider a new method involving the formation of a differential equation. As a preliminary the following must first be considered.

3.3. Leibnitz's Theorem

History records that Leibnitz, a contemporary of Sir Isaac Newton, spent many years endeavouring to find the formula for differentiating once the product of two functions. Posterity has rewarded him by giving his name to a theorem for differentiating n times the product of two functions.

If u, v are two functions of x, differentiable n times, let

$$y = uv$$

then

$$y_1 = uv_1 + u_1v$$

and

$$y_2 = uv_2 + 2u_1v_1 + u_2v$$

$$y_3 = uv_3 + 3u_1v_2 + 3u_2v_1 + u_3v.$$

We can deal with this in the following manner. Let P be an operator which differentiates u only leaving v unchanged whilst Q acts similarly on v only, then we may represent these results as

$$y_1 = (P + Q)uv$$

for this is

$$= P(uv) + Q(uv)$$

$$= u_1v + uv_1$$

since $P(uv) = vPu = vu_1$ and $Q(uv) = uQv = uv_1$.

Similarly,

$$y_2 = (P + Q)^2uv$$

$$= (P^2 + 2PQ + Q^2)uv$$

for this is

$$= P^2(uv) + 2PQ(uv) + Q^2(uv)$$

$$= u_2v + 2u_1v_1 + uv_2$$

since, for example, $2PQ(uv) = 2P \cdot Q(uv)$

$$= 2P(uv_1)$$

$$= 2u_1v_1$$

and generally

$$y_n = (P + Q)^n uv$$

$$= (P^n + nP^{n-1}Q + \frac{n(n-1)}{1 \cdot 2} P^{n-2}Q^2 + \ldots + Q^n)uv$$

$$= u_nv + nu_{n-1}v_1 + \frac{n(n-1)}{1 \cdot 2} u_{n-2}v_2 + \ldots uv_n.$$

In most applications we are able to take as u the function whose nth differential coefficient can be written down easily and for v a function which gives zero after a small number of differentiations.

Example 3. Find the nth differential coefficient of x^3e^{2x}.

If

$$u = e^{2x}, \quad u_1 = 2e^{2x}, \quad u_2 = 2^2e^{2x}, \quad \ldots \quad u_n = 2^ne^{2x}.$$

$$v = x^3, \quad v_1 = 3x^2, \quad v_2 = 6x, \quad v_3 = 6$$

and v_4 and all further differentiations give zero.

By the formula

$$y_n = (2^ne^{2x})x^3 + n(2^{n-1}e^{2x})(3x^2) + \frac{n(n-1)}{1 \cdot 2} (2^{n-2}e^{2x})(6x) +$$

$$\frac{n(n-1)(n-2)}{1 \cdot 2 \cdot 3} (2^{n-3}e^{2x})(6).$$

$$= 2^{n-3}e^{2x}\{8x^3 + 12nx^2 + 6n(n-1)x + n(n-1)(n-2)\}.$$

Example 4. Differentiate n times $(1 + x^2)y_2 + 2xy_1 = 7y$.

By the formula

$$D^n\{(1 + x^2)y_2\} = (1 + x^2)y_{n+2} + n(2x)y_{n+1} + \frac{n(n-1)}{1 \cdot 2}(2)y_n$$
$$D^n\{2xy_1\} = (2x)y_{n+1} + n(2)y_n$$
$$D^n\{7y\} = 7y_n.$$

The result required is

$$(1 + x^2)y_{n+2} + 2(n + 1)xy_{n+1} + (n^2 + n - 7)y_n = 0.$$

3.4. The Value of $(y_n)_0$

If we assume that as an expansion in powers of x

$$y = a_0 + a_1x + \frac{a_2x^2}{2!} + \frac{a_3x^3}{3!} + \ldots \frac{a_nx^n}{n!} + \ldots$$

putting $x = 0$ gives

$$(y)_0 = a_0.$$

Differentiate :

$$y_1 = a_1 + \frac{a_2x}{1!} + \frac{a_3x^2}{2!} + \ldots \frac{a_nx^{n-1}}{(n-1)!} + \ldots$$

with $x = 0$,

$$(y_1)_0 = a_1 \quad . \quad . \quad . \quad . \quad . \quad . \quad (1)$$

Proceeding in this way by successive differentiation of the series and then putting $x = 0$ we obtain, for all n

$$(y_n)_0 = a_n \quad . \quad . \quad . \quad . \quad . \quad . \quad (2)$$

where we have assumed for the expansion

$$y = \Sigma a_n \frac{x^n}{n!} \quad . \quad . \quad . \quad . \quad . \quad (3)$$

Note the factorial in the denominator, since the above relation (2) is not true if we assume

$$y = \Sigma a_n x^n.$$

It then appears that if we can obtain an expression involving differential coefficients, we can put $x = 0$ in it, and each y_n becomes a_n if y and a_n are connected by (3).

3.5. Formation of the Differential Equation

Suppose

$$y = (\sin^{-1} x)^2$$

then

$$y_1 = \frac{2(\sin^{-1} x)}{\sqrt{(1 - x^2)}}$$

and

$$\sqrt{(1 - x^2)}y_1 = 2(\sin^{-1} x)$$

differentiate :

$$\sqrt{(1 - x^2)}y_2 + y_1 \cdot \frac{-x}{\sqrt{(1 - x^2)}} = \frac{2}{\sqrt{(1 - x^2)}}$$

or

$$(1 - x^2)y_2 - xy_1 = 2 \quad . \quad . \quad . \quad . \quad (4)$$

We have now eliminated the special function concerned, $(\sin^{-1} x)^2$, and obtained an equation involving powers of x and orders of y_n only. If this is differentiated n times by Leibnitz's Theorem :

$$(1 - x^2)y_{n+2} + n(-2x)y_{n+1} + \frac{n(n-1)}{2}(-2)y_n$$
$$- xy_{n+1} + n(-1)y_n = 0.$$

Putting $x = 0$ in this and using (2) above

$$a_{n+2} - n(n-1)a_n - na_n = 0$$

or
$$a_{n+2} = n^2 a_n . \quad . \quad . \quad . \quad . \quad . \quad (5)$$

a relation between coefficients in the expansion of

$$(\sin^{-1} x)^2 = \Sigma \frac{a_n x^n}{n!}.$$

It must be noticed that this relation (5) is only true for $n > 0$, since we obtained it by differentiating (4) n times. We cannot put $n = 0$ in it to find $a_2 = 0$, but we may put $n = 1$ to obtain a_3 in terms of a_1.

3.6. Expansion in Series

Consider the above example

$$y = (\sin^{-1} x)^2 = \sum_{n=0}^{\infty} \frac{a_n x^n}{n!}.$$

Putting $x = 0$:
$$0 = a_0$$

from
$$y_1 = \frac{2 \sin^{-1} x}{\sqrt{(1 - x^2)}}$$

$x = 0$ gives :
$$(y_1)_0 = a_1 = 0$$

from
$$(1 - x^2)y_2 - xy_1 = 2$$

$x = 0$ gives :
$$a_2 = 2.$$

using (5) above :
$$a_{n+2} = n^2 a_n$$

$n = 1$ gives :
$$a_3 = 1^2 a_1 = 0$$

$n = 2$,,
$$a_4 = 2^2 a_2 = 2^2 . 2$$

$n = 3$,,
$$a_5 = 3^2 a_3 = 0$$

$n = 4$,,
$$a_6 = 4^2 a_4 = 4^2 . 2^2 . 2.$$

It is clear that the coefficients of odd powers of x are zero. This could have been seen in advance, since y is an even function of x. The expansion is

$$(\sin^{-1} x)^2 = \frac{2}{2!} x^2 + \frac{2 . 2^2}{4!} x^4 + \frac{2 . 2^2 . 4^2}{6!} x^6 + \cdots$$
$$+ \frac{2 . 2^2 . 4^2 . 6^2 \cdots (2n-2)^2}{(2n)!} x^{2n} + \cdots$$

Example 5. Expand $y = (\sinh^{-1} x)/\sqrt{(1 + x^2)}$ as a power series in x.

We have
$$y\sqrt{(1 + x^2)} = \sinh^{-1} x \quad \cdots \cdots \quad (6)$$

and if
$$y = \Sigma \frac{a_n x^n}{n!}$$

$x = 0$ in (6) gives $\quad a_0 = 0.$

Differentiate:
$$y \frac{x}{\sqrt{(1 + x^2)}} + \sqrt{(1 + x^2)}y_1 = \frac{1}{\sqrt{(1 + x^2)}} \quad \cdots \cdots \quad (7)$$

$x = 0$: $\quad a_1 = 1.$

From (7) $\quad xy + (1 + x^2)y_1 = 1$

differentiate: $\quad xy_1 + y + (1 + x^2)y_2 + y_1 2x = 0 \quad \cdots \cdots \quad (8)$

$x = 0$: $\quad a_0 + a_2 = 0,$

therefore $\quad a_2 = 0.$

(8) is $\quad y + (1 + x^2)y_2 + 3xy_1 = 0.$

Differentiate n times:
$$y_n + (1 + x^2)y_{n+2} + n(2x)y_{n+1} + \frac{n(n - 1)}{2}(2)y_n + 3xy_{n+1} + 3n(1)y_n = 0$$

$x = 0$: $\quad a_n + a_{n+2} + n(n - 1)a_n + 3na_n = 0$

or $\quad a_{n+2} + (n^2 + 2n + 1)a_n = 0 \quad \cdots \cdots \quad (9)$

in (9) $n = 1$: $\quad a_3 = -2^2 a_1 = -2^2$

$n = 2$: $\quad a_4 = -3^2 a_2 = 0.$

By changing the sign of x in the given function for y we find that it is an odd function. All even coefficients will therefore be zero.

$n = 3$: $\quad a_5 = -4^2 a_3 = 2^2 \cdot 4^2$

$n = 5$: $\quad a_7 = -6^2 a_5 = -2^2 \cdot 4^2 \cdot 6^2.$

The expansion is
$$y = x - \frac{2^2}{3!}x^3 + \frac{2^2 \cdot 4^2}{5!}x^5 - \frac{2^2 \cdot 4^2 \cdot 6^2}{7!}x^7 + \cdots$$

$$= x - \frac{2}{3}x^3 + \frac{2 \cdot 4}{3 \cdot 5}x^5 - \frac{2 \cdot 4 \cdot 6}{3 \cdot 5 \cdot 7}x^7 + \cdots$$

The student will note the method of obtaining the first two or three coefficients from the given function and its derivatives. But to obtain the general relation between successive coefficients, the differential equation must be formed, differentiated n times and the procedure above followed.

3.7. Approximate Integration

We will confine ourselves to applications of Simpson's Rule only. The student will know that if over a range $-h$ to $+h$ it is assumed that a function may be represented as
$$f(x) = a + bx + cx^2 + dx^3,$$

then the area under it between the ordinates is given by
$$\tfrac{1}{3}h\{f(-h) + 4f(0) + f(h)\},$$

a formula requiring the evaluation of three ordinates. An interval of any length may be divided into an even number of equal parts, each of length h, and on applying this rule to each succession of three ordinates, the formula becomes:

total area under curve

$$= \frac{h}{3} \text{[first ordinate + last ordinate + 4 (sum of even ordinates)}$$
$$+ 2 \text{ (sum of } \textit{other} \text{ odd ordinates)].}$$

A tabular layout ensures that ordinates are covered once only, as demonstrated in the following examples.

Example 6. The net accelerative force acting on a train of mass 400 tonnes varies with the speed, being $F(v) \times 10^{-4}$ N when the speed is v m/s (1 tonne = 1000 kg).

If $F(v)$ is given by the table below, find the time taken to reach a speed of 54 km/h (= 15 m/s).

V	0	2·5	5	7·5	10	12·5	15	m/s
$F(v)$	8·93	8·67	8·36	7·99	7·56	7·07	6·52	$\times 10^4$ N

$$F = ma$$

If v denotes the speed in m/s, $a = \dfrac{dv}{dt}$

$$F(v) \times 10^4 = 400 \times 10^3 \frac{dv}{dt}$$

$$dt = \frac{40}{F(v)} \, dv$$

$$t = 40 \int \frac{1}{F(v)} \, dv$$

$$\text{and } t = 40 \int_0^{15} \frac{1}{F(v)} \, dv$$

This integral will be evaluated by Simpson's rule.

$$\frac{1}{F(v)} = 0{\cdot}1120, \ 0{\cdot}1153, \ 0{\cdot}1196, \ 0{\cdot}1252, \ 0{\cdot}1323, \ 0{\cdot}1414 \text{ and } 0{\cdot}1534$$

	First and last	Even	Remaining odd
Width of strip = 2·5	0·1120	0·1153	0·1196
		0·1252	0·1323
	0·1534	0·1414	
	0·2654	0·3819	0·2519

$$t = 40 \int \frac{1}{F(v)} \, dv = 40 \left\{ \frac{2{\cdot}5}{3} \left[0{\cdot}2654 + 4(0{\cdot}3819) + 2(0{\cdot}2519) \right] \right\}$$

$$= \frac{100}{3} [0{\cdot}2654 + 1{\cdot}5276 + 0{\cdot}5038]$$

$$= \frac{100 \times 2{\cdot}2968}{3} = \underline{76{\cdot}6 \text{ seconds}}$$

Example 7. Evaluate

$$\int_0^{1.5} \frac{x^3}{e^x - 1} \, dx$$

by using Simpson's Rule on the interval divided into 6 equal parts. [This integral is required in certain problems in Physics.]

x	x^3	$e^x - 1$	$x^3/(e^x - 1)$
0	0	0	0
0·25	0·0156	0·2840	0·0550
0·50	0·1250	0·6487	0·1927
0·75	0·4219	1·1170	0·3777
1·00	1·0000	1·7183	0·5820
1·25	1·9531	2·4903	0·7843
1·50	3·3750	3·4817	0·9694

	First and last	Even	Remaining odd
Width of strip = 0·25	0	0·0550	0·1927
		0·3777	0·5820
		0·7843	
	0·9694		
	0·9694	1·2170	0·7747

$$I = \frac{0 \cdot 25}{3} \left\{ 0 \cdot 9694 + 4(1 \cdot 2170) + 2(0 \cdot 7747) \right\}$$

$$= \frac{0 \cdot 25 \times 7 \cdot 3868}{3} = 0 \cdot 6156$$

There are many other formulae for approximate integration, but for the amount of work required Simpson's Rule is perhaps the most accurate.

3.8. Other Methods

We have already dealt with the methods in which the integrand is expanded in a power series (by the differential equation method 3.6 if no easier way is available) and integrated term by term to the required degree of accuracy.

Example 8.

$$\int_0^{\frac{1}{2}} \frac{\log (1 + x)}{\sqrt{x}} \, dx = \int_0^{\frac{1}{2}} \frac{1}{\sqrt{x}} \left(x - \frac{x^2}{2} + \frac{x^3}{3} - \frac{x^4}{4} + \frac{x^5}{5} \cdots \right) dx$$

$$= \int_0^{\frac{1}{2}} \left(x^{1/2} - \frac{1}{2} x^{3/2} + \frac{1}{3} x^{5/2} - \frac{1}{4} x^{7/2} + \frac{1}{5} x^{9/2} \cdots \right) dx$$

$$= \left(\frac{2}{3} x^{3/2} - \frac{1}{5} x^{5/2} + \frac{2}{21} x^{7/2} - \frac{1}{18} x^{9/2} \cdots \right)_0^{\frac{1}{2}}$$

$$= \frac{1}{\sqrt{2}} \left(\frac{2}{3} \cdot \frac{1}{2} - \frac{1}{5} \cdot \frac{1}{4} + \frac{2}{21} \cdot \frac{1}{8} - \frac{1}{18} \cdot \frac{1}{16} \cdots \right)$$

$$= 0 \cdot 207 \text{ to 3 sig. figures.}$$

In some cases we can find values between which the integral must lie, even if it cannot be evaluated exactly.
Thus for $x > 1$, clearly

$$e^{-x^2} < xe^{-x^2},$$

and therefore regarding the definite integrals as areas, it is seen that

$$\int_1^\infty e^{-x^2} dx < \int_1^\infty xe^{-x^2} dx.$$

The right-hand integral is

$$\left(-\tfrac{1}{2} e^{-x^2} \right)_1^\infty = \frac{1}{2e},$$

so that
$$\int_1^\infty e^{-x^2}dx < \frac{1}{2e}.$$

A rough sketch of the integrand shows that in the range 0 to 1 the area under it is certainly less than 1. We then have

$$\int_0^1 e^{-x^2}dx + \int_1^\infty e^{-x^2}dx = \int_0^\infty e^{-x^2}dx < 1 + \frac{1}{2e},$$

giving a value for an upper bound to an integral requiring double integration for its exact evaluation (see 15.5).

As another example, for $0 < x < 1$

$$1 < 1 + x^4 < 1 + x^2$$

$$\therefore \quad 1 > \frac{1}{1 + x^4} > \frac{1}{1 + x^2}$$

and
$$\int_0^1 1dx > \int_0^1 \frac{dx}{1 + x^4} > \int_0^1 \frac{dx}{1 + x^2}$$

or
$$1 > \int_0^1 \frac{dx}{1 + x^4} > \frac{\pi}{4},$$

a case where rough upper and lower bounds are easily obtained.

EXERCISE 7

1. Expand $f(x) = 3 + x^4 + 2x^6$ in powers of $(x - 2)$ by Taylor's Theorem, and hence put $f(x)/(x - 2)^4$ into partial fractions.

2. Find
$$\int \frac{x^4 + x^3 + 3x^2 - 2x + 1}{(x - 1)^3}\,dx.$$

3. Expand $\cos x$ in powers of $x - \frac{\pi}{2}$. Hence find $\cos 91°$ correct to 4 decimals.

4. Expand: (a) $\log x$ in powers of $x - 1$; (b) e^x in powers of $x - 2$.

5. If a is near to the root of the equation $f(x) = 0$ so that if $a + h$ is the exact root h is small show that

$$0 = f(a + h) \simeq f(a) + hf'(a).$$

Hence obtain (Newton's formula)

$$a - f(a)/f'(a)$$

as the next approximation to a as a root of $f(x) = 0$.

Given that $x = 4$ is a good approximation to a root of $x^3 - 8x - 40 = 0$, find this root correct to two decimal places.

6. If $y = \log (x + a)$, show that $y_n = (-1)^{n-1}(n-1)!/(x + a)^n$.

7. If $y = \sin (ax + b)$, show that $y_n = a^n \sin (ax + b + \frac{1}{2}n\pi)$.

8. If $y = e^x \cos x$, show that $y_1 = 2^{\frac{1}{2}}e^x \cos (x + \frac{1}{4}\pi)$, $y_2 = 2^{2/2}e^x \cos (x + \frac{2}{4}\pi)$ and find y_n. Hence expand $y = e^x \cos x$ in ascending powers of x.

9. Expand $y = e^x \sin x$ in ascending powers of x.

10. By means of Taylor's Theorem obtain the expansion

$$\tan (\tfrac{1}{4}\pi + x) = 1 + 2x + 2x^2 + \tfrac{8}{3}x^3 + \dots$$

and find also the coefficient of x^4.

Use the expansion to calculate $\tan 46° 6'$ to four decimal places. [L.U.]

11. Use Maclaurin's Theorem to show that as far as the term in x^4

$$\log_e [(1 + x)^{(e^x - 1)}] = x^2 + \tfrac{1}{4}x^4.$$

Hence or otherwise find an approximation to two decimal places of the real roots of

$$\log_e (1 + x)^{16} = (e^x - 1)^{-1} \qquad \text{[L.U.]}$$

12. Prove $D^{n+1}(xy) = (n + 1)D^n y + xD^{n+1}y$, where $D \equiv \dfrac{d}{dx}$. By taking $y = x^{n-1}e^{1/x}$ prove by induction that

$$D^n(x^{n-1}e^{1/x}) = (- 1)^n \frac{e^{1/x}}{x^{n+1}}. \qquad \text{[L U]}$$

13. State Leibnitz's theorem for the nth derivative of a product.

If $$\frac{d^2y}{dx^2} = xy,$$

show that $$\frac{d^{n+2}y}{dx^{n+2}} = x \frac{d^n y}{dx^n} + n \frac{d^{n-1}y}{dx^{n-1}}.$$

If y is a function satisfying these equations for which $y = 0$ and $dy/dx = 1$ when $x = 0$, find the first four non-zero terms in the expansion of y as a series of powers of x. [L.U.]

14. By first expanding the derivative or otherwise obtain a series of powers of x for $\cos^{-1}(1 - 2x)$, giving the formula for the general term. Use the series to evaluate $\cos^{-1}(0.98)$ correct to four decimals. [L.U.]

15. Find the value of $\cos 10°$. What is the maximum value of the error involved if terms after that in x^6 are neglected?

16. If $y = \dfrac{2}{\sqrt{3}} \tan^{-1}\left(\dfrac{x\sqrt{3}}{2 + x}\right)$, prove that $\dfrac{dy}{dx} = \dfrac{1}{1 + x + x^2}$.

If $-1 < x < 1$, by assuming that y can be expanded in a series of ascending powers of x and using the equation

$$(1 - x^3) \frac{dy}{dx} = 1 - x$$

prove that

$$y = x - \frac{x^2}{2} + \frac{x^4}{4} - \frac{x^5}{5} + \frac{x^7}{7} \cdots + \frac{x^{3n+1}}{3n+1} - \frac{x^{3n+2}}{3n+2} + \cdots$$
$$\text{[L.U.]}$$

17. If $y = e^{a \sin^{-1} x}$ show that :

(i) $(1 - x^2)y_2 - xy_1 - a^2 y = 0$;

(ii) $(1 - x^2)y_{n+2} - (2n + 1)xy_{n+1} - (n^2 + a^2)y_n = 0$.

Hence or otherwise expand y into a series of ascending powers of x, giving explicitly the coefficients of x^{2r} and x^{2r+1}. [L.U.]

18. If $y = (1 - x^2)^{\frac{1}{2}} \sin^{-1} x$ show :

(i) $(1 - x^2)y_1 + xy = 1 - x^2$;

(ii) $(1 - x^2)y_{n+1} - (2n - 1)xy_n - n(n - 2)y_{n-1} = 0 \quad (n > 2)$.

Hence or otherwise expand y in a series of ascending powers of x as far as the term in x^7. [L.U.]

19. If $y = (x^2 - 1)^n$, prove that

$$(x^2 - 1)y_{n+2} + 2xy_{n+1} - n(n + 1)y_n = 0.$$

If $P = \dfrac{1}{2^n n!} \dfrac{d^n}{dx^n}(x^2 - 1)^n$, show that P satisfies the equation

$$(1 - x^2)y_2 - 2xy_1 + n(n + 1)y = 0. \qquad \text{[L.U.]}$$

20. If $y = [\log_e \{x + (a^2 + x^2)^{\frac{1}{2}}\}]^2$, show that

$$(a^2 + x^2)y_2 + xy_1 = 2.$$

Differentiate this equation n times, and deduce or find by any other means the expansion for y in terms of positive integral powers of x, giving the general term. [L.U.]

21. Show that

$$2 \cos x \cosh x = \sum_0^\infty 2^{\frac{1}{2}n}\{1 + (-1)^n\} \frac{x^n}{n!} \cos \frac{n\pi}{4}.$$ [L.U.]

22. If $y = (\sin^{-1} x)^2 + a \sin^{-1} x$ where a is a constant, prove that :

$$(1 - x^2)y_2 - xy_1 - 2 = 0.$$

Find the expansion of y in positive integral powers of x and give the coefficients of the terms in x^{2p} and x^{2p+1}.   [L.U.]

23. A function y is such that when $x = 0$, $y = 1$ and $dy/dx = 0$. A differential equation formed from it is

$$xy_2 + y_1 + xy = 0.$$

Find an expression for y as an infinite series in powers of x. [This is $J_0(x)$, Bessel's function of order zero.]

24. Sketch the graph of $y(1 + x^3) = 12$. By the use of Simpson's Rule with nine ordinates calculate the area between the curve, the x axis and the ordinates $x = 1$, $x = 3$. Give the answer correct to two places of decimals.

[L.U.]

25. The velocity v of a particle at distance s from a point on its path is given in the table.

s	0	10	20	30	40	50	60 m
v	47	58	64	65	61	52	38 m/s

Show that the time t is given by $t = \int ds/v$ and by drawing the graph of $1/v$ against s, or otherwise estimate the time taken to traverse the 60 ft.   [L.U.]

26. The co-ordinates of a point on a curve are

x	0	1	2	3	4	5	6
y	8	7·9	7·6	7·1	6·4	5·5	4·4

The portion between $x = 0$ and 6 rotates round the x axis. Find the volume of solid formed and the position of its centre of gravity.

27. A reservoir discharging through sluices at a depth h below the water surface has a surface area A for various values of h as given below :

h	10	11	12	13	14 m
A	950	1070	1200	1350	1530 m²

If t denotes time in minutes, the rate of fall of the surface is given by

$$\frac{dh}{dt} = -\frac{48\sqrt{h}}{A}.$$

Using some graphical or numerical method estimate the time taken for the water level to fall from 14 to 10 m above the sluices.   [L.U.]

28. A body of mass M tonnes is acted upon by a variable force of F kN. It acquires a speed of v km/h after travelling y m. Show that, if M = 600

$$v^2 = \frac{54}{1250} \int F\,dy$$

If F is given by the table below, estimate the velocity at the end of the 400 m from a standing start.

y	0	50	100	150	200	300	400 m
F	900	620	450	340	260	150	80 kN

(1 tonne = 1000 kg)

29. For a certain material the following readings of flux density B and magnetising force H were obtained:

(a) H increasing—

B	.	.	.	0	5600	8000	9700	10 700
H	.	.	.	2	3	4	5	6

(b) H decreasing—

B	.	.	0	5300	7800	8900	9600	10 100	10 400	10 600	10 700
H	.	.	−2	−1	0	1	2	3	4	5	6

Find the area contained by the two (B, H) curves with the H axis horizontal.

30. For all x in the range $(0, \frac{1}{2})$

$$x^2 > x^n > 0 \quad \text{(for } n > 2\text{)}$$

and therefore

$$1 - x^2 < 1 - x^n < 1.$$

Use this to prove that

$$0.524 > \int_0^{\frac{1}{2}} \frac{dx}{\sqrt{(1 - x^n)}} > \frac{1}{2} \quad (n > 2).$$

31. By division show that for $z \neq -1$

$$(1 + z)^{-1} = 1 - z + z^2 - \frac{z^3}{1 + z}$$

so that for $0 < z < 1$, since $z^3/(1 + z) < z^3$

$$1 - z + z^2 - z^3 < \frac{1}{1 + z} < 1 - z + z^2.$$

Deduce that

$$x - \tfrac{1}{2}x^2 + \tfrac{1}{3}x^3 - \tfrac{1}{4}x^4 < \log(1 + x) < x - \tfrac{1}{2}x^2 + \tfrac{1}{3}x^3.$$

32. By considering the graph of $\sin x$ show that

$$\sin x \geqslant \frac{2x}{\pi} \quad \text{for} \quad 0 \leqslant x \leqslant \frac{\pi}{2}.$$

Use this to show that as R increases without limit through positive values, the value of

$$R \int_0^{\pi/2} e^{-R \sin x} dx$$

never exceeds $\pi/2$.   [L.U.]

33. The rate at which a black body radiates heat is given by

$$\int_0^\infty \frac{c_1 dx}{x^5(e^{c_2/Tx} - 1)}$$

where c_1, c_2, T are positive constants. Transform this to the form

$$\left(\frac{c_1 T^4}{c_2^4}\right) \int_0^{-\infty} \frac{z^3 e^z dz}{1 - e^z}$$

where $e^z < 1$, and thence obtain its value as

$$\left(\frac{c_1 T^4}{c_2^4}\right) \int_0^{-\infty} z^3(e^z + e^{2z} + e^{3z} + \ldots) dz$$

$$= \left(\frac{6 c_1 T^4}{c_2^4}\right) \left(1 + \frac{1}{2^4} + \frac{1}{3^4} + \frac{1}{4^4} + \ldots\right).$$

CHAPTER 4

DETERMINANTS

> " In the middle of the eighteenth century one of the independent discoverers of the fundamental idea, Cramer (in 1750), was fortunate enough to attract attention to it and it soon became the common property of mathematicians in France and elsewhere. . . Cauchy . . . Jacobi . . . Cayley . . . Sylvester and Hermite contributed. . . . In 1850 correspondence appeared which showed that the fundamental idea had been clear to Leibnitz half a century before Cramer's time."
>
> T. Muir, *The Theory of Determinants.*

4.1. Second-order Determinants

We will begin with the problem in which Leibnitz first built up determinant theory, but modern notation will be used.

The solution of the pair of first-degree equations

$$\left. \begin{array}{l} a_1 x + b_1 y = k_1 \\ a_2 x + b_2 y = k_2 \end{array} \right\} \quad \cdot \quad \cdot \quad \cdot \quad \cdot \quad \cdot \quad (1)$$

is given by the expressions

$$x = \frac{k_1 b_2 - k_2 b_1}{a_1 b_2 - a_2 b_1}, \quad y = \frac{k_2 a_1 - k_1 a_2}{a_1 b_2 - a_2 b_1} \quad \cdot \quad \cdot \quad (2)$$

where we assume that $a_1 b_2 - a_2 b_1 \neq 0$.

We can simplify the solution of the above system of equations and, more importantly, of systems of first-degree equations in three or more unknowns by the introduction of *determinants*.

By a determinant of the second order we mean the symbol

$$\begin{vmatrix} a_1 & b_1 \\ a_2 & b_2 \end{vmatrix} \equiv a_1 b_2 - a_2 b_1$$

which is evaluated as shown. A second-order determinant has two (horizontal) rows and an *equal* number of (vertical) columns. With this notation we may now write the solution of the above equations as

$$x = \begin{vmatrix} k_1 & b_1 \\ k_2 & b_2 \end{vmatrix} \Big/ \begin{vmatrix} a_1 & b_1 \\ a_2 & b_2 \end{vmatrix} = \begin{vmatrix} k_1 & b_1 \\ k_2 & b_2 \end{vmatrix} / D,$$

$$y = \begin{vmatrix} a_1 & k_1 \\ a_2 & k_2 \end{vmatrix} \Big/ \begin{vmatrix} a_1 & b_1 \\ a_2 & b_2 \end{vmatrix} = \begin{vmatrix} a_1 & k_1 \\ a_2 & k_2 \end{vmatrix} / D.$$

The common denominator D is called the determinant of the coefficients, and is the left-hand side of the equations with the variables x, y missed out and the determinant bars inserted. Thus the expressions

$$\begin{array}{l} a_1 x + b_1 y \\ a_2 x + b_2 y \end{array} \quad \text{give} \quad \begin{vmatrix} a_1 & b_1 \\ a_2 & b_2 \end{vmatrix} \equiv D$$

as the determinant of the coefficients. The respective numerators are the above determinant with the column of values $\begin{matrix} k_1 \\ k_2 \end{matrix}$ substituted for the column of coefficients associated with the variable x or y whose value is being found. Thus to find x we change the column $\begin{matrix} a_1 \\ a_2 \end{matrix}$ in D into $\begin{matrix} k_1 \\ k_2 \end{matrix}$ to obtain the determinant which divided by D gives x. To find y we act similarly with a determinant in which the column $\begin{matrix} b_1 \\ b_2 \end{matrix}$ in D has been changed into $\begin{matrix} k_1 \\ k_2 \end{matrix}$.

Example 1. Solve the pair of equations $3x + 2y - 1 = 0$, $7y = -4x - 3$. Write these :

$$3x + 2y = 1$$
$$4x + 7y = -3.$$

Here
$$D = \begin{vmatrix} 3 & 2 \\ 4 & 7 \end{vmatrix} = 21 - 8 = 13,$$

$$\therefore \quad 13x = \begin{vmatrix} 1 & 2 \\ -3 & 7 \end{vmatrix} = 13 \text{ and } x = 1.$$

$$13y = \begin{vmatrix} 3 & 1 \\ 4 & -3 \end{vmatrix} = -13 \text{ and } y = -1.$$

This method shows that the pair of equations (1) will always have a definite finite solution provided $a_1b_2 - a_2b_1 \neq 0$. If this relation is zero it may be written

$$\frac{a_1}{b_1} = \frac{a_2}{b_2}$$

showing that the two lines (1) whose point of intersection we are finding are in fact parallel.

EXERCISE 8

1. Evaluate the following determinants :

(a) $\begin{vmatrix} 3 & 4 \\ 1 & 2 \end{vmatrix}$ (b) $\begin{vmatrix} 2 & -1 \\ 4 & -3 \end{vmatrix}$ (c) $\begin{vmatrix} 1 & 2 \\ 2 & 4 \end{vmatrix}$ (d) $\begin{vmatrix} \cos\theta & -\sin\theta \\ \sin\theta & \cos\theta \end{vmatrix}$

2. If j denotes $\sqrt{-1}$ show that

(a) $\begin{vmatrix} 1 + 3j & 5 + 4j \\ -5 + 4j & 1 - 3j \end{vmatrix} = 1^2 + 3^2 + 4^2 + 5^2;$

(b) $\begin{vmatrix} a + jb & c + jd \\ -c + jd & a - jb \end{vmatrix} = a^2 + b^2 + c^2 + d^2.$

3. Solve the following sets of simultaneous equations :

(a) $3x - 2y = 5$ (b) $4a + 2b = 1$ (c) $i_1 + i_2 = 0$
     $x + y = 4$ $a - b = 4$ $3i_1 - 4i_2 = 5$

4. Solve when possible :

(a) $x - 3y = 2$ (b) $x + 2y = 4$ (c) $y - x = -4$
     $2x + 4y = 5$ $3x + 6y = 5$ $7x - 7y = 28$

5. Express as a ratio of $x : z$ and $y : z$ and hence solve the pair of equations to find $x : y : z$:

$$7x + y - z = 0;$$
$$x + 2y + 4z = 0.$$

6. Solve the equations :

(a) $\begin{vmatrix} 3 - x & 2 \\ 4 & 3 + x \end{vmatrix} = 0;$ (b) $\begin{vmatrix} x & -2 \\ x + 2 & 3x + 2 \end{vmatrix} = 3.$

7. Show that if the equations $ax^2 + bx + c = 0$, $lx^2 + mx + n = 0$ have a common root, then

$$\begin{vmatrix} c & a \\ n & l \end{vmatrix}^2 = \begin{vmatrix} b & c \\ m & n \end{vmatrix} \times \begin{vmatrix} a & b \\ l & m \end{vmatrix}.$$

4.2. Third-order Determinants

A third-order determinant consists of three rows and three columns. It is evaluated in terms of the first row as shown:

$$\begin{vmatrix} a_1 & b_1 & c_1 \\ a_2 & b_2 & c_2 \\ a_3 & b_3 & c_3 \end{vmatrix} = a_1 \begin{vmatrix} b_2 & c_2 \\ b_3 & c_3 \end{vmatrix} - b_1 \begin{vmatrix} a_2 & c_2 \\ a_3 & c_3 \end{vmatrix} + c_1 \begin{vmatrix} a_2 & b_2 \\ a_3 & b_3 \end{vmatrix}$$
$$= a_1(b_2c_3 - b_3c_2) - b_1(a_2c_3 - a_3c_2) + c_1(a_2b_3 - a_3b_2).$$

In order to find the coefficient of say b_1 we imagine the row and column containing b_1 erased and the determinant of the remaining coefficients put down as they stand. For the sign we count forward element by element, starting with a_1 as $+$ and changing the sign alternately to $-$ and $+$ with every element passed over until we reach the element concerned. Thus $a_1(+)$, $b_1(-)$, $c_1(+)$.

Although we will usually multiply out in this way, we can do so in terms of any row or column. Thus to use the second row we would have, since $a_1(+)$, $a_2(-)$, the determinant (denoted by D) is

$$D = - a_2 \begin{vmatrix} b_1 & c_1 \\ b_3 & c_3 \end{vmatrix} + b_2 \begin{vmatrix} a_1 & c_1 \\ a_3 & c_3 \end{vmatrix} - c_2 \begin{vmatrix} a_1 & b_1 \\ a_3 & b_3 \end{vmatrix}$$
$$= - a_2(b_1c_3 - b_3c_1) + b_2(a_1c_3 - a_3c_1) - c_2(a_1b_3 - a_3b_1),$$

and is the same as the expression found above.

Example 2. Evaluate the determinant using: (a) the first row; (b) the first column; (c) the second column. Find also the coefficient of (the element) 7.

$$\begin{vmatrix} 3 & 2 & -1 \\ 4 & 5 & 7 \\ 6 & 3 & 8 \end{vmatrix}.$$

(a) $3 \begin{vmatrix} 5 & 7 \\ 3 & 8 \end{vmatrix} - 2 \begin{vmatrix} 4 & 7 \\ 6 & 8 \end{vmatrix} - 1 \begin{vmatrix} 4 & 5 \\ 6 & 3 \end{vmatrix}$

$= 3(40 - 21) - 2(32 - 42) - 1(12 - 30)$
$= 57 + 20 + 18 = 95.$

(b) $3 \begin{vmatrix} 5 & 7 \\ 3 & 8 \end{vmatrix} - 4 \begin{vmatrix} 2 & -1 \\ 3 & 8 \end{vmatrix} + 6 \begin{vmatrix} 2 & -1 \\ 5 & 7 \end{vmatrix}$

$= 3(40 - 21) - 4(16 + 3) + 6(14 + 5)$
$= 57 - 76 + 114 = 95.$

(c) $- 2 \begin{vmatrix} 4 & 7 \\ 6 & 8 \end{vmatrix} + 5 \begin{vmatrix} 3 & -1 \\ 6 & 8 \end{vmatrix} - 3 \begin{vmatrix} 3 & -1 \\ 4 & 7 \end{vmatrix}$

$= - 2(32 - 42) + 5(24 + 6) - 3(21 + 4)$
$= 20 + 150 - 75 = 95.$

Counting down to 7 by means of $3(+)$, $4(-)$, $5(+)$, $7(-)$ we see that the sign associated with the determinant is $-$. Therefore the coefficient is

$$- \begin{vmatrix} 3 & 2 \\ 6 & 3 \end{vmatrix} = 3.$$

4.3. The Solution of Linear Equations in Three Unknowns

The method used for two unknowns can be used for three or more unknowns. Thus to solve the system of equations

$$a_1x + b_1y + c_1z = k_1$$
$$a_2x + b_2y + c_2z = k_2$$
$$a_3x + b_3y + c_3z = k_3$$

we evaluate each variable as the ratio of two determinants. The determinant of the coefficients is

$$\begin{vmatrix} a_1 & b_1 & c_1 \\ a_2 & b_2 & c_2 \\ a_3 & b_3 & c_3 \end{vmatrix} \equiv D,$$

which we assume is not zero.

If we substitute the column of k's for the column of a's (the coefficients of x) then the value of x is found as

$$x = \begin{vmatrix} k_1 & b_1 & c_1 \\ k_2 & b_2 & c_2 \\ k_3 & b_3 & c_3 \end{vmatrix} \Big/ \begin{vmatrix} a_1 & b_1 & c_1 \\ a_2 & b_2 & c_2 \\ a_3 & b_3 & c_3 \end{vmatrix} = \begin{vmatrix} k_1 & b_1 & c_1 \\ k_2 & b_2 & c_2 \\ k_3 & b_3 & c_3 \end{vmatrix} \Big/ D.$$

Similarly,

$$y = \begin{vmatrix} a_1 & k_1 & c_1 \\ a_2 & k_2 & c_2 \\ a_3 & k_3 & c_3 \end{vmatrix} \Big/ D, \qquad z = \begin{vmatrix} a_1 & b_1 & k_1 \\ a_2 & b_2 & k_2 \\ a_3 & b_3 & k_3 \end{vmatrix} \Big/ D.$$

Example 3. Solve the system of simultaneous equations:

$$2x + 3y + 4z = 5;$$
$$x + 2y - 3z = 7;$$
$$x + y + 5z = -8.$$

$$D = \begin{vmatrix} 2 & 3 & 4 \\ 1 & 2 & -3 \\ 1 & 1 & 5 \end{vmatrix} = 2(10 + 3) - 3(5 + 3) + 4(1 - 2)$$
$$= 26 - 24 - 4 = -2$$

$$\therefore \quad -2x = \begin{vmatrix} 5 & 3 & 4 \\ 7 & 2 & -3 \\ -8 & 1 & 5 \end{vmatrix} = 5(10 + 3) - 3(35 - 24) + 4(7 + 16)$$
$$= 65 - 33 + 92 = 124$$

$$x = -62$$

$$-2y = \begin{vmatrix} 2 & 5 & 4 \\ 1 & 7 & -3 \\ 1 & -8 & 5 \end{vmatrix} = 2(35 - 24) - 5(5 + 3) + 4(-8 - 7)$$
$$= 22 - 40 - 60 = -78$$

$$y = 39$$

$$-2z = \begin{vmatrix} 2 & 3 & 5 \\ 1 & 2 & 7 \\ 1 & 1 & -8 \end{vmatrix} = 2(-16 - 7) - 3(-8 - 7) + 5(1 - 2)$$
$$= -46 + 45 - 5 = -6$$

$$z = 3.$$

We have therefore found the solutions as $x = -62$, $y = 39$, $z = 3$ (which should be checked by substitution).

4.4. *Note* 1.
In addition to the above method for multiplying out *third-order* determinants, the *Rule of Sarrus* may be used: (a) Repeat

the first two columns of the determinant by writing them to the right of it. (*b*) Find the value of the continued product of the elements of each cross diagonal containing three elements.

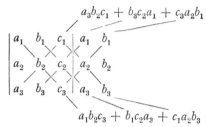

(*c*) The value of the determinant is then the sum of the lower values less the sum of the upper values. Here

$$D = (a_1b_2c_3 + b_1c_2a_3 + c_1a_2b_3) - (a_3b_2c_1 + b_3c_2a_1 + c_3a_2b_1).$$

In Example 3 to find D proceed :

$$8 - 6 + 15 = 17$$

$$\begin{vmatrix} 2 & 3 & 4 \\ 1 & 2 & -3 \\ 1 & 1 & 5 \end{vmatrix} \begin{matrix} 2 & 3 \\ 1 & 2 \\ 1 & 1 \end{matrix}$$

$$20 - 9 + 4 = 15$$

$$\therefore \quad D = 15 - 17 = -2.$$

Note 2. In solving equations as above it is often quicker to find one of the variables *x*, suppose, by the determinant method and then to substitute for *x* in any two of the equations. The resulting equations in two unknowns are then easily solved, *e.g.*, in Example 3, having found $x = -62$, we could substitute in the first two equations and find

$$3y + 4z = 129,$$
$$2y - 3z = 69.$$

For these two equations

$$D = \begin{vmatrix} 3 & 4 \\ 2 & -3 \end{vmatrix} = -17,$$

$$\therefore \quad -17y = \begin{vmatrix} 129 & 4 \\ 69 & -3 \end{vmatrix} = -663,$$

$$-17z = \begin{vmatrix} 3 & 129 \\ 2 & 69 \end{vmatrix} = -51,$$

therefore, as found above

$$y = 39, \quad z = 3.$$

Note 3. It will be shown later that the equation $ax + by + cz = d$ represents a plane. Since two planes generally meet in a line which will generally meet a third plane in a single point, we expect the solution of three simultaneous equations such as those considered to give a single unique solution, the single point common to the three planes. We could, however, have the three planes parallel or two parallel and cut by the third in two parallel lines. In each case D, the determinant of the coefficients, is zero and the values of x, y, z are infinite. We could also have three planes which coincide or else have a line in common. In these cases both numerator and denominator in the expressions for x, y, z would be zero. The values are therefore indeterminate and there are an infinite number of solutions:—any triplet of numbers (x, y, z) which denotes a point on the plane or on the line.

EXERCISE 9

1. Evaluate:

(a) $\begin{vmatrix} 3 & 11 & 1 \\ 11 & 13 & 66 \\ -12 & -65 & 40 \end{vmatrix}$ (b) $\begin{vmatrix} 1 & 4 & 27 \\ 2 & 9 & 64 \\ 3 & 16 & 125 \end{vmatrix}$ (c) $\begin{vmatrix} 25 & 3 & 35 \\ 16 & 10 & -18 \\ 34 & 6 & 38 \end{vmatrix}$

2. Solve the equations:

(a) $\begin{vmatrix} x & 6 & 3 \\ 1 & x & 1 \\ -2 & 4 & x \end{vmatrix} = 0$ (b) $\begin{vmatrix} 3x+1 & 5x & 6x+1 \\ 2x+1 & 3x-1 & 4x \\ 5x & 7x-1 & 10x-1 \end{vmatrix} = 0$

Solve the following sets of simultaneous equations:

3. $3x + 2y + 4z = 3$;
$x + y + z = 2$;
$2x - y + 3z = -3$.

4. $7x - 3y + 5z = 21$;
$2x + 5y - z = 12$;
$x + 6y + 3z = 2$.

5. $x + y + z = 1$;
$11x + 12y - 7z = 11$;
$37x + 40y + 24z = 38$.

6. $2x - 3y + z = 1$;
$x + y + z = 5$;
$3x - 2y + 3z = 6$.

7. In the solution of an electrical network problem the following equations were obtained. Find the values of i_1, i_2, i_3.

$$3(i_1 + i_2) + 1(i_1 + i_3) + 2(i_1 + i_3) = 5$$
$$2i_2 + 3(i_1 + i_2) + 3(i_1 - i_3) + 4i_2 = 0$$
$$1(i_1 + i_2) + 2(i_1 + i_3) + 2i_3 + 3(i_3 - i_2) = 0.$$

8. The following set of three linear equations in four unknowns is homogeneous, *i.e.*, contains no constant terms. Divide by i_4 and solve the resulting set of three equations in the three unknowns, the ratios $\frac{i_1}{i_4}, \frac{i_2}{i_4}, \frac{i_3}{i_4}$. If the results are expressed as

$$\frac{i_1}{D_1} = \frac{i_2}{D_2} = \frac{i_3}{D_3} = \frac{i_4}{D_4}$$

find D_1, D_2, D_3, D_4.

$$3i_1 + 3i_2 - i_3 + 4i_4 = 0$$
$$i_1 \quad\quad + i_3 + 3i_4 = 0$$
$$i_2 + i_3 + 3i_4 = 0.$$

9. Show that the following systems of equations have no unique solution and give the reason:

(a) $3x + 2y + 5z = 1$,
$x - y + z = 4$,
$6x + 4y + 10z = 7$;

(b) $x + y + z = 2$,
$2x + 4y + 5z = 7$,
$3x + 5y + 6z = 9$.

10. If $2x = a^2 + b^2 + c^2$ show that

$$a^2b^2c^2 \begin{vmatrix} 1 & \dfrac{x-c^2}{ab} & \dfrac{x-b^2}{ca} \\ \dfrac{x-c^2}{ab} & 1 & \dfrac{x-a^2}{bc} \\ \dfrac{x-b^2}{ca} & \dfrac{x-a^2}{bc} & 1 \end{vmatrix} = 4(x-a^2)(x-b^2)(x-c^2).$$

11. The solution of a circuit by the node method gives the equations:

$$10 \cdot 1V_1 - 0 \cdot 10V_2 = 20;$$
$$-0 \cdot 10V_1 + 0 \cdot 35V_2 - 0 \cdot 050V_3 = 0;$$
$$-0 \cdot 05V_2 + 5 \cdot 05V_3 = 20.$$

Show that
$$V_1 = 1 \cdot 99, \quad V_2 = 1 \cdot 14, \quad V_3 = 3 \cdot 96 \text{ volts.}$$

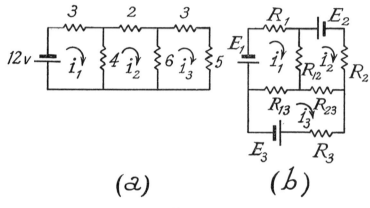

Fig. 7.

12. Using the loop-current method on the circuit shown (Fig. 7(a)) obtain and solve the equations:

$$7i_1 - 4i_2 = 12;$$
$$-4i_1 + 12i_2 - 6i_3 = 0;$$
$$-6i_2 + 14i_3 = 0.$$

13. Obtain and solve the equations for the loop currents i_1, i_2, i_3 in the circuit of Fig. 7(b).

$$(R_1 + R_{12} + R_{13})i_1 - R_{12}i_2 - R_{13}i_3 = E_1,$$
$$-R_{12}i_1 + (R_2 + R_{12} + R_{23})i_2 - R_{23}i_3 = E_2,$$
$$-R_{13}i_1 - R_{23}i_2 + (R_3 + R_{13} + R_{23})i_3 = E_3,$$

when $R_1 = 1$, $R_2 = 9$, $R_3 = 2$, $R_{12} = 2$, $R_{13} = 1$, $R_{23} = 3$, $E_1 = 10$, $E_2 = 5$, $E_3 = 2$.

14. The diagram shows a Wheatstone bridge where E is the e.m.f. and r_1 the total resistance external to the bridge. Obtain the equations:

$$i_1 r_1 + (i_1 - i_2)R_4 + (i_1 - i_3)R_3 = E,$$
$$(i_1 - i_3)R_3 - i_3 R_1 - (i_3 - i_2)r_2 = 0,$$
$$(i_1 - i_2)R_4 + (i_3 - i_2)r_2 - i_2 R_2 = 0,$$

where i_1, i_2, i_3 are the assumed loop currents as shown in the figure. Solve for i_2 and i_3, and hence obtain the usual relation

$$R_1R_4 = R_2R_3$$

when there is no current through the galvanometer.

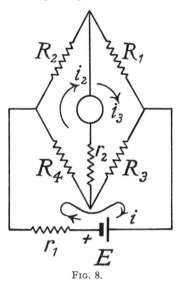

FIG. 8.

15. In the theory of vibrations we encounter equations whose roots give information about the stability of the vibration being studied. Such an equation is said to be stable if its real roots are all negative and its complex roots all have negative real parts. It can be shown that the equation

$$a_0x^4 + a_1x^3 + a_2x^2 + a_3x + a_4 = 0 \quad (a_0 > 0)$$

is stable if each of the following determinants is positive in value:

$$|a_1| \qquad \begin{vmatrix} a_1 & a_0 \\ a_3 & a_2 \end{vmatrix} \qquad \begin{vmatrix} a_1 & a_0 & 0 \\ a_3 & a_2 & a_1 \\ 0 & a_4 & a_3 \end{vmatrix} \qquad \begin{vmatrix} a_1 & a_0 & 0 & 0 \\ a_3 & a_2 & a_1 & a_0 \\ 0 & a_4 & a_3 & a_2 \\ 0 & 0 & 0 & a_4 \end{vmatrix}$$

(a) Test for stability the equation

$$x^4 + 4x^3 + 6x^2 + 5x + 2 = 0.$$

(b) Given that two of the roots are small integers, find all four roots and check the above definition of stability.

4.5. Some Important Properties

The following properties are of great use in decreasing the work required in the evaluation of determinants.

(1) The value of a determinant is unaltered if it is rewritten with the three rows as columns, in the same consecutive order.

That is

$$\begin{vmatrix} a_1 & b_1 & c_1 \\ a_2 & b_2 & c_2 \\ a_3 & b_3 & c_3 \end{vmatrix} = \begin{vmatrix} a_1 & a_2 & a_3 \\ b_1 & b_2 & b_3 \\ c_1 & c_2 & c_3 \end{vmatrix}$$

This could be proved by expanding the first by the first row and the second by the first column.

(2) The interchange of any two columns or of any two rows multiplies the value of the determinant by -1.

Thus we could show that

$$\begin{vmatrix} a_1 & b_1 & c_1 \\ a_2 & b_2 & c_2 \\ a_3 & b_3 & c_3 \end{vmatrix} = - \begin{vmatrix} c_1 & b_1 & a_1 \\ c_2 & b_2 & a_2 \\ c_3 & b_3 & a_3 \end{vmatrix}$$

where the third and first columns of the first have been interchanged to form the second determinant.

(3) If a determinant has two rows (or columns) identical its value is zero.

By (2) the interchange of the two rows changes the sign of the determinant. But if they are identical the value must be unchanged. Therefore

$$D = - D$$

or the value is zero.

(4) A common factor of any row or column can be taken out to multiply the remaining determinant value.

Thus clearly

$$\begin{vmatrix} kla_1 & lb_1 & lc_1 \\ ka_2 & b_2 & c_2 \\ ka_3 & b_3 & c_3 \end{vmatrix} = kl \begin{vmatrix} a_1 & b_1 & c_1 \\ a_2 & b_2 & c_2 \\ a_3 & b_3 & c_3 \end{vmatrix}$$

(5) If each element of any row (or column) is the sum of two or more terms, the determinant may be expressed as the sum of two or more determinants. Thus by multiplying out and regrouping, it can be shown that

$$\begin{vmatrix} a_1 + l_1 & b_1 & c_1 \\ a_2 + l_2 & b_2 & c_2 \\ a_3 + l_3 & b_3 & c_3 \end{vmatrix} = \begin{vmatrix} a_1 & b_1 & c_1 \\ a_2 & b_2 & c_2 \\ a_3 & b_3 & c_3 \end{vmatrix} + \begin{vmatrix} l_1 & b_1 & c_1 \\ l_2 & b_2 & c_2 \\ l_3 & b_3 & c_3 \end{vmatrix}$$

and similarly when there are more terms in each element.

(6) The value of a determinant is unaltered if the same multiple of the elements of any other column is added to a given column. The same is true for rows.

Thus since, by (5)

$$\begin{vmatrix} a_1 + lb_1 & b_1 & c_1 \\ a_2 + lb_2 & b_2 & c_2 \\ a_3 + lb_3 & b_3 & c_3 \end{vmatrix} = \begin{vmatrix} a_1 & b_1 & c_1 \\ a_2 & b_2 & c_2 \\ a_3 & b_3 & c_3 \end{vmatrix} + l \begin{vmatrix} b_1 & b_1 & c_1 \\ b_2 & b_2 & c_2 \\ b_3 & b_3 & c_3 \end{vmatrix}$$

and the value of the last determinant is zero. The value of the first determinant in which l times the second column has been added to the first column is therefore equal to the original value. More generally, using (a_1, b_1, c_1) to denote the determinant we could show that

$$(a_1 + lb_1 + mc_1, \ b_1 + nc_1, \ c_1) = (a_1, \ b_1, \ c_1),$$

the proof requiring only that one column should remain unchanged. In this case it is the third, containing the elements c_1, c_2, c_3. The same is true for rows.

Example 4. Evaluate

$$\begin{vmatrix} 80 & 37 & 2 \\ 60 & 19 & 3 \\ 57 & 21 & 5 \end{vmatrix}$$

If we subtract twice the second column plus three times the third from the first, we obtain

$$D = \begin{vmatrix} 0 & 37 & 2 \\ 13 & 19 & 3 \\ 0 & 21 & 5 \end{vmatrix} \qquad (c_1 - 2c_2 - 3c_3)$$

where the obvious notation on the right is used to show what has been done. Multiplying out in terms of the first column

$$D = -13(37 \times 5 - 21 \times 2) = -1859.$$

Example 5. Solve the equation

$$\begin{vmatrix} 8 + 2x & 13 & 11 \\ 9 + 4x & 19 & 8 \\ 17 - 2x & 12 & 14 \end{vmatrix} = 0.$$

Subtract twice the first row from the second and add the first row to the third. The determinant, which is unchanged in value, becomes

$$\begin{vmatrix} 8 + 2x & 13 & 11 \\ -7 & -7 & -14 \\ 25 & 25 & 25 \end{vmatrix} = 0.$$

Divide both sides by the common factors -7, 25,

$$\begin{vmatrix} 8 + 2x & 13 & 11 \\ 1 & 1 & 2 \\ 1 & 1 & 1 \end{vmatrix} = 0.$$

Subtract the third row from the second,

$$\begin{vmatrix} 8 + 2x & 13 & 11 \\ 0 & 0 & 1 \\ 1 & 1 & 1 \end{vmatrix} = 0.$$

Multiply out in terms of the second row,

$$\begin{vmatrix} 8 + 2x & 13 \\ 1 & 1 \end{vmatrix} = 0.$$

$$\therefore \quad (8 + 2x) - 13 = 0, \quad x = 2 \cdot 5.$$

4.6. Minors and Co-factors

The determinant obtained by deleting the row and column through an element is called the *minor* of the element. Thus in the above determinant (a_1, b_1, c_1), the minor of b_2 is

$$\begin{vmatrix} a_1 & c_1 \\ a_3 & c_3 \end{vmatrix}.$$

We can clearly multiply out the determinant, writing it as, for example, by using the first row,

$$a_1 \text{ (minor of } a_1) - b_1 \text{ (minor of } b_1) + c_1 \text{ (minor of } c_1).$$

These changes of sign are a nuisance, and to avoid them we introduce the term *co-factor*, which denotes a minor with the proper sign (plus or minus) attached to it. If A_r denotes the co-factor of a_r, B_r of b_r and C_r of c_r, then the value of the determinant is either

$$a_1A_1 + b_1B_1 + c_1C_1$$
or
$$a_2A_2 + b_2B_2 + c_2C_2$$
or
$$a_1A_1 + a_2A_2 + a_3A_3$$

with similar expressions for the expansion in terms of other rows or columns. In short, *the sum of the elements of any row (or column) each multiplied by its co-factor is the value of the determinant.*

On the other hand, consider, for example,

$$\begin{vmatrix} a_2 & b_2 & c_2 \\ a_2 & b_2 & c_2 \\ a_3 & b_3 & c_3 \end{vmatrix}.$$

The value of this is zero, since it has two rows identical. But it differs only from D in having a_2, b_2, c_2 instead of a_1, b_1, c_1 in the first row, so that its expansion in terms of the first row is

$$a_2A_1 + b_2B_1 + c_2C_1,$$

and this is zero. In the same way we may show that the sum of the elements of any row multiplied by the corresponding co-factors of *another row* is zero. The same holds for columns. Thus, for example,

$$a_1A_3 + b_1B_3 + c_1C_3 = 0$$
and
$$b_1A_1 + b_2A_2 + b_3A_3 = 0.$$

4.7. The Solution of Linear Equations

We can now repeat the previous work of 4.3 and provide a proof of the rule given there for the solution of linear equations. We will consider only three equations in three unknowns, *e.g.*, the system

$$a_1x + b_1y + c_1z = k_1 \quad \cdots \quad (1)$$
$$a_2x + b_2y + c_2z = k_2 \quad \cdots \quad (2)$$
$$a_3x + b_3y + c_3z = k_3 \quad \cdots \quad (3)$$

Let capital letters as above stand for the co-factors of the small letters in the determinant of the coefficients, so that, for example,

$$\left. \begin{array}{l} a_1A_1 + a_2A_2 + a_3A_3 = D \\ b_1A_1 + b_2A_2 + b_3A_3 = 0 \\ c_1A_1 + c_2A_2 + c_3A_3 = 0 \end{array} \right\} \quad \cdots \quad (4)$$

whilst since k_1, k_2, k_3 are not elements of the determinant of the coefficients we cannot say in advance what symmetrical expressions involving the k's are in value and must evaluate each. Thus we will have to evaluate

$$k_1A_1 + k_2A_2 + k_3A_3,$$

which is the above determinant D with the column of k's substituted for the column of a's, and is therefore

$$\begin{vmatrix} k_1 & b_1 & c_1 \\ k_2 & b_2 & c_2 \\ k_3 & b_3 & c_3 \end{vmatrix}.$$

Multiply the equations (1), (2), (3) respectively by the co-factors A_1, A_2, A_3 and add; this gives

$$x(a_1A_1 + a_2A_2 + a_3A_3) + y(b_1A_1 + b_2A_2 + b_3A_3)$$
$$+ z(c_1A_1 + c_2A_2 + c_3A_3) = (k_1A_1 + k_2A_2 + k_3A_3).$$

The coefficients of y and z are each zero by (4), so that the equation reduces to

$$x\mathrm{D} = \begin{vmatrix} k_1 & b_1 & c_1 \\ k_2 & b_2 & c_2 \\ k_3 & b_3 & c_3 \end{vmatrix}.$$

Similarly, by multiplying the equations (1)–(3) by the co-factors B_1, B_2, B_3 and adding,

$$x(a_1B_1 + a_2B_2 + a_3B_3) + y(b_1B_1 + b_2B_2 + b_3B_3)$$
$$+ z(c_1B_1 + c_2B_2 + c_3B_3) = (k_1B_1 + k_2B_2 + k_3B_3).$$

The first and third coefficients are zero, and this reduces to

$$y\mathrm{D} = \begin{vmatrix} a_1 & k_1 & c_1 \\ a_2 & k_2 & c_2 \\ a_3 & k_3 & c_3 \end{vmatrix}$$

where the column of k's has been substituted for the column of b's in D to give the determinant on the right. Similarly, it can be shown

$$z\mathrm{D} = \begin{vmatrix} a_1 & b_1 & k_1 \\ a_2 & b_2 & k_2 \\ a_3 & b_3 & k_3 \end{vmatrix}.$$

The method clearly applies to a system as above but of any number n equations in n unknowns, since determinants of the fourth and higher orders are evaluated similarly to the above.

4.8. Consistency and Elimination

We may have the two equations in one unknown

$$a_1x + b_1 = 0$$
$$a_2x + b_2 = 0.$$

If the value of x obtained from each is the same, *i.e.*, the equations are *consistent*,

$$-\frac{b_1}{a_1} = -\frac{b_2}{a_2} \quad \text{or} \quad a_1b_2 - a_2b_1 = 0.$$

This can be written in the determinant form

$$\begin{vmatrix} a_1 b_1 \\ a_2 b_2 \end{vmatrix} = 0$$

and it will be seen that we have *eliminated* the variable x and obtained the determinant of the coefficients.

Similarly, given *three* equations in *two* unknowns

$$a_1 x + b_1 y + c_1 = 0$$
$$a_2 x + b_2 y + c_2 = 0$$
$$a_3 x + b_3 y + c_3 = 0.$$

We can solve the first two equations for x and y, and if the three equations are consistent these values must satisfy the third. Performing the calculations, we have

$$\frac{x}{\begin{vmatrix} b_1 c_1 \\ b_2 c_2 \end{vmatrix}} = \frac{y}{-\begin{vmatrix} a_1 c_1 \\ a_2 c_2 \end{vmatrix}} = \frac{1}{\begin{vmatrix} a_1 b_1 \\ a_2 b_2 \end{vmatrix}}.$$

Substitution in the third equation gives

$$a_3 \begin{vmatrix} b_1 c_1 \\ b_2 c_2 \end{vmatrix} - b_3 \begin{vmatrix} a_1 c_1 \\ a_2 c_2 \end{vmatrix} + c_3 \begin{vmatrix} a_1 b_1 \\ a_2 b_2 \end{vmatrix} = 0,$$

which is, as above, the determinant, of the coefficients and in this case is

$$\begin{vmatrix} a_1 b_1 c_1 \\ a_2 b_2 c_2 \\ a_3 b_3 c_3 \end{vmatrix} = 0.$$

Once again we have *eliminated* the unknowns by means of the determinant of the coefficients and expressed that the three equations all relate to the same values of x and y, *i.e.*, are consistent.

Example 6. Find the equation of the line through the points $(1, 2)$, $(4, -3)$. If the equation to the line is

$$ax + by + c = 0 \quad . \quad . \quad . \quad . \quad . \quad . \quad . \quad (3)$$

then since $(1, 2)$ is on it

$$a + 2b + c = 0.$$

Similarly

$$4a - 3b + c = 0.$$

We have to solve these two equations for two ratios such as a/c, b/c and substitute in (3), *i.e.*, we have to eliminate $a : b : c$ from the three equations. By the above, this eliminant is given immediately by the determinant

$$\begin{vmatrix} x & y & 1 \\ 1 & 2 & 1 \\ 4 & -3 & 1 \end{vmatrix} = 0$$

which gives $5x + 3y - 11 = 0$ as the equation of the line determined by the two points.

Example 7. Express by means of a determinant the condition that the three equations

$$(3 + \lambda)x + (2 + 2\lambda)y + \lambda - 2 = 0$$
$$(2\lambda - 3)x + (2 - \lambda)y + 3 = 0$$
$$3x + 7y - 1 = 0.$$

should be consistent. Hence determine the possible values of λ. [L.U.]

The condition is

$$\begin{vmatrix} 3 + \lambda & 2 + 2\lambda & \lambda - 2 \\ 2\lambda - 3 & 2 - \lambda & 3 \\ 3 & 7 & -1 \end{vmatrix} = 0.$$

This can be written

$$\begin{vmatrix} 5 & 6 & \lambda - 2 \\ 2\lambda - 6 & -\lambda - 4 & 3 \\ 4 & 9 & -1 \end{vmatrix} = 0$$

by subtracting the last column from the first and twice the last from the second column.

or

$$\begin{vmatrix} 17 & 6 & \lambda - 2 \\ -14 & -(\lambda + 4) & 3 \\ 22 & 9 & -1 \end{vmatrix} = 0$$

by adding twice the second column to the first.

This is evaluated as

$$17(\lambda - 23) - 6(-52) + (\lambda - 2)(22\lambda - 38) = 0,$$

i.e.,

$$22\lambda^2 - 65\lambda - 3 = 0,$$
$$(22\lambda + 1)(\lambda - 3) = 0,$$
$$\therefore \quad \lambda = -\tfrac{1}{22} \text{ or } 3,$$

giving two values for λ, each of which when substituted in the three equations will give a set of three consistent equations.

EXERCISE 10

1. (*a*) Express the determinant

$$\begin{vmatrix} 1 & a & bc \\ 1 & b & ca \\ 1 & c & ab \end{vmatrix}$$

as the product of simple factors.

(*b*) Solve by determinants the equations

$$4x - 3y + 2z + 7 = 0,$$
$$6x + 2y - 3z - 33 = 0,$$
$$2x - 4y - z + 3 = 0. \qquad \text{[L.U.]}$$

2. (*a*) Evaluate the determinant

$$\begin{vmatrix} 2 & 1 + j & 3 \\ 1 - j & 0 & 2 + j \\ 3 & 2 - j & 1 \end{vmatrix}$$

(where $j \equiv \sqrt{-1}$) and explain how you could conclude without expansion that its value is real.

(*b*) If

$$x + 3y + 6z = 5$$
$$3 \cdot 1x + 2 \cdot 3y + 2 \cdot 6z = 6 \cdot 5$$
$$2 \cdot 01x + 5 \cdot 03y + 3 \cdot 06z = 2 \cdot 05$$

find the ratio x/y. [L.U.]

3. Find the values of λ for which the equations

$$(2 - \lambda)x + 2y + 3 = 0$$
$$2x + (4 - \lambda)y + 7 = 0$$
$$2x + 5y + 6 - \lambda = 0$$

are consistent and find the values of x and y corresponding to each of these values of λ. [L.U.]

4. Solve the equation

$$\begin{vmatrix} x + 1 & x + 2 & 3 \\ 2 & x + 3 & x + 1 \\ x + 3 & 1 & x + 2 \end{vmatrix} = 0.$$

[L.U.]

5. (a) Solve the equation

$$\begin{vmatrix} 15 - 2x & 11 & 10 \\ 11 - 3x & 17 & 16 \\ 7 - x & 14 & 13 \end{vmatrix} = 0.$$

(b) Express as a determinant the result of eliminating x, y, z from the equations

$$x + y + z = 0, \quad ax + by + cz = 0, \quad a^2x + b^2y + c^2z = 0,$$

and deduce the relations between a, b, c which are necessary for the elimination to be possible. [L.U.]

6. Find by means of a determinant the equation of the line through the points (2, 3), (4, 1). Find also the equation of the circle through these points and the point (3, − 1).

7. (a) If $x = r \cos \theta$, $y = r \sin \theta$ prove that

$$\begin{vmatrix} \dfrac{\partial x}{\partial r} & \dfrac{\partial x}{\partial \theta} \\ \dfrac{\partial y}{\partial r} & \dfrac{\partial y}{\partial \theta} \end{vmatrix} = r.$$

(b) If $x = r \sin \theta \cos \phi$, $y = r \sin \theta \sin \phi$, $z = r \cos \theta$, prove that

$$\begin{vmatrix} \dfrac{\partial x}{\partial r} & \dfrac{\partial x}{\partial \theta} & \dfrac{\partial x}{\partial \phi} \\ \dfrac{\partial y}{\partial r} & \dfrac{\partial y}{\partial \theta} & \dfrac{\partial y}{\partial \phi} \\ \dfrac{\partial z}{\partial r} & \dfrac{\partial z}{\partial \theta} & \dfrac{\partial z}{\partial \phi} \end{vmatrix} = r^2 \sin \theta.$$

8. By means of a determinant eliminate x, y, z and w from the equations

$$\begin{aligned} \lambda x + a(y + z + w) &= 0, \\ \lambda y + b(z + w + x) &= 0, \\ \lambda z + c(w + x + y) &= 0, \\ \lambda w + d(x + y + z) &= 0. \end{aligned}$$

Find the condition that the result shall contain no term in λ^2. [L.U.]

9. The preliminary steps in the construction of a nomogram require the elimination of u and v from the equations

$$\begin{aligned} u - 16 \log N &= 0 \\ v - S \log n &= 0 \\ \frac{u}{16} - \frac{v}{S} - R &= 0. \end{aligned}$$

Show that the eliminant may be written

$$\begin{vmatrix} 1 & 0 & -16 \log N \\ 0 & 1 & -S \log n \\ \dfrac{1}{16} & -\dfrac{1}{S} & -R \end{vmatrix} = 0.$$

10. Eliminate x from the equations

$$x^2 + ax + b, \quad x^2 + px + q = 0.$$

Show that the eliminant is equivalent to

$$\begin{vmatrix} 1 & a & b & 0 \\ 0 & 1 & a & b \\ 0 & 1 & p & q \\ 1 & p & q & 0 \end{vmatrix} = 0.$$

[L.U.]

(Multiply each equation by x to form a total of four equations. Eliminate x^3, x^2 and x from these.)

11. A problem in which two masses m_1, m_2 oscillate in line affected each by a spring and joined by a coupling spring gives rise to the differential equations

$$m_1\ddot{x}_1 + (p_1 + p_3)x_1 - p_3x_2 = 0$$
$$m_2\ddot{x}_2 + (p_2 + p_3)x_2 - p_3x_1 = 0.$$

Assume that both masses oscillate with the same (but unknown) frequency ω and different amplitudes so that

$$x_1 = a_1 \sin \omega t, \quad x_2 = a_2 \sin \omega t.$$

Substitute these in the given differential equations and obtain the simultaneous equations

$$a_1(- m_1\omega^2 + p_1 + p_3) - p_3a_2 = 0$$
$$- p_3a_1 + a_2(- m_2\omega^2 + p_2 + p_3) = 0.$$

Eliminate the ratio a_1/a_2 and hence obtain a " frequency equation " giving two values for ω^2.

12. A problem of a bar resting on two springs gives rise to the equations

$$m\ddot{x} + p(x + l\theta) + 2p(x - l\theta) = 0$$
$$\tfrac{1}{3}ml^2\ddot{\theta} + pl(x + l\theta) - 2pl(x - l\theta) = 0,$$

assume $x = x_0 \sin \omega t$, $\theta = \theta_0 \sin \omega t$ and obtain

$$(- m\omega^2 + 3p)x_0 - pl\theta_0 = 0$$
$$- plx_0 + (- \tfrac{1}{3}ml^2\omega^2 + 3pl^2)\theta_0 = 0.$$

Obtain the frequency equation in ω^2 and solve it in the special case when $p/m = \sqrt{3}$.

13. A problem concerning a shaft carrying three equal pulleys gives as the differential equations governing the rotation of the pulleys

$$- I\ddot{q}_1 = c(2q_1 - q_2)$$
$$- I\ddot{q}_2 = c(- q_1 + 2q_2 - q_3)$$
$$- I\ddot{q}_3 = c(- q_2 + q_3).$$

Assume $q_r = a_r \sin \omega t$ ($r = 1, 2, 3$), and by eliminating the ratios $a_1 : a_2 : a_3$ obtain the frequency equation

$$\lambda^3 - 5\lambda^2 + 6\lambda - 1 = 0,$$

where $\lambda = I\omega^2/c$.

14. Show that in any triangle

$$b = c \cos A + a \cos C.$$

Deduce, by the elimination of the ratios $a : b : c$ from this and two similar expressions that

$$\begin{vmatrix} -1 & \cos C & \cos B \\ \cos C & -1 & \cos A \\ \cos B & \cos A & -1 \end{vmatrix} = 0.$$

Hence show that in any triangle with angles A, B, C

$$\cos^2 A + \cos^2 B + \cos^2 C + 2 \cos A \cos B \cos C = 1.$$

4.9. Multiplication of Determinants

We will state the rule for multiplying together two determinants of the *same* order : (*a*) The *rows* of the first determinant are multiplied element by element into the *columns* of the second determinant, the first element in a row combining with the first element in the column, the second with the second and so on.   Thus for

$$\begin{vmatrix} a_1 & b_1 & c_1 \\ a_2 & b_2 & c_2 \\ a_3 & b_3 & c_3 \end{vmatrix} \times \begin{vmatrix} A_1 & B_1 & C_1 \\ A_2 & B_2 & C_2 \\ A_3 & B_3 & C_3 \end{vmatrix}$$

an element in the resulting determinant would be obtained by

$$\begin{vmatrix} a_2 b_2 c_2 \\ \longrightarrow \end{vmatrix} \times \begin{vmatrix} B_1 \\ B_2 \\ B_3 \end{vmatrix}\bigg\downarrow$$

and is
$$a_2 B_1 + b_2 B_2 + c_2 B_3.$$

(b) An element obtained by multiplying the ith row into the jth column appears in the ith row and jth column of the resulting determinant. In the illustration above it is in the position where the horizontal arrow meets the vertical arrow.

Example 8. Find the product

$$\begin{vmatrix} 2 & 1 & 3 \\ -1 & 5 & 4 \\ -2 & -1 & 3 \end{vmatrix} \times \begin{vmatrix} 0 & 1 & 6 \\ 2 & 5 & 3 \\ 4 & 7 & 2 \end{vmatrix}$$

In the product: *1st row.*

1st element: $(2 \times 0) + (1 \times 2) + (3 \times 4) = 14.$
2nd element: $(2 \times 1) + (1 \times 5) + (3 \times 7) = 28.$
3rd element: $(2 \times 6) + (1 \times 3) + (3 \times 2) = 21.$

2nd row.

1st element: $(-1 \times 0) + (5 \times 2) + (4 \times 4) = 26.$
2nd element: $(-1 \times 1) + (5 \times 5) + (4 \times 7) = 52.$
3rd element: $(-1 \times 6) + (5 \times 3) + (4 \times 2) = 17.$

3rd row.

1st element: $(-2 \times 0) + (-1 \times 2) + (3 \times 4) = 10.$
2nd element: $(-2 \times 1) + (-1 \times 5) + (3 \times 7) = 14.$
3rd element: $(-2 \times 6) + (-1 \times 3) + (3 \times 2) = -9.$

Therefore the required product is

$$\begin{vmatrix} 14 & 28 & 21 \\ 26 & 52 & 17 \\ 10 & 14 & -9 \end{vmatrix}.$$

(i) Since the first determinant in the product is evaluated as **66**, the second as -28 and the resulting product determinant as -1848, the result is checked.

(ii) The student should reverse the order of the determinants and check that the product is the same.

(iii) We are not confined to multiplying determinants of the same order by this method, since in the product

$$\begin{vmatrix} a_1 b_1 c_1 \\ a_2 b_2 c_2 \\ a_3 b_3 c_3 \end{vmatrix} \times \begin{vmatrix} A_1 B_1 \\ A_2 B_2 \end{vmatrix}$$

the second determinant may be written

$$\begin{vmatrix} A_1 & B_1 & 0 \\ A_2 & B_2 & 0 \\ 0 & 0 & 1 \end{vmatrix}.$$

It is now of the same order as the other, and the above method may be used.

EXERCISE 11

Express as a single determinant Numbers 1–5.

1. $\begin{vmatrix} 2 & 3 \\ 1 & 4 \end{vmatrix}$ $\begin{vmatrix} -1 & 2 \\ 3 & 5 \end{vmatrix}$ 2. $\begin{vmatrix} 1 & 2 \\ 4 & 3 \end{vmatrix}^2$ 3. $\begin{vmatrix} 2 & 3 & 1 \\ -1 & 0 & 2 \\ 5 & 3 & 1 \end{vmatrix}$ $\begin{vmatrix} 0 & 2 & 3 \\ 1 & 4 & 7 \\ 6 & 4 & 2 \end{vmatrix}$

4. $\begin{vmatrix} a & h & g \\ h & b & f \\ g & f & c \end{vmatrix}$ $\begin{vmatrix} 1 & 0 & 0 \\ 0 & 1 & 0 \\ 0 & 0 & 1 \end{vmatrix}$ 5. $\begin{vmatrix} a & h & g \\ h & b & f \\ g & f & c \end{vmatrix}$ $\begin{vmatrix} 0 & 0 & 1 \\ 0 & 1 & 0 \\ 1 & 0 & 0 \end{vmatrix}$

6. Show that if $j \equiv \sqrt{-1}$

$$\begin{vmatrix} 1 + 2j & 3 + 4j \\ -3 + 4j & 1 - 2j \end{vmatrix} \begin{vmatrix} 1 + 3j & 5 + 7j \\ -5 + 7j & 1 - 3j \end{vmatrix} = \begin{vmatrix} -48 + 6j & 6 + 12j \\ -6 + 12j & -48 - 6j \end{vmatrix}$$

Deduce that

$$(1^2 + 2^2 + 3^2 + 4^2)(1^2 + 3^2 + 5^2 + 7^2) = 6^2 + 6^2 + 12^2 + 48^2.$$

(The student will note that this method can be used to show that the product of two sums, each of four squares, is itself the sum of four squares.)

7. If $x = r \cos \theta$, $y = r \sin \theta$ show that

$$\begin{vmatrix} \dfrac{\partial x}{\partial r} & \dfrac{\partial x}{\partial \theta} \\ \dfrac{\partial y}{\partial r} & \dfrac{\partial y}{\partial \theta} \end{vmatrix} \begin{vmatrix} \dfrac{\partial r}{\partial x} & \dfrac{\partial \theta}{\partial x} \\ \dfrac{\partial r}{\partial y} & \dfrac{\partial \theta}{\partial y} \end{vmatrix} = 1.$$

8. If capital letters stand for the co-factors of the small letters, show that

$$\begin{vmatrix} a_1 b_1 c_1 \\ a_2 b_2 c_2 \\ a_3 b_3 c_3 \end{vmatrix} \begin{vmatrix} A_1 B_1 C_1 \\ A_2 B_2 C_2 \\ A_3 B_3 C_3 \end{vmatrix} = D^3$$

and hence that the value of the second determinant is D^2.

9. Prove that the vertices of the triangle formed by the three lines

$$a_r x + b_r y + c_r = 0 \qquad (r = 1, 2, 3)$$

are

$$(A_1/C_1, \ B_1/C_1), \quad (A_2/C_2, \ B_2/C_2), \quad (A_3/C_3, \ B_3/C_3),$$

and show that the area of the triangle is

$$D^2/2C_1 C_2 C_3.$$

CHAPTER 5

MAXIMA AND MINIMA. DIFFERENTIATION UNDER THE INTEGRAL SIGN. CURVILINEAR OR LINE INTEGRALS

" The true meaning of a term is to be found by observing what a man does with it not what he says about it."

H. Bridgeman, *The Logic of Modern Physics.*

5.1. Maxima and Minima

We have already dealt with this for a function of one variable [Vol. I (5.9)], and will here consider functions of two or more variables. We require the values of x and y that make

$$z = f(x, y)$$

a maximum or minimum. It is clear that at a genuine maximum the function will have a maximum when x alone varies and y is kept constant or y varies and x is kept constant. We must therefore have

$$\frac{\partial z}{\partial x} = 0, \quad \frac{\partial z}{\partial y} = 0 \quad . \quad . \quad . \quad . \quad (1)$$

at any point where a genuine turning value occurs. It will be seen later that $z = f(x, y)$ is a surface, and if a maximum is regarded as a mountain peak, it will remain so even when we move over it parallel to the y axis (or the x axis). Hence the need for equations (1).

Just as the solution of $dy/dx = 0$ gives those values of x at which possible turning values occur so the solution of the above pair of equations, treated as a pair of simultaneous equations, provide the pairs of values (x, y) at which turning values may occur.

If there are more than 2 independent variables we have an equation as in (1) for each variable. There are certain methods of testing whether the pair of values (x, y) obtained from (1) provides a genuine maximum or minimum but we will not give these. In the problems which occur here it is evident from the problem that a genuine turning value has been found and its nature.

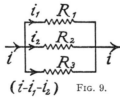

$(i\text{-}i_1\text{-}i_2)$ Fig. 9.

Example 1. The diagram shows 3 wires of resistance R_1, R_2, R_3 respectively connected in parallel. If they are part of a circuit containing a current i find the distribution of i between the respective resistances.

64

We must know that the heat generated in a resistance R by a current i is Ri^2, and it is a physical principle that the current will so distribute itself in the network that the total heat generated is a minimum.

If i_1 is the current through R_1, i_2 through R_2 so that $i - i_1 - i_2$ is the current through R_3, then H, the total heat generated, is given by

$$H = R_1 i_1^2 + R_2 i_2^2 + R_3(i - i_1 - i_2)^2.$$

Since this is a minimum

$$\frac{\partial H}{\partial i_1} = 2R_1 i_1 - 2R_3(i - i_1 - i_2) = 0$$

$$\frac{\partial H}{\partial i_2} = 2R_2 i_2 - 2R_3(i - i_1 - i_2) = 0.$$

We have to solve the above simultaneous equations for i_1 and i_2. They are

$$i_1(R_1 + R_3) + i_2 R_3 - R_3 i = 0$$
$$i_1 R_3 + i_2(R_2 + R_3) - R_3 i = 0$$

so that

$$i_1 = \frac{iR_2R_3}{R_1R_2 + R_2R_3 + R_3R_1} \qquad i_2 = \frac{iR_3R_1}{R_1R_2 + R_2R_3 + R_3R_1}$$

and

$$i_3 = i - i_1 - i_2 = \frac{iR_1R_2}{R_1R_2 + R_2R_3 + R_3R_1}$$

values which can, of course, be more easily obtained in the usual way.

Example 2. Prove that the rectangular solid of maximum volume which can be inscribed in a given sphere is a cube.

Let x, y, z be the length, breadth and height of the solid with d the diameter of the sphere. A diagonal of the solid will be a diameter of the sphere so that

$$x^2 + y^2 + z^2 = d^2.$$

If V is the volume of the box

$$V = xyz$$
$$= xy\sqrt{(d^2 - x^2 - y^2)},$$
$$V^2 = x^2y^2(d^2 - x^2 - y^2).$$

Using this value:

$$\frac{\partial(V^2)}{\partial x} = 2d^2xy^2 - 4x^3y^2 - 2xy^4 = 0,$$

$$\frac{\partial(V^2)}{\partial y} = 2d^2x^2y - 2x^4y - 4x^2y^3 = 0.$$

Since $x = y = 0$ will give a minimum, neglecting these values and dividing by xy^2 and x^2y respectively, we have to solve

$$2x^2 + y^2 - d^2 = 0,$$
$$x^2 + 2y^2 - d^2 = 0.$$

The solution is $x = y = d/\sqrt{3},$

giving $z = \sqrt{(d^2 - x^2 - y^2)} = d/\sqrt{3}.$

The required solid is therefore a cube of edge $d/\sqrt{3}$.

5.2. Turning Values With Restrictive Conditions

The problem often arises of finding the turning values of an expression $z = f(x, y)$, where x and y are connected by an equation $F(x, y) = 0$. If we could solve $F(x, y)$ for, say, y and substitute in $f(x, y)$, z would then be a function of one variable x and could be dealt with in the usual way. When this cannot be done the following artifice is required, and it often helps to a quicker solution, even when $F(x, y) = 0$ can be solved easily for y.

We form the artificial function

$$\phi(x, y) = f(x, y) + \lambda F(x, y),$$

where it will be noted $F(x, y)$ is zero for the values concerned. We now find

$$\frac{\partial \phi}{\partial x} = \frac{\partial f}{\partial x} + \lambda \frac{\partial F}{\partial x} = 0$$

$$\frac{\partial \phi}{\partial y} = \frac{\partial f}{\partial y} + \lambda \frac{\partial F}{\partial y} = 0$$

and we have

$$F(x, y) = 0$$

a total of three equations to determine the values of λ and (x, y) at the turning values.

Example 3. Solve the problem of Example 1 taking the currents as i_1, i_2, i_3 with the restrictive condition

$$i_1 + i_2 + i_3 - i = 0.$$

We form

$$\phi = R_1 i_1{}^2 + R_2 i_2{}^2 + R_3 i_3{}^2 + \lambda(i_1 + i_2 + i_3 - i)$$

and find

$$\frac{\partial \phi}{\partial i_1} = 2R_1 i_1 + \lambda = 0$$

$$\frac{\partial \phi}{\partial i_2} = 2R_2 i_2 + \lambda = 0$$

$$\frac{\partial \phi}{\partial i_3} = 2R_3 i_3 + \lambda = 0.$$

Substituting in

$$i_1 + i_2 + i_3 = i$$

from these equations

$$-\frac{\lambda}{2}\left(\frac{1}{R_1} + \frac{1}{R_2} + \frac{1}{R_3}\right) = i$$

$$\lambda = \frac{-2i R_1 R_2 R_3}{\Sigma R_1 R_2}.$$

Therefore

$$i_1 = -\frac{\lambda}{2R_1} = \frac{i R_2 R_3}{\Sigma R_1 R_2}$$

and similar values for i_2 and i_3 as found above.

EXERCISE 12

1. A piece of metal of width 12 mm has a width of x mm from each edge turned up at an angle θ. Find x and θ if the resultant trough is to have a maximum volume.

2. Find the values of x and y for which $z = y^2 + x^2 + 6x + 2$ has a minimum.

3. Divide a number a into 3 parts such that the product of them is a maximum.

4. Find the volume of the largest rectangular box that can be inscribed in the ellipsoid

$$x^2/a^2 + y^2/b^2 + z^2/c^2 = 1.$$

5. A tent on a square base of side $2a$ has four vertical sides of height b and on top of this a regular pyramid of height h (but base $2a \times 2a$). If the volume enclosed is V show that the area of canvas in the tent is

$$\frac{2V}{a} - \frac{8}{3}ah + 4a\sqrt{(h^2 + a^2)}.$$

Hence show that if a and h can both vary, the least area of canvas for a given volume V is given by

$$a = \tfrac{1}{2}\sqrt{5h}, \quad b = \tfrac{1}{2}h.$$

6. There are n points $P_r \equiv (x_r, y_r, z_r)$ $(r = 1, 2 \ldots n)$. Show that a point $P \equiv (x, y, z)$ chosen so that the sum of the squares of its distances from each P_r is a minimum has co-ordinates

$$x = \frac{1}{n}\Sigma x_r, \quad y = \frac{1}{n}\Sigma y_r, \quad z = \frac{1}{n}\Sigma z_r.$$

7. To fit a line $y = ax + b$ to the n points (x_r, y_r) $(r = 1, 2, \ldots n)$ we find a and b so as to make a minimum of

$$S = \Sigma(y_r - ax_r - b)^2.$$

Show that the equations for a and b are

$$a\Sigma x_r^2 + b\Sigma x_r - \Sigma x_r y_r = 0$$
$$a\Sigma x_r + bn - \Sigma y_r = 0.$$

Turn the following equation into the linear form and hence find a law of the form $y = ax^2 + b$ for the data

x	. . .	10	20	30	40	50
y	. . .	8	10	15	21	30

8. A quadrilateral ABCD is formed by four wires of resistance $AB = 5$, $BC = 5$, $CD = 5$, $DA = 3$ ohms. Also B and D are joined by a wire of resistance 8 ohms. Find the equivalent resistance of the network for a current entering at A and leaving at C.

9. A tetrahedron frame ABCD is formed by six wires the resistance of opposite edges being equal. Prove that the whole resistance of the frame for a current entering at A and leaving at D is

$$\frac{r_3(r_1 r_3 + 2r_1 r_2 + r_2 r_3)}{2(r_1 + r_3)(r_2 + r_3)}$$

where AB, CD each have resistance r_1, AC, BD each have resistance r_2 and r_3 is the resistance of AD or BC.

10. A cube consists of 12 equal wires forming its edges. A current i enters at one corner and leaves at another corner nearest to it. If each wire has a resistance r, find the currents along the 3 edges forming the corner where i enters.

11. The capacity of a condenser formed from four concentric conducting spherical shells of radii a, x, y, b $(a < x < y < b)$, of which those of radii a and y are connected and insulated and those of radii x and b are connected to earth is

$$\frac{ax}{x - a} + \frac{xy}{y - x} + \frac{yb}{b - y}.$$

Verify that this capacity is a minimum when

$$x = \frac{3ab}{2b + a}, \qquad y = \frac{3ab}{2a + b}$$

assuming that a and b are fixed and x, y allowed to vary independently.

5.3. Differentiation Under the Integral Sign

Since

$$\int_0^1 3(x + \alpha)^2 dx = (x + \alpha)^3 \Big]_0^1$$

$$= (1 + \alpha)^3 - \alpha^3 = 1 + 3\alpha + 3\alpha^2 \quad . \quad . \quad . \quad (1)$$

and is a function of α only we will denote it by $F(\alpha)$. We have then

$$F(\alpha) = \int_0^1 3(x + \alpha)^2 dx = 1 + 3\alpha + 3\alpha^2,$$

therefore
$$\frac{dF}{d\alpha} = 3 + 6\alpha \quad . \quad . \quad . \quad . \quad . \quad . \quad (2)$$

Before evaluating the integral, this must be written

$$\frac{dF}{d\alpha} = \frac{d}{d\alpha} \int_0^1 3(x + \alpha)^2 dx$$

and the question arises whether we can interchange the order on the right-hand side—integration and then differentiation into differentiation and then integration. Using the notation of partial differentiation (to show that x is also a variable) the right-hand side becomes, if we may effect the interchange,

$$\int_0^1 3 \frac{\partial}{\partial \alpha} (x + \alpha)^2 dx = \int_0^1 6(x + \alpha) dx = 3 + 6\alpha$$

which is the same as (2), showing that in this case the interchange is legitimate. A general proof follows.

Let $f(x, \alpha)$ be a continuous function of x and α which we have to integrate between the limits a and b where a and b are independent of α. The integral, as above, will be a function of α and we have

$$F(\alpha) = \int_a^b f(x, \alpha) dx.$$

If α changes to $\alpha + \Delta\alpha$

$$F(\alpha + \Delta\alpha) = \int_a^b f(x, \alpha + \Delta\alpha) dx$$

from which we obtain

$$\frac{F(\alpha + \Delta\alpha) - F(\alpha)}{\Delta\alpha} = \int_a^b \frac{f(x, \alpha + \Delta\alpha) - f(x, \alpha)}{\Delta\alpha} dx.$$

Proceeding to the limit as $\Delta\alpha \longrightarrow 0$, and provided $\frac{\partial f}{\partial \alpha}$ is also a continuous function of x and α

$$\frac{dF}{d\alpha} = \int_a^b \frac{\partial f}{\partial \alpha} dx$$

or
$$\frac{d}{d\alpha} \int_a^b f(x, \alpha) dx = \int_a^b \frac{\partial f}{\partial \alpha} dx$$

showing that when certain conditions are fulfilled, the derivative of the integral is equal to the integral of the derivative.

Example 4. Find

$$\int_{0}^{\infty} \frac{dx}{x^2 + 1}$$

and deduce: (i) $\displaystyle\int_{0}^{\infty} \frac{dx}{(x^2 + 1)^3}$; (ii) $\displaystyle\int_{0}^{\infty} \frac{x^2 dx}{(x^2 + 1)^3}$.

We will find

$$\int_{0}^{\infty} \frac{dx}{x^2 + a^2} = \frac{1}{a} \tan^{-1}\left(\frac{x}{a}\right)\Bigg]_{0}^{\infty} = \frac{\pi}{2a} \quad \cdots \quad (3)$$

From this, using the above theorem,

$$\int_{0}^{\infty} \frac{\partial}{\partial a}\left(\frac{1}{x^2 + a^2}\right) dx = \frac{d}{da}\left(\frac{\pi}{2a}\right),$$

$$\int_{0}^{\infty} \frac{-2a\,dx}{(x^2 + a^2)^2} = -\frac{\pi}{2a^2},$$

or

$$\int_{0}^{\infty} \frac{dx}{(x^2 + a^2)^2} = \frac{\pi}{4a^3} \quad \cdots \cdots \cdots \quad (4)$$

Note that once we have differentiated with respect to a, we revert to using that a is a constant with respect to x.

Applying the theorem again

$$\int_{0}^{\infty} \frac{\partial}{\partial a}\left(\frac{1}{(x^2 + a^2)^2}\right) dx = \int_{0}^{\infty} \frac{-4a}{(x^2 + a^2)^3} dx = -\frac{3\pi}{4a^4}$$

or

$$\int_{0}^{\infty} \frac{dx}{(x^2 + a^2)^3} = \frac{3\pi}{16}\frac{1}{a^5} \quad \cdots \cdots \quad (5)$$

Putting $a = 1$ in this, we have for (i)

$$\int_{0}^{\infty} \frac{dx}{(x^2 + 1)^3} = \frac{3\pi}{16}.$$

For (ii)

$$\int_{0}^{\infty} \frac{x^2 dx}{(x^2 + 1)^3} = \int_{0}^{\infty} \frac{(x^2 + 1) - 1}{(x^2 + 1)^3} dx$$

$$= \int_{0}^{\infty} \frac{1}{(x^2 + 1)^2} dx - \int_{0}^{\infty} \frac{dx}{(x^2 + 1)^3}$$

using (4) and (5):

$$= \frac{\pi}{4} - \frac{3\pi}{16} = \frac{\pi}{16}.$$

5.4. Application to Beam Problems

The following example illustrates an important use of the theorem. We need, however, to know :

(a) If M is the bending moment at any point x of a beam the total potential energy stored in the beam is given by

$$U = \int \frac{M^2}{2EI}\, dx.$$

integrated over the length of the beam.

(b) If one of the forces P acting on the beam makes a small displacement δ in the direction in which it is acting

$$\delta = \frac{\partial U}{\partial P}.$$

For a couple C and a small rotation θ, $\quad \theta = \dfrac{\partial U}{\partial C}.$

(c) In equilibrium the total potential energy of the system is a minimum.

Example 5. A light uniform cantilever is deflected by a vertical force P at the free end. To find the amount of the deflection.

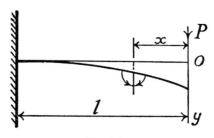

FIG. 10.

The B.M. at a distance x from the free end is given by

$$M = Px.$$

For the energy stored in the beam, by (a)

$$U = \frac{1}{2EI} \int_0^l M^2 dx.$$

If δ is the displacement due to P

$$\delta = \frac{\partial U}{\partial P} = \frac{1}{2EI} \frac{\partial}{\partial P} \int_0^l M^2 dx = \frac{1}{2EI} \int_0^l 2M \frac{\partial M}{\partial P}\, dx \qquad \text{(by 5.3)}$$

$$= \frac{1}{EI} \int_0^l Px^2 dx = \frac{Pl^3}{3EI}.$$

In some cases it would be easier to use (c). In this case since the force P has descended a distance δ and done work $P\delta$, the total energy of the system is

$$U = \frac{1}{2EI} \int_0^l M^2 dx - P\delta$$

Since this is a minimum,

$$\frac{\partial U}{\partial P} = \frac{1}{2EI}\int_0^l 2M\frac{\partial M}{\partial P}\,dx - \delta = 0$$

$$\delta = \frac{1}{2EI}\int_0^l 2M\frac{\partial M}{\partial P}\,dx = \underline{\frac{Pl^3}{3EI}}$$

as found by (a).

5.5. Curvilinear or Line Integrals

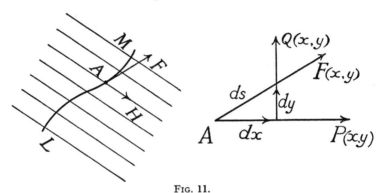

FIG. 11.

Let LM (Fig. 11) be a continuous curve in a field of electric force whose intensity at a point A on the curve is H. If the component of H along the curve at A is F, then the work done in a small displacement ds by the field on a unit charge at A is Fds and the total work done by the field is

$$\int_{LM} F ds$$

where we integrate along the curve LM along which the unit charge is supposed to move. If the components of F, which generally is a function of x and y, along the axes x and y are P(x, y) and Q(x, y) respectively, then instead of a movement ds along the curve we can imagine a movement dx in which P(x, y)dx is the total work done and a movement dy, in which Q(x, y)dy is the total work done. The total work done in the displacement $\overline{ds} = \overline{dx} + \overline{dy}$ is then

$$\int_{LM} [P(x, y)dx + Q(x, y)dy]$$

where again we must integrate along the curve. In this way it is seen that an integral along a curve, called a curvilinear or line integral, is of great importance in applied mathematics.

5.6. Evaluation of Line Integrals

If an ordinate cuts the curve in only one point and if the x value of L in Fig. 11 is x_1 and that of M is x_2 then the integral may be written

$$\int F(x, y)ds = \int_{x_1}^{x_2} F(x, y) \sqrt{\left\{1 + \left(\frac{dy}{dx}\right)^2\right\}} dx$$

since
$$\frac{ds}{dx} = \sqrt{\left\{1 + \left(\frac{dy}{dx}\right)^2\right\}}$$
[Vol. I (22.5)],

and we will put $y = f(x)$ the equation to the curve LM and so reduce the integral to one in a single variable x.

It is often easier to express x and y in terms of a single parameter t, and if $F(x, y)$ becomes $\phi(t)$ we will have to evaluate

$$\int_{t_1}^{t_2} \phi(t) \sqrt{\left\{\left(\frac{dx}{dt}\right)^2 + \left(\frac{dy}{dt}\right)^2\right\}} dt$$

where t_1, t_2 are the values in terms of t of x_1, x_2 respectively.

Both these methods are often inconvenient, and the best way is to evaluate the integral as

$$\int [Pdx + Qdy]$$

where, of course, we still need to know y in terms of x as $y = f(x)$ or both x and y in terms of a parameter.

Example 6. Find $\int_{(0, 0)}^{(1, 1)} [(xy + 2y^2)dx + (3x^2 + y)dy]$

taken along the paths: (a) $y = x^2$; (b) the line $y = x$.

(a) Using the equation $y = x^2$ the integral may be written

$$\int_0^1 [(x . x^2 + 2x^4)dx + (3x^2 + x^2)2xdx]$$

$$= \int_0^1 (2x^4 + 9x^3)dx = \frac{53}{20}.$$

In this example it is easy to invent and introduce the parametric representation $x = t$, $y = t^2$. The integral is then

$$\int_0^1 [(t . t^2 + 2t^4)dt + (3t^2 + t^2)2tdt] = \int_0^1 (9t^3 + 2t^4)dt = \frac{53}{20}.$$

(b) Along $y = x$ the integral is

$$\int_0^1 [(x . x + 2x^2)dx + (3x^2 + x)dx] = \int_0^1 (6x^2 + x)dx = \underline{2 \cdot 5}.$$

5.7. Dependence on Path of Integration

The above example illustrates what is expected, that the evaluation of the integral along different paths gives different answers. There is

one important case when the result of the integration is independent
of the path joining the beginning and end points. Suppose $Pdx + Qdy$
is an exact differential, *i.e.*, as it stands it is $d\phi(x, y)$, where, when
required, we may use

$$d\phi(x, y) = \frac{\partial \phi}{\partial x} dx + \frac{\partial \phi}{\partial y} dy.$$

In this case

$$\int_{x_1 y_1}^{x_2 y_2} (Pdx + Qdy) = \int_{x_1 y_1}^{x_2 y_2} d\phi(x, y) = \phi(x, y) \Big]_{x_1 y_1}^{x_2 y_2}$$

$$= \phi(x_2, y_2) - \phi(x_1, y_1),$$

and therefore depends only on the value of the function at the end
points, and not in any way on the curve joining them.

It will be noticed that if the integral along ABC (Fig. 12(a)) equals
that along ADC then the integral along ABCDA is the integral along
ABC minus that along ADC, and is zero. Therefore in the case of a
perfect differential, the line integral along any closed path is zero.

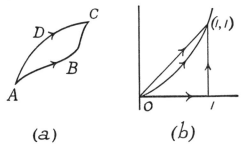

Fig. 12.

Example 7. If
$$\phi(x, y) = x^2 y^3$$
$$d\phi = 2xy^3 dx + 3x^2 y^2 dy$$

and is a perfect differential. Integrate this along the three different paths given,
without using the fact that it is a perfect differential. In each case the path
begins at (0, 0) and ends at (1, 1). Fig. 12 (b).

(a) Along $y = x$.

$$\int (Pdx + Qdy) = \int_{(0, 0)}^{(1, 1)} (2xy^3 dx + 3x^2 y^2 dy)$$

$$= \int_0^1 (2x . x^3 dx + 3x^2 . x^2 dx)$$

$$= \int_0^1 5x^4 dx = 1 \quad . \quad . \quad . \quad . \quad . \quad . \quad (a)$$

(b) Along $y = x^2$ the integral is

$$\int_0^1 (2x \cdot x^6 dx + 3x^2 \cdot x^4 \cdot 2x dx)$$

$$= \int_0^1 8x^7 dx = 1 \quad \ldots \ldots \ldots \ldots \quad (b)$$

(c) Along $y = 0$ from $x = 0$ to 1 and then along $y = 1$ from $y = 0$ to $y = 1$. In this case, since y and dy are zero on the first section of the path and $x = 1$, $dx = 0$, on the second section, the integral is

$$\int_0^1 0 dx + \int_0^1 3y^2 dy = 1 \quad \ldots \ldots \ldots \quad (c)$$

The three values (a), (b), (c) are the same, and equal to the value

$$\int_{(0,\,0)}^{(1,\,1)} d(x^2y^3) = \left[x^2y^3 \right]_{(0,\,0)}^{(1,\,1)} = 1.$$

The student will remember from his elementary Mechanics that when there are no dissipative forces such as friction in a field the work done, say, in raising a weight from one level to another, is independent of the path travelled. In such a field, called a *conservative* field of force,

$$\int_C (P dx + Q dy)$$

along every closed path C in the field is zero and the integrand is a perfect (or exact) differential.

EXERCISE 13

1. Show $\int_0^\infty e^{-ax} dx = \dfrac{1}{a}$, $(a > 0)$ and hence show that

$$\int_0^\infty x^n e^{-ax} dx = \frac{n!}{a^{n+1}}.$$

2. Find

$$\int_0^\infty e^{-ax} \cos bx \, dx.$$

Hence find: (i) $\int_0^\infty x e^{-ax} \cos bx \, dx$;   (ii) $\int_0^\infty x e^{-ax} \sin bx \, dx$.

3. Find

$$\int_0^\infty \frac{dx}{(x + a)(x + b)} \quad \text{if } a > b > 0.$$

Deduce

$$\int_0^\infty \frac{dx}{(x + a)^2(x + b)}.$$

4. Find

$$\int_0^\infty \frac{dx}{(x^2 + a^2)(x^2 + b^2)}$$

and deduce

$$\int_0^\infty \frac{dx}{(x^2 + a^2)^2(x^2 + b^2)}.$$

5. Given
$$f(a) = \int_{-\infty}^{0} \frac{e^{-ax} - e^{x}}{x} \, dx$$

show that $df/da = \dfrac{1}{a}$. Solve this simple differential equation to find $f(a)$, the value of the integral using $f(1) = 0$ to evaluate the constant of integration.

6. Show that for $a > b$
$$\int_{0}^{\pi} \frac{dx}{a + b \cos x} = \frac{\pi}{\sqrt{(a^2 - b^2)}}.$$

Deduce the values of:

$$(a) \int_{0}^{\pi} \frac{dx}{(a + b \cos x)^2}; \qquad (b) \int_{0}^{\pi} \frac{\cos x \, dx}{(a + b \cos x)^2};$$

$$(c) \int_{0}^{\pi} \frac{a_1 + b_1 \cos x}{(a + b \cos x)^2} \, dx.$$

7. If
$$I = \int_{0}^{\infty} e^{-x^2} \cos 2bx \, dx$$

show
$$\frac{dI}{db} + 2bI = 0.$$

Solve this differential equation and given that
$$\int_{0}^{\infty} e^{-x^2} dx = \tfrac{1}{2}\sqrt{\pi}$$

show that
$$I = \tfrac{1}{2}\sqrt{\pi} e^{-b^2}.$$

8. Show that
$$\int_{0}^{\infty} x^n e^{-x} dx = n \int_{0}^{\infty} x^{n-1} e^{-x} dx$$

and deduce that when n is a positive integer the value of the integral is $n\,!$
Find the value of a for which
$$\int_{0}^{\infty} (1 - ax^2)^2 e^{-x} dx$$

is a minimum.   [L.U.]

9. When a beam of length l clamped at one end and supported at the other carries a uniformly distributed load of intensity w, the strain energy U is given by
$$2EIU = \int_{0}^{l} (\tfrac{1}{2}wx^2 - Rx)^2 dx$$

and the correct value of the reaction R is the one which makes U a minimum. Find the value of R and the value of U for this value of R.   [L.U.]

10. A light uniform girder is simply supported at each end in a horizontal position and carries a central load W. Show that the central deflection is given by $\delta = Wl^3/48EI$.

11. Find the value of

$$\int_{(0,0)}^{(1,1)} [(x+y)dx + (x-y)dy]$$

along the curves:

(a) the line $x = t$, $y = t$;

(b) the parabola $x = t^2$, $y = t$;

(c) the parabola $x = t$, $y = t^2$;

(d) the curve $x = t$, $y = t^3$.

12. Find

$$\int_{(0,2)}^{(3,0)} [(x-2y)dx + xydy]$$

along the curves:

(a) the line joining the points;

(b) the ellipse $x = 3\cos\theta$, $y = 2\sin\theta$;

(c) the parabola $2(x-3)^2 = 9y$.

13. Find the value of

$$\int_{(-a,0)}^{(a,0)} (xdy + ydx)$$

along the upper half of the circle $x^2 + y^2 = a^2$.

14. If

$$Pdx + Qdy \equiv d\phi$$

show that

$$P = \frac{\partial\phi}{\partial x}, \qquad Q = \frac{\partial\phi}{\partial y}$$

and hence that

$$\frac{\partial P}{\partial y} = \frac{\partial Q}{\partial x}$$

is a necessary condition for $Pdx + Qdy$ to be a perfect differential. [It is also a sufficient condition.] In this case show that

$$\int (Pdx + Qdy)$$

taken along a curve joining two given points is independent of the path joining the points.

15. Show that the value of each of the following is independent of the path of integration and evaluate by any convenient method:

(a) $\displaystyle\int_{(0,1)}^{(1,2)} [(x^2 + 3y^2)dx + 6xydy]$;   (b) $\displaystyle\int_{(-1,0)}^{(2,1)} (x^2 + 4y)dx + (4x - y)dy]$;

(c) $\displaystyle\int_{(0,0)}^{(1,1)} \left[\frac{1-y^2}{(1+x)^3} dx + \frac{y}{(1+x)^2} dy\right]$;

(d) $\displaystyle\int_{(1,3)}^{(3,1)} [(e^x + 3y)dx + (3x + \log y)dy]$.

16. Show that the line integral

$$\int \left(\frac{-ydx}{x^2 + y^2} + \frac{xdy}{x^2 + y^2}\right)$$

taken along a square of side unity parallel to the axes and centre the origin has a value 2π although the integrand is a perfect differential.

[This is $\displaystyle\int d\theta$ where $\theta = \tan^{-1}(y/x)$ and for any path enclosing the origin θ is not a one-valued function, but increases by 2π for every circuit of the origin.

If, however, the curve does not include the origin, θ returns to its original value and is a one-valued function.]

17. Show that if $u(x, y)$, $v(x, y)$ are functions which satisfy a certain differential equation, then $\int (u\,dx + v\,dy)$ as a line integral between points A and B is independent of the path AB.

Show that the condition is not satisfied when

$$u = x \sin y, \qquad v = y \cos x$$

and evaluate

$$\int (x \sin y\,dx + y \cos x\,dy):$$

(i) Along the straight line $y = mx$; and

(ii) Along the straight lines $y = 0$ and $x = x_1$ in each case from $(0, 0)$ to (x_1, y_1). [L.U.]

18. Evaluate

$$\int \{(x + y)dx + x^2 dy\}$$

between $(0, 0)$ and $(2a, 0)$ along the upper half of the circle whose centre is at the point $(a, 0)$. [L.U.]

[Show that any point P on the curve is $x = 2a \cos^2 \theta$, $y = 2a \cos \theta \sin \theta$ where OP is inclined to the x axis at angle θ.]

THE CATENARY

"When the overhead line (1·3 miles long and 600 ft high at the pylons) was erected across the river it was found that the clearance for ships was some 13 ft less than the calculated value. At first this was believed due to the expansion of the line in the heat. Later it was found that the calculation was made on the parabolic assumption which did not hold in this case."

A Report on a Grid System.

6.1. In Vol. I (13.12) we showed that when a hanging string is so tightly stretched that the centre of gravity of any section may be assumed to act through the mid point of its horizontal projection, the curve assumed by the string is a parabola. We will now not make this assumption and find the shape assumed by a heavy string hanging from fixed supports.

6.2. The Cartesian Equation

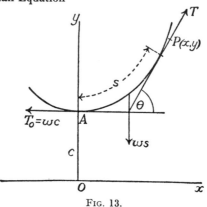

Fig. 13.

A uniform rope of weight w N/m is suspended from two fixed points and hangs vertically. If A is the lowest point draw the y axis vertically through A. The position of the x axis will be settled later.

Consider the forces acting on AP a length of s m where P is a point (x, y) on the curve. The forces acting on this length are T_0 the tension at the point A, T the tension at the point P and ws the weight of AP. We note that a tension acts along the join of two consecutive elements of the rope and so is tangential to it. Let the tension T make angle θ with the horizontal, whilst T_0 at the lowest point where the gradient is zero will act horizontally. It is convenient to let

$T_0 = wc$ so that T_0 is equal to the weight of a length c of the rope. This length c is called the *parameter* of the curve.

For the equilibrium of the length AP we have by resolving horizontally

$$T \cos \theta = wc \quad (= T_0)$$

resolving vertically
$$T \sin \theta = ws$$
$$\therefore \quad \tan \theta = s/c$$
$$s = c \tan \theta . \quad . \quad . \quad . \quad . \quad . \quad . \quad (1)$$

Since $\tan \theta = \dfrac{dy}{dx}$ and

$$\frac{ds}{dx} = \sqrt{\left(1 + \left(\frac{dy}{dx}\right)^2\right)} \qquad \text{[Vol. I (22.1)]},$$

$$\therefore \quad \frac{ds}{dx} = \sqrt{\left(1 + \frac{s^2}{c^2}\right)},$$

$$\therefore \quad \frac{ds}{\sqrt{(c^2 + s^2)}} = \frac{dx}{c}.$$

Integrating this

$$\sinh^{-1}\left(\frac{s}{c}\right) = \frac{x}{c} + D.$$

But $s = 0$ when $x = 0$ since the y axis passes through A so that $D = 0$

and
$$s = c \sinh\left(\frac{x}{c}\right) \quad . \quad . \quad . \quad . \quad . \quad (2)$$

an equation connecting the x co-ordinate of any point with the length of the rope *measured from where dy/dx is zero* up to that point.

Returning to (1) we have

$$\frac{dy}{dx} = \frac{s}{c} = \sinh\left(\frac{x}{c}\right) \text{ by (2)}$$

Integrate :

$$y = c \cosh\left(\frac{x}{c}\right) + E.$$

We now choose the x axis a distance c below A so that when $x = 0$, $y = c$. This gives $E = 0$

and
$$y = c \cosh\left(\frac{x}{c}\right) \quad . \quad . \quad . \quad . \quad . \quad (3)$$

This (x, y) equation to the curve shows that the rope hangs in the form of the hyperbolic cosine which we have already considered [Vol. I (29)].

From (3)
$$y^2 = c^2 \mathrm{ch}^2\left(\frac{x}{c}\right) = c^2\left[1 + \mathrm{sh}^2\left(\frac{x}{c}\right)\right]$$

$$= c^2\left[1 + \frac{s^2}{c^2}\right] \text{ by (2)},$$

$$\therefore \quad y^2 = c^2 + s^2 \quad . \quad . \quad . \quad . \quad . \quad . \quad . \quad (4)$$

an equation showing that y, the height of P above the x axis (often for this curve called the directrix), the arc length AP and the length 0A ($= c$) can form the sides of a right-angled triangle.

If we square and add the equations for the equilibrium of AP we find

$$T^2(\cos^2\theta + \sin^2\theta) = w^2(c^2 + s^2)$$
$$\therefore \quad T^2 = w^2y^2 \quad \text{by (4)}$$
$$T = wy \quad . \quad . \quad . \quad . \quad . \quad (5)$$

showing that the tension at any point (x, y) on the rope is equal to the weight of the length of the same rope hanging from that point down to the x axis. Since $T_0 = wc$, this also holds for the lowest point.

The equations (1)–(5) must be known and the appropriate ones used in solving a given problem.

Example 1. One end of a uniform chain of length 870 mm is fixed at A. It passes over a small smooth peg B, which is vertically 60 mm higher than A, and 450 mm of chain hang freely from B. Show that the horizontal distance of B from A is 360 log 3 mm (Fig. 14(a)).

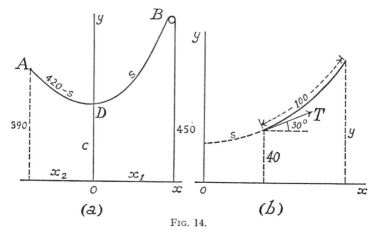

Fig. 14.

The tension at B is equal to the weight of 450 mm of the string and, on the other side of the smooth peg also equal to the weight of a length hanging from B to the directrix (x axis). Therefore the directrix is 450 mm below B and 390 mm below A.

Using
$$y^2 = s^2 + c^2$$
for DB
$$450^2 = s^2 + c^2,$$
AD
$$390^2 = (420 - s)^2 + c^2,$$
subtract and find
$$s = 270,$$
$\therefore$ from either equation $c = 360$

Using $y = c \operatorname{ch}(x/c)$
$$x = c \operatorname{ch}^{-1}(y/c)$$
$$= 360 \log \left(\frac{y + \sqrt{(y^2 - 360^2)}}{12} \right),$$

for x_1, $y = 450$
$$x_1 = 360 \log 2$$
x_2, $y = 390$
$$x_2 = 360 \log (3/2),$$
therefore
$$x_1 + x_2 = 360 \log 2 + 360 \log (3/2)$$
$$= \underline{360 \log 3 \text{ mm}}$$

Example 2. A kite is flown at the end of 100 m of string. The tension at the hand is equal to the weight of 40 m of string, and is inclined at 30° to the horizontal. Show that the kite is about 85 m above the hand.

The hand (Fig 14 (b)) is 40 m above the directrix of the catenary in which the string hangs and the tension at the hand is 40 w.

∴ resolving horizontally $40 w \cos 30 = wc (= T_0)$,

$$c = 20\sqrt{3}.$$

If s is the length required from the hand to the lowest point of the curve, since $y^2 = s^2 + c^2$

$$40^2 = s^2 + (20\sqrt{3})^2.$$
$$s = 20.$$

If y is the height of the kite above the directrix

$$y^2 = (20 + 100)^2 + (20\sqrt{3})^2$$
$$= 15\,600$$
$$y = 20\sqrt{39} = 125 \text{ m approx.}$$

∴ height above hand $= 125 - 40 = \underline{85 \text{ m}}$

Example 3. A uniform chain is stretched between 2 points at the same level 66 m apart. If the sag at the centre is 3 m prove that the parameter is approximately 182 m.

The co-ordinates of the right-hand end are $x = 33$ m, $y = 3 + c$ m. Using $y = c \operatorname{ch} (x/c)$

$$3 + c = c \operatorname{ch} (33/c)$$
$$\therefore \quad 1 + 3/c = \operatorname{ch} (33/c)$$

put $33/c = z$: $\qquad 1 + \dfrac{z}{11} = \operatorname{ch} z$ . . . . (6)

which must be solved graphically or by Newton's method.

Since c is to be 182, z is about $\frac{1}{6}$, which is small enough to justify the use of

$$\operatorname{ch} z = 1 + \frac{z^2}{2!} + \frac{z^4}{4!} + \cdots$$

$$= 1 + \frac{z^2}{2!} \text{ approx.}$$

(6) gives

$$1 + \frac{z}{11} = 1 + \frac{z^2}{2},$$

or as an approximate value $z = \dfrac{2}{11}$.

Using this approximate value of z

$$\frac{33}{c} = z = \frac{2}{11}, \qquad \therefore \quad c = 181 \cdot 5$$
$$= \underline{182 \text{ approx.}}$$

As an exercise we will solve (6) more accurately by Newton's method.

Let $\qquad f(z) = 1 + \dfrac{z}{11} - \operatorname{ch} z,$

$$\therefore \quad f(0 \cdot 1) = 1 + 0 \cdot 0091 - 1 \cdot 0050 = + 0 \cdot 0041,$$
$$f(0 \cdot 2) = 1 + 0 \cdot 0182 - 1 \cdot 0201 = - 0 \cdot 0019,$$
$$f(0 \cdot 15) = 1 + 0 \cdot 0136 - 1 \cdot 0113 = + 0 \cdot 0023,$$

∴ the root is between 0·15 and 0·2.

Since $\qquad f''(z) = - \operatorname{ch} z$

and is negative throughout the range, we must choose the value of z that makes

$f(z)$ negative, *i.e.*, $z_1 = 0.2$, as our first approximation. For the second approximation

$$z_2 = z_1 - \frac{f(z_1)}{f'(z_1)} = 0.2 - \frac{-0.0019}{(\frac{1}{11} - \text{sh } z)_{0.2}}$$

$$= 0.2 - \frac{-0.0019}{-0.1104} = 0.18.$$

Again

$$z_3 = 0.18 - \frac{0.000\,12}{-0.09} = 0.181,$$

$$\therefore \quad c = \frac{33}{z_3} = \underline{182.3}.$$

EXERCISE 14

1. A wire hangs in the form $y = 100 \text{ ch } (x/100)$. Find the length of the wire and the sag at the centre if the span between the ends at the same level is 50 m.

2. A chain hangs from two points at the same level. If the inclination at either end is $45°$, find the sag if the total length of chain is $2l$.

3. A chain 10 m long and weighing 60 N hangs from two points at the same level, the sag at the centre being 3 m. Find the tension at the lowest point and at each support. If a support is a pole 20 m high, find the maximum bending moment acting on it.

4. A chain 13 m long rests in limiting equilibrium on a rough horizontal table for which the coefficient of friction $\mu = \frac{2}{5}$. It lies stretched out straight with one end held 5 m above the table. Find the length on the table.

5. A uniform string has one end fixed and rests in equilibrium passing over a small smooth peg with its other end hanging freely. The length of the vertical portion of the string is 2 m. The fixed end and the lowest point of the catenary are respectively $2/\sqrt{3}$ m and 1 m above the free end. Find the angles which the string makes with the horizontal at the fixed end and at the peg and also find the total length of the string.

6. An endless uniform chain hangs in equilibrium over a smooth pulley, and is in contact with it over three-quarters of the circumference. Show that the length of the free portion of the string is $\sqrt{2}/\log (1 + \sqrt{2})$ times the radius of the pulley.

7. Show that the length of an endless chain which will hang over a circular pulley of radius a so as to be in contact with $\frac{2}{3}$ of the circumference of the pulley is

$$a \left[\frac{3}{\log (2 + \sqrt{3})} + \frac{4\pi}{3} \right].$$

8. A kite flies at the end of 120 m of string, and the pull on the hand is equal to the weight of 20 m of the string inclined at $30°$ upwards to the horizontal. Find the vertical height of the kite above the hand.

9. A uniform flexible chain of length l and weight W hangs between two fixed points at the same level, and a weight W_1 is attached to its mid point. If k is the sag in the middle, prove that the pull on either point of support is

$$\frac{k\text{W}}{2l} + \frac{l\text{W}_1}{4k} + \frac{l\text{W}}{8k}.$$

[L.U.]

(Consider half the string with $\frac{1}{2}\text{W}_1$ at its end. The vertical component of the tension in the catenary at this point supports this weight.)

10. A uniform chain hangs over two smooth pegs at different levels so that there is a length l of chain between the pegs and lengths l_1, l_2 hanging vertically on the two sides. Prove that the lowest point of the curved part of the chain divides this part in the ratio

$$l^2 + l_1^2 - l_2^2 : l^2 + l_2^2 - l_1^2$$

and that the parameter c of the catenary is given by

$$4l^2c^2 = (l + l_1 + l_2)(l + l_2 - l_1)(l + l_1 - l_2)(l_1 + l_2 - l).$$

[L.U.]

11. A heavy uniform chain of length l is lying in a straight line on a horizontal floor. One end of the chain is slowly raised vertically until one-half of the chain is clear of the floor, the remainder being on the point of slipping. If the coefficient of friction is $\frac{1}{3}$, show that the height above the floor of the raised end is $\frac{1}{6}l(\sqrt{10}-1)$ and that the horizontal distance of the mid point of the chain from that end is $\frac{1}{6}l\log(\sqrt{10}+3)$.

12. A chain 52 m long is suspended from 2 points at the same level 50 m apart. Find the sag.

13. A chain 102 m long is suspended from 2 points at the same level 100 m apart. Show the sag is about 8·7 m.

14. A chain 155 m long is suspended from 2 points, at the same level and 150 m apart. Show that the tension at the lowest point is nearly 1·08 times the weight of the chain.

15. Obtain the formulae $s = c\tan\theta$, $y = c\sec\theta$ for the common catenary.

One end of a uniform chain of weight W, length l, slides on a smooth straight wire OA inclined at 45° below the horizontal. The other end is attached to a small ring of weight W which slides on a smooth vertical wire OB. Prove that the catenary formed has a parameter $2l$ and find the difference in height of the ends. [L.U.]

16. One end of a rough uniform chain of length l is fastened to a point on a vertical wall at a height h above the ground. Show that the greatest distance from the wall at which the free end of the chain can rest on level ground is

$$\frac{c}{\mu}\left\{1 + \mu\log\left(\frac{h+l}{c}+1-\frac{1}{\mu}\right)\right\},$$

where μ is the coefficient of friction and c is the parameter of the catenary given by

$$c := \mu(l+\mu h) - \mu\{(\mu^2+1)h^2 + 2\mu lh\}^{\frac{1}{2}}.$$

6.3. Further Equations

In some problems it is useful to use the equation of the catenary in a somewhat different form. We will now obtain the equations giving the value of x, y at any point in terms of θ, the slope of the curve at the point.

We have

$$s = c\tan\theta \qquad\qquad [6.2\ (1)]$$

$$\therefore\ \frac{ds}{d\theta} = c\sec^2\theta.$$

We will relate this to $\qquad \dfrac{dx}{ds} = \cos\theta \qquad$ [Vol. I (22.5)]

and write $\qquad \dfrac{dx}{d\theta} = \dfrac{dx}{ds}\cdot\dfrac{ds}{d\theta}$

$$= \cos\theta\cdot c\sec^2\theta = c\sec\theta,$$

integrate: $\qquad x = c\displaystyle\int\frac{d\theta}{\cos\theta}$

$$= c\log(\sec\theta + \tan\theta) + C$$

with the usual axis $\theta = 0$ when $x = 0$, so that $C = 0$

and $\qquad\qquad \underline{x = c\log(\sec\theta + \tan\theta)} \quad \cdot\ \cdot\ \cdot\ (7)$

Also
$$\frac{dy}{d\theta} = \frac{dy}{ds} \cdot \frac{ds}{d\theta}$$
$$= \sin\theta \cdot c \sec^2\theta,$$
$$\therefore \quad y = c \int \frac{\sin\theta}{\cos^2\theta}\, d\theta + D$$
$$= c \sec\theta + D,$$

$y = c$ when $\theta = 0$ so that $D = 0$

and
$$y = c \sec\theta \qquad \cdot \quad \cdot \quad \cdot \quad \cdot \quad \cdot \quad \cdot \quad (8)$$

From (7)
$$e^{x/c} = \sec\theta + \tan\theta$$

also
$$1 = \sec^2\theta - \tan^2\theta,$$
$$\therefore \quad e^{-x/c} = \sec\theta - \tan\theta,$$
$$\therefore \quad \frac{e^{x/c} + e^{-x/c}}{2} = \sec\theta = \frac{y}{c} \quad \text{by (8)}$$

or
$$y = c \operatorname{ch}(x/c) \qquad \cdot \quad \cdot \quad \cdot \quad \cdot \quad \cdot \quad (9)$$

From (8)
$$y^2 = c^2 \sec^2\theta = c^2[1 + \tan^2\theta]$$
$$= c^2 + s^2 \qquad \cdot \quad \cdot \quad \cdot \quad \cdot \quad \cdot \quad (10)$$
$$\therefore \quad s^2 = y^2 - c^2 = c^2 \operatorname{ch}^2(x/c) - c^2$$
$$= c^2 \operatorname{sh}^2(x/c)$$

or
$$s = c \operatorname{sh}(x/c) \qquad \cdot \quad \cdot \quad \cdot \quad \cdot \quad \cdot \quad (11)$$

Equations (7) and (8) are the new equations which we will now use.

Example 4. A chain AB of length l hangs under gravity with the end A fixed and the other end B attached by a light inextensible string BC to a fixed point C at the same level as A. The lengths of the string and chain are such that the tangent to the catenary at A makes an angle of 45° with the horizontal, and the vertex divides the length of chain in the ratio of 2 : 1. Find the ratio of the length of string to the length of the chain and the horizontal distance between A and B.

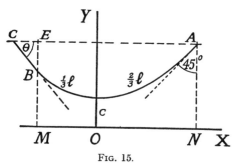

FIG. 15.

Using $\qquad s = c \tan\theta$
for LA $\qquad \frac{2}{3}l = c \tan 45,$ $\qquad \therefore \qquad c = \frac{2}{3}l,$
for BL $\qquad \frac{1}{3}l = c \tan\theta,$ $\qquad \therefore \qquad \tan\theta = \frac{1}{2},$
$\qquad \sec\theta = \frac{1}{2}\sqrt{5}.$

Using
$$y = c \sec \theta$$
$$BM = \tfrac{2}{3}l \times \tfrac{1}{2}\sqrt{5} = \frac{\sqrt{5}}{3}l$$

also
$$AN = \tfrac{2}{3}l \sec 45 = \frac{2\sqrt{2}}{3}l,$$

$$\therefore \quad BE = AN - BM$$
$$= \frac{2\sqrt{2}}{3}l - \frac{\sqrt{5}}{3}l,$$

$$\therefore \quad BC = BE \cosec \theta = \frac{l}{3}(2\sqrt{2} - \sqrt{5})\sqrt{5}$$
$$= \frac{l}{3}(2\sqrt{10} - 5),$$

$$\therefore \quad \frac{BC}{BA} = \frac{2\sqrt{10} - 5}{3} \quad . \quad . \quad . \quad . \quad . \quad . \quad . \quad (12)$$

Since
$$x = c \log (\sec \theta + \tan \theta)$$
$$MO = \tfrac{2}{3}l \log (\tfrac{1}{2}\sqrt{5} + \tfrac{1}{2})$$
$$ON = \tfrac{2}{3}l \log (\sqrt{2} + 1),$$

$$\therefore \quad MN = \tfrac{2}{3}l \log \left\{ \left(\frac{\sqrt{5} + 1}{2} \right) \left(\sqrt{2} + 1 \right) \right\} \quad . \quad . \quad . \quad (13)$$

EXERCISE 15

1. A chain 100 m long is suspended from 2 points at the same level, and the ends are inclined at 60° to the horizontal. Find the span and the sag at the lowest point. [L.U.]

2. One end of a uniform flexible chain of length a and weight W is attached to a fixed point, and the other end is drawn aside by a horizontal force. If this force is equal to W, show that the horizontal deflection of the lower end is $a \log (1 + \sqrt{2})$ and find the tension at the support. [L.U.]

3. A chain of length L has its ends attached to light rings which slide on a rough horizontal rod. If μ is the coefficient of friction, show that the greatest distance apart at which the rings can rest in equilibrium is
$$\mu L \log [\{1 + \sqrt{(1 + \mu^2)}\}/\mu]. \qquad [L.U.]$$

4. One end of a uniform heavy chain is fastened to one end of a light in-extensible string of length l, and the other ends of the chain and string are fastened to points A and B respectively in the same horizontal line. If the inclination of the chain at A and of the string at B to the horizontal are 60° and 30° respectively prove that the length of the chain is
$$\tfrac{1}{2}(\sqrt{3} + 1)l$$
and find the distance AB. [L.U.]

5. A uniform chain is stretched between a point on the ground and a point 300 m above the ground. The tension at the ground is equal to a weight of 500 m of the chain, and is inclined at $\cos^{-1}(\tfrac{4}{5})$ to the horizontal. Show that the length of the chain is nearly 393 m.

6. Prove that the centroid of the arc of length s of the catenary $y = c \operatorname{ch} \dfrac{x}{c}$ included between the origin and any point (x, y) is at a height $\tfrac{1}{2}y + \dfrac{cx}{2s}$ above the directrix.

A uniform flexible chain of length $2l$ and weight w per unit length hangs with one end A attached to a fixed point and the other end B supported close to A. Prove that the work done in moving B horizontally until the chain is inclined at 45° to the vertical at each end is
$$wl^2[1 - \sqrt{2} + \log (1 + \sqrt{2})]. \qquad [L.U.]$$

CHAPTER 7

DIFFERENTIAL EQUATIONS

7.1. Introduction

In Vol. I (34) we saw that by the solution of an ordinary differential equation we mean a relation between the two variables concerned, free from differential coefficients. The *general* solution of such an equation must contain a number of *independent* unknowns equal to the *order* of the equation solved. Thus to solve

$$\frac{dy}{dx} = x$$

as
$$y + A = \tfrac{1}{2}x^2 + B$$

is not to obtain a solution with two independent unknowns, since A and B are effectively the one unknown $C = B - A$, a method of statement we learn to adopt at once. A solution with less than the required number of constants is called a particular solution or integral.

For convenience, differential equations are divided into types, each having a standard method of solution.

7.2. Variables Separable

An equation of this type can be written as

$$P(x)dx + Q(y)dy = 0,$$

where $P(x)$ involves x only and $Q(y)$ involves y only. The solution is

$$\int P(x)dx + \int Q(y)dy = C.$$

Example 1. A tank contains 50 litres of brine containing 1 kg of salt. Another brine solution with 0·2 kg of salt per litre runs in at 3 litres/min and the mixture runs out at the same rate. Find the amount of salt present after t min.

If Q kg is the amount present after t minutes and this is increased by ΔQ in the next Δt minute,

$$\Delta Q = 0·6\,\Delta t + \frac{3Q}{50}\,\Delta t$$
$$\text{(in)} \qquad \text{(out)}$$

Dividing by Δt and proceeding to the limit:

$$\frac{dQ}{dt} = 0·6 - \frac{3Q}{50} = \frac{30 - 3Q}{50}.$$

Separating the variables:

$$\frac{dQ}{30 - 3Q} = \frac{dt}{50}$$

$$-\tfrac{1}{3}\log(30 - 3Q) = \frac{t}{50} + C$$

but $Q = 1$ when $t = 0$,

$$-\tfrac{1}{3}\log 27 = C.$$

Subtracting:
$$-\tfrac{1}{3}\log\left(\frac{30 - 3Q}{27}\right) = \frac{t}{50},$$

or
$$Q = 10 - 9e^{\frac{-3t}{50}}$$

giving the amount present after t min, and showing that Q increases with time to a maximum of 10 kg, a maximum which is never attained.

7.3. Linear Equation, First Order

The type
$$L(x)\frac{dy}{dx} + M(x)y = f(x),$$

where, as shown L, M, f contain x only, is called a linear equation since dy/dx and y occur in the first power only. To solve it we divide first by $L(x)$ and rewrite it in the standard form

$$\frac{dy}{dx} + Py = Q \quad \cdots \quad \cdots \quad (1)$$

where P and Q are functions of x only.

This type suggests that we had at first (if $R \equiv R(x)$)

$$\frac{d}{dx}(Ry) = S \quad \cdots \quad \cdots \quad (2)$$

and obtained

$$R\frac{dy}{dx} + y\frac{dR}{dx} = S \quad \cdots \quad \cdots \quad (3)$$

or
$$\frac{dy}{dx} + y\frac{dR/dx}{R} = \frac{S}{R} \quad \cdots \quad \cdots \quad (4)$$

If (1), the equation which we are given, is the same as (4), then considering the left-hand side only

$$\frac{dR}{R} = Pdx \quad \text{or} \quad \log R = \int Pdx + C.$$

and
$$R = Ae^{\int Pdx} \qquad [A = e^C].$$

Multiplying both sides of (1) by this *integrating factor* to remake (3), it becomes

$$Ae^{\int Pdx}\frac{dy}{dx} + APe^{\int Pdx}y = AQe^{\int Pdx}$$

which by (2) can be written, after cancelling the constant A,

$$\frac{d}{dx}\left(ye^{\int Pdx}\right) = Qe^{\int Pdx}$$

or
$$\frac{d}{dx}(yR_1) = QR_1,$$

where $R_1 = e^{\int Pdx}$ which has been evaluated (and simplified as far as possible).

It follows that

$$yR_1 = \int QR_1dx + D,$$

which is the general solution.

(i) Since the constant A cancels we can omit it from the beginning and take as the integrating factor

$$R_1 \equiv e^{\int Pdx}$$

(ii) It will be noticed that we concentrate on obtaining the left-hand side in an " integrable form "

$$\frac{d}{dx}\left(ye^{\int Pdx}\right) \quad \text{or} \quad \frac{d}{dx}(yR_1).$$

The right-hand side contains x only and in theory gives no trouble in integration.

Example 2. Solve

$$x\frac{dy}{dx} + (1 + x)y = e^x.$$

Rewriting this as

$$\frac{dy}{dx} + \left(\frac{1}{x} + 1\right)y = e^x/x$$

we note that it is of the type solved above and the integrating factor is

$$e^{\int Pdx}, \quad \text{where} \quad \int Pdx = \int \left(\frac{1}{x} + 1\right) dx$$

$$= \log x + x$$

so that $e^{\log x + x} = e^{\log x} \cdot e^x = xe^x.$

Multiplying both sides by this, the equation becomes

$$xe^x\frac{dy}{dx} + (1 + x)e^xy = e^{2x},$$

which we know, from the method, to be

$$\frac{d}{dx}(yxe^x) = e^{2x},$$

therefore

$$yxe^x = \tfrac{1}{2}e^{2x} + C$$

and

$$y = \frac{1}{x}(\tfrac{1}{2}e^x + Ce^{-x}).$$

(Note the necessity for evaluating the integrating factor and expressing it in the simplest form.)

Example 3. A tank has initially 50 litres of brine containing 5 kg of salt. 2 litres of fresh water per minute enter A, and the mixture passes at the same rate from A into B, which initially has 50 litres of fresh water. The resulting mixture leaves B at the same rate of 2 litres/min. Find the amount of salt in B after 1 h.

Suppose there is Q kg in A at time t min and this increases by ΔQ in the next interval of Δt min.

$$\Delta Q = -2\Delta t\,\frac{Q}{50} \quad \text{or} \quad \frac{\Delta Q}{\Delta t} = -\frac{2Q}{50}$$

therefore

$$\frac{dQ}{dt} = -\frac{2Q}{50} \quad \text{or} \quad \frac{dQ}{Q} = -\frac{1}{25}\,dt.$$

Integrate:

$$\log Q = -\frac{1}{25}t + C$$

$Q = 5, t = 0$:

$$\log 5 = 0 + C$$

$$\log \frac{Q}{5} = -\frac{1}{25}t \quad \text{or} \quad Q = 5e^{-t/25},$$

giving the amount at time t min.

Now suppose there is P kg in B at time t, which increases by ΔP kg in the next Δt min.

$$\Delta P = 2\Delta t\,\frac{Q}{50} - \frac{P}{20}\cdot 2\Delta t$$

$$= 0{\cdot}2\Delta t\,e^{-t/25} - \frac{P}{50}\cdot 2\Delta t$$

therefore

$$\frac{\Delta P}{\Delta t} = 0{\cdot}2e^{-t/25} - \frac{P}{25}$$

or in the limit when $\Delta t \longrightarrow 0$,

$$\frac{dP}{dt} = 0{\cdot}2e^{-t/25} - \frac{P}{25} \text{ giving } \frac{dP}{dt} + \frac{1}{25}P = 0{\cdot}2e^{-t/25}.$$

The integrating factor is $e^{\int 1/25dt} = e^{t/25}$; multiply by this to obtain

$$\frac{d}{dt}(Pe^{t/25}) = 0{\cdot}2$$

or

$$Pe^{t/25} = 0{\cdot}2 + D$$

$P = 0, t = 0$, giving

$$D = 0.$$

Finally,

$$P = 0{\cdot}2te^{-t/25}$$

$t = 60$ min:

$$= 0{\cdot}2te^{-2{\cdot}4} = \underline{1{\cdot}09 \text{ kg}}$$

7.4. Exact Equations

Given

$$\phi(x, y) = C . \quad \cdot \quad \cdot \quad \cdot \quad \cdot \quad \cdot \quad (1)$$

we have by total differentiation

$$\frac{\partial \phi}{\partial x}dx + \frac{\partial \phi}{\partial y}dy = 0 . \quad \cdot \quad \cdot \quad \cdot \quad \cdot \quad (2)$$

giving the first order equation

$$\frac{\partial \phi}{\partial x} + \frac{\partial \phi}{\partial y}\frac{dy}{dx} = 0.$$

Since (2) may be integrated as it stands, it is called an exact equation, (1) being its solution. Because we may have divided all through (2) by some function in x and y, in general the equation

$$Pdx + Qdy = 0 \quad \cdot \quad \cdot \quad \cdot \quad \cdot \quad \cdot \quad (3)$$

is not exact. If it is it must be the same as (2) and so

$$P = \frac{\partial \phi}{\partial x}, \qquad Q = \frac{\partial \phi}{\partial y},$$

from which we obtain

$$\frac{\partial P}{\partial y} = \frac{\partial^2 \phi}{\partial y \partial x}, \qquad \frac{\partial Q}{\partial x} = \frac{\partial^2 \phi}{\partial x \partial y}$$

or since

$$\frac{\partial^2 \phi}{\partial y \partial x} = \frac{\partial^2 \phi}{\partial x \partial y} \qquad \text{[see Vol. I (27.2)]}$$

$$\frac{\partial P}{\partial y} = \frac{\partial Q}{\partial x}$$

is a necessary condition for an equation such as (3) to be exact.

Example 4. Solve $(y^2 + 2xy + 1)dx + (2xy + x^2)dy = 0$.

If this is

$$P dx + Q dy = 0$$

$$\frac{\partial P}{\partial y} = 2y + 2x, \qquad \frac{\partial Q}{\partial x} = 2y + 2x,$$

and the equation is exact. If the integral is $\phi(x, y) = 0$, where ϕ may include a constant

$$P = \frac{\partial \phi}{\partial x} = y^2 + 2xy + 1$$

$$\therefore \quad \phi = y^2 x + x^2 y + x + f(y).$$

[Note this form $f(y)$ instead of the usual constant of integration, since $\frac{\partial f(y)}{\partial x} = 0$].

From this

$$\frac{\partial \phi}{\partial y} = 2yx + x^2 + f'(y).$$

But

$$Q = \frac{\partial \phi}{\partial y},$$

$$\therefore \quad 2xy + x^2 = 2yx + x^2 + f'(y),$$

$$f'(y) = 0, \qquad f(y) = A.$$

The general solution is

$$y^2 x + x^2 y + x + A = 0.$$

The importance of exact equations is that the integral between 2 points $(x_1 y_1)$ $(x_2 y_2)$ is independent of the path travelled between the points by the variable point (x, y) of the integrand. This is a characteristic of a conservative field of force (see 5.7).

7.5. Some Miscellaneous Types

We will illustrate these by examples.

Example 5. Solve $\frac{dy}{dx} = \cos(x + y)$.

If

$$x + y = z, \qquad 1 + \frac{dy}{dx} = \frac{dz}{dx}.$$

Equation becomes

$$\frac{dz}{dx} - 1 = \cos z$$

or

$$dx = \frac{dz}{1 + \cos z}$$

therefore

$$x + C = \int \frac{dz}{1 + \cos z}$$

$$= \tan\left(\tfrac{1}{2} z\right)$$

$$= \underline{\tan \tfrac{1}{2}(x + y)}$$

Example 6. Solve $\frac{d^2y}{dx^2} = 2(y^3 + y)$ if $y = 0$ and $\frac{dy}{dx} = 1$ when $x = 0$. [L.U.]

An equation of this type has the characteristic that x does not appear (except in the differential coefficient).

Put
$$\frac{dy}{dx} = p,$$

$$\therefore \quad \frac{d^2y}{dx^2} = \frac{dp}{dx} = \frac{dy}{dx} \cdot \frac{dp}{dy} = p \frac{dp}{dy}.$$

The equation becomes a first-order equation

$$p\frac{dp}{dy} = 2(y^3 + y),$$

$$pdp = 2(y^3 + y)dy,$$

$$\tfrac{1}{2}p^2 + C = \tfrac{1}{2}y^4 + y^2,$$

$p = 1,\ y = 0:$
$$\tfrac{1}{2} + C = 0.$$

Solving for
$$p = \frac{dy}{dx}.$$

$$\frac{dy}{dx} = \pm \sqrt{(y^4 + 2y^2 + 1)}$$

$$= (y^2 + 1) \text{ [since } p = 1, y = 0.]$$

Therefore
$$\frac{dy}{y^2 + 1} = x$$

$$\tan^{-1} y + C = x$$

$y = 0,\ x = 0:$
$$C = 0.$$

Solution is :
$$\underline{\tan^{-1} y = x.}$$

Example 7. Solve $(x^2 - 1)\frac{d^2y}{dx^2} + x\frac{dy}{dx} = 0$.

In this type y does not appear except in the differential coefficients.

Put
$$\frac{dy}{dx} = p, \quad \text{so that} \quad \frac{d^2y}{dx^2} = \frac{dp}{dx}.$$

The equation becomes a first-order equation

$$(x^2 - 1)\frac{dp}{dx} + xp = 0,$$

$$\frac{dp}{p} + \frac{xdx}{x^2 - 1} = 0,$$

$$\log p + \tfrac{1}{2}\log(x^2 - 1) = \log A.$$

$$p\sqrt{x^2 - 1} = A,$$

$$dy = \frac{Adx}{\sqrt{x^2 - 1}},$$

$$\underline{y = A\ \mathrm{ch}^{-1} x + B.}$$

7.6. Orthogonal Trajectories

The equation $x^2 + y^2 = r^2$ denotes a circle centre at the origin, and as r varies we get a *family* of such circles. The equation $y = mx$ denotes a line through the origin, and whatever m is, this line will be a radius of each circle of the family and so cut it at right angles. Two families of curves each of which cuts every member of the other orthogonally are called orthogonal trajectories.

If the slope of one curve at a point is dy/dx, the slope of the orthogonal curve at that point is $-1/dy/dx$ (since $m_1 m_2 = -1$ for two perpendicular lines), from which we expect to obtain its equation.

Example 8. Find the orthogonal trajectories of the family of circles

$$x^2 + y^2 = kx.$$

From this equation

$$2x + 2y \frac{dy}{dx} = k$$

and we must eliminate k to obtain the differential equation governing the whole family. This is seen to be

$$2x + 2y \frac{dy}{dx} = \frac{x^2 + y^2}{x}$$

or

$$\frac{dy}{dx} = \frac{y^2 - x^2}{2xy}.$$

Every member of the orthogonal trajectories must then be governed by

$$\frac{dy}{dx} = \frac{-2xy}{y^2 - x^2}$$

To solve this, a homogeneous equation, put $y = vx$.

Since

$$\frac{dy}{dx} = v + x \frac{dv}{dx}$$

the equation gives

$$v + x \frac{dv}{dx} = -\frac{2v}{v^2 - 1}$$

$$x \frac{dv}{dx} = -\frac{v^3 + v}{v^2 - 1}$$

$$\frac{v^2 - 1}{v(v^2 + 1)} dv = -\frac{dx}{x}$$

$$\left(-\frac{1}{v} + \frac{2v}{v^2 + 1} \right) dv = -\frac{dx}{x}$$

$$-\log v + \log (v^2 + 1) = -\log x + \log C$$

$$\frac{v^2 + 1}{v} = \frac{C}{x}$$

finally,

$$x^2 + y^2 = Cy.$$

EXERCISE 16

Solve the following differential equations.

1. $x^4 \frac{dy}{dx} = x^2 + 3x - 7.$

2. $\cot x \frac{dy}{dx} = 4.$

3. $\frac{dy}{dx} = \frac{y^2 + 4}{x^2 + 9}.$

4. $\frac{dy}{dx} = \frac{y^2 + 4}{2y}.$

5. $\frac{dy}{dx} = \frac{x(1 + y^2)}{y(1 + x^2)}.$

6. $e^x \sqrt{(1 - y^2)} dx + \frac{y}{x} dy = 0.$

7. $(1 + x^2) \frac{dy}{dx} + xy = x.$

8. $x^2(1 + y)dy + y^2(x - 1)dx = 0.$

9. Water is added drop by drop to a beaker of volume V filled to the brim with a salt solution of relative density d, the excess being allowed to overflow. Show that the relative density d_1 of the solution when a volume v of water has been added is given by

$$d_1 = 1 + (d - 1)e^{-v/V}$$

10. The amount x of a substance present in a certain chemical action after time t is given by

$$\frac{dx}{dt} = k(a - x)(b - x)$$

where k, a, b are constant and x is zero at $t = 0$. If $x = 3$ when $t = 10$ min find the value of x after 20 min when $a = 6$ and $b = 9$.

11. It is required to find the area of cross-section of a vertical column such that the pressure p per unit area of cross-section is constant throughout. Consider a section of area A distant x from the top ($x = 0$) of area A_0. Let the area of the section $x + \Delta x$ from the top be $A + \Delta A$. By equating the forces keeping this volume between x and $x + \Delta x$ in equilibrium obtain the equation

$$p \frac{dA}{dx} = wA$$

where w is the weight per unit volume and hence the relation

$$A = A_0 e^{wx/p}.$$

12. The ideal cable for a barrage balloon would taper so that the stress across every section has the same value. If w is the weight per unit volume of the material of the cable, f the constant stress and r the radius at a distance x below the balloon, show that

$$\frac{dr}{dx} = -\frac{wr}{2f}.$$

Find the formula for r in terms of w, f and the tension T at $x = 0$ and show that the total weight W of the cable of length L is given by

$$W = T(1 - e^{-wL/f}). \tag{L.U.}$$

13. A cylindrical tank of cross-section A m² at water level lies with its axis horizontal and has a hole of area B m² in the bottom. If h m is the depth of water in the tank at time t seconds then the rate of outflow is given by

$$A \frac{dh}{dt} = -0.6B \sqrt{(2gh)}.$$

Find the time to empty the tank when full if its height is equal to the diameter d of its round ends and $1200B = d$.

14. A 50 litre tank is filled with water containing 0.5 kg of salt. 2 litres/min of brine each litre holding 0.1 kg of salt runs in and the mixture runs out at the same rate. Find the amount of salt present at time t min and the amount ultimately present.

15. A tank contains 100 litres of fresh water. 2 litres/min of brine run in, each litre containing 0.1 kg of salt and the mixture runs out at the rate of 1 litre/min. Find the amount of salt present when the tank contains 150 litres of brine.

16. A motor under load generates heat at a constant rate and radiates it at a rate proportional to the temperature excess θ so that $d\theta/dt = K - k\theta$, where t denotes time and K, k are constants. The initial rate of rise is 10°C/min and after 10 min the rise is 50°C. Show that the ultimate rise is $(10/k)$°C and that k is a root of the equation

$$e^{-10k} = 1 - 5k. \tag{L.U.}$$

Solve the equations:

17. $xy_1 + y = x^3$.

18. $y_1 + y \cos x = \cos x$.

19. $\sin x \cdot y_1 + y \cos n = x^2$.

20. $y_1 = (xy - 6)/2x^2$.

21. A circuit contains a resistance R ohms and an inductance L henries in series with an e.m.f. e. If i is the current at time t after the e.m.f. has been introduced obtain the equation

$$L\frac{di}{dt} + Ri = e.$$

Solve this when $e = E_0 \sin \omega t$ if $i = 0$ when $t = 0$.

22. If the above circuit contains a capacitance C farads instead of the inductance L obtain the equations:

(i) $e - q/C = Ri$,

where q is the charge on the condenser;

(ii) $i = dq/dt$, where i is the current charging the condenser;

(iii) $R\frac{di}{dt} + \frac{i}{C} = \frac{de}{dt}$. Solve this latter when $e = E_0 \sin \omega t$ and $i_0 = 0$.

23. A tank has initially 20 litres of brine containing 0·5 kg of salt. 2 litres/min of brine enter, each litre containing 0·1 kg of salt, and the mixture passes at the same rate into another tank of fresh water containing initially 20 litres. The mixture leaves this second tank at the same rate. Obtain the equation

$$\frac{dP}{dt} + \frac{P}{10} = 2 - 1·5e^{-t/10}$$

for P kg, the amount of salt in the second tank at time t and hence the maximum amount of salt it will contain during the process.

24. A heavy uniform chain of weight w N/m is lying on a rough vertical circle of radius r. If it is just on the point of slipping, the tension T at a point where the slope of the circle is θ is given by

$$\frac{dT}{d\theta} - \mu T = wr (\sin \theta - \mu \cos \theta)$$

where μ is the coefficient of friction. If $T = 0$ when $\theta = \frac{1}{2}\pi$ show that the solution is given by

$$T = \frac{wr}{1 + \mu^2} [(\mu^2 - 1) \cos \theta - 2\mu \sin \theta] + \frac{2\mu wr}{1 + \mu^2} e^{\mu\left(\theta - \frac{\pi}{2}\right)}$$

25. A substance M changes into a substance N at a rate p times the amount of M present; N changes into P at a rate q times the amount of N present. If initially only M is present and its amount is a show that the amount of P present at time t is

$$a + a(qe^{-pt} - pe^{-qt})/(p - q).$$

Show that the following (26)–(28) are exact differential equations and solve.

26. $(e^x + 1)dx + dy = 0$.

27. $2x \log y dx + \frac{x^2}{y} dy = 0$.

28. $y(1 + x^2)^{-1}dx + (\tan^{-1} x)dy = 0$.

29. Find the indices α, β so that when

$$(x^2y^2 - y)dx + (2x^3y + x)dy = 0$$

is multiplied by $x^\alpha y^\beta$ it becomes an exact equation. Hence integrate it.

Solve the equations:

30. $\frac{dy}{dx} = ax + by$.

31. $\frac{dy}{dx} = \frac{y}{x} + \tan\left(\frac{y}{x}\right)$.

32. $x + y\frac{dy}{dx} = x^2 + y^2$.

33. $\left(\frac{dy}{dx}\right)^2 + y^2 = 1$.

34. $\dfrac{d^2y}{dx^2} + y = 0.$ **35.** $x\dfrac{d^2y}{dx^2} + \dfrac{dy}{dx} = 4x.$

Find the orthogonal trajectories to:

36. $x^2 = y^2 + ky^3.$ **37.** $xy = a.$

38. $y^2 = 4ax.$ **39.** $x^2 + ky^2 = 1.$

40. A spherical raindrop initially of radius a falls freely from rest. While falling, its volume increases continuously through condensation at a rate equal to λ times its surface at any instant. Show that its velocity v at time t satisfies the equation

$$\frac{dv}{dt} = g - \frac{3\lambda}{r}v,$$

where $r = a + \lambda t.$

Solve this equation for v and so show that at time $t = a/\lambda,\ v = 15ga/32\lambda.$

[L. U.]

[By Newton's second law of motion

$$\frac{d}{dt}\{\rho\tfrac{4}{3}\pi r^3 v\} = \rho g\tfrac{4}{3}\pi r^3,$$

where ρ is the density of water and v the velocity at time t when the radius is r. Also

$$\frac{d}{dt}\{\tfrac{4}{3}\pi r^3\} = \lambda 4\pi r^2.]$$

CHAPTER 8

OPERATIONAL METHODS AND LINEAR DIFFERENTIAL EQUATIONS

" Shall I refuse my dinner because I do not fully understand the process of digestion?"

O. Heaviside.

8.1. The Operator D

For convenience and as an essential shorthand leading to useful extensions D is used to stand for d/dx so that Dy and dy/dx are two ways of stating the same thing. This is extended so that $(D + a)y$ and $\dfrac{dy}{dx} + ay$ are also the same, as are

$$(D^2 + aD + b)y \quad \text{and} \quad \frac{d^2y}{dx^2} + a\frac{dy}{dx} + by.$$

Generally we will treat D and expressions containing it as an algebraic operator. Thus since

$$(1 - D)(1 + D + D^2 + D^3 + \ldots)x^2$$
$$= (1 - D)(x^2 + 2x + 2)$$
$$= (x^2 + 2x + 2) - D(x^2 + 2x + 2)$$
$$= x^2 + 2x - 2 - 2x - 2 = x^2$$

we regard $(1 - D)(1 + D + D^2 + D^3 \ldots)$ as identical with 1 and from

$$(1 - D)(1 + D + D^2 + \ldots) \equiv 1,$$

$$1 + D + D^2 \ldots \equiv \frac{1}{1 - D},$$

which is the binomial expansion of the right-hand side. In other words, so long as the result is finite we will act with D and functions containing D as though they are algebraic operators obeying the usual laws.

Theorem 1

$$F(D)e^{ax} = e^{ax}F(a)$$

if $F(a)$ is finite. This means that if we are acting on e^{ax} with an expression containing D we can substitute a for D as a quick way of evaluating the operation,

e.g.,
$$(D^2 + 2D + 3)e^{4x} = e^{4x}(4^2 + 2 \cdot 4 + 3)$$
$$= 27e^{4x},$$

96

as can be checked by simple differentiation. This theorem is often used on expressions such as

$$\frac{1}{D-a}e^{bx},$$

which is evaluated as

$$\frac{1}{b-a}e^{bx},$$

provided b is not equal to a so that the result is finite. In this case we can easily check the result. It may be considered that we have solved

$$y = \frac{1}{D-a}e^{bx},$$

which was

$$(D-a)y = e^{bx}.$$

This is the type of 7.3 and the integrating factor is e^{-ax}, so that the equation can be written

$$\frac{d}{dx}(ye^{-ax}) = e^{(b-a)x}$$

and

$$ye^{-ax} = \int e^{(b-a)x}dx = \frac{e^{(b-a)x}}{b-a}$$

or

$$y = \frac{e^{bx}}{b-a}$$

as found by the operator method. We have omitted the constant of integration. This will be dealt with later.

Theorem 2.

$$F(D)[e^{ax}V(x)] = e^{ax}F(D+a)V(x)$$

In this theorem we are operating on a product e^{ax} times $V(x)$, another function of x. We carry e^{ax} through the operator and multiply it into the result of acting on $V(x)$ with $F(D+a)$.

As a simple example, by the usual method

$$(D+4)e^{2x}\cos 3x = e^{2x}(-3\sin 3x) + \cos 3x \,.\, 2e^{2x} + 4e^{2x}\cos 3x$$
$$= e^{2x}(6\cos 3x - 3\sin 3x),$$

but by the theorem this is evaluated as

$$e^{2x}(D+2+4)\cos 3x$$
$$= e^{2x}(D+6)\cos 3x$$
$$= e^{2x}(-3\sin 3x + 6\cos 3x).$$

8.2. Linear Equations with Constant Coefficients

We are now in a position to repeat the methods of Vol. I (34.7, 8, 9).

Example 1. To solve $(D^2 - 2D - 63)y = 0$.
Let $y = Ae^{mx}$ be a trial solution so that

$$(D^2 - 2D - 63)Ae^{mx} = 0.$$

$$Ae^{mx}(m^2 - 2m - 63) = 0 \qquad \text{(by Theorem 1)}$$

$A, e^{mx} \neq 0$, $\qquad \therefore \quad m^2 - 2m - 63 = 0$

$$m = 9 \text{ and } - 7$$

and $\qquad \underline{y = Ae^{9x} + Be^{-7x}}.$

Example 2. To solve $(D^2 + 6D + 13)y = 0$.
As above if Ae^{mx} is a trial solution

$$m^2 + 6m + 13 = 0$$

$$m = -3 \pm 2j$$

$$y = Ae^{(-3 + 2j)x} + Be^{(-3 - 2j)x}$$

$$= \underline{e^{-3x}(L \cos 2x + M \sin 2x)}.$$

Example 3. To solve $(D^2 + 4D + 4)y = 0$.
The usual trial solution $y = Ae^{mx}$ gives

$$(m + 2)^2 = 0,$$

from which only one solution $y = Ae^{-2x}$ is obtained. We will try $y = e^{-2x}V(x)$, where $V(x)$ is to be found.
Substituting this solution

$$(D^2 + 4D + 4)[e^{-2x}V(x)] = 0$$

$$e^{-2x}(D + 2 - 2)^2 . V(x) = 0 \qquad \text{(by Theorem 2)}$$

and $\qquad\qquad D^2V(x) = 0,$

integrating twice: $\qquad V(x) = a + bx,$

giving as a solution $\qquad \underline{y = e^{-2x}(a + bx)},$

a solution with two independent constants.

From this method we see that for say 3 equal roots each equal to p we would have

$$y = e^{px}(A + Bx + Cx^2)$$

as the general solution. The student should check that this is a solution of

$$(D - p)^3y = 0,$$

or $\qquad \dfrac{d^3y}{dx^3} - 3p\dfrac{d^2y}{dx^2} + 3p^2\dfrac{dy}{dx} - p^3y = 0.$

Theorem 3.

$$\frac{1}{D} f(x) \equiv \int f(x)dx.$$

This is an obvious extension of the use of D as an algebraic operator, since

$$D\left(\frac{1}{D}f(x)\right) = f(x).$$

To fit in with other methods of solution we must omit the constant of integration. We will thus have, for example,

$$\frac{1}{D} x = \tfrac{1}{2}x^2$$

$$\frac{1}{D^2} x = \frac{1}{D} \left(\frac{1}{D} x\right)$$

$$= \frac{1}{D} \left(\tfrac{1}{2}x^2\right) = \tfrac{1}{6}x^3.$$

We can now deal with a difficulty that arises. In finding

$$\frac{1}{D - a} e^{ax}$$

Theorem 1 cannot be used, since this would give an infinite answer. We can employ Theorem 2 by means of a trick method. Write the expression as

$$\frac{1}{D - a} (e^{ax} . 1),$$

where 1 is a special case of the $V(x)$ of Theorem 2, and proceed

$$= e^{ax} \frac{1}{D + a - a} . 1 = e^{ax} \frac{1}{D} . 1 = e^{ax}x.$$

This can be checked by solving by the usual method

$$y = \frac{1}{D - a} e^{ax}$$

or $$(D - a)y = e^{ax}.$$

Example 4. Find $\int e^{3x}x^4 dx.$

We could integrate by parts. The operational method is

$$\frac{1}{D} (e^{3x}x^4) = e^{3x} \frac{1}{D + 3} . x^4$$

$$= \frac{e^{3x}}{3} \left(1 + \frac{D}{3}\right)^{-1} x^4$$

$$= \frac{e^{3x}}{3} \left(1 - \frac{D}{3} + \frac{D^2}{9} - \frac{D^3}{27} + \frac{D^4}{81} \cdots\right) x^4$$

$$= \frac{e^{3x}}{3} \left(x^4 - \frac{4x^3}{3} + \frac{12x^2}{9} - \frac{24x}{27} + \frac{24}{81}\right)$$

and a constant of integration will now be added on.

Theorem 4.

$$F(D^2) \frac{\cos ax}{\sin ax} = F(- a^2) \frac{\cos ax}{\sin ax}.$$

This formula means that when acting on a sine or cosine of ax with *even* powers of D, we may put $- a^2$ for D^2.

We can easily check by the usual method of evaluation that

$$(D^4 - 2D^3 + D^2 + D - 2) \sin 3x$$
$$= [(-3^2)^2 - 2(-3^2)D + (-3^2) + D - 2] \sin 3x$$
$$= (19D + 70) \sin 3x$$
$$= 57 \cos 3x + 70 \sin 3x.$$

More often we require an evaluation such as

$$y = \frac{1}{D^4 - 3D^3 + 2} \cos 4x \quad \cdot \quad \cdot \quad \cdot \quad \cdot \quad \cdot \quad \cdot \quad (1)$$

$$= \frac{1}{(-4^2)^2 - 3(-4^2)D + 2} \cos 4x$$

$$= \frac{1}{48D + 258} \cos 4x.$$

In this way the powers of D in the denominator are reduced to first degree powers only. In order to be able to continue with Theorem 4 we adopt the trick of multiplying above and below by $48D - 258$ to obtain

$$\frac{1}{48D + 258} \times \frac{48D - 258}{48D - 258} \cdot \cos 4x$$

$$= \frac{48D - 258}{48^2D^2 - 258^2} \cdot \cos 4x$$

$$= \frac{48D - 258}{48^2(-4^2) - 258^2} \cos 4x.$$

Having cleared the denominator of D's the method continues

$$= -\frac{1}{17\,238 \times 6} (48D - 258) \cos 4x$$

$$= -\frac{1}{17\,238 \times 6} (48[-4 \sin 4x] - 258 \cos 4x)$$

$$= \frac{1}{17\,238} (32 \sin 4x + 43 \cos 4x),$$

a result which can be checked as a particular solution of (1) written as

$$(D^4 - 3D^3 + 2)y = \cos 4x.$$

8.3. The General Solution of $F(D)y = f(x)$

Let $y = u$ be the general solution of $F(D)y = 0$, *i.e.*, the solution containing the required number of arbitrary constants. If we put $y = u + v$, where v like u is a function of x

$$F(D)[u + v] = f(x)$$
or
$$F(D)u + F(D)v = f(x).$$

Since $F(D)u = 0$, this becomes

$$F(D)v = f(x)$$

and we have seen how to solve

$$v = \frac{1}{F(D)} f(x)$$

to obtain a solution with no arbitrary constants. From this we obtain the method of solution of $F(D)y = f(x)$.

(1) Solve $F(D)y = 0$ to obtain the *complementary function* (C.F.). This will contain the required number of constants equal to the order of $F(D)$.

(2) Obtain a *particular integral* (P.I.) of

$$\frac{1}{F(D)} f(x)$$

by operational or other methods. This will not contain any unknown constants.

(3) The general solution is the sum of these two, *i.e.*,

$$y = \text{C.F.} + \text{P.I.}$$

and contains the required number of arbitrary constants.

Writing it as

$$y = u + \frac{1}{F(D)} \cdot f(x)$$

and operating on both sides with $F(D)$:

$$F(D)y = F(D)\left[u + \frac{1}{F(D)} \cdot f(x)\right]$$
$$= F(D)u + f(x)$$
$$= f(x) \qquad\qquad [F(D)u = 0]$$

showing that the original differential equation is satisfied.

Example 5. Solve $Dy = x$.
The solution is obviously

$$y = \tfrac{1}{2}x^2 + A.$$

By the operational method :

C.F.	$Dy = 0$
Try Ae^{mx} :	$m = 0$
	C.F. is $Ae^{0x} = A$
P.I.	$\dfrac{1}{D} x = \tfrac{1}{2}x^2$
general solution	$\underline{y = A + \tfrac{1}{2}x^2.}$

8.4. Worked Examples

A number of examples with different forms of functions of x in the particular integral will now be considered.

Example 6. Solve $\frac{d^2y}{dx^2} + 4\frac{dy}{dx} + 3y = e^{2x}$.

C.F. $\qquad\qquad (D^2 + 4D + 3)y = 0$

Ae^{mx}: $\qquad\qquad m^2 + 4m + 3 = 0$

$\qquad\qquad\qquad m = -3 \text{ and } -1$

$\qquad\qquad\qquad \underline{Ae^{-3x} + Be^{-x}.}$

P.I.

$$\frac{1}{D^2 + 4D + 3}e^{2x} = \frac{1}{4 + 8 + 3}e^{2x}$$

$$= \tfrac{1}{15}e^{2x}$$

general solution $\qquad\qquad y = Ae^{-3x} + Be^{-x} + \tfrac{1}{15}e^{2x}.$

Note 1. We are concerned with finding a " particular " integral of the above differential equation. Any such integral (or solution) will do. It is clear that a function of the form Le^{2x} will do, and we may therefore avoid the operational method of finding the P.I. and using $y = Le^{2x}$ proceed

$$(D^2 + 4D + 3)Le^{2x} = e^{2x},$$

$$Le^{2x}(4 + 8 + 3) = e^{2x},$$

and, as above $\qquad\qquad L = 1/15.$

If, however, the differential equation is

$$(D^2 + 4D + 3)y = e^{-3x}$$

the index -3 of e^{-3x} being a root of the equation in m, $m^2 + 4m + 3$, then this method of assuming the P.I. to be Le^{-3x} will not provide the value of L. The operational method does so, giving

$$\frac{1}{(D + 3)(D + 1)}e^{-3x}$$

$$= \frac{1}{D + 3} \cdot e^{-3x} \cdot \frac{1}{(-3 + 1)}$$

by taking the exponential through the function that does not vanish

$$= -\frac{1}{2}\frac{1}{D + 3}(e^{-3x} \cdot 1)$$

$$= -\tfrac{1}{2}e^{-3x}\frac{1}{D} \cdot 1 = \underline{-\tfrac{1}{2}e^{-3x}x.}$$

This operational method may again be avoided if we accept the rule that when the obvious P.I. such as $y = Le^{-3x}$ will not provide the value of L, then $y = Lxe^{-3x}$ will be tried. This gives

$$(D + 3)(D + 1)Lxe^{-3x} = e^{-3x}$$

$$Le^{-3x}D(D - 2)x = e^{-3x}$$

$$LD(D - 2)x = 1$$

$$L(-2) = 1$$

and, as above $\qquad\qquad L = -\tfrac{1}{2}.$

Note 2. If the right-hand side is a constant in an equation such as

$$(aD^2 + bD + c)y = A$$

the P.I. may be found as

$$\frac{1}{a\mathrm{D}^2 + b\mathrm{D} + c} A e^{0x}$$

$$= \frac{A}{c} e^{0x} = \frac{A}{c} \quad \text{(Theorem 1).}$$

Here again we can avoid this by guessing the obvious P.I., $y = L$, which gives

$$(a\mathrm{D}^2 + b\mathrm{D} + c)L = A$$

$$cL = A, \quad L = A/c \text{ as above.}$$

If c is zero the P.I. is

$$\frac{1}{\mathrm{D}(a\mathrm{D} + b)} A e^{0x}$$

$$= A \cdot \frac{1}{\mathrm{D}} e^{0x} \cdot \frac{1}{b}$$

$$= \frac{A}{b} \frac{1}{\mathrm{D}} \cdot 1 = \frac{A}{b} x,$$

and this again may be obtained by the rule above that since $y = L$ will not give a value to L when used as a P.I. for

$$(a\mathrm{D}^2 + b\mathrm{D})y = A$$

we will **try** $y = Lx$ to obtain

$$(a\mathrm{D}^2 + b\mathrm{D})Lx = A.$$

$$L(b) = A, \quad L = A/b \text{ as above.}$$

Example 7. Solve $(\mathrm{D}^2 + 4\mathrm{D} + 3)y = x^3 + x^2 + 2.$

$C.F.$ as before this is $A e^{-3x} + B e^{-x}.$

P.I. $\dfrac{1}{(\mathrm{D} + 3)(\mathrm{D} + 1)} (x^3 + x^2 + 2)$

$$= \frac{1}{2} \left(\frac{1}{\mathrm{D} + 1} - \frac{1}{\mathrm{D} + 3} \right) (x^3 + x^2 + 2)$$

$$= \left[\frac{1}{2} (1 + \mathrm{D})^{-1} - \frac{1}{6} \left(1 + \frac{\mathrm{D}}{3} \right)^{-1} \right] (x^3 + x^2 + 2)$$

$$= \tfrac{1}{2}(1 - \mathrm{D} + \mathrm{D}^2 - \mathrm{D}^3 \ . \ .)(x^3 + x^2 + 2)$$
$$- \frac{1}{6} \left(1 - \frac{\mathrm{D}}{3} + \frac{\mathrm{D}^2}{9} - \frac{\mathrm{D}^3}{27} \cdot \cdot \right) (x^3 + x^2 + 2)$$

$$= \tfrac{1}{2}\{(x^3 + x^2 + 2) - (3x^2 + 2x) + (6x + 2) - 6\}$$
$$- \tfrac{1}{6}\{(x^3 + x^2 + 2) - \tfrac{1}{3}(3x^2 + 2x) + \tfrac{1}{9}(6x + 2) - \tfrac{1}{27} \cdot 6\}$$

$$= \tfrac{1}{2}\{x^3 - 2x^2 + 4x - 2\} - \tfrac{1}{6}\{x^3 + 2\}$$

$$= \tfrac{1}{3}x^3 - x^2 + 2x - 4/3.$$

The solution is $\underline{y = A e^{-3x} + B e^{-x} + \tfrac{1}{3}x^3 - x^2 + 2x - 4/3.}$

(i) In many cases the partial fractions are not necessary, and we may expand $(\mathrm{D}^2 + 4\mathrm{D} + 3)^{-1}$ as $\dfrac{1}{3} \left(1 + \dfrac{\mathrm{D}^2 + 4\mathrm{D}}{3} \right)^{-1}$ to as many terms as are required.

(ii) In a simple case such as this it is easier to assume that the P.I. is

$$Lx^3 + Mx^2 + Nx + P.$$

Substitution for y gives

$$(\mathrm{D}^2 + 4\mathrm{D} + 3)(Lx^3 + Mx^2 + Nx + P) \equiv x^3 + x^2 + 2.$$

This gives

$$3Lx^3 + x^2(3M + 12L) + x(3N + 8M + 6L)$$
$$+ 2M + 4N + 3P \equiv x^3 + x^2 + 2.$$

Equating coefficients:

$$3L = 1,$$
$$3M + 12L = 1,$$
$$3N + 8M + 6L = 0,$$
$$2M + 4N + 3P = 2.$$

Therefore $L = \frac{1}{3}$, $M = -1$, $N = 2$, $P = -4/3$, giving the same P.I. as obtained by the operator method.

Example 8. Find the P.I. of $(D^2 + 4D + 5)y = 2 \cos 2x$.
This is

$$\frac{2}{D^2 + 4D + 5} \cos 2x$$

$$= \frac{2}{-4 + 4D + 5} \cos 2x$$

$$= \frac{2}{4D + 1} \times \frac{4D - 1}{4D - 1} \cos 2x$$

$$= \frac{8D - 2}{16D^2 - 1} \cos 2x$$

$$= \frac{8D - 2}{16(-4) - 1} \cos 2x$$

$$= -\frac{1}{65}(8D - 2) \cos 2x$$

$$= \frac{1}{65}(16 \sin 2x + 2 \cos 2x).$$

It is often easier to assume that $A \cos 2x + B \sin 2x$ is the P.I. Substitution in the equation gives

$$(D^2 + 4D + 5)(A \cos 2x + B \sin 2x) = 2 \cos 2x$$

or $- 4A \cos 2x - 8A \sin 2x + 5A \cos 2x$

$$- 4B \sin 2x + 8B \cos 2x + 5B \sin 2x = 2 \cos 2x,$$
$$\cos 2x(A + 8B) + \sin 2x(B - 8A) = 2 \cos 2x,$$
$$\therefore \quad A + 8B = 2,$$
$$- 8A + B = 0,$$

from which $A = 2/65$ $B = 16/65$ as above.

Example 9. Find the P.I. of $(D^2 + 2D + 5)y = e^{2x} \cos 2x$. This is

$$\frac{1}{D^2 + 2D + 5} e^{2x} \cos 2x$$

$$= e^{2x} \frac{1}{(D + 2)^2 + 2(D + 2) + 5} \cos 2x$$

$$= e^{2x} \frac{1}{D^2 + 6D + 13} \cos 2x$$

$$= e^{2x} \frac{1}{6D + 9} \times \frac{6D - 9}{6D - 9} \cos 2x$$

$$= -\frac{e^{2x} (6D - 9)}{225} \cos 2x$$

$$= \frac{e^{2x}}{225} (12 \sin 2x + 9 \cos 2x).$$

In this case it is again often simpler to assume $e^{2x}(A \cos 2x + B \sin 2x)$ as the P.I. and find A and B as in the previous example.

8.5. The Complex Number Method for the P.I.

Let y_1, y_2 be the solutions respectively of
$$F(D)y = A \cos nx, \qquad F(D)y = A \sin nx,$$
so that
$$F(D)y_1 = A \cos nx,$$
$$F(D)y_2 = A \sin nx.$$
From these with
$$j \equiv \sqrt{(-1)}$$
$$F(D)[y_1 + jy_2] = A[\cos nx + j \sin nx]$$
$$= A e^{jnx}.$$

We can therefore solve in the usual way, for the particular integral,
$$F(D)y = A e^{jnx}$$
to obtain

$$y = \frac{1}{F(D)} A e^{jnx}$$

$$= \frac{A}{F(jn)} e^{jnx} \qquad \text{(if } F(jn) \neq 0),$$

and the real part of this gives the solution to
$$F(D)y = A \cos nx,$$
whilst the imaginary part is the solution to
$$F(D)y = A \sin nx.$$

Example 10. Find the P.I. for $(D^2 + 3D + 4)y = A \sin 2x.$

Solve
$$(D^2 + 3D + 4)y = A e^{j2x}$$
to obtain
$$y = \frac{A e^{j2x}}{(D^2 + 3D + 4)_{D=2j}} = \frac{-Aje^{j2x}}{6},$$

$\therefore$ P.I. is
$$\frac{-jA}{6}(\cos 2x + j \sin 2x).$$

The P.I. required is the j component or $-\frac{1}{6}A \cos 2x$.

When the right-hand side is $f(x) \sin nx$ or $f(x) \cos nx$ this method is applied to $f(x)e^{jnx}$.

8.6. The Resonance Condition

It will be seen later that an important condition in mechanical and electrical circuits arises when the frequency of the external force or e.m.f. equals the "natural" frequency of the circuit. In such a case, ignoring any mechanical or electrical resistance, the equation to be solved is of the form

$$(D^2 + n^2)x = A \cos nt. \qquad \left(D \equiv \frac{d}{dt}\right)$$

The P.I. cannot be obtained as

$$\frac{A}{D^2 + n^2} \cos nt$$

since this gives an infinite answer.   Using the form
$$\frac{A}{D^2 + n^2} e^{jnt}$$
we proceed
$$= \frac{A}{(D - jn)(D + jn)} e^{jnt}$$
$$= \frac{A}{D - jn} e^{jnt} \frac{1}{jn + jn}$$
$$= \frac{A}{2jn} \frac{1}{D - jn} (e^{jnt} . 1)$$
$$= \frac{A}{2jn} e^{jnt} \frac{1}{D + jn - jn} . 1$$
$$= \frac{A}{2jn} e^{jnt} . t$$

The real part is
$$\frac{At}{2n} \sin nt,$$
the j part is
$$- \frac{At}{2n} \cos nt.$$

We have then
$$\frac{1}{D^2 + n^2} (A \cos nt) = \frac{At}{2n} \sin nt,$$
$$\frac{1}{D^2 + n^2} (A \sin nt) = - \frac{At}{2n} \cos nt,$$

so that the solution to
$$(D^2 + n^2)x = A \cos nt$$
is
$$x = L \cos nt + M \sin nt + \frac{At}{2n} \sin nt$$

and to
$$(D^2 + n^2)x = A \sin nt$$
is
$$x = L \cos nt + M \sin nt - \frac{At}{2n} \cos nt$$

showing the exact manner in which the displacement x becomes infinite with the time.

Again we can avoid this operational method by trying
$$x = Lte^{jnt}$$
as the P.I. for
$$(D^2 + n^2)x = Ae^{jnt}$$
to obtain
$$L = \frac{A}{2jn}, \quad x = \frac{A}{2jn} te^{jnt}$$
$$= - \frac{At}{2n} (j \cos nt - \sin nt)$$

and on separation the above results are obtained.

Example 11. An inductance L is in series with a condenser C and acted on by an e.m.f. $E_0 \cos \omega t$ in a circuit of negligible resistance. Find the voltage across the condenser at time t.

If at time t, i is the current, Q the charge caused by it on the condenser and V the p.d. of the plates of the condenser

$$E_0 \cos \omega t - L \frac{di}{dt} - V = 0.$$

But $i = \frac{dQ}{dt} = C \frac{dV}{dt}$ so that $\frac{di}{dt} = C \frac{d^2V}{dt^2}$.

Substitution for $\frac{di}{dt}$ gives

$$LC \frac{d^2V}{dt^2} + V = E_0 \cos \omega t$$

or

$$\frac{d^2V}{dt^2} + \frac{1}{LC} V = \frac{E_0}{LC} \cos \omega t.$$

Suppose now that $\frac{1}{LC} = \omega^2$. We have to solve

$$(D^2 + \omega^2)V = \frac{E_0}{LC} \cos \omega t.$$

Using the above, the solution is

$$V = A \cos \omega t + B \sin \omega t + \frac{E_0}{2\omega LC} \cdot t \sin \omega t$$

showing that the voltage across the plates of the condenser consists of two parts, one a wave of constant amplitude $\sqrt{(A^2 + B^2)}$ and the other a wave of steadily increasing amplitude so that a breakdown must occur.

In practice, owing to the presence of some resistance electrical or mechanical (in mechanical problems) this stage need not be reached, but the potential difference or amplitude of the mechanical oscillations may become too large for safety. Because of this factor of resonance soldiers crossing a bridge endeavour to walk out of step, the rolling of a ship will become excessive if the period of the waves hitting it tends to coincide with its " natural " period of rolling, and the vertical oscillations of a railway carriage may become large if the speed is such that it travels the length of a rail (and so receives a slight jolt) in a time equal to that of its vertical oscillation on its springs.

8.7. Homogeneous Linear Equations

This name is given to the type such as

$$5x^2 \frac{d^2y}{dx^2} + 7x \frac{dy}{dx} + 3y = f(x),$$

which is like the usual type of linear differential equation with constant coefficients that we have just solved, but associated with each differential coefficient is a power of x equal to its order. Thus the derivative d^3y/dx^3 would occur as $x^3 d^3y/dx^3$, d^2y/dx^2 occurs as $x^2 d^2y/dx^2$, and so on.

By taking a new variable t, say, where $x = e^t$ we can reduce this homogeneous form to the usual form with constant coefficients and so solve by the methods already used.

If

$$x = e^t, \quad dx/dt = e^t = x$$

and

$$\frac{dy}{dx} = \frac{dy}{dt} \cdot \frac{dt}{dx} = \frac{dy}{dt} \frac{1}{x}$$

so that
$$x \frac{dy}{dx} = \frac{dy}{dt} \qquad \cdots \cdots \cdots \cdots \quad (1)$$

also
$$\frac{d^2y}{dx^2} = \frac{d}{dx}\left(\frac{dy}{dx}\right) = \frac{d}{dt}\left(\frac{1}{x}\frac{dy}{dt}\right)\frac{dt}{dx}$$

$$= \left[\frac{1}{x}\frac{d^2y}{dt^2} - \left(\frac{1}{x^2}\frac{dx}{dt}\right)\frac{dy}{dt}\right]\frac{dt}{dx}$$

$$= \frac{1}{x}\frac{d^2y}{dt^2}\frac{dt}{dx} - \frac{1}{x^2}\frac{dy}{dt}$$

$$= \frac{1}{x^2}\left(\frac{d^2y}{dt^2} - \frac{dy}{dt}\right). \qquad \cdots \cdots \cdots \quad (2)$$

If θ is used to denote d/dt (1) and (2) above can be written

$$xDy = \theta y,$$

$$x^2D^2y = \theta^2 y - \theta y = \theta(\theta - 1)y.$$

Similarly, it can be shown that

$$x^3D^3y = \theta(\theta - 1)(\theta - 2)y,$$

and the form for the higher derivatives can be easily seen.

Example 12. Solve $x^2\dfrac{d^2y}{dx^2} + 4x\dfrac{dy}{dx} + 2y = x^2$.

With the substitution $x = e^t$, the equation becomes

$$[\theta(\theta - 1) + 4\theta + 2]y = e^{2t},$$

$$(\theta^2 + 3\theta + 2)y = e^{2t}.$$

Proceeding with θ as with the D operator, the solution is

$$y = Ae^{-2t} + Be^{-t} + \tfrac{1}{12}e^{2t}$$

$$= \frac{A}{x^2} + \frac{B}{x} + \frac{1}{12}x^2.$$

EXERCISE 17

Using operational methods find:

1. $(D^2 + 2D - 1)e^{3x}$.

2. $(D^3 - 3D^2 + 2)\{e^{2x}x^2\}$.

3. $(D^4 - 8D^3 + D)\{e^{4x}\sin 2x\}$.

4. $\displaystyle\int e^{ax}\cos bx\,dx$.

5. $\displaystyle\int e^{px}\sin qx\,dx$.

6. $\displaystyle\int e^{4x}x^3\,dx$.

7. $\dfrac{1}{D - 4}\cos 3x$.

8. $\dfrac{1}{D^2 + 2D + 1}\sin x$.

9. $\dfrac{1}{D^2 + 9}\sin 3x$.

10. $\dfrac{1}{D^2 + 4}e^{2x}\cos 3x$.

11. $\dfrac{1}{D^2 + D + 1}(x^2 + 3x + 2)$.

12. $\dfrac{1}{D^2 - 5D + 6}(x^3 + x^2 + 1)$.

Find the particular integrals of:

13. $(D^2 + 2D + 1)y = e^{-2x}$.

14. $(6D^2 + 11D - 6)y = \sinh 5x$.

15. $(D^2 - 5D + 1)y = x^2 + 2x + 7$.

16. $(D^3 + D^2)y = x^2 + 2$.

17. $(D^2 - 3D + 2)y = x^2 - e^x$.

18. $(D^2 + 1)y = \sin x$.

19. $(D^2 + 1)y = x^2e^{2x}$.

20. $(D^2 - 2D + 2)y = e^{3x}\sin 4x$.

Solve completely the following equations:

21. $(D^2 - 2D - 1)y = 4 \sin x.$ 22. $(2D^2 + D)y = \cos 2x.$
23. $(D^2 - 4)y = x \sin x.$ 24. $(D^2 - 2D + 2)y = e^{-t} \sin 2t.$
25. $(D^2 + 2D + 3)y = x^2 + 2x.$ 26. $(D^2 - 4D + 4)y = \sin 3x.$

27. $4\dfrac{d^2y}{dx^2} - 4\dfrac{dy}{dx} + 5y = e^x$, given that $y = 1$, $\dfrac{dy}{dx} = 1$ when $x = 0$.

[L.U.]

28. $3\dfrac{d^2x}{dt^2} - 2\dfrac{dx}{dt} - x = 2t - 1$, given that $x = 7$, $\dfrac{dx}{dt} = 2$ when $t = 0$.

[L.U.]

29. $\dfrac{d^2s}{dt^2} + 4\dfrac{ds}{dt} + 5s = 2e^{-2t}$, given that $s = 1$, $\dfrac{ds}{dt} = -2$ when $t = 0$.

[L.U.]

30. $\dfrac{d^2y}{dt^2} - 2\dfrac{dy}{dt} + 5y = \sin 3t$, given that $y = 0$, $\dfrac{dy}{dt} = 0$ when $t = 0$.

[L.U.]

31. Find a particular integral of:

(a) $\dfrac{d^4y}{dx^4} - y = x \sin x$;

(b) $\dfrac{d^4y}{dx^4} + 8\dfrac{d^2y}{dx^2} + 16y = \sin 2x.$ [L.U.]

32. Find a particular integral of

$$\dfrac{d^2y}{dx^2} + 8\dfrac{dy}{dx} + 16y = 12x^2 - 16 \sin 4x.$$

Write down the general solution and determine the arbitrary constants if $y = 0$, $\dfrac{dy}{dx} = 0$ when $x = 0$.

Show that the graph of y for large values of x is independent of the initial values of y and $\dfrac{dy}{dx}$.

[L.U.]

33. Solve:

(a) $\dfrac{d^3y}{dx^3} + 4\dfrac{dy}{dx} = 6 \sin x$;

(b) $\dfrac{d^4y}{dx^4} - y = \cos x \cosh x$;

(c) $\dfrac{d^2y}{dx^2} + 2a\dfrac{dy}{dx} + a^2y = 12x^2 e^{-ax}$;

given that $y = 1$ and $\dfrac{dy}{dx} = 0$ when $x = 0$.

[L.U.]

34. Solve:

(a) $\dfrac{d^2y}{dx^2} + 2\dfrac{dy}{dx} + 5y = 1 + x + x^2$;

(b) $\dfrac{d^4y}{dx^4} + 20\dfrac{d^2y}{dx^2} + 64y = 2 \cos x \cos 3x$;

(c) $\dfrac{d^4y}{dx^4} - 16y = 4e^{-2x} \cos 2x.$

[L.U.]

35. Solve $4\dfrac{d^2y}{dt^2} + 4\dfrac{dy}{dt} + 37y = 6e^{-\frac{1}{2}t} \sin 3t$ subject to the conditions that $y = 1$ and $dy/dt = -1 \cdot 5$ when $t = 0$.

[L.U.]

36. Show that the solution of the equation

$$\frac{d^2y}{dt^2} + 2n\frac{dy}{dt} + n^2y = A\cos pt$$

for which y and $\frac{dy}{dt}$ both vanish when $t = 0$ can be written

$$y = A\{\cos(pt - \phi) - e^{-nt}(nt + \cos\phi)\}/(n^2 + p^2)$$

where $\tan\phi = 2np/(n^2 - p^2)$. [L.U.]

37. Obtain the general solutions of the equations:

 (i) $d^2y/dx^2 + 3dy/dx + 2y = e^{-x}\sin 2x$;

 (ii) $d^2y/dx^2 + 2ady/dx + a^2y = x^2e^{-ax}$. [L.U.]

38. Solve the equation

$$\frac{d^4y}{dx^4} - 16y = 15\cos x$$

subject to the conditions that

when $\qquad x = 0, \quad y = 0, \quad \frac{dy}{dx} = 2$

when $\qquad x = \tfrac{1}{2}\pi, \quad y = -1, \quad \frac{dy}{dx} = -1,$

showing that the solution is purely periodic in x. [L.U.]

Solve the equations:

39. $x^2\dfrac{d^2y}{dx^2} - 6y = \log x.$

40. $x^2y_2 - 2xy_1 - 4y = x^2.$

41. $x^2y_2 - xy_1 - 3y = x^2\log x.$

42. $x^2y_2 - xy_1 + y = 2x\log x$, where $y = 1$, $y_1 = 0$ when $x = 1$.

43. By means of the substitution $x = e^z$ reduce the equation

$$x^2\frac{d^2y}{dx^2} - x\frac{dy}{dx} + y = 4x^3$$

to one in which the coefficients are constants. Hence or otherwise solve it.

[L.U.]

44. Solve $\qquad x^2\dfrac{d^2y}{dx^2} + 5x\dfrac{dy}{dx} + 3y = \left(1 + \dfrac{1}{x}\right)\log x.$

[L.U.]

45. There is a steady flow of heat through the walls of a hollow cylindrical pipe carrying steam where r_1, r_2 are the inner and outer radii respectively. Show that the temperature θ at a distance r from the axis in the wall of the pipe is given by

$$r\frac{d^2\theta}{dr^2} + \frac{d\theta}{dr} = 0.$$

If θ_1, θ_2 are the inner and outer temperatures of the wall find θ in terms of r.

46. If the above is a hollow sphere, r_1, r_2 its inner and outer radii respectively show that

$$r\frac{d^2\theta}{dr^2} + 2\frac{d\theta}{dr} = 0 \qquad (r_1 < r < r_2)$$

and again find θ in terms of r_1, r_2, θ_1, θ_2.

[Each of the above equations can be solved more simply than by making it homogeneous.]

47. Show that the constant n may be chosen so that by the substitution $y = x^nz$, the differential equation

$$x^2\frac{d^2y}{dx^2} + 4x(x + 1)\frac{dy}{dx} + (8x + 2)y = \cos x$$

reduces to the form

$$\frac{d^2z}{dx^2} + a\frac{dz}{dx} + bz = \cos x,$$

where a, b are constant. Hence or otherwise solve the given equation.

[L.U.]

48. Show that there is a value of m for which $y = x^m$ is a solution of the equation

$$2x^3 \frac{d^2y}{dx^2} - x \frac{dy}{dx} + y = 0$$

and hence or otherwise obtain the complete solution. [L.U.]

49. If $t = x^2$ prove that

$$\frac{dy}{dx} = 2t^{\frac{1}{2}} \frac{dy}{dt}, \qquad \frac{d^2y}{dx^2} = 4t \frac{d^2y}{dt^2} + 2 \frac{dy}{dt}.$$

Change the independent variable from x to t in the equation

$$x \frac{d^2y}{dx^2} - \frac{dy}{dx} + 4x^3 y = 0$$

and hence find y in terms of x.

50. Show that $x = \sinh z$ transforms

$$(1 + x^2) \frac{d^2y}{dx^2} + x \frac{dy}{dx} - y = x^2$$

into

$$\frac{d^2y}{dz^2} - y = \sinh^2 z.$$

Hence show that the complete solution is

$$y = Ax + B\sqrt{(1 + x^2)} + \tfrac{1}{3}(x^2 + 2).$$

APPLICATIONS OF LINEAR DIFFERENTIAL EQUATIONS. SIMULTANEOUS DIFFERENTIAL EQUATIONS. SOLUTION IN SERIES

9.1. The Equation $EIy_2 = M$

In Vol. I (34.14) we arrived at the following rule for using the bending moment equation. Choose the axes Ox, Oy along and perpendicular to the undisplaced beam and at any point (x, y) on the beam at which the equation is to be applied draw a line from the beam in the positive direction of the y axis. Put in the arrows as shown on the figure below and any forces to the left of the point (x, y) or to the right turning in the direction shown by the arrow for that side are exerting a positive moment in the equation

$$EIy_2 = M.$$

[The arrows can be reversed and $EIy_2 = -M$, used instead.]

Example 1. A cantilever l m long and weight w N/m is subjected to a horizontal compressive force P N at the free end. Find the deflection at this end.

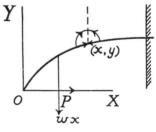

Fig. 16.

Choosing axes at the free end as shown and using the above rule, we note that the moment of P about the point (x, y) is in the opposite direction to that indicated by the left-hand arrow, and the moment is therefore $- Py$. Similarly, for the moment exerted by the weight wx acting at the mid point of the length x. The B.M. equation is

$$EIy_2 = - (Py + \tfrac{1}{2}wx^2)$$

or

$$(D^2 + m^2)y = - \frac{w}{2EI} x^2 \qquad \left(m^2 = \frac{P}{EI}\right).$$

The C.F. is

$$A \cos mx + B \sin mx.$$

112

The P.I. is
$$-\frac{w}{2EI} \cdot \frac{1}{D^2 + m^2} \cdot x^2$$
$$= -\frac{w}{2EI} \cdot \frac{1}{m^2}\left(1 + \frac{D^2}{m^2}\right)^{-1} \cdot x^2$$
$$= -\frac{w}{2EI} \cdot \frac{EI}{P}\left(1 - \frac{D^2}{m^2} \cdot \cdot\right)x^2$$
$$= -\frac{w}{2P}\left(x^2 - \frac{2}{m^2}\right).$$

The general solution is
$$y = A\cos mx + B\sin mx - \frac{w}{2P}\left(x^2 - \frac{2}{m^2}\right)$$

$y = 0$, $x = 0$ at the free end:
$$0 = A + \frac{w}{Pm^2},$$
$$\therefore \quad A = -\frac{w}{Pm^2} = -\frac{wEI}{P^2}.$$

Also
$$\frac{dy}{dx} = -mA\sin mx + Bm\cos mx - \frac{wx}{P}$$

$\frac{dy}{dx} = 0$ when $x = l$ (at the fixed end):
$$0 = -mA\sin ml + Bm\cos ml - \frac{wl}{P}$$
$$\therefore \quad B = A\tan ml + \frac{wl}{Pm}\sec ml$$
$$= -\frac{wEI}{P^2}\tan ml + \frac{wl}{Pm}\sec ml.$$

The vertical displacement y at any point is
$$y = -\frac{wEI}{P^2}\cos mx + \left(\frac{wl}{Pm}\sec ml - \frac{wEI}{P^2}\tan ml\right)\sin mx - \frac{w}{2P}\left(x^2 - \frac{2}{m^2}\right).$$

At the fixed end, where $x = l$ this gives for its deflection δ above the origin at the free end
$$\delta = -\frac{wEI}{P^2}\cos ml + \left(\frac{wl}{Pm}\sec ml - \frac{wEI}{P^2}\tan ml\right)\sin ml - \frac{w}{2P}\left(l^2 - \frac{2}{m^2}\right)$$
$$= \frac{wEI}{P^2}[1 - \sec\theta + \theta\tan\theta - \tfrac{1}{2}\theta^2] \qquad \text{where } \theta \equiv lm.$$

9.2. Electrical Circuits

To obtain the differential equation governing the instantaneous value of the current and voltage in a circuit we proceed as follows.

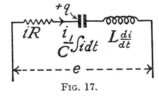

FIG. 17.

Let i in a chosen direction be the value of the current at time t, so that $\int_0^t idt$ is the charge (initially zero) on the condenser. The voltage drops in an L, R, C circuit in series are : (1) Ri across the resistance; (2) L$\frac{di}{dt}$ across the inductance; and (3) $\frac{1}{C}\int_0^{'} idt$ across the plates of the condenser. If e is the external e.m.f. at that instant applied to this circuit

$$e = iR + L\frac{di}{dt} + \frac{1}{C}\int_0^{'} idt \quad . \quad . \quad . \quad . \quad . \quad (1)$$

Since the current is charging the condenser

$$i = \frac{dq}{dt} \quad . \quad . \quad . \quad . \quad . \quad . \quad . \quad (2)$$

where q is the charge on the condenser at that instant of time. [If the current is caused by the condenser discharging, then

$$i = -\frac{dq}{dt} \quad . \quad . \quad . \quad . \quad . \quad . \quad (3)$$

since q is now decreasing with time.]

Equations (1), (2), (3) provide the necessary data for dealing with most problems.

Example 2. A condenser C and initial charge Q_0 is discharged through a resistance R and an inductance L in series. Prove that if $R^2C < 4L$ the current at time t is

$$- Q_0 e^{-ht}\left(k + \frac{h^2}{k}\right)\sin kt,$$

where $- h \pm jk$ are the roots of

$$CLx^2 + CRx + 1 = 0 \qquad\qquad \text{[L.U.]}$$

The student will draw the usual circuit diagram and show i flowing towards the $+q$ plate. [If he takes i in the other direction the answer without the minus sign will be obtained.]

Since the total drop in potential is zero

$$L\frac{di}{dt} + Ri + \frac{q}{C} = 0 \quad \text{where } i = \frac{dq}{dt}$$

Differentiate the left-hand equation and substitute .or $\frac{dq}{dt}$. This gives

$$LC\frac{d^2i}{dt^2} + RC\frac{di}{dt} + i = 0$$

The usual trial solution $i = Ae^{xt}$ gives

$$CLx^2 + CRx + 1 = 0 \quad . \quad . \quad . \quad . \quad . \quad . \quad (1)$$

Since $R^2C < 4L$ the roots of this are complex. If they are $- h \pm jk$, the solution is

$$i = e^{-ht}[L\cos kt + M \sin kt].$$

$i = 0$ when $t = 0$, showing that L = 0. Therefore

$$i = Me^{-ht}\sin kt \quad . \quad . \quad . \quad . \quad . \quad . \quad . \quad (2)$$

From the first equation above, at time $t = 0$,

$$\left(L\frac{di}{dt}\right)_{t=0} + (Ri)_{t=0} + \left(\frac{q}{C}\right)_{t=0} = 0$$

or, since $i_0 = 0$

$$L \left(\frac{di}{dt}\right)_0 + \frac{Q_0}{C} = 0 \quad . \quad . \quad . \quad . \quad . \quad . \quad . \quad (3)$$

From (2),

$$\frac{di}{dt} = Me^{-ht} [k \cos kt - h \sin kt]$$

and at $t = 0$;

$$\left(\frac{di}{dt}\right)_0 = Mk.$$

By (3) $$M = - \frac{Q_0}{LCk} = - \frac{Q_0}{k}(h^2 + k^2)$$

since $$(- h + jk)(- h - jk) = h^2 + k^2 = 1/LC \text{ by (1).}$$

Finally, $i = Me^{-ht} \sin kt = - \left(k + \frac{h^2}{k}\right) Q_0 e^{-ht} \sin kt.$

9.3. Forced Vibrations

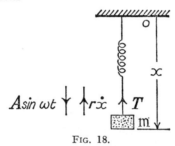

FIG. 18.

Consider a mass m kg oscillating vertically at the end of a spring which exerts a tension of $k \times$ (extension of spring in metres) newtons. On the mass of m kg is acting a force of $A \sin wt$ newtons. Apart from the tension in the spring there are other resistances to motion, and for many instruments these may be summed up, fairly accurately, by assuming a total resistance of $r \times$ (velocity in metres per second) newtons.   In this, k, A and r are all constant.

The displacement x of the mass can be measured from any fixed point.   We will choose as our origin the fixed point to which the spring is attached

The effect of gravity on the mass is mg newtons.   If l is the natural length of the spring, the tension due to its extension is $k(x - l)$ newtons and the equation for the motion is

$$m\ddot{x} = mg + A \sin wt - r\dot{x} - k(x - l)$$

or $$m\ddot{x} + r\dot{x} + kx = mg + kl + A \sin wt$$

If we put $$x - \frac{mg}{k} - l = z$$

since $$\dot{x} = \dot{z} \text{ and } \ddot{x} = \ddot{z}$$

the equation becomes

$$m\ddot{z} + r\dot{z} + kz = A \sin wt$$

where z is now the displacement of m from the point where it would hang in static equilibrium with only the forces due to gravity and the tension acting on it.

Rewrite the equation as

$$\ddot{z} + 2K\dot{z} + N^2z = F \sin wt \quad . \quad . \quad . \quad . \quad . \quad (1)$$

where
$$\frac{r}{m} = 2K, \quad \frac{k}{m} = N^2 \quad \frac{A}{m} = F$$

The complementary function part of the general solution requires the solution of

$$(D^2 + 2KD + N^2)z = 0.$$

which is for

(i) $N > K$
$$e^{-Kt}\{L \cos t\sqrt{(N^2 - K^2)} + M \sin t\sqrt{(N^2 - K^2)}\};$$

(ii) $N = K$
$$e^{-Kt}\{L + Mt\};$$

(iii) $N < K$
$$Le^{\{-K - \sqrt{(K^2 - N^2)}\}t} + Me^{\{-K + \sqrt{(K^2 - N^2)}\}t}.$$

In each case the exponential term has a negative index, and as t increases this component soon dies out, hence its name of the transient component of the solution.

The particular integral can be evaluated by finding

$$\frac{1}{D^2 + 2KD + N^2} (F \sin \omega t)$$

or by putting $z = a \sin \omega t + b \cos \omega t$ in (1) and so finding a and b. Either way we find the P.I. to be

$$z = F \frac{(N^2 - \omega^2) \sin \omega t - 2K\omega \cos \omega t}{(N^2 - \omega^2)^2 + 4K^2\omega^2}$$

$$= \frac{F \sin (\omega t - \alpha)}{\sqrt{\{(N^2 - \omega^2)^2 + 4K^2\omega^2\}}}$$

where
$$\tan \alpha = 2K\omega/(N^2 - \omega^2).$$

This is the persistent component of the solution or the *steady state* solution, showing that the period of this forced vibration (arising solely from the action of the force F sin ωt) is the same as that of the force. If instead of the force A sin wt acting on the mass m, the point of support of the spring has this motion where A is now an amplitude the equation of motion of the mass m is

$$m\ddot{x} = mg - r\dot{x} - k(x - l - A \sin wt)$$

and the same results as above are obtained for the transients and the period of the forced vibration.

It is important to consider the variation in the amplitude of the forced vibration when N, K and F are constant but ω varies, *i.e.*, we apply a periodic force to the system and consider the effect on the amplitude of the motion of varying the frequency. The amplitude is

$$F/\sqrt{\{(N^2 - \omega^2)^2 + 4K^2\omega^2\}}.$$

Let

$$y = (N^2 - \omega^2)^2 + 4K^2\omega^2$$

then

$$\frac{dy}{d\omega} = 4\omega[2K^2 - N^2 + \omega^2]$$

$$\frac{d^2y}{d\omega^2} = 12\omega^2 + 8K^2 - 4N^2.$$

The turning values are given by

$$\omega = 0, \ \sqrt{(N^2 - 2K^2)}$$

the latter value existing only when $N^2 > 2K^2$.

From these we find :

(i) When $N^2 > 2K^2$ the amplitude is a maximum when $\omega = \sqrt{(N^2 - 2K^2)}$ and a minimum when $\omega = 0$.

(ii) When $N^2 < 2K^2$ the amplitude is a maximum at $\omega = 0$.

The diagram shows the manner in which the amplitude varies with the frequency for various degrees of damping, the less the damping, the greater the amplitude and the amount of variation in it as the frequency varies.

In most cases K is very small compared to N and then $\omega \simeq N$. In this case or in the case above where $\omega = \sqrt{(N^2 - 2K^2)}$ the applied

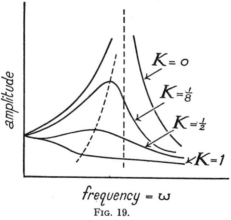

FIG. 19.

Variation of amplitude with frequency for different degrees of damping.

force is said to be in *resonance* with the free oscillations of the body, since the amplitude is then a maximum.

The amplitude of the forced vibration is, as above,

$$F/\sqrt{\{(N^2 - \omega^2)^2 + 4K^2\omega^2\}},$$

which for $\omega \simeq N$ becomes

$$F/\sqrt{(4K^2N^2)} = F/(2KN).$$

This is large, since K is small, showing that in such a case ($\omega \simeq N$ and K small) the oscillations can be dangerously large.

9.4. The Complex Variable Method of Solution

Based on the above and the method of 8.5 we may now repeat the method of obtaining the magnitude and phase of the *steady state solution* or particular integral of an equation such as

$$m\ddot{x} + k\dot{x} + hx = A \cos \omega t \text{ or } A \sin \omega t.$$

Consider

$$m\ddot{x} + k\dot{x} + hx = Ae^{j\omega t} \quad . \quad . \quad . \quad . \quad . \quad . \quad (1)$$

By the method of 8.5 and even more so since the particular integral is of the same frequency as the impressed force (by 9.3), we try as a solution

$$x = Fe^{j\omega t} \quad . \quad . \quad . \quad . \quad . \quad . \quad (2)$$

where F may be complex and equal to $|F| e^{j\theta}$, $|F|$ being its amplitude and θ its phase.

The substitution of (2) in (1) gives

$$(-m\omega^2 + j\omega k + h)Fe^{j\omega t} = Ae^{j\omega t}$$

$$\therefore \quad F = \frac{A}{(h - m\omega^2) + j\omega k}$$

$$= \frac{A(h - m\omega^2 - j\omega k)}{(h - m\omega^2)^2 + \omega^2 k^2}$$

$$\therefore \quad |F| = \frac{A}{\sqrt{\{(h - m\omega^2)^2 + \omega^2 k^2\}}}, \quad \tan \theta = -\frac{\omega k}{h - m\omega^2}$$

and with these values the steady-state solution is,

$$\begin{matrix} \text{real part} \\ \text{or} \\ \text{imaginary part} \end{matrix} \quad \text{of} \quad |F| e^{j(\omega t + \theta)} = \begin{matrix} |F| \cos(\omega t + \theta) \\ |F| \sin(\omega t + \theta) \end{matrix}.$$

As an illustration of this consider a circuit containing L, R, C in series with an applied voltage $E_m \sin \omega t$. If q is the charge on the plate of the condenser and i the current charging the condenser at time t

$$E_m \sin \omega t = iR + L\frac{di}{dt} + \frac{1}{C}\int_0^t i\,dt$$

$$i = \frac{dq}{dt}.$$

From the first equation

$$\omega E_m \cos \omega t = R \frac{di}{dt} + L \frac{d^2i}{dt^2} + \frac{i}{C}$$

or $\qquad (CLD^2 + RCD + 1)i = \omega CE_m \cos \omega t.$

The solution is, therefore, for the steady state above,

$$i = \frac{E_m \cos (\omega t + \theta)}{\sqrt{\{R^2 + (L\omega - 1/C\omega)^2\}}}.$$

The amplitude of this current is greatest when $R^2 + (L\omega - 1/C\omega)^2$ is least, *i.e.*, *resonance* occurs when $L\omega = 1/C\omega$ or $\omega = 1/\sqrt{(LC)}$, or when the period of the applied voltage is the same as the period of the free oscillations when there is no resistance in the circuit. The diagram shows the variation of the amplitude of i_{max} with ω.

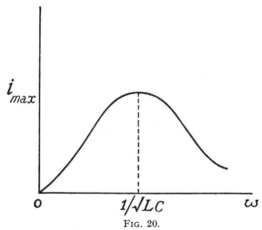

FIG. 20.

It is simple to establish a routine method of applying these principles. Thus for the usual equation

$$Ri + L \frac{di}{dt} + \frac{1}{C} \int_0^t i\,dt = E_m \sin \omega t$$

we may write

$$\left(R + LD + \frac{1}{CD}\right) i = E_m \sin \omega t,$$

so that inductance may in future be considered as a resistance of value LD and capacitance as a resistance $1/CD$. We then use Ohm's law on the circuit for the *total* resistance thus obtained. Also since the operator D^2 is to be replaced by $-\omega^2$, we will replace D by $j\omega$. However complicated an expression in D we may have on the left-hand side above, from it we obtain

$$i = \frac{a + jb}{c + jd} E_m \sin \omega t$$

after replacing D^2 and D by $-\omega^2$ and $j\omega$ respectively. We will now put $e^{j\omega t}$ for $\sin \omega t$ and pick out the j component when the rationalisation, etc., has been performed. Or, since

$$a + jb = \sqrt{(a^2 + b^2)}e^{j\phi}, \text{ where } \tan \phi = b/a;$$
$$c + jd = \sqrt{(c^2 + d^2)}e^{j\theta}, \text{ where } \tan \theta = d/c;$$

it follows that

$$i = \frac{\sqrt{(a^2 + b^2)}}{\sqrt{(c^2 + d^2)}} e^{j(\omega t + \phi - \theta)},$$

so that from the j component

$$i = \sqrt{\frac{(a^2 + b^2)}{(c^2 + d^2)}} \sin (\omega t + \phi - \theta),$$

giving the lag or lead of the current on the applied e.m.f. and the maximum value of i.

Example 3. An e.m.f. $E \sin \omega t$ is applied through a resistance R to two branches each containing a resistance R, one a condenser of capacitance C and the other a coil of inductance L. Determine the peak value and the phase lag or lead of the total current supplied. Show that if $\omega^2 LC = 1$ the circuit behaves as a pure resistance of amount $(3R^2 + \omega^2 L^2)/2R$. [L.U.]

If i is the total current, i_1 that through the R, C branch and i_2 that through the R, L branch

$$i = i_1 + i_2 \quad . \quad . \quad . \quad . \quad . \quad . \quad . \quad (1)$$

and using the equivalent resistance values for each circuit

$$\frac{E \sin \omega t - Ri}{R + 1/CD} = i_1, \quad \frac{E \sin \omega t - Ri}{R + LD} = i_2,$$

therefore from (1)

$$i\left\{\frac{R}{R + LD} + \frac{R}{R + 1/CD} + 1\right\} = \left\{\frac{1}{R + LD} + \frac{1}{R + 1/CD}\right\} E \sin \omega t,$$

$$i = \frac{1 + 2RCD + LCD^2}{2RC + D(3R^2C + L) + 2RCLD^2} E \sin \omega t$$

$$= \frac{(1 - LC\omega^2) + j2RC\omega}{2R(1 - LC\omega^2) + j(3R^2C + L)\omega} \cdot E \sin \omega t.$$

If $\omega^2 LC = 1$ this reduces to

$$i = \frac{2RC}{3R^2C + L} E \sin \omega t,$$

showing that the circuit behaves as a pure resistance of value

$$\frac{3R^2C + L}{2RC} = \frac{3R^2 + L/C}{2R}$$

$$\left(\text{or with } \frac{1}{C} = \omega^2 L\right) \qquad = \underline{\frac{3R^2 + \omega^2 L^2}{2R}}$$

as required.

EXERCISE 18

1. A uniform horizontal strut of length l and weight w per unit length is freely supported at the ends, and is also subject to a thrust P at each end. Show that with the origin at one end the deflection is given by the equation

$$EI\frac{d^2y}{dx^2} + Py = -\tfrac{1}{2}wx(l - x).$$

Integrate this equation and obtain an expression for the maximum deflection.

2. A light rod is clamped horizontally at one end $(x = 0)$, is freely hinged at the other end $(x = l)$ and is subject to a horizontal thrust P. Show that the deflection y satisfies the equation

$$EI \frac{d^2y}{dx^2} + Py = G \left(1 - \frac{x}{l}\right),$$

where G is the couple applied at the clamped end. Show further that the rod can take up a curved form given by

$$y = \frac{G}{P} \left(\frac{\sin nx - nx}{nl} + 1 - \cos nx\right)$$

provided $\tan nl = nl$, where $n^2 = P/EI$.

3. For a horizontal strut length l freely hinged at its ends and subjected to an axial thrust P, the bending moment under a given system of loading is determined by the equation

$$EI \frac{d^2y}{dx^2} + Py = - \frac{wl^2}{8} \sin \left(\frac{\pi x}{l}\right),$$

where E, I have their usual meanings. Solve the equation and show that the bending moment at the centre is

$$\frac{wl^2}{8} \left(\frac{Q}{Q - P}\right), \quad \text{where} \quad Ql^2 = EI\pi^2.$$

4. A light horizontal bar is clamped at one end. At the other end there is a vertical load W and a horizontal thrust P. Prove that the deflection at the free end is

$$\frac{W}{nP} (\tan nl - nl)$$

where $n^2 = P/EI$ and l is the length of the bar.

5. A strut of length l has both ends hinged and carries an axial load P. There is in addition a transverse load W applied at the centre of the strut. If y is the deflection at a distance x from one end, the equation which determines y is

$$EI \frac{d^2y}{dx^2} + Py = - \frac{Wx}{2}.$$

Integrate this equation and show that the deflection d at the centre is given by

$$d = \frac{W}{2P} \sqrt{\left(\frac{EI}{P}\right)} \tan \left(\frac{l}{2}\sqrt{\frac{P}{EI}}\right) - \frac{Wl}{4P}.$$

Find also the maximum bending moment.

6. A uniform strut of length l is freely hinged at each end; the ends are at the same level and are at the points $x = \pm \frac{1}{2}l$, Ox being horizontal. The strut carries a load w per unit length and is subject to a thrust P at each end. Show that the deflection y at any point satisfies

$$EI \frac{d^2y}{dx^2} + Py = - \frac{1}{2}w(\tfrac{1}{4}l^2 - x^2).$$

Solve this equation and show that the deflection at the middle of the strut is

$$\frac{wEI}{P^2} \left[\sec \left(\tfrac{1}{2}l \sqrt{\frac{P}{EI}}\right) - 1\right] - \frac{wl^2}{8P}.$$

7. A uniform beam is supported at its ends and carries a uniformly distributed load along the middle half. Show that the additional deflection due to the load is 57/64 times the additional deflection had the load been concentrated at the mid-point. [L.U.]

8. A naturally straight uniform elastic rod of length $2l$ is laid across a rigid horizontal table of breadth $2a$ with equal lengths $l - a$ projecting beyond the

table. If a straight portion AB of the rod is in contact with the table, show that the bending moments at A and B are zero and that

$$\frac{l}{a} < 1 + \frac{1}{\sqrt{2}}$$

[L.U.]

9. The lower end of a uniform light cantilever of length l and flexural rigidity EI is clamped at angle θ to the vertical. A vertical load W is applied to the upper end. Obtain the differential equation for y the deflection (from the undeflected position) of a point on the strut distant x from the lower end in terms of the end deflection a. Show that at the free end the value of dy/dx is $\tan \theta$ (sec $nl - 1$), where $EIn^2 =$ W cos θ. [L.U.]

[Since the deflection at the free end is a and the gradient at the origin is zero, two values can be obtained for an unknown constant.]

10. A beam of uniform cross-section and length $2l$ m has its ends built in horizontally at the same level and carries a distributed load. The load intensity at any point is proportional to the distance of the point from the end nearer to it and at the centre it is p N/m. Assuming the formula $EI \dfrac{d^4y}{dx^4} = w$, find the deflection at the beam of the centre and the fixing couple at each end.

[L.U.]

11. A light vertical pole AB of strength EI and height h is clamped vertically at its base A and has a light cord attached to its free top end B. The cord is pulled downwards at an angle of $45°$ to the horizontal with a force F, and produces at B a horizontal displacement δ.

If y is the deflection at a height x above the ground show that

$$\frac{d^2y}{dx^2} + m^2y = m^2(h + \delta - x)$$

where $m^2 = F/(EI \sqrt{2})$.

Hence show that $\delta = (\tan mh - mh)/m$, and that if F is small $\delta = Fh^3/(3\sqrt{2}EI)$ approx. [L.U.]

12. Integrate the equation

$$\ddot{x} + 2hn\dot{x} + n^2x = 0$$

given that $0 < h < 1$ and $x = 0$, $\dot{x} = v$ when $t = 0$. To measure h the ratio r of the amplitude of successive swings (i.e., on opposite sides of the equilibrium position) is found. Prove that $h^2 = (\log r)^2/\{n^2\pi^2 + (\log r)^2\}$.

13. A flywheel of moment of inertia I about its axis is acted on by a variable couple of moment G sin$^2 pt$, and is subject to a retarding couple k times the angular velocity. Show that ultimately its angular velocity is

$$\tfrac{1}{2}G\left\{\frac{1}{k} - \frac{2Ip \sin 2pt + k \cos 2pt}{4I^2p^2 + k^2}\right\}.$$

[L.U.]

14. A body of mass 8 kg hangs from the end of a spiral spring which elongates 1 m for a pull of 1000 N. Its vibrations are damped, the damping being proportional to the speed and equal to 80 N for a speed of 1 m/s. What is the time of a complete vibration?

If the upper end of the spring is moved up and down with a simple harmonic motion of amplitude 0·03m and with one-half the natural frequency, what is the amplitude of the forced vibration after some time has elapsed?

15. Find the condition that the motion represented by

$$\ddot{x} + 2p\dot{x} + qx = 0$$

may be oscillatory, and find the period.

A mass of 8 kg hangs from a spiral spring whose stiffness is such that an axial force of 1600 N elongates it 1 m. The motion is resisted by fluid friction, which in N is numerically equal to 160 times the velocity. The point of support has an axial motion, the displacement y m at time t s being given by $y = \frac{1}{3} \sin 9t$. Find the motion of the mass when its natural vibration has been destroyed by friction.

16. A 2 kg mass hangs at rest on a spring, producing in the spring an extension of 0·3 m. The upper end of the spring is now made to execute a vertical simple harmonic oscillation $x = \frac{1}{3} \sin 4t$, x being measured vertically downwards in metres. If the mass is subject to a frictional resistance whose magnitude in newtons is three times its velocity in m/s, obtain the differential equation for the motion of the mass and find the expression for its displacement at time t when t is large.

17. A particle of mass m is attached to the lower end of a light elastic string of unstretched length a and modulus λ. The upper end of the string is made to move in a vertical line with simple harmonic motion of amplitude c and period $2\pi/n$. Prove that x the extension of the string at time t satisfies the differential equation

$$\frac{d^2x}{dt^2} + \frac{\lambda x}{am} = g + cn^2 \sin (nt + \beta)$$

where β is a constant.

Find the general solution of this equation and prove that if $n = (\lambda/ma)^{\frac{1}{2}}$, the amplitude of the oscillations of the particle will increase indefinitely with the time t. [L.U.]

18. A condenser of capacity 2 microfarads is charged by a battery of 100 volts through a coil of resistance 40 ohms and inductance 2 henrys. Find the potential difference between the plates of the condenser at time t after closing the circuit.

19. An uncharged condenser of capacity C is charged by applying an e.m.f. of E $\sin (t/\sqrt{LC})$ through leads of self-inductance L and negligible resistance. Prove that at time t the charge on one of the plates is

$$\tfrac{1}{2}EC \left[\sin \frac{t}{\sqrt{LC}} - \frac{t}{\sqrt{LC}} \cos \frac{t}{\sqrt{LC}} \right].$$ [L.U.]

20. An electric circuit consists of a generator of voltage 200 sin 100t, an open switch, a resistance of 300 ohms, an inductance of 1 henry and an uncharged condenser of 50 microfarads, all in series. At time $t = 0$ the switch is closed. If q is the charge on a plate of the condenser at time t, show that

$$\frac{d^2q}{dt^2} + 300 \frac{dq}{dt} + 20\,000\,q = 200 \sin 100t.$$

Hence obtain a complete expression for the current flowing at time t. [L.U.]

21. A circuit consists of an inductance L and a capacity C in series. An alternating e.m.f. E $\sin nt$ is applied to the circuit at time $t = 0$, the initial current and charge on the condenser being zero. Prove that the current at time t is

$$I = \frac{nE}{L(n^2 - \omega^2)} \{\cos \omega t - \cos nt\}, \quad \text{where} \quad CL\omega^2 = 1.$$

Prove also that if $n = \omega$ the current at time t is $\dfrac{Et \sin \omega t}{2L}$. [L.U.]

22. A condenser of capacity C is discharged through a circuit of resistance R and self-inductance L and the current x satisfies the equation

$$L\ddot{x} + R\dot{x} + \frac{x}{C} = 0.$$

If $R^2 > 4L/C$ and the potential V of the condenser at any time is given by $V = L\dot{x} + Rx$, and if initially $V = V_0$, $x = 0$ show that

$$x = \frac{2V_0}{(R^2 - 4L/C)^{\frac{1}{2}}} e^{-Rt/2L} \sinh \left\{ t \sqrt{\left(\frac{R^2}{4L^2} - \frac{1}{CL} \right)} \right\}.$$

23. An e.m.f. E cos pt is switched across an electric circuit consisting of a coil of inductance L henrys, resistance R ohms, in series with a condenser of capacity C farads. State the differential equation for the charge q in the condenser at any time t and the form of the general solution when the resistance is just sufficient to prevent natural oscillations. If $L = 0.0001$, $R = 2$, find the values of C for this condition to hold and calculate the amplitude of the steady current for an imposed peak voltage of 100 at 50 hertz.   [L.U.]

24. Two condensers of capacities C_1 and C_2 have one plate of each earthed and the two insulated plates joined by a wire of resistance R and self-inductance L. Initially the first condenser has a charge Q and the second is uncharged.

Prove that the charge on the second condenser after a time t is

$$\frac{QC_2}{C_1 + C_2} \left[\frac{\beta e^{-\alpha t} - \alpha e^{-\beta t}}{\alpha - \beta} + 1 \right]$$

where α and β are the roots of

$$L\lambda^2 - R\lambda + \frac{1}{C_1} + \frac{1}{C_2} = 0.$$ [L.U.]

25. An e.m.f. $E_0 \sin \omega t$ is applied to a series circuit of R, L and C. Show that if $1 = \omega^2 CL$, then the steady state current is in phase with the e.m.f. and

$$i = \frac{E_0}{R} \sin \omega t.$$

26. A circuit consists of two branches, a coil (R, L) in parallel with a condenser of capacity C. If an e.m.f. $E_0 \sin \omega t$ is imposed upon this, show that the steady state (total) current is

$$\frac{\sqrt{\{(1 - \omega^2 CL)^2 + \omega^2 C^2 R^2\}}}{\sqrt{\{R^2 + \omega^2 L^2\}}} E_0 \sin (\omega t + \alpha - \beta)$$

where $\tan \alpha = \dfrac{\omega CR,}{1 - \omega^2 CL}$, $\tan \beta = \dfrac{\omega L}{R}.$

Find the condition that this total current is in phase with the e.m.f. and show that when this condition is satisfied

$$i_{(max.)} = \frac{E_0}{R} \cos^2 \beta.$$

27. A coil of resistance R_1 and inductance L_1 is in parallel with a non-inductive resistance R_2. Find the total steady-state current when an e.m.f. $E_0 \sin \omega t$ is applied.

28. A coil (R, L) is in parallel with a condenser C. If the total current (steady state) is $I_0 \cos \omega t$ and the charge on the condenser is in phase with this current, prove that $C(R^2 + \omega^2 L^2) = L$ and that the charge is $\dfrac{LI_0}{R} \cos \omega t.$

29. Two points A, B are joined by a wire of resistance R without self-induction; B is joined to a third point C by two wires each of resistance R, of which one is without self-induction and the other has a coefficient of self-induction L. If the ends A, C are kept at a potential difference E cos ωt and if there are no mutual inductances, prove that the current in AB is

$$\frac{E}{R} \sqrt{\left(\frac{4R^2 + L^2\omega^2}{9R^2 + 4L^2\omega^2} \right)} \cos (\omega t - \alpha)$$

where $\tan \alpha = RL\omega/(6R^2 + 2L^2\omega^2)$. Find also the difference of potential of B and C.   [L.U.]

30. An alternating e.m.f. of amplitude E and frequency $\omega/2\pi$ is supplied to a coil of inductance L and resistance R. Write down the differential equation for the current in the coil and solve it, indicating the transient term.

If the coil is shunted by a condenser of capacitance C and resistance S, show that the circuit can be replaced (as far as permanent current is concerned) by a non-inductive resistance provided that

$$CR^2 - L = \omega^2 CL(CS^2 - L) \qquad \text{[L.U.]}$$

9.5. Simultaneous Differential Equations

We will solve some examples before considering the way in which these equations arise.

Example 4. Solve

$$\frac{d^2x}{dt^2} + 8x + 2y = 24 \cos 4t$$

$$\frac{d^2y}{dt^2} + 2x + 5y = 0$$

given that when $t = 0$, x, y, $\frac{dx}{dt}$, $\frac{dy}{dt}$ are all zero. [L.U.]

These equations can be written

$$(D^2 + 8)x + 2y = 24 \cos 4t \quad . \quad . \quad . \quad . \quad . \quad (1)$$
$$2x + (D^2 + 5)y = 0 \quad . \quad . \quad . \quad . \quad . \quad (2)$$

To eliminate x operate on the first equation with the factor 2 and on the second equation with the factor $D^2 + 8$, to obtain

$$2(D^2 + 8)x + 4y = 48 \cos 4t,$$
$$2(D^2 + 8)x + (D^2 + 8)(D^2 + 5)y = 0.$$

Subtract:

$$[(D^2 + 8)(D^2 + 5) - 4]y = -48 \cos 4t,$$
$$(D^4 + 13D^2 + 36)y = -48 \cos 4t$$

or
$$(D^2 + 9)(D^2 + 4)y = -48 \cos 4t.$$

The solution of this is

$$y = A \cos 3t + B \sin 3t + C \cos 2t + E \sin 2t - \tfrac{4}{7} \cos 4t.$$

(a) In this example there is a simple method for finding x. From (2)

$$2x = -(D^2 + 5)y$$
$$= -(D^2 + 5)[A \cos 3t + B \sin 3t + C \cos 2t + E \sin 2t - \tfrac{4}{7} \cos 4t]$$
$$= 4A \cos 3t + 4B \sin 3t - C \cos 2t - E \sin 2t - \tfrac{44}{7} \cos 4t$$
$$x = 2A \cos 3t + 2B \sin 3t - \tfrac{1}{2}C \cos 2t - \tfrac{1}{2}E \sin 2t - \tfrac{22}{7} \cos 4t.$$

Since $x = y = 0$ when $t = 0$,

$$0 = 2A - \tfrac{1}{2}C - \tfrac{22}{7},$$
$$0 = A + C - \tfrac{4}{7},$$
$$\therefore \quad A = \tfrac{48}{35}, \; C = -\tfrac{28}{35}.$$

Also $\dfrac{dx}{dt} = -6A \sin 3t + 6B \cos 3t + C \sin 2t - E \cos 2t + \tfrac{88}{7} \sin 4t,$

$$\frac{dy}{dt} = -3A \sin 3t + 3B \cos 3t - 2C \sin 2t + 2E \cos 2t + \tfrac{16}{7} \sin 4t,$$

and since $\dfrac{dx}{dt} = \dfrac{dy}{dt} = 0$ when $t = 0$,

$$0 = 6B - E,$$
$$0 = 3B + 2E,$$
$$\therefore \quad B = E = 0.$$

The solution is therefore

$$x = \tfrac{96}{35} \cos 3t + \tfrac{2}{5} \cos 2t - \tfrac{22}{7} \cos 4t,$$
$$y = \tfrac{48}{35} \cos 3t - \tfrac{4}{5} \cos 2t - \tfrac{4}{7} \cos 4t.$$

(b) If this simple method in (a) did not exist it might be necessary to solve (1) and (2) for x, proceeding as in the following method.

From (1) and (2)

$$(D^2 + 5)(D^2 + 8)x + 2(D^2 + 5)y = 24(D^2 + 5) \cos 4t,$$
$$4x + 2(D^2 + 5)y = 0.$$

Subtract:

$$(D^2 + 9)(D^2 + 4)x = -264 \cos 4t,$$
$$\therefore \quad x = L \cos 3t + M \sin 3t + P \cos 2t + Q \sin 2t - \tfrac{22}{7} \cos 4t.$$

These unknowns L, M, P, Q are not independent of A, B, C, D, since x and y must satisfy the given equations (1) and (2).

Substituting in (2),

$$2L \cos 3t + 2M \sin 3t + 2P \cos 2t + 2Q \sin 2t - \tfrac{44}{7} \cos 4t$$
$$+ (D^2 + 5)[A \cos 3t + B \sin 3t + C \cos 2t + E \sin 2t - \tfrac{4}{7} \cos 4t] = 0.$$

Equating the separate coefficients to zero,

$$2L - 4A = 0,$$
$$2M - 4B = 0,$$
$$2P + C = 0,$$
$$2Q + E = 0,$$

giving $L = 2A$, $M = 2B$, $P = -\tfrac{1}{2}C$, $Q = -\tfrac{1}{2}E$, from which x is obtained in terms of A, B, C, E as previously. Since initial conditions are given, the method continues as above to find A, B, C, E.

(c) This matter of the number of independent unknowns in the general solution should be noted. The maximum number is equal to the sum of the orders of the differential equations. In this case since each equation (1) and (2) is of the second order we expect four unknowns in the solution. There may be less, and the exact number is the degree of the determinant of the coefficients. From (1) and (2) this is

$$\begin{vmatrix} D^2 + 8 & 2 \\ 2 & D^2 + 5 \end{vmatrix}$$

which multiplied out is of the fourth degree in D showing that there are four arbitrary constants in the solution.

9.6. A Special Method

Some pairs of equations are in such a form that a special method may be used.

Example 5. Solve the equations

$$m\ddot{x} + He\dot{y} = Ee$$
$$m\ddot{y} - He\dot{x} = 0$$

given that $x = y = \dot{x} = \dot{y} = 0$ when $t = 0$. [This is the mathematics of J. J. Thomson's famous experiment to find m/e the ratio of the mass of an electron to its charge.]

The equations are such that adding j times the second to the first, gives

$$m(\ddot{x} + j\ddot{y}) + He(\dot{y} - j\dot{x}) = Ee$$

or

$$mD^2(x + jy) - jHeD(x + jy) = Ee$$

with $x + jy = z$, this is

$$D^2 z - j\frac{He}{m} Dz = Ee/m$$

or

$$D\left(D - j\frac{He}{m}\right) z = Ee/m.$$

The solution is, as usual

$$z = A + Be^{j\frac{He}{m}t} + \frac{jE}{H}t,$$

where the constants may be complex.   Hence

$$x + jy = a + jb + (c + jd)(\cos \omega t + j \sin \omega t) + \frac{jE}{H}t,$$

where $\omega = \dfrac{He}{m}$ and a, b, c, d are real.   From this equation

$$x = a + c \cos \omega t - d \sin \omega t,$$

$$y = b + d \cos \omega t + c \sin \omega t + \frac{E}{H}t.$$

From $x = y = 0$ when $t = 0$:

$$0 = a + c,$$
$$0 = b + d,$$

from $\dot{x} = \dot{y} = 0$ when $t = 0$:

$$0 = d,$$

$$0 = \omega c + \frac{E}{H},$$

$$\therefore \quad b = d = 0, \ c = -\frac{E}{\omega H}, \ a = \frac{E}{\omega H}.$$

The solution is therefore

$$x = \frac{E}{\omega H}(1 - \cos \omega t),$$

$$y = \frac{E}{\omega H}(\omega t - \sin \omega t),$$

the parametric equations to a cycloid.

9.7. Applications

We have seen that a body, *i.e.*, machine part on which an alternating force is acting, may make oscillations of dangerously large amplitude especially when conditions are near or at resonance.   One method of eliminating this is as follows :

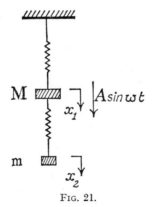

FIG. 21.

Consider a mass M oscillating from a spring of stiffness K attached to a fixed support.   To the mass M is attached a smaller mass m by means of a spring of stiffness k.   Let the displacements x_1 and x_2 be measured from the points where M and m would hang in statical equilibrium so that it is only the further extensions x_1 and x_2 that enter into the equations of motion. These are

$$M\ddot{x}_1 + (K + k)x_1 - kx_2 = A \sin \omega t,$$

$$m\ddot{x}_2 + k(x_2 - x_1) = 0.$$

We could solve this pair of simultaneous equations in the usual way. However, here we are only concerned with the forced vibrations which we know to be of the same period as the applied force. We will therefore try as a solution,

$$x_1 = a_1 \sin \omega t \qquad x_2 = a_2 \sin \omega t$$

where a_1, a_2 the amplitudes of these motions are required.

Making the substitution and cancelling the factor $\sin \omega t$,

$$a_1 \left(- M\omega^2 + K + k \right) - ka_2 = A$$

$$- ka_1 + a_2 \left(- m\omega^2 + k \right) = 0$$

The elimination of a_2 gives

$$a_1 = \frac{A \left(k - m\omega^2 \right)}{k^2 + (k - m\omega^2)(K + k - M\omega^2)}$$

The amplitude of the motion of a_1 will then be zero, *i.e.*, despite the periodic force $A \sin \omega t$, M will be at rest if

$$k - m\omega^2 = 0$$

an equation for the stiffness k of the lower spring.

9.8. Coupled Electrical Circuits

Suppose the primary has a resistance R_1, inductance L_1 and an e.m.f. of e_1. Similarly, for the secondary let these be R_2, L_2 and e_2, with M as the mutual inductance of the coils.

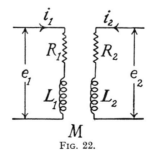

Fig. 22.

If the coils are so linked that each current describes a right-hand screw round the magnetic flux due to the other, then each current causes an e.m.f. $- M di/dt$ in the other coil.

By Ohm's law for each coil

$$R_1 i_1 = e_1 - L_1 \frac{di_1}{dt} - M \frac{di_2}{dt},$$

$$R_2 i_2 = e_2 - L_2 \frac{di_2}{dt} - M \frac{di_1}{dt}.$$

We will solve completely for the simple case of an induction coil where $e_1 = E$, a constant, $e_2 = 0$ and $R_1 = R_2 = R$, $L_1 = L_2 = L$. In this case the equations are

$$(LD + R)i_1 + MDi_2 = E \quad \cdots \quad (1)$$
$$MDi_1 + (LD + R)i_2 = 0 \quad \cdots \quad (2)$$

The elimination of i_2 gives

$$\{(L^2 - M^2)D^2 + 2RLD + R^2\}i_1 = RE,$$

the solution of which is

$$i_1 = \frac{E}{R} + Ae^{-Rt/(L+M)} + Be^{-Rt/(L-M)}. \quad \cdots \quad (3)$$

$L \times (1)$ minus $M \times (2)$ gives

$$\{(L^2 - M^2)D + LR\}i_1 - MRi_2 = LE$$

and after substituting for i_1 from (3)

$$i_2 = Ae^{-Rt/(L+M)} - Be^{-Rt/(L-M)}.$$

Since $i_1 = 0 = i_2$ when $t = 0$,

$$A = B = - E/2R,$$

$$\therefore \quad i_1 = \frac{E}{R}\{1 - \tfrac{1}{2}e^{-Rt/(L+M)} - \tfrac{1}{2}e^{-Rt/(L-M)}\}.$$

$$i_2 = \frac{E}{2R}\{e^{-Rt/(L-M)} - e^{-Rt/(L+M)}\}.$$

These equations show that

$$i_1 \longrightarrow E/R, \quad i_2 \longrightarrow 0 \text{ as } t \longrightarrow \infty$$

and i_2 is negative, *i.e.*, in the opposite direction to that assumed.

9.9. Solution in Series

Unfortunately the methods described above are not sufficient for all the types of differential equations that occur in applied mathematics. When a differential equation does not fall—or cannot be transformed—into one of the standard types dealt with above we try to find a solution in the form of an infinite power series.

The simplest type of solution assumed is

$$y = a_0 + a_1 x + a_2 x^2 + \ldots a_n x^n + \ldots \quad \cdots \quad (1)$$

and by substitution in the given differential equation we hope to find the coefficients $a_0, a_1, \ldots a_n, \ldots$ and then the range of values of x for which the series is convergent. Within this range the power series will then be a solution.

As an example we will solve by this method

$$(D^2 - 3D + 2)y = 0 \quad . \quad . \quad . \quad . \quad . \quad . \quad (2)$$

where the solution is clearly

$$y = Ae^x + Be^{2x} \quad . \quad . \quad . \quad . \quad . \quad . \quad (3)$$

From (1)

$$\frac{dy}{dx} = a_1 + 2a_2x + 3a_3x^2 + \ldots na_nx^{n-1} \ldots$$

$$\frac{d^2y}{dx^2} = 2a_2 + 6a_3x + 12a_4x^2 + \ldots n(n-1)a_nx^{n-2} + \ldots$$

Substitution in (2) gives

$$[2a_2 + 6a_3x + 12a_4x^2 + \ldots n(n-1)a_nx^{n-2} + \ldots]$$
$$- 3[a_1 + 2a_2x + 3a_3x^2 + \ldots na_nx^{n-1} + \ldots]$$
$$+ 2[a_0 + a_1x + a_2x^2 + \ldots a_nx^n + \ldots] = 0$$

The coefficients of each power of x are separately zero. This gives

x^0:	$2a_2 - 3a_1 + 2a_0 = 0$
x:	$6a_3 - 6a_2 + 2a_1 = 0$
x^2:	$12a_4 - 9a_3 + 2a_2 = 0$
x^3:	$20a_5 - 12a_4 + 2a_3 = 0$, etc.

Apparently we cannot find a_0, a_1 but can obtain the other coefficients in terms of them. This gives:

$$a_2 = \tfrac{1}{2}(3a_1 - 2a_0)$$
$$a_3 = \tfrac{1}{6}(6a_2 - 2a_1)$$
$$= \tfrac{1}{6}(7a_1 - 6a_0)$$
$$a_4 = \tfrac{1}{12}(9a_3 - 2a_2)$$
$$= \tfrac{1}{12}(4a_1 - 7a_0)$$
$$a_5 = \tfrac{1}{20}(12a_4 - 2a_3)$$
$$= \tfrac{1}{4}(\tfrac{1}{3}a_1 - a_0).$$

The solution is

$$y = a_0 + a_1x + \tfrac{1}{2}(3a_1 - 2a_0)x^2 + \tfrac{1}{6}(7a_1 - 6a_0)x^3$$
$$+ \tfrac{1}{12}(4a_1 - 7a_0)x^4 + \tfrac{1}{4}(\tfrac{1}{3}a_1 - a_0)x^5 + \ldots$$
$$= a_0(1 - x^2 - x^3 - \tfrac{7}{12}x^4 - \tfrac{1}{4}x^5 + \ldots)$$
$$+ a_1(x + \tfrac{3}{2}x^2 + \tfrac{7}{6}x^3 + \tfrac{1}{3}x^4 + \tfrac{1}{12}x^5 + \ldots)$$

This is the same as the solution (3) if

$$A + B = a_0, \quad A + 2B = a_1.$$

A general solution containing two unknown constants has therefore been found by this method.

This method will not give a solution to such a simple equation as

$$\frac{dy}{dx} = \frac{1}{x},$$

since the solution is $y = \log x + c$ and $\log x$ cannot be expanded as a power series in x.

9.10. Method of Frobenius

A more general method is to assume as a trial solution

$$y = x^c(a_0 + a_1x + a_2x^2 + \ldots a_nx^n + \ldots) \quad . \quad . \quad (1)$$

where a_0 is the first coefficient that is not zero in the expansion. In addition to the coefficients a_n, c also has to be found. We will use this method to solve Bessel's equation, an equation that occurs in the vibration of chains and membranes, the flow of heat and the study of the propagation of electricity in conductors, *e.g.*, the skin effect.

The equation is

$$x^2 y_2 + x y_1 + (x^2 - n^2) y = 0,$$

where n is a constant.

Assuming y as in (1) we find

$$x^2 y_2 = c(c-1)a_0 x^c + c(c+1)a_1 x^{c+k} + \ldots$$
$$+ a_k(c+k)(c+k-1)x^{c+k} + \ldots$$
$$x y_1 = c a_0 x^c + (c+1)a_1 x^{c+1} + \ldots (c+k)a_k x^{c+k} + \ldots$$
$$x^2 y = a_0 x^{c+2} + \ldots a_{k-2} x^{c+k} + \ldots$$
$$- n^2 y = - n^2 a_0 x^c - n^2 a_1 x^{c+1} + \ldots - n^2 a_k x^{c+k} + \ldots$$

Equate successive coefficients to zero:

x^c:
$$(c^2 - n^2)a_0 = 0 . \quad . \quad . \quad . \quad . \quad . \quad . \quad (2)$$
$$a_0 \neq 0 \qquad \therefore \quad c = \pm n.$$

The equation (2) which gives c, by equating to zero the coefficient of the lowest power of x is called the *indicial equation*, since it gives the value of the index c.

x^{c+1}:
$$[(c+1)^2 - n^2]a_1 = 0 \quad . \quad . \quad . \quad . \quad (3)$$
x^{c+k}:
$$[(c+k)^2 - n^2]a_k + a_{k-2} = 0 \quad . \quad . \quad . \quad . \quad (4)$$

giving the general relation

$$a_k = \frac{-a_{k-2}}{(c+k)^2 - n^2} \quad . \quad . \quad . \quad . \quad . \quad . \quad (5)$$

From (3) $a_1 = 0$ therefore by (5), $a_3 = a_5 = \ldots = 0$.
For the other coefficients

$$a_2 = - a_0/2(2n+2)$$
$$a_4 = - a_2/4(2n+4) = a_0/2 . 4(2n+2)(2n+4).$$

In this way we find for $c = n$

$$y = a_0 x^n \left[1 - \frac{x^2}{2(2n+2)} + \frac{x^4}{2 . 4(2n+2)(2n+4)} \cdots \right]$$

for $c = -n$

$$y = a_0 x^{-n} \left[1 - \frac{x^2}{2(2n-2)} + \frac{x^4}{2 . 4(2n-2)(2n-4)} \cdots \right]$$

For $n = 0$ the two solutions are identical. For n a negative integer the first solution is meaningless, but the second holds and vice versa when n is a positive integer. For other values of n we have two distinct solutions and if y_1 denotes the first series and y_2 the second then

$$A y_1 + B y_2$$

is the general solution.

There are a number of difficulties with this method, and the student is referred to *Piaggio : Differential Equations* for a more detailed treatment.

EXERCISE 19

Solve the simultaneous differential equations:

1. $\dfrac{dy}{dx} - 2y + 3z = 1, \quad y - \dfrac{dz}{dx} - z = 2.$
2. $Dy + 2z = x, \quad D^2 y + (2D+1)z = 2.$
3. $(D-1)x + Dy = 4, \quad Dx + (D-1)y = 0.$

4. $\dfrac{dx}{dt} = 3x - y$, $\dfrac{dy}{dt} = x + y$.

5. $\dot{x} - 2y = \cos t$, $\dot{y} + 2x = \sin t$,
given that $x = 1$, $y = 2$ when $t = 0$.

6. The free oscillations of the currents in the primary and secondary circuits of a transformer are given by:

$$L_1 \frac{di_1}{dt} + M \frac{di_2}{dt} + R_1 i_1 = 0,$$

$$M \frac{di_1}{dt} + L_2 \frac{di_2}{dt} + R_2 i_2 = 0,$$

where $M^2 < L_1 L_2$. If L_1, L_2, $M = 2$, 4, 2 respectively and $R_1 = R_2 = 3$, $i_1 = 2$, $i_2 = 3$ at $t = 0$, find the currents at time t.

7. Solve the differential equations:

$$dx/dt + 2x + y = 0, \quad dy/dt + x + 2y = 0,$$

subject to the conditions that $x = 1$ and $y = 0$ at $t = 0$. [L.U.]

8. Solve the equations

$$6d^2x/dt^2 = 2(y - x) = - 3d^2y/dt^2,$$

given that when $t = 0$, $dx/dt = V$, and x, y and dy/dt all vanish.

Find the smallest value of t for which $(y - x)$ again vanishes and the values of dx/dt and dy/dt then. [L.U.]

9. Solve the simultaneous differential equations

$$\frac{dx}{dt} + 3x - 2y = 1,$$

$$\frac{dy}{dt} - 2x + 3y = e^t,$$

given that when $t = 0$, $x = y = 0$. [L.U.]

10. Solve the equations

$$\frac{dx}{dt} + y = \sin t, \quad dy/dt + x = \cos t,$$

subject to the conditions $x = 2$, $y = 0$ when $t = 0$. [L.U.]

11. Solve the simultaneous differential equations

$$\frac{d^2x}{dt^2} + x + y = 0,$$

$$4 \frac{d^2y}{dt^2} - x = 0,$$

subject to the conditions that when $t = 0$, $x = 2a$, $y = - a$, $dx/dt = 2b$, $dy/dt = - b$, and show that the solution is then purely periodic.

12. In a " heat exchanger " the temperatures θ and T of the two liquids satisfy the equations

$$m \frac{d\theta}{dx} = k(T - \theta) = \frac{dT}{dx}$$

where m and k are constants. Find the general solutions of these equations. If $m = 2$, $k = 0.5$ and $\theta = 20$ when $x = 0$; $T = 100$ when $x = 3$, find θ when $x = 3$ and T when $x = 0$. [L.U.]

13. The currents x and y in two coupled circuits are given by

$$L \frac{dx}{dt} + Rx + R(x - y) = E, \quad L \frac{dy}{dt} + Ry - R(x - y) = 0,$$

where L, R and E are constants. Find x and y in terms of t, given that $x = y = 0$ at $t = 0$. [L.U.]

14. The acceleration components of a particle moving in a plane are given by

$$\frac{d^2x}{dt^2} = n \frac{dy}{dt}, \quad \frac{d^2y}{dt^2} = a - n \frac{dx}{dt},$$

where a and n are constant. Solve these equations assuming that the particle is at rest at the origin at $t = 0$ and show that the particle describes a cycloid.

[L.U.]

15. Solve the simultaneous differential equations

$$\frac{d^2x}{dt^2} + x = \frac{dy}{dt}, \quad 4\frac{dx}{dt} + 2x = \frac{dy}{dt} + 2y$$

If x, y are the cartesian co-ordinates of a point in a plane, and if, when $t = 0$

$$x = 0, \quad y = 1, \quad \frac{dx}{dt} = 2,$$

prove that the point lies on the parabola $(5x - 2y)^2 = 4(y - 2x)$. [L.U.]

16. If
$$6\frac{d^2x}{dz^2} - 3\frac{dx}{dz} + 6x + \frac{dy}{dz} = z$$

and
$$\frac{d^2x}{dz^2} + 8x + y = 0,$$

find x and y in terms of z.

If when $z = 0$, $x = \frac{11}{36}$, $y = -\frac{22}{9}$, $\frac{dx}{dz} = \frac{1}{6}$, show that the locus of the point (x, y) is a straight line through the origin. [L.U.]

17(a). Solve the simultaneous equations:

$$(D - 1)u - 2v = e^{-x};$$
$$- 2u + (D - 1)v = 1.$$

(b) Solve:

$$\frac{dx}{dt} + 6\frac{dy}{dt} + x = 0,$$

$$- 2\frac{dx}{dt} + 2\frac{dy}{dt} - 3y = 0,$$

given that $x = 0$, and $y = a$ when $t = 0$. [L.U.]

18. A double pendulum is performing oscillations under gravity. The angles θ and ϕ made with the vertical at time t by the two parts of the pendulum satisfy the equations:

$$5a\frac{d^2\theta}{dt^2} + 12a\frac{d^2\phi}{dt^2} + 6g\theta = 0;$$

$$5a\frac{d^2\theta}{dt^2} + 16a\frac{d^2\phi}{dt^2} + 6g\phi = 0.$$

Find θ and ϕ in terms of t, given that $\theta = \frac{7\alpha}{4}$, $\phi = \alpha$, $\frac{d\theta}{dt} = 0 = \frac{d\phi}{dt}$, when $t = 0$. [L.U.]

[Eliminate, say, θ and, using the given conditions, obtain

$$\phi = A \cos lt + B \sin lt + (\alpha - A) \cos mt - \frac{B}{\sqrt{20}} \sin mt,$$

where
$$l = \sqrt{\frac{3g}{10a}}, \quad m = \sqrt{\frac{6g}{a}}.$$

From the given differential equations, by subtraction obtain

$$4a\frac{d^2\phi}{dt^2} + 6g\phi = 6g\theta.$$

This gives θ in terms of A, B. Finally, A $= \frac{5}{4}\alpha$, B $= 0$,

$$\theta = \alpha \cos lt + 3\frac{\alpha}{4} \cos mt,$$

$$\phi = \frac{5}{4}\alpha \cos lt - \frac{\alpha}{4} \cos mt \Big].$$

19. A mass M_2 hangs from a fixed support by a spring of stiffness k_2 N/m. From M_2 is suspended another mass M_1 by means of a spring of stiffness k_1 N/m. If x is the displacement of M_1 and y of M_2 measured in each case from its position of statical equilibrium, show that the equations of motion are:

$$M_1 \ddot{x} = - k_1(x - y);$$

$$M_2 \ddot{y} = k_1(x - y) - k_2 y.$$

By assuming $x = a_1 \sin \omega t$, $y = a_2 \sin \omega t$, find the periods of vibration when $M_1 = 3$ kg, $M_2 = 18$ kg, $k_1 = 18$, $k_2 = 108$.

20. An axle free to rotate horizontally carries a heavy fly wheel at each end of moments of inertia, J_1, J_2 respectively. If each receives a displacement θ_1, θ_2 respectively in the same sense, show that, if C is the torsional stiffness of the shaft, the equations of rotation are:

$$J_1 \ddot{\theta}_1 = - C(\theta_1 - \theta_2),$$

$$J_2 \ddot{\theta}_2 = - C(\theta_2 - \theta_1).$$

Find the period of the free oscillation and show that the flywheels move in opposite directions throughout the motion.

21. A shaft of stiffness C_1 Nm/rad fixed at one end carries a rotor of moment of inertia I_1. Fastened to this and in line with the shaft is another shaft of stiffness C_2 Nm/rad carrying a rotor of moment of inertia I_2. If the rotor I_1 is acted on by a torque F, show that, θ_1, θ_2 being the respective angular displacements,

$$F - C_1 \theta_1 + C_2(\theta_2 - \theta_1) = I_1 \frac{d^2\theta_1}{dt^2},$$

$$- C_2(\theta_2 - \theta_1) = I_2 \frac{d^2\theta_2}{dt^2}.$$

When $F = F_0 \sin pt$ assume $\theta_1 = A \sin pt$, $\theta_2 = B \sin pt$ and obtain the equations giving A, B the amplitudes of the forced vibrations.

22. The currents i_1, i_2 in the primary and secondary windings of a transformer are given by

$$L \frac{di_1}{dt} + M \frac{di_2}{dt} - R i_1 = E e^{jpt}$$

$$M \frac{di_1}{dt} + N \frac{di_2}{dt} - S i_2 = 0$$

where L, M, N, R, S, E and p are given constants and t is the time. If for sufficiently large values of t it can be assumed that $i_1 = A e^{jpt}$, $i_2 = B e^{jpt}$ where A and B are complex constants, show that

$$\left| \frac{B}{A} \right| = \frac{Mp}{\sqrt{(S^2 + N^2 p^2)}}$$

Find A, B when $R = 0$ and $2LN = M^2$. [L.U.]

23. A mass m is supported on a horizontal platform to which it is attached by a spring of stiffness λ, and its vibration is damped by a damper which applies a force kv when the velocity of the mass relative to the platform is v. If the platform oscillates horizontally its displacement at any time being y, while the displacement of the mass at the same time is x, show that

$$m\ddot{x} + k\dot{x} + \lambda x = k\dot{y} + \lambda y.$$

If $y = a \sin pt$, determine the amplitude of the steady oscillation of the mass and show that it attains its maximum value when

$$p^2 = \lambda^2 \{\sqrt{(1 + 2k^2/m\lambda)} - 1\}/k^2. \qquad [L.U.]$$

24. An induction coil has two equal coils each resistance R, inductance L, the mutual inductance being M. At $t = 0$, an e.m.f. E_0 is applied to the primary coil. Obtain the equations

$$(LD + R)i_1 + MDi_2 = E$$

$$MDi_1 + (LD + R)i_2 = 0$$

for i_1, i_2 the currents in the primary and secondary. Prove that i_2 will reach its maximum value after a time

$$\frac{L^2 - M^2}{2MR} \log \left(\frac{L + M}{L - M}\right)$$

and that this maximum current is

$$\frac{E}{2R} \left\{ \left(\frac{L - M}{L + M}\right)^{\frac{L + M}{2M}} - \left(\frac{L - M}{L + M}\right)^{\frac{L - M}{2M}} \right\}.$$

25. By using $z = u + jv$ or otherwise solve the simultaneous equations

$$m \frac{du}{dt} = eE - evH,$$

$$m \frac{dv}{dt} = euH,$$

where m, e, E, H are constants.

If $u = dx/dt$, $v = dy/dt$, find x and y as functions of t the time, from the solution in z, and show further that if

$$x = y = u = v = 0 \text{ when } t = 0,$$

$$x = \frac{E}{\omega H} (1 - \cos \omega t) \qquad y = \frac{E}{\omega H} (\omega t - \sin \omega t)$$

where $\omega = \dfrac{eH}{m}$.

[L.U.]

26(a). Two exactly similar solenoids are wrapped on the same core. The terminals of the primary circuit are connected to the plates of a condenser of capacity C, whilst those of the secondary are joined by a wire of negligible resistance and inductance. Show that if R is the resistance and L the self-induction of either solenoid (and also the mutual induction) and the condenser has an initial charge, the charge Q on the positive plate of the condenser and the current i in the secondary circuit at time t are given by

$$L \frac{d^2Q}{dt^2} - L \frac{di}{dt} - Ri = 0,$$

$$CR \frac{dQ}{dt} + CRi + Q = 0.$$

Show that the discharge is oscillatory if

$$(3 - 2\sqrt{2})L < CR^2 < (3 + 2\sqrt{2})L.$$

(b) Two points A, B are connected by two circuits; the first contains a coil of resistance R and self induction L, the second contains an equal coil connected in series with a condenser of capacity C. The coefficient of mutual induction between the circuits is M and between A and B an alternating potential difference of frequency $p/2\pi$ is maintained. If the phase difference between the currents in the two circuits is $\pi/2$, prove that

$$C\{R^2 + (L - M)^2 p^2\} = L - M.$$

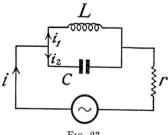

FIG. 23.

27. An alternating e.m.f. $E_0 \sin \omega t$ is applied to the choke circuit shown. Show that the currents i_1 and i_2 are given by:

$$L\frac{di_1}{dt} + ri_1 + ri_2 = E_0 \sin \omega t,$$

$$LC\frac{d^2i_1}{dt^2} - i_2 = 0.$$

Show that the current consists of a transient part, if $r > \sqrt{(L/4C)}$, and its frequency is

$$\frac{1}{2\pi}\sqrt{\left(\frac{4r^2C - L}{4Lr^2C^2}\right)}$$

and a steady part of frequency $2\pi/\omega$. Show that the impedance of the whole circuit to the current $(i_1 + i_2)$ is a maximum when $\omega = 1/\sqrt{LC}$.

28. Find a power series solution to

$$\frac{dy}{dx} - xy = 1 + x$$

and verify the solution by solving the equation in another way.

29. Find a power series solution of

$$4xy_2 + 2y_1 - y = 0$$

and show that only one solution is obtained. Then assume the more general form of solution and find the two solutions.

30. Find a power series solution of

$$y_2 - xy_1 + y = 0.$$

31. Find the values of c so that $y = \overset{\infty}{\underset{0}{\Sigma}} a_r x^{r+c}$ may be a solution of

$$(2x - x^2)y_2 + (5 - 7x)y_1 - 3y = 0.$$

Show that one of the resultant series reduces to a single term and find the other series as far as the term in x^4.

32. Find the general solution to the differential equation

$$3xy_2 + y_1 + y = 0.$$

33. Solve by a series

$$\frac{d^2y}{dx^2} + x\frac{dy}{dx} + y = 0.$$

34. Find a solution in series of

$$\frac{d^2y}{dx^2} + (x - 1)y = 0.$$

CHAPTER 10

LAPLACE TRANSFORMS

" The modern form of this (Heaviside) operational calculus consists of the use of the Laplace transformation. This is a mathematical procedure which not only yields the rules of the operational calculus in a straightforward manner but which demonstrates at the same time conditions under which the rules are valid. In addition to this, the theory of the Laplace transformation introduces a large number of additional rules and methods that are important in the analysis of problems in engineering and physics."

R. V. Churchill, *Modern Operational Mathematics in Engineering.*

10.1. In this chapter we will consider an alternative method for the solution of the linear differential equations discussed in the previous chapter. Instead of operational formulae, all that will be needed is a knowledge of partial fractions and the ability to use a *table of transforms*.

10.2. Definition of the Laplace Transformation

Given a function of the time $F(t)$ multiply it by e^{-st} (where s is a parameter, *i.e.*, a constant introduced by us) and integrate the product between zero and infinity. The result, if it exists, is denoted by $L\{F(t)\}$ and is called the Laplace transform of $F(t)$. In symbols

$$L\{F(t)\} = \int_0^\infty e^{-st}F(t)dt.$$

Examples.

(a)
$$L\{1\} = \int_0^\infty e^{-st}.1.dt$$

$$= \frac{e^{-st}}{-s}\Big]_0^\infty = \frac{1}{s} \quad . \quad . \quad . \quad . \quad . \quad (1)$$

provided $s > 0$, so that the value of the exponential is zero when $t = \infty$. (It will be seen that in our applications this restriction is no handicap.)

(b)
$$L\{e^{at}\} = \int_0^\infty e^{-st}.e^{at}dt$$

$$= \int_0^\infty e^{-(s-a)t}dt$$

$$= \frac{1}{s-a} \quad . \quad . \quad . \quad . \quad . \quad . \quad (2)$$

provided $s > a$.

(c) $$L\{\sin at\} = \int_0^\infty e^{-st} \sin at\, dt$$

[see Vol. I (19.6)] $$= \left[\frac{e^{-st}}{s^2 + a^2}(-s\sin at - a\cos at)\right]_0^\infty$$

$$= \frac{a}{s^2 + a^2} \quad \cdot \quad \cdot \quad \cdot \quad \cdot \quad \cdot \quad \cdot \quad \cdot \quad (3)$$

(d) We can similarly find $L\{\cos at\}$. An alternative way is :

$$L\{\cos at\} = \int_0^\infty e^{-st}\left(\frac{e^{jat} + e^{-jat}}{2}\right) dt,$$

and, assuming that we can proceed as usual even when the index is complex, by (2) this is

$$\frac{1}{2}\left(\frac{1}{s - ja} + \frac{1}{s + ja}\right) = \frac{s}{s^2 + a^2} \quad \cdot \quad \cdot \quad \cdot \quad (4)$$

(e) $$L\{t\} = \int_0^\infty e^{-st}t\,dt$$

$$= \left(t\,\frac{e^{-st}}{-s}\right)_0^\infty - \int_0^\infty \frac{e^{-st}}{-s}\,dt$$

$$= \frac{1}{s^2} \quad \cdot \quad \cdot \quad \cdot \quad \cdot \quad \cdot \quad \cdot \quad \cdot \quad \cdot \quad (5)$$

since for $s > 0$ $\underset{t \to \infty}{\mathcal{L}t.}\ (te^{-st}) = 0$ [Vol. I (19.6)]

We now have

$$\frac{1}{s^2} = \int_0^\infty e^{-st}t\,dt.$$

Differentiating both sides with respect to s,

$$-\frac{2}{s^3} = \frac{d}{ds}\int_0^\infty e^{-st}t\,dt$$

$$= \int_0^\infty \frac{\partial}{\partial s}(e^{-st})t\,dt \qquad \text{(see 5.3)}$$

$$= \int_0^\infty -te^{-st}t\,dt,$$

so that $$\frac{2}{s^3} = \int_0^\infty e^{-st}t^2\,dt,$$

giving the transform of t^2,

Differentiating again,

$$\frac{3!}{s^4} = \int_0^\infty e^{-st}t^3 dt,$$

and finally we have

$$\frac{n!}{s^{n+1}} = \int_0^\infty e^{-st}t^n dt = L\{t^n\} \quad \cdot \quad \cdot \quad \cdot \quad \cdot \quad (6)$$

when n is a positive integer.

The student should, by the above methods, obtain

$$L\{\sinh at\} = \frac{a}{s^2 - a^2} \quad \cdot \quad \cdot \quad \cdot \quad (7)$$

$$L\{\cosh at\} = \frac{s}{s^2 - a^2} \quad \cdot \quad \cdot \quad \cdot \quad (8)$$

$$L\left\{\frac{t}{2a} \sin at\right\} = \frac{s}{(s^2 + a^2)^2} \quad \cdot \quad \cdot \quad (9)$$

$$L\left\{\frac{1}{2a^3} (\sin at - at \cos at)\right\} = \frac{1}{(s^2 + a^2)^2} \quad \cdot \quad \cdot \quad (10)$$

These transforms, and others, constitute the *table of transforms* to which reference will be made.

They may be generalized by means of the following :

Theorem

If $f(s)$ denotes the transform of $F(t)$ then $f(s + a)$ is the transform of $e^{-at}F(t)$.

The proof is clear, since if

$$f(s) = \int_0^\infty e^{-st}F(t)dt,$$

$$f(s + a) = \int_0^\infty e^{-(s + a)t}F(t)dt$$

$$= \int_0^\infty e^{-st}\{e^{-at}F(t)\}at,$$

so that $f(s + a)$ is the transform of $e^{-at}F(t)$.

Immediate applications of this are :

(*a*) since $\qquad\qquad L\{t^n\} = \dfrac{n!}{s^{n+1}}$

therefore $\qquad\qquad L\{e^{-at}t^n\} = \dfrac{n!}{(s + a)^{n+1}} \quad \cdot \quad \cdot \quad \cdot \quad \cdot \quad (11)$

(*b*) since $\qquad\qquad L\{\cos at\} = \dfrac{s}{s^2 + a^2}$

therefore $\qquad\qquad L\{e^{-bt} \cos at\} = \dfrac{s + b}{(s + b)^2 + a^2} \quad \cdot \quad \cdot \quad \cdot \quad (12)$

E.g., later when dealing with an expression in *s* such as

$$\frac{s-3}{s^2+4s+13}$$

we will arrange this as

$$\frac{s-3}{(s+2)^2+3^2} = \frac{s+2}{(s+2)^2+3^2} - \frac{5}{(s+2)^2+3^2},$$

and from (12) and (3) above this is

$$= L\{e^{-2t}\cos 3t\} - L\{\tfrac{5}{3}\,e^{-2t}\sin 3t\}.$$

10.3. The Transforms of Derivatives

From the definition, if $F'(t)$ denotes the derivative of $F(t)$

$$L\{F'(t)\} = \int_0^\infty e^{-st}F'(t)dt.$$

Integrating by parts,

$$= e^{-st}F(t)\Big]_0^\infty - \int_0^\infty - se^{-st}F(t)dt.$$

For the functions with which we deal $\underset{t\to\infty}{\mathcal{L}t.}\ \{e^{-st}F(t)\} = 0$ so that the above then gives, with $f(s)$ denoting the transform of $F(t)$,

$$L\{F'(t)\} = -F(0) + sf(s) \qquad \cdot \quad \cdot \quad \cdot \quad \cdot \quad \cdot \quad (1)$$

Similarly, substituting $F'(t)$ for $F(t)$

$$L\{F''(t)\} = -F'(0) + sL\{F'(t)\}$$
$$= -F'(0) + s[-F(0) + sf(s)]$$

by using (1).

$$= s^2f(s) - sF(0) - F'(0).$$

Continuing with this we find

$$L\{F'''(t)\} = s^3f(s) - s^2F(0) - sF'(0) - F''(0)$$

and generally

$$L\{F^n(t)\} = s^nf(s) - s^{n-1}F(0) - s^{n-2}F'(0) \ldots - F^{(n-1)}(0).$$

These formulae are fundamental in the solution of differential equations, and we will repeat them using the usual notation employed in differential equations.

When $t = 0$ let $x = x_0$, $\dfrac{dx}{dt} = x_1$, and $\dfrac{d^nx}{dt^n} = x_n$,

then denoting the transform of x by $\bar{x}$, we have :

$$L\{x\} = \bar{x},$$

$$L\left\{\frac{dx}{dt}\right\} = s\bar{x} - x_0,$$

$$L\left\{\frac{d^2x}{dt^2}\right\} = s^2\bar{x} - sx_0 - x_1,$$

$$L\left\{\frac{d^3x}{dt^3}\right\} = s^3\bar{x} - s^2x_0 - sx_1 - x_2,$$

and generally

$$L\left\{\frac{d^n x}{dt^n}\right\} = s^n \bar{x} - s^{n-1} x_0 - s^{n-2} x_1 \ldots - s x_{n-2} - x_{n-1}.$$

10.4. The Inverse

If $\qquad\qquad L\{F(t)\} = f(s)$

we write $\qquad\qquad F(t) = L^{-1}\{f(s)\},$

so that, for example,

$$e^{at} = L^{-1}\left\{\frac{1}{s-a}\right\}.$$

As far as our applications are concerned, the inverse relation is unique, *i.e.*, given a function of t it has a definite transform and given a function of s it has a definite inverse or was derived from a unique finite function of t, as shown in a complete table of transforms.

10.5. Applications to Differential Equations

Example 1. To solve

$$L\frac{di}{dt} + Ri = E_0 \quad \text{(given that } i = 0 \text{ when } t = 0\text{)}.$$

In this case L denotes inductance, so using $\mathcal{L}$ to denote the operation of taking the transform, we have by taking the transform of the equation

$$\mathcal{L}\left(L\frac{di}{dt}\right) + \mathcal{L}(Ri) = \mathcal{L}(E_0),$$

which means that we have multiplied each term in the equation by e^{-st} and integrated from 0 to ∞ .

Since L is a constant independent of t, the first term gives

$$L\mathcal{L}\left(\frac{di}{dt}\right) = L(s\bar{i} - i_0)$$

by using the formula for the transform of a derivative, $\bar{i}$ denoting the transform of i and i_0 the value of i at $t = 0$,

$$\mathcal{L}(Ri) = R\mathcal{L}(i) = R\bar{i},$$

$$\mathcal{L}(E_0) = E_0\mathcal{L}(1) = E_0\frac{1}{s}.$$

The transformed equation becomes

$$L(s\bar{i} - i_0) + R\bar{i} = E_0/s,$$

which, after a little practice, can be written down directly from the given differential equation.

Since $i_0 = 0$, solving for $\bar{i}$

$$\bar{i} = \frac{E_0}{s(Ls + R)}$$

$$= \frac{E_0}{R}\left(\frac{1}{s} - \frac{1}{s + R/L}\right)$$

by putting the expression for s into partial fractions.   Taking the inverse of each term,

$$\underline{i = \frac{E_0}{R}\left(1 - e^{-Rt/L}\right)}$$

(by using numbers (1) and (2) in the table of transforms).

Although other examples will generally involve more work with the partial fractions section, the above example contains the essence of the matter. (1) Apply the formula of 10.3 to the differential equation, inserting the initial conditions. (2) Solve the transform involved for $\bar{x}$ or $\bar{\imath}$, etc. (3) Rearrange the expression by partial fractions to give the standard forms shown in the table of transforms. (4) Take the inverse of each term to obtain the solution of the differential equation.

Example 2. To solve $(D^2 - 3D + 2)x = 0$, given that when $t = 0$, $x = 4$ and $dx/dt = 3$.

Taking the transform:
$$(s^2\bar{x} - sx_0 - x_1) - 3(s\bar{x} - x_0) + 2\bar{x} = 0.$$
Inserting the given values for x_0 and x_1:
$$(s^2\bar{x} - 4s - 3) - 3(s\bar{x} - 4) + 2\bar{x} = 0.$$
Solving for $\bar{x}$,
$$\bar{x} = \frac{4s - 9}{(s - 2)(s - 1)}$$
$$= \frac{5}{s - 1} - \frac{1}{s - 2}$$
therefore
$$x = \underline{5e^t - e^{2t}}.$$

Example 3. To solve $\qquad (D^2 - 3D + 2)x = e^{2t}$.

Since initial conditions are not given, we must work with the constants x_0, x_1. In the usual way
$$(s^2\bar{x} - sx_0 - x_1) - 3(s\bar{x} - x_0) + 2\bar{x} = \frac{1}{s - 2},$$
$$\bar{x}(s^2 - 3s + 2) = \frac{1}{s - 2} + (sx_0 + x_1 - 3x_0),$$
$$\bar{x} = \frac{1 + x_0 s(s - 2) + (x_1 - 3x_0)(s - 2)}{(s - 1)(s - 2)^2}$$
$$= \frac{(1 + 2x_0 - x_1)}{s - 1} + \frac{A}{s - 2} + \frac{1}{(s - 2)^2}.$$

The numerators of the first and third have been found by the " cover-up " method [Vol. I (4.1)]. Multiplying up
$$1 + x_0 s(s - 2) + (x_1 - 3x_0)(s - 2) = (1 + 2x_0 - x_1)(s - 2)^2$$
$$+ A(s - 1)(s - 2) + (s - 1)$$
Equating coefficients of x^2
$$x_0 = (1 + 2x_0 - x_1) + A.$$
We now have
$$\bar{x} = \frac{(1 + 2x_0 - x_1)}{s - 1} + \frac{(x_1 - x_0 - 1)}{s - 2} + \frac{1}{(s - 2)^2}.$$
Using transforms numbered (2) and (11)
$$x = \underline{(1 + 2x_0 - x_1)e^t + (x_1 - x_0 - 1)e^{2t} + te^{2t}}.$$

Example 4. To solve $\qquad (D^2 + \omega^2)x = \cos nt$

with $x = x_0$, $Dx = x_1$ when $t = 0$.

The transformed equation is
$$(s^2\bar{x} - sx_0 - x_1) + \omega^2\bar{x} = \frac{s}{s^2 + n^2},$$
$$\bar{x} = \frac{s}{(s^2 + n^2)(s^2 + \omega^2)} + \frac{sx_0 + x_1}{s^2 + \omega^2}$$
$$= \frac{1}{\omega^2 - n^2}\left(\frac{s}{s^2 + n^2} - \frac{s}{s^2 + \omega^2}\right) + \frac{sx_0 + x_1}{s^2 + \omega^2}.$$

Using transforms (3), (4)

$$x = \frac{1}{\omega^2 - n^2} (\cos nt - \cos \omega t) + x_0 \cos \omega t + \frac{x_1}{\omega} \sin \omega t.$$

This differential equation arises when we consider the motion caused by a periodic force acting on a mass hanging from a spring. When $n \neq \omega$, i.e., the period of the force is not equal to the "natural" period of vibration of the mass on its spring we see that the result is a compound of two vibrations. It is evident that as $n \to \omega$ the factor $1/(\omega^2 - n^2)$ becomes very large, and hence the amplitude x will be dangerously large. We could evaluate the indeterminate form

$$\lim_{n \to \omega} \left[\frac{1}{\omega^2 - n^2} (\cos nt - \cos \omega t) \right]$$

as in Chapter 1, to find what happens when $n = \omega$. The next example considers this problem independently.

Example 5. To solve $\qquad (D^2 + n^2)x = \cos nt$
with $x = x_0$, $Dx = x_1$ when $t = 0$.
   The transformed equation is

$$(s^2\bar{x} - sx_0 - x_1) + n^2\bar{x} = \frac{s}{s^2 + n^2},$$

$$\bar{x} = \frac{s}{(s^2 + n^2)^2} + \frac{sx_0 + x_1}{(s^2 + n^2)}.$$

Using transforms (9), (4), (3),

$$x = \frac{t}{2n} \sin nt + x_0 \cos nt + \frac{x_1}{n} \sin nt.$$

The first term shows that the amplitude will increase to infinity with the time t (see 8.6).

Example 6. To solve the simultaneous equations

$$\frac{dx}{dt} + 4x + 5y = 7 \cos 2t,$$

$$\frac{dy}{dt} - 4y - 5x = 8 \cos 2t,$$

given that when $t = 0$, $x = y = 0$.
   The transformed equations are:

$$s\bar{x} + 4\bar{x} + 5\bar{y} = \frac{7s}{s^2 + 4},$$

$$s\bar{y} - 4\bar{y} - 5\bar{x} = \frac{8s}{s^2 + 4}.$$

These are:

$$\bar{x}(s + 4) + 5\bar{y} = 7s/(s^2 + 4),$$
$$- 5\bar{x} + \bar{y}(s - 4) = 8s/(s^2 + 4).$$

Eliminating $\bar{x}$ gives

$$\bar{y} = \frac{s(8s + 67)}{(s^2 + 9)(s^2 + 4)}.$$

Noting that $s^2 + 9 = (s - 3j)(s + 3j)$, this

$$= -\frac{(67 + 24j)}{10(s - 3j)} - \frac{(67 - 24j)}{10(s + 3j)} + \frac{(67 + 16j)}{10(s - 2j)} + \frac{(67 - 16j)}{10(s + 2j)},$$

where each numerator has been obtained by the cover-up method, although the second and fourth may be obtained from their conjugates by changing the sign of j. Combining these in pairs

$$\bar{y} = -\frac{67s + 72}{5(s^2 + 9)} + \frac{67s - 32}{5(s^2 + 4)},$$

from which

$$y = -\tfrac{67}{5}\cos 3t + \tfrac{24}{5}\sin 3t + \tfrac{67}{5}\cos 2t - \tfrac{16}{5}\sin 2t.$$

In some cases it is easier to assume

$$\frac{s(8s + 67)}{(s^2 + 9)(s^2 + 4)} = \frac{As + B}{s^2 + 9} + \frac{Cs + D}{s^2 + 4},$$

so that

$$s(8s + 67) = (As + B)(s^2 + 4) + (Cs + D)(s^2 + 9).$$

Equating coefficients:

s^3: $\qquad\qquad\qquad 0 = A + C$

s^2: $\qquad\qquad\qquad 8 = B + D$

s^1: $\qquad\qquad\qquad 67 = 4A + 9C$

s^0: $\qquad\qquad\qquad 0 = 4B + 9D.$

$$\therefore \quad C = \tfrac{67}{5} = -A, \ D = -\tfrac{32}{5}, \ B = \tfrac{72}{5}$$

which are the coefficients found above.

We can now eliminate $\bar{y}$ from the given differential equations and proceed similarly with $\bar{x}$. But in this case the second differential equation gives

$$5x = \frac{dy}{dt} - 4y - 8\cos 2t$$

$$= (\tfrac{201}{5}\sin 3t + \tfrac{72}{5}\cos 3t - \tfrac{134}{5}\sin 2t - \tfrac{32}{5}\cos 2t)$$
$$\qquad - 4(-\tfrac{67}{5}\cos 3t + \tfrac{24}{5}\sin 3t + \tfrac{67}{5}\cos 2t - \tfrac{16}{5}\sin 2t) - 8\cos 2t$$

$$= 21\sin 3t + 68\cos 3t - 14\sin 2t - 68\cos 2t,$$

$$x = \tfrac{68}{5}\cos 3t + \tfrac{21}{5}\sin 3t - \tfrac{68}{5}\cos 2t - \tfrac{14}{5}\sin 2t.$$

Note: In all the above examples the initial conditions have been given at the time $t = 0$ but this is not really necessary.

As an example to solve

$$\frac{d^2x}{dt^2} + 3\frac{dx}{dt} + 2x = \cos 4t$$

given that $x = 1$, $\dfrac{dx}{dt} = 4$ when $t = 2$, we proceed by putting $T = t - 2$. Since $dt = dT$ and $(dt)^2 = (dT)^2$ the equation becomes

$$\frac{d^2x}{dT^2} + 3\frac{dx}{dT} + 2x = \cos 4(T + 2)$$

with $x = 1$, $\dfrac{dx}{dT} = 4$ when $T = 0$ and we proceed with the solution in the usual manner. The alternative is to solve the original equation in terms of the unknown constants x_0, x_1 which are brought in when its transform is found. From the solution of x in terms of x_0, x_1 and t the time, $\dfrac{dx}{dt}$ is found and x_0, x_1 determined by using the initial data.

EXERCISE 20

Show, by using the table of transforms, that:

1. $L(a + bt + ct^2) = \dfrac{1}{s^3}(as^2 + bs + 2c).$

2. $L(e^{at} - e^{bt}) = (a - b)/(s - a)(s - b).$

3. $L\left(\dfrac{1}{a}\sin at - \dfrac{1}{b}\sin bt\right) = (b^2 - a^2)/(s^2 + a^2)(s^2 + b^2)$.

4. $L(\cos at - \cos bt) = (b^2 - a^2)s/(s^2 + a^2)(s^2 + b^2)$.

5. $L(\operatorname{ch} at - \operatorname{ch} bt) = (a^2 - b^2)s/(s^2 - a^2)(s^2 - b^2)$.

6. $L\{\cos(at + \alpha)\} = \dfrac{s}{s^2 + a^2}\cos\alpha - \dfrac{a}{s^2 + a^2}\sin\alpha$.

7. $L\{\sin(bt - \alpha)\} = \dfrac{b}{s^2 + b^2}\cos\alpha - \dfrac{s}{s^2 + b^2}\sin\alpha$.

8. $L\{e^{-2t}(\cos t + 6\sin t)\} = \dfrac{s + 8}{s^2 + 4s + 5}$.

9. Using the formula on $F(t) = \sin t$, of
$$L\{F''(t)\} = s^2 f(s) - sF(0) - F'(0),$$
find the transform of $\sin t$. Find similarly that of $\cos t$.

10. Using the general formula for the transform of a derivative applied to $F^{n+1}(t)$ with $F(t) = t^n$, find the transform of t^n where n is a positive integer.

11. Prove that if $0 < \lambda < p$
$$\int_0^\infty e^{-px}e^{\lambda x}dx = \frac{1}{p - \lambda}.$$

Hence or otherwise prove that
$$\int_0^\infty e^{-px}\cos ax\,dx = \frac{p}{p^2 + a^2}, \quad \int_0^\infty e^{-px}\sin ax\,dx = \frac{a}{p^2 + a^2}.$$

By differentiating with respect to a, deduce that
$$\int_0^\infty e^{-px}x\sin ax\,dx = \frac{2ap}{(p^2 + a^2)^2}, \quad \int_0^\infty e^{-px}x\cos ax\,dx = \frac{p^2 - a^2}{(p^2 + a^2)^2}.$$

[L.U.]

12. In example (6) above we had to put into partial fractions
$$s(8s + 67)/(s^2 + 9)(s^2 + 4).$$

Assume
$$\frac{s(8s + 67)}{(s^2 + 9)(s^2 + 4)} = \frac{As + B}{s^2 + 9} + \frac{Cs + D}{s^2 + 4}.$$

Hence
$$\frac{s(8s + 67)}{s^2 + 9} = \frac{(As + B)(s^2 + 4)}{s^2 + 9} + Cs + D,$$

put $s = 2j$:
$$\frac{2j(16j + 67)}{5} = 2jC + D.$$

Hence obtain C and D. Repeat this to find A and B.

13. If $h(p) = \displaystyle\int_0^\infty e^{-pt}f(t)dt$ find $h(p)$

if $\qquad$ (i) $f(t) = e^{-at}$; (ii) $f(t) = t$.

Show that if $f(t)$ is finite at infinity
$$\int_0^\infty e^{-pt}f'(t)dt = ph(p) - f(0).$$

If $f'(t) + 2f(t) = e^{-t}$ and $f(0) = 0$ show that
$$h(p) = 1/(p + 1) - 1/(p + 2)$$
and by inverting the result (i) determine $f(t)$ in this case.

[L.U.]

Find the Inverses of 14–27.

14. $1/s(s + 1)(s + 2)$.   15. $1/s(s^2 + a^2)$.   16. $1/s^2(s^2 + a^2)$.   17. $1/s^2(s + b)$.

18. $s/(s + a)(s + b)$.   19. $1/(s + a)^2$.   20. $s/(s + a)^2$.

21. $1/(s^2 + a^2)^2$.   22. $s/(s^2 + a^2)^2$.   23. $s^2/(s^2 + a^2)^2$.

24. $(s^2 - a^2)/(s^2 + a^2)^2$.   25. $1/s^2(s^2 + 1)(s^2 + 4)$.   26. $3a^2/(s^3 + a^3)$.

27. $1/(s^4 - a^4)$.

28. Show that

$$L\{\sin (at) \text{ ch } (at) - \cos (at) \text{ sh } (at)\} = \frac{4a^3}{s^4 + 4a^4}.$$

29. Show that

$$\frac{2s + 1}{(s^2 + 6s + 13)^2} = \frac{2(s + 3) - 5}{[(s + 3)^2 + 4]^2}$$

and hence using transforms (9), (10) that its inverse is

$$\tfrac{1}{2}te^{-3t} \sin 2t - \tfrac{5}{16}e^{-3t}(\sin 2t - 2t \cos 2t).$$

Solve the differential equations 30–48.

30. $R\dfrac{dq}{dt} + \dfrac{q}{C} = 0$ given $q = EC$ when $t = 0$.

31. $\dfrac{dq}{dt} + \dfrac{1}{RC}q = \dfrac{E}{R}$ given that $q = 0$ when $t = 0$.

32. $\dfrac{dT}{d\theta} - \mu T = 0$ given that $T = T_0$ when $\theta = 0$.   (There is no need for the independent variable to be t, the time.)

In the following, the conditions are all given at $t = 0$ except where otherwise stated.

33. $4\dfrac{di}{dt} + 2i = 100 \cos t$, given that $i = 0$.

34. $(D^2 + D - 20)x = 0$, given that $x = 1$, $Dx = 2$.

35. $\dfrac{d^2x}{dt^2} = 25x$, given that $x = 2$, $dx/dt = 1$.

36. $\dfrac{d^2r}{d\theta^2} + 4\dfrac{dr}{d\theta} + 3r = 0$, $r = 2$, $\dfrac{dr}{d\theta} = 3$ when $\theta = 0$.

37. $(D^2 + 4)x = 0$, $x = 0$, $\dot{x} = 1$.

38. $(D^2 + D + 1)x = 0$, $x = 2$, $\dot{x} = -1$.

39. $(D^2 + 6D + 9)x = 0$, $x = 2$, $\dot{x} = 4$.

40. $(D^2 + 9x) = t$, $x = 1$, $\dot{x} = 2$.

41. $(D^2 + 4)x = 5 \cos 2t$, $x = 1$, $\dot{x} = 3$.

42. $(D^2 + 6D + 9)x = 4e^{2t}$, $x = 0$, $\dot{x} = 0$.

43. $(D^2 + 6D + 9)x = e^{3t}$, $x = 1$, $\dot{x} = -2$.

44. $\left.\begin{array}{l} (3D + 2)x + Dy = 1 \\ Dx + (4D + 3)y = 0 \end{array}\right\}$ given $x = 0 = y$ at $t = 0$.

45. $\left.\begin{array}{l} 2\dfrac{dx}{dt} + 2x + \dfrac{dy}{dt} = \cos t \\[2mm] \dfrac{dx}{dt} + 2\dfrac{dy}{dt} + y = 0 \end{array}\right\}$ $x = 0 = y$ at $t = 0$.

46. $\left.\begin{array}{l} \dot{x} + 2x - 3y = t \\ \dot{y} - 3x + 2y = e^{2t} \end{array}\right\}$ $x = 1$, $y = 2$ at $t = 0$.

47. $\left.\begin{array}{l} \dot{x} = x + y + e^{-t} \\ \dot{y} = 3x - y \end{array}\right\}$ $y = -1$, $x = 4$ at $t = 0$.

48. $\left.\begin{array}{l} \ddot{x} - 3x - 4y = 0 \\ \ddot{y} + x + y = 0 \end{array}\right\}$.

49. An e.m.f. of $10 \cos t$ is applied to the primary of a transformer network which is dead at $t = 0$. The constants are such that the currents i_1, i_2 in the primary and secondary are given by

$$2\frac{di_1}{dt} - \frac{di_2}{dt} + i_1 = 10 \cos t$$

$$2\frac{di_2}{dt} - \frac{di_1}{dt} + i_2 = 0.$$

Find i_1 and i_2 at time t.

50. Solve the simultaneous differential equations

$$\frac{dx}{dt} + \frac{dy}{dt} + 2x + y = e^{-3t}$$

$$\frac{dy}{dt} + 5x + 3y = 5e^{-2t}$$

given that when $t = 0$, $x = -1$ and $y = 4$. [L.U.]

51. If $L\frac{dx}{dt} - Ry = E$, $RCy + q = 0$, $x + y = \frac{dq}{dt}$

where L, R, C, E are constants, find the differential equation relating q and t and show that q oscillates only if $L < 4CR^2$. If this condition is satisfied and if $q = x = 0$ at $t = 0$, show that

$$q = EC - \frac{E}{2Rp} e^{(-t/2RC)}\{\sin pt + 2RCp \cos pt\},$$

where

$$p^2 = \frac{1}{LC} - \frac{1}{4R^2C^2}.$$ [L.U.]

52. Solve

$$5a\frac{d^2\theta}{dt^2} + 12a\frac{d^2\phi}{dt^2} + 6g\theta = 0,$$

$$5a\frac{d^2\theta}{dt^2} + 16a\frac{d^2\phi}{dt^2} + 6g\phi = 0,$$

given that $\theta = \frac{7\alpha}{4}$, $\phi = \alpha$, $\frac{d\theta}{dt} = 0 = \frac{d\phi}{dt}$ when $t = 0$. [L.U.]

10.6. The (R, L, C) Electrical Circuit

As in 9.2 we have the two equations

$$L\frac{di}{dt} + Ri + \frac{q}{C} = e \quad \cdots \quad (1)$$

$$i = \frac{dq}{dt} \quad \cdots \quad (2)$$

Taking transforms and using $\bar{i}$, $\bar{q}$ to denote the transforms and i_0, q_0 the initial values

$$(Ls + R)\bar{i} + \frac{1}{C}\bar{q} = Li_0 + \bar{e} \quad \cdots \quad (3)$$

$$\bar{q} = \frac{\bar{i}}{s} + \frac{q_0}{s} \quad \cdots \quad (4)$$

Substitution for $\bar{q}$ gives

$$\left(Ls + R + \frac{1}{Cs}\right)\bar{i} = \bar{e} + Li_0 - \frac{q_0}{Cs} \quad \cdots \quad (5)$$

the fundamental equation for $\bar{i}$, $\bar{q}$ may then be obtained from (4).

(a) If e has the constant value E_0 and $i_0 = q_0 = 0$, (5) gives

$$\left(Ls + R + \frac{1}{Cs}\right)i = \frac{E_0}{s},$$

$$i = \frac{E_0}{L[(s + \mu)^2 + n^2]}$$

where $\qquad \mu = \frac{R}{2L}, \quad n^2 = \frac{1}{LC} - \frac{R^2}{4L^2}$

From the table of transforms we find

$$i = \frac{E_0}{nL} e^{-\mu t} \sin nt \quad \text{if} \quad n^2 > 0$$

$$= \frac{E_0}{L} t e^{-\mu t} \quad \text{if} \quad n^2 = 0$$

$$= \frac{E_0}{kL} e^{-\mu t} \sinh kt \quad \text{if} \quad n^2 = -k^2 < 0.$$

(b) In the case of a condenser charged to a voltage E_0 and discharged at $t = 0$ through the (L, R) inductive resistance

$$q_0 = CE, \quad i_0 = 0, e = 0$$

and (5) becomes

$$\left(Ls + R + \frac{1}{Cs}\right)i = -\frac{E_0}{s}.$$

Therefore by (4)

$$\bar{q} = \frac{CE}{s} - \frac{CE}{s(LCs^2 + RCs + 1)}$$

$$= \frac{CE(s + 2\mu)}{(s + \mu)^2 + n^2}$$

using the abbreviations of (a) above. We may consider the three cases as above, but for the usual case of $n^2 > 0$

$$q = CEe^{-\mu t}\left\{\cos nt + \frac{\mu}{n}\sin nt\right\}$$

10.7. When dealing with electrical examples we often need to find the transform of $\frac{1}{C}\int_0^t i\,dt$. We have avoided this so far by dealing with $\frac{1}{C}q$ where $i = \frac{dq}{dt}$. A simple theorem enables us to deal directly with such a case.

Let $\qquad G(t) = \int_0^t F(T)dT,$

then $G(0) = 0$ and $G'(t) = F(t)$.

As usual

$$L\{F(t)\} = L\{G'(t)\}$$
$$= s\bar{G} - G(0)$$

or

$$f(s) = s\bar{G} \qquad . \quad . \quad . \quad . \quad . \quad (1)$$

where

$$\bar{G} = L\left\{\int_0^t F(T)dT\right\}, \quad f(s) = L\{F(t)\}.$$

Therefore if

$$f(s) \text{ is the transform of } F(t)$$

$$(\bar{G} =) \frac{1}{s}f(s) \text{ is the transform of } \int_0^t F(T)dT.$$

In particular, since

$$\frac{a}{s^2 + a^2} \text{ is the transform of } \sin at$$

$$\frac{a}{s(s^2 + a^2)} \text{ is the transform of } \int_0^t \sin at\, dt$$

$$= \frac{1}{a}\left(1 - \cos at\right)$$

and since

$$\frac{i}{C} \text{ is the transform of } \frac{i}{C},$$

$$\frac{i}{Cs} \text{ is the transform of } \frac{1}{C}\int_0^t i dt.$$

[There is no need to distinguish between t and T when the theorem has been proved and there is no risk of confusion.]

Example 8. Show that the effect of the circuits R, L and R, C in parallel as shown in the diagram is that of a resistance only, if $L = CR^2$. [L.U.]

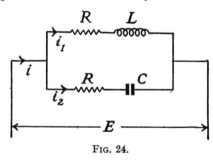

FIG. 24.

The circuit equations are:

$$Ri_1 + L\frac{di_1}{dt} = E,$$

$$Ri_2 + \frac{1}{C}\int i_2 dt = E,$$

$$i_1 + i_2 = i.$$

Taking the transform of each equation,

$$\bar{\imath}_1(R + Ls) = \bar{E},$$

$$\bar{\imath}_2\left(R + \frac{1}{Cs}\right) = \bar{E} \qquad \text{(by the above theorem)},$$

$$\bar{\imath}_1 + \bar{\imath}_2 = \bar{\imath},$$

therefore

$$\bar{\imath} = \frac{\bar{E}}{R + Ls} + \frac{\bar{E}}{R + \dfrac{1}{Cs}}$$

$$= \bar{E}\left[\frac{LCs^2 + 2CRs + 1}{CRLs^2 + (CR^2 + L)s + R}\right]$$

For a pure resistance this expression must be independent of s, *i.e.*, the numerator must be a constant factor of the denominator or

$$LCs^2 + 2CRs + 1 = k[CRLs^2 + (CR^2 + L)s + R],$$

$$\therefore \quad \frac{LC}{CRL} = \frac{2CR}{CR^2 + L} = \frac{1}{R} = k,$$

from which we obtain

$$L = CR^2.$$

10.8. Applications to Beam Problems

Example 9. Find the deflection for a uniform beam clamped horizontally at each end and carrying a uniform load w per foot.

The usual equation for such a case is

$$EID^4y = w$$

where $y = 0 = y_1$ at $x = 0$ and also at $x = l$. From the given equation, by transforming,

$$EI(s^4\bar{y} - s^3y_0 - s^2y_1 - sy_2 - y_3) = w/s,$$

or since $y_0 = y_1 = 0$ $\left(\text{note that } y_0 = (y)_{x=0} \text{ and } y_1 = \left(\dfrac{dy}{dx}\right)_{x=0}\right)$,

$$EI(s^4\bar{y} - sy_2 - y_3) = w/s,$$

$$\bar{y} = \frac{1}{s^4}\left(\frac{w}{EIs} + sy_2 + y_3\right)$$

$$= \frac{w}{EIs^5} + \frac{y_2}{s^3} + \frac{y_3}{s^4},$$

$$\therefore \quad y = \frac{w}{EI}\cdot\frac{x^4}{4!} + \frac{x^2}{2!}y_2 + \frac{x^3}{3!}y_3,$$

at $x = l$, $y = 0$:

$$0 = \frac{w}{EI}\frac{l^4}{24} + \frac{l^2}{2}y_2 + \frac{l^3}{6}y_3,$$

$$\frac{dy}{dx} = 0: \qquad 0 = \frac{w}{EI}\frac{l^3}{6} + ly_2 + \frac{l^2}{2}y_3,$$

giving

$$y_3 = -\frac{wl}{12EI}, \quad y_2 = -\frac{wl^2}{12EI}.$$

The expression for y is then

$$y = \frac{wx^2(l - x)^2}{24EI}.$$

It will be noticed that when initial values are given for values of x other than $x = 0$, we work with the constants (in this case y_2 and y_3) and find them at the end of the process.

To apply the transform method to the case of discontinuous or concentrated loads requires further theorems, which will now be considered.

10.9. Heaviside's Unit Function H(t)

The further development of the subject is helped by a new function defined as, t being the time,

$$H(t) = 0 \text{ for } t \leqslant 0,$$
$$= 1 \text{ for } t > 0.$$

On this basis $H(t - a)$ is a function which is zero for $t - a \leqslant 0$ or $t \leqslant a$, and is unity for $t - a > 0$ or $t > a$.

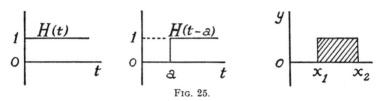

FIG. 25.

Graphically $H(t)$, $H(t - a)$ are as shown in the diagram.

Similarly, when dealing with problems on beams we use $H(x)$ where x is the distance from the origin. Thus a uniform load of w per foot from the point x_1 onwards to infinity would be represented by

$$wH(x - x_1),$$

and the load w per foot in the figure between x_1 and x_2 is represented by

$$w[H(x - x_1) - H(x - x_2)].$$

10.10. The Translation of F(t)

Instead of finding the transform of $H(t - a)$ we will deal with a more general function $F(t)$ defined for $t > 0$ and zero for negative values of t. Then clearly $F(t - a)$ is $F(t)$ translated a distance a to the right as shown.

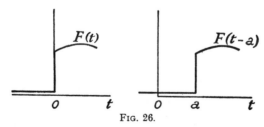

FIG. 26.

If $f(s)$ is the transform of $F(t)$ so that

$$f(s) = \int_0^\infty e^{-st}F(t)dt,$$

then the transform of $F(t - a)$ is

$$\int_0^\infty e^{-st}F(t - a)dt$$

$$= \int_0^a e^{-st} \cdot 0dt + \int_a^\infty e^{-st}F(t - a)dt.$$

The substitution $t - a = T$ changes this to

$$\int_0^\infty e^{-s(T + a)}F(T)dT = e^{-as}\int_0^\infty e^{-sT}F(T)dT$$

$$= e^{-as}f(s).$$

We therefore have the general theorem that if $f(s)$ is the transform of $F(t)$, $e^{-as}f(s)$ is the transform of the function which is $F(t)$ translated a distance a to the right, $F(t)$ being assumed zero for negative values of t. The inverses should be noted. Thus, since

$$L^{-1}\left(\frac{1}{s^2}\right) = t,$$

by the above theorem,

$$L^{-1}\left(\frac{e^{-as}}{s^2}\right) = 0 \quad (0 < t < a)$$

$$= t - a \quad (t > a).$$

Similarly,

$$L^{-1}\left(\frac{se^{-as}}{s^2 + k^2}\right) = 0 \quad (0 < t < a),$$

$$= \cos k(t - a) \quad (t > a).$$

Since the transform of $H(t)$ is

$$\int_0^\infty e^{-st} \cdot 1 \cdot dt = \frac{1}{s}$$

that of $H(t - a)$ is e^{-as}/s, and we have the important inverse

$$L^{-1}\left\{\frac{e^{-as}}{s}\right\} = H(t - a) = 0 \quad (0 < t < a)$$

$$= 1 \quad (t > a).$$

Example 10. During the time interval t_1 to t_2 seconds a constant e.m.f. E_0 acts on an R, C circuit (in series). To find the current at time t.
If i denotes the current at time t

$$Ri + \frac{1}{C}\int_0^t idt = E,$$

where the e.m.f. can be written

$$E_0[H(t - t_1) - H(t - t_2)].$$

We then have

$$Ri + \frac{1}{C}\int_0^t idt = E_0[H(t - t_1) - H(t - t_2)].$$

Taking transforms:

$$R i + \frac{1}{Cs} i = E_0 \left[\frac{e^{-st_1}}{s} - \frac{e^{-st_2}}{s} \right],$$

$$i = \frac{E_0}{R \left(s + \frac{1}{RC} \right)} [e^{-st_1} - e^{-st_2}],$$

therefore $\qquad i = \frac{E_0}{R} [e^{(t-t_1)/RC} H(t - t_1) - e^{(t-t_2)/RC} H(t - t_2)].$

This result can be written at length as

$$i = 0 \qquad\qquad \text{for} \qquad\qquad (0 < t < t_1)$$

$$i = \frac{E_0}{R} e^{(t-t_1)/RC} \qquad\qquad (t_1 < t < t_2)$$

$$i = \frac{E_0}{R} [e^{(t-t_1)/RC} - e^{(t-t_2)/RC}] \qquad (t_2 < t)$$

Example 11. A light horizontal beam of length l is clamped at each end and carries a load w N/m over the legnth x_1 to x_2 measured from the left-hand end. To find the deflection

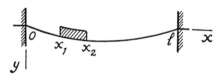

FIG. 27.

The differential equation here is

$$EI \frac{d^4 y}{dx^4} = wH(x - x_1) - wH(x - x_2).$$

Taking transforms,

$$EI[s^4 \bar{y} - s^3 y_0 - s^2 y_1 - s y_2 - y_3] = w \left[\frac{e^{-sx_1}}{s} - \frac{e^{-sx_2}}{s} \right].$$

At $x = 0$, $y_0 = 0$, $y_1 = 0$ and the equation reduces to

$$s^4 \bar{y} - s y_2 - y_3 = \frac{w}{EIs} (e^{-sx_1} - e^{-sx_2}),$$

$$\bar{y} = \frac{y_2}{s^3} + \frac{y_3}{s^4} + \frac{w}{EIs^5} (e^{-sx_1} - e^{-sx_2}),$$

$$y = y_2 \frac{x^2}{2} + y_3 \frac{x^3}{6} + \frac{w}{24EI} [(x - x_1)^4 H(x - x_1) - (x - x_2)^4 H(x - x_2)].$$

Since $y = 0$ and $\frac{dy}{dx} = 0$ when $x = l$,

$$0 = \tfrac{1}{2} y_2 l^2 + \tfrac{1}{6} y_3 l^3 + \frac{w}{24EI} [(l - x_1)^4 - (l - x_2)^4],$$

$$0 = y_2 l + \tfrac{1}{2} y_3 l^2 + \frac{w}{6EI} [(l - x_1)^3 - (l - x_2)^3],$$

two equations for the two unknowns y_2, y_3. These values are then substituted in the expression for y to give a single expression for the deflection y instead of as in the usual method, three separate expressions.

10.11. The Impulse Function $\delta(t)$

In order to deal with concentrated loads acting on beams and impulsive actions in circuits (electrical or mechanical) another invented function must be investigated.

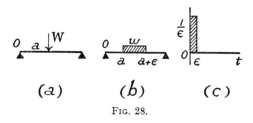

FIG. 28.

To find the transform of a concentrated load W we will find the transform of a uniform load w per unit length over the range a to $a + \varepsilon$ where ε is a small positive quantity which will ultimately be zero. To effect the trick we must clearly let $\varepsilon \to 0$ and $w \to \infty$ in such a way that

$$\lim_{\substack{\varepsilon \to 0 \\ w \to \infty}} w\varepsilon = W \quad \cdots \quad \cdots \quad (1)$$

Now the transform of such a load is (Fig. 28(a), (b)).

$$\int_a^{a+\varepsilon} e^{-sx} w\, dx = \frac{w}{s}\left(e^{-sa} - e^{-s(a+\varepsilon)}\right)$$

$$= \frac{w}{s} e^{-sa}(1 - e^{-s\varepsilon})$$

$$= \frac{w}{s} e^{-sa}\left\{1 - \left(1 - s\varepsilon + \frac{s^2\varepsilon^2}{2!} \cdots\right)\right\}$$

$$= \frac{w}{s} e^{-sa}\left\{s\varepsilon - \frac{s^2\varepsilon^2}{2!} \cdots\right\}$$

$$= e^{-sa}\left\{(w\varepsilon) - s\frac{(w\varepsilon)}{2!}\varepsilon + \cdots\right\}$$

Using the limit (1) it appears that every term within the brackets except the first gives zero value and the transform of the concentrated load is

$$e^{-sa}W \quad \cdots \quad \cdots \quad \cdots \quad (2)$$

where W is the load concentrated at the point $x = a$ on the beam.

It will soon be seen that if $\delta(x)$ is the delta function (defined below) then a concentrated load W at $x = a$ can be written

$$W\delta(x - a).$$

We will show that

$$L\{\delta(x - a)\} = e^{-sa},$$

so that

$$\int_0^\infty e^{-st}\delta(x-a)dx = e^{-sa},$$

and

$$L\{W\delta(x-a)\} = We^{-sa}.$$

Example 12. A light beam of length l is clamped horizontally at each end and carries a concentrated load W at $x = a$. To find the deflection at any point

The differential equation is

$$EI\frac{d^4y}{dx^4} = W\delta(x-a),$$

with $y = \dfrac{dy}{dx} = 0$ at $x = 0$ and $x = l$.

Taking transforms,

$$EI\{s^4\bar{y} - s^3y_0 - s^2y_1 - sy_2 - y_3\} = We^{-sa},$$

which reduces to

$$s^4\bar{y} = \frac{W}{EI}e^{-sa} + sy_2 + y_3.$$

From this

$$\bar{y} = \frac{W}{EI}\frac{e^{-sa}}{s^4} + \frac{y_2}{s^3} + \frac{y_3}{s^4},$$

$$y = \frac{W}{6EI}(x-a)^3H(x-a) + y_2\frac{x^2}{2} + y_3\frac{x^3}{6}.$$

At $x = l$, $y = 0 = \dfrac{dy}{dx}$ and $H(x-a) = 1$. Using these conditions

$$0 = \frac{W}{6EI}(l-a)^3 + \tfrac{1}{2}l^2y_2 + \tfrac{1}{6}l^3y_3,$$

$$0 = \frac{W}{2EI}(l-a)^2 + ly_2 + \tfrac{1}{2}l^2y_3.$$

From these

$$y_2 = \frac{Wa(l-a)^2}{EIl^2}, \quad y_3 = -\frac{W(l-a)^2(l+2a)}{EIl^3}.$$

Finally,

$$EIy = \frac{W}{6}(x-a)^3H(x-a) + \frac{Wa(l-a)^2x^2}{2l^2} - \frac{W(l-a)^2(l+2a)x^3}{6l^3}.$$

We now define the delta function $\delta(t)$ as

$$\delta(t) = 0 \quad t \leqslant 0,$$

$$= \frac{1}{\varepsilon} \quad 0 < t < \varepsilon,$$

$$= 0 \quad t > \varepsilon,$$

where ε may be as small as we please and in the limit is zero. The function is represented diagrammatically in Fig. 28(c), and since the area under the curve is unity for all ε

$$\int_{-\infty}^{+\infty} \delta(t)dt = 1.$$

We may now express an impulsive force P as $P\delta(t)$, since it will thus be very large over a very small interval of time ϵ, and the integral over the time t is by the above

$$\int_{-\infty}^{+\infty} P\delta(t)dt = P.$$

An impulsive voltage whose time integral is E may be written as $E\delta(t)$, an impulsive current which transfers a charge Q as $Q\delta(t)$. In terms of x the distance, we have already stated that a concentrated load W at the point $x = a$ can be represented as $W\delta(x - a)$.

The transform of $\delta(t)$ is the limit as $\epsilon \to 0$ of

$$\int_0^\infty e^{-st}\delta(t)dt = \int_0^\epsilon e^{-st} \cdot \frac{1}{\epsilon}\, dt$$

$$= \frac{1 - e^{-\epsilon s}}{\epsilon s}.$$

By expanding the exponential form or evaluating the indeterminate form, this limit is seen to be unity, so that

$$L\{\delta(t)\} = 1.$$

Hence by the general theorem

$$L\{\delta(t - a)\} = e^{-sa}$$

as used above.

Example 13. An inductance L in series with a capacitance C is acted on by an e.m.f. in the form of an impulsive voltage E_0. If initially current and charge are zero, find the current at time t.

The differential equation governing the current i is

$$L\frac{di}{dt} + \frac{q}{C} = E_0\delta(t)$$

where

$$i = \frac{dq}{dt}.$$

Taking transforms,

$$L(s\bar{i} - i_0) + \frac{1}{C}\bar{q} = E_0,$$

$$\bar{i} = s\bar{q} - q_0,$$

where $i_0 = q_0 = 0$.

Eliminating $\bar{q}$,

$$\bar{i}\left(Ls + \frac{1}{Cs}\right) = E_0,$$

$$\bar{i} = E_0 \frac{s}{Ls^2 + 1/C}$$

$$= \frac{E_0}{L} \frac{s}{s^2 + \dfrac{1}{LC}},$$

$$\therefore \quad i = \frac{E_0}{L} \cos\left(\frac{t}{\sqrt{LC}}\right).$$

10.12. It should be noticed that the H and δ functions can be used without Laplace Transforms and help considerably in the solution of problems on the deflection of beams by the usual methods. However, certain integrals must first be obtained.

Since $H(x - a)$ is zero for $x < a$ and unity for $x > a$

$$\int_0^x H(X - a)dX = 0 \qquad (x < a)$$

$$= x - a \quad (x > a)$$

which values can be combined into one expression $(x - a)H(x - a)$ so that

$$\int_0^x H(X - a)dX = (x - a)H(x - a).$$

Similarly

$$\int_0^x (X - a)^n H(X - a)dX = \frac{(x - a)^{n+1}}{n + 1} H(x - a) . \quad . \quad (1)$$

Also from the definition of $\delta(x - a)$,

$$\int_0^\infty \delta(x)dx = 1 \qquad \int_0^x \delta(X - a)dX = H(x - a) . \quad . \quad (2)$$

Example 14. A light beam of length l rests horizontally on supports at each end and carries a load W at a point $\frac{1}{3}l$ from the left-hand end. Find the deflection at the load and at the mid point of the beam.

With the orgin at the left-hand end and the y axis measured vertically downwards, since the reaction at $x = 0$ is 2W/3,

$$EIy_2 = W\left(x - \frac{l}{3}\right) H\left(x - \frac{l}{3}\right) - \frac{2W}{3} x.$$

Integrate :

$$EIy_1 = \tfrac{1}{2}W\left(x - \frac{l}{3}\right)^2 H\left(x - \frac{l}{3}\right) - \tfrac{1}{3}Wx^2 + A$$

and

$$EIy = \tfrac{1}{6}W\left(x - \frac{l}{3}\right)^3 H\left(x - \frac{l}{3}\right) - \tfrac{1}{9}Wx^3 + Ax + B.$$

$y = 0$ at $x = 0$ where $H(x - \tfrac{1}{3}l) = 0$ and $y = 0$ at $x = l$ where $H(x - \tfrac{1}{3}l) = 1$. From these values

$$B = 0, \ A = \frac{5}{81} Wl^2,$$

$$EIy = \tfrac{1}{6}W(x - \tfrac{1}{3}l)^3 H(x - \tfrac{1}{3}l) - \tfrac{1}{9}Wx^3 + \frac{5}{81} Wl^2 x.$$

At $x = \tfrac{1}{3}l$:

$$EIy = -\tfrac{1}{9}W(\tfrac{1}{3}l)^3 + \frac{5}{81} Wl^2 (\tfrac{1}{3}l) = \underline{\frac{4}{243} Wl^3}$$

At $x = \tfrac{1}{2}l$:

$$EIy = \tfrac{1}{6}W(\tfrac{1}{6}l)^3 - \tfrac{1}{9}W(\tfrac{1}{2}l)^3 + \frac{5}{81} Wl^2(\tfrac{1}{2}l) = \underline{\frac{23}{1296} Wl^3}$$

Using the δ function, the method of solution is

$$EID^4y = W\delta(x - \tfrac{1}{3}l)$$

integrating twice :

$$EID^3y = WH(x - \tfrac{1}{3}l) + A$$
$$EID^2y = W(x - \tfrac{1}{3}l)H(x - \tfrac{1}{3}l) + Ax + B.$$

At a freely supported end the B.M. and hence D^2y is zero so that for $x = 0$ and $x = l$, $D^2y = 0$. This gives $B = 0$, $A = -\frac{2}{3}W$.

Using these values and integrating again,

$$EIDy = \tfrac{1}{2}W(x - \tfrac{1}{3}l)^2 H(x - \tfrac{1}{3}l) - \tfrac{1}{3}Wx^2 + C$$
$$EIy = \tfrac{1}{6}W(x - \tfrac{1}{3})^3 H(x - \tfrac{1}{3}l) - \tfrac{1}{9}Wx^3 + Cx + E.$$

$y = 0$ when $x = 0$ and l so that

$$E = 0, \qquad C = \frac{5}{81} Wl^2$$

and we have the same expression for y as above.

EXERCISE 21

[Further problems are provided in Exercises 17–19.]

1. A beam (l) is fixed horizontally at each end and carries a load w per metre. Assuming a couple C Nm at each end keeping the end horizontal and a vertical force $\frac{1}{2}lw$, obtain the equation

$$EIy_2 = -\tfrac{1}{2}lwx + \tfrac{1}{2}wx^2 + C,$$

and hence find y, the deflection at any point distant x from the left-hand end.

2. A uniform beam (l) is fixed horizontally at each end and carries a load equal to the distance from one end. From the equation

$$EID^4y = x$$

obtain the deflection at any point distant x from the end.

3. An e.m.f. $E_0 \cos \omega t$ is applied at $t = 0$ to a circuit of a capacity C and inductance L in series. If initially the current and charge are zero show that at time t, if $n^2 = 1/LC$,

$$i = \frac{E}{L(\omega^2 - n^2)} [\omega \sin \omega t - n \sin nt].$$

4. An e.m.f. $E_0 \sin \omega t$ acts at $t = 0$ on an inductive resistance L, R. If the initial current is zero, show that at time t

$$i = \frac{E}{\sqrt{(R^2 + L^2\omega^2)}} [e^{-Rt/L} \sin \theta + \sin (\omega t - \theta)],$$

where $\tan \theta = L\omega/R$.

5. A uniform beam of length l is fixed horizontally at one end and freely supported at the other. If w is the weight per unit length, show that the fixing couple is $wl^2/8$ and that the reaction at the free end is $3wl/8$. Show that the deflection is a maximum at a point distant $0.58l$ approximately from the fixed end. [L.U.]

6. A light rod is clamped horizontally at one end ($x = 0$), is freely hinged at the other ($x = l$) and is subject to a horizontal thrust P at $x = 0$. Show that, if G is the couple applied at the clamped end, the deflection satisfies the equation

$$EI \frac{d^2y}{dx^2} + Py = G\left(1 - \frac{x}{l}\right).$$

Show also that the rod can take the curved form

$$y = \frac{G}{P} \left(\frac{\sin nx - nx}{nl} + 1 - \cos nx\right),$$

provided that $\tan nl = nl$, where $n^2 = P/EI$.

7. A uniform strut of length $2a$ and weight w per unit length is subject to a thrust P at each end which is freely hinged, these ends being the points $x = \pm a$, Ox being horizontal. Show that the deflection y at any point is given by

$$EI \frac{d^2y}{dx^2} + Py = \tfrac{1}{2}w(x^2 - a^2).$$

Solve this equation and show that if
$$\sec \theta = 1 + \tfrac{1}{3}\theta^2$$
where $\theta = a(\mathrm{P/EI})^{\frac{1}{2}}$, then there is no deflection at the mid point of the strut.

8. If an electron is projected into a uniform magnetic field perpendicular to its direction of motion, its path is given by
$$m\frac{d^2y}{dt^2} = -\frac{He}{c}\frac{dx}{dt}, \quad m\frac{d^2x}{dt^2} = \frac{He}{c}\frac{dy}{dt}$$
with $x = 0$, $\dfrac{dx}{dt} = u$, $y = 0 = \dfrac{dy}{dt}$, all at $t = 0$. Find the values of x and y at time t and show that the path of the electron is a circle.

9. Show that if θ satisfies the equation
$$\ddot{\theta} + 2k\dot{\theta} + n^2\theta = 0 \quad (k < n)$$
and if when $t = 0$, $\theta = \alpha$, $\dot{\theta} = 0$ then
$$\theta = e^{-kt}\Big(\alpha \cos pt + \frac{k\alpha}{p}\sin pt\Big),$$
where $p^2 = n^2 - k^2$.

The complete period of small oscillation of a simple pendulum is 2 seconds, and the angular retardation due to air resistance is $0.04 \times$ (angular velocity of the pendulum). The bob is held at rest so that the string makes a small angle $\alpha = 1°$ with the downward vertical and let go. Show that after 10 complete oscillations, the string will make an angle of about 40′ with the vertical.

[L.U.]

10. A unit e.m.f. is applied between the terminals A and B of the network shown (Fig. 29(a)). at an instant when both condensers are uncharged. Show

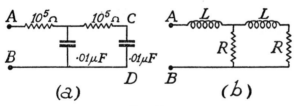

(a)   (b)

FIG. 29.

that the difference of potential between C and D after an interval of t milliseconds is
$$1 - e^{-3t/2}\Big\{\cosh\frac{\sqrt{5}}{2}t + \frac{3}{\sqrt{5}}\sinh\frac{\sqrt{5}}{2}t\Big\}.$$

11. A unit e.m.f. is applied at $t = 0$ to the input terminals A, B of the network shown (Fig. 29(b)), which is dead at time $t = 0$. Show that the input current at time t is
$$\frac{2}{R}\Big\{1 - e^{-\frac{3Rt}{2L}}\Big(\cosh\frac{\sqrt{5}R}{2L}t + \frac{2}{\sqrt{5}}\sinh\frac{\sqrt{5}R}{2L}t\Big)\Big\}.$$

12. Show that the indicial admittance (ie., current caused by the application at $t = 0$ of unit e.m.f.) between the terminals A and B of the network shown (Fig. 30(a)) is approximately
$$\frac{1}{L}\Big\{\frac{2}{3}t - \frac{1}{18}\frac{R}{L}t^2\Big\}$$
for small values of t. [Obtain $\bar{\imath}$, the transform of i in terms of s. Expand in powers of $1/s$.]

160

13. In Fig. 30(b) the inductance L has the value $\frac{1}{2}R^2C$. Show that if the condensers are initially uncharged and a battery of unit e.m.f. is connected to the input terminals at time $t = 0$, the input current is approximately

$$\frac{1}{R}(1 - t^3/R^3C^3)$$

for small values of t and that the current in the resistance R is approximately

$$\frac{1}{R}\left(1 - 2\frac{t}{RC} + \frac{t^2}{R^2C^2} - \frac{1}{3}\frac{t^3}{R^3C^3}\right).$$

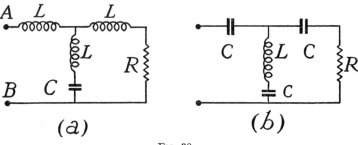

Fig. 30.

14. An impulsive force P acts on a mass m at rest at time $t = 0$. Obtain the equation

$$m\frac{d^2x}{dt^2} = P\delta(t)$$

and the solution

$$x = \frac{P}{m}t.$$

15. A mass M is fastened to a spring of stiffness ω^2M and is set in motion at time $t = 0$ from its equilibrium position by an impulse P. Obtain the equation

$$M\frac{d^2x}{dt^2} + M\omega^2x = P\delta(t)$$

and the solution

$$x = \frac{P}{M\omega}\sin \omega t.$$

16. An impulsive voltage is applied to an L, R, C circuit in series at time $t = 0$ when the circuit has $i_0 = q_0 = 0$. If $\mu = \frac{R}{2L}$, $n^2 = \frac{1}{LC} - \mu^2$, show that for the case of $n^2 > 0$

$$i = \frac{E}{Ln}e^{-\mu t}\{n\cos nt - \mu\sin nt\}.$$

17. A uniform light beam (l) is clamped horizontally at one end and free at the other. It carries a concentrated load P at $x = a$. Show that the deflection is given by

$$y = \frac{Px^2}{EI}(\tfrac{1}{2}a - \tfrac{1}{6}x) \qquad\qquad 0 < x < a$$

$$= \frac{Pa^2}{EI}(\tfrac{1}{2}x - \tfrac{1}{6}a) \qquad\qquad a < x < l$$

18. A uniform light beam is clamped horizontally at each end. It carries a

constant load w over the length $0 < x < \frac{1}{2}l$ only, the load being zero over the rest of the beam. Show that the deflection at any point is given by

$$EIy = \frac{11wl^2x^2}{384} - \frac{13wlx^3}{192} + \frac{wx^4}{24} - \frac{w(x - \frac{1}{2}l)^4}{24} \, H(x - \tfrac{1}{2}l)$$

19. A leaky condenser of capacity C_1 and leakage conductance G_1 is put in series with a similar condenser (C_2, G_2) at time $t = 0$ when the charges on the condensers are Q_1 and zero (on C_2) respectively. Show that the current in the circuit at time t is given by

$$\left[\frac{1}{G_1 + C_1s} + \frac{1}{G_2 + C_2s}\right] i = -\frac{Q_1}{G_1 + C_1s}$$

and that

$$i = -\frac{Q_1C_2}{C_1 + C_2} \, \delta(t) + \frac{(G_1C_2 - G_2C_1)}{(C_1 + C_2)^2} \, e^{-\frac{(G_1 + G_2)t}{C_1 + C_2}}.$$

Explain the meaning of this expression.

20. A voltage of constant value E_0 for the time interval t_1 to t_2 is imposed on a circuit consisting of R, C in series. The condenser at time $t = 0$ has a charge Q_0. Show that for the current at time t

$$i = -\frac{Q_0}{RC} e^{-t/RC} + \frac{E_0}{R} [e^{-(t - t_1)/RC} H(t - t_1) - e^{-(t - t_2)/RC} H(t - t_2)]$$

21. A " square " wave Fig. 31(a) is given by $y = E$ for $0 < t < T$, $y = -E$, $(T < t < 2T)$, $y = E$, $(2T < t < 3T)$ and so on. Show that

$$y = E\{H(t) - 2H(t - T) + 2H(t - 2T) \ldots \}.$$
$$\bar{y} = \frac{E}{s} \{1 - 2e^{-sT} + 2e^{-2sT} \ldots \}$$
$$= \frac{E}{s} \frac{1 - e^{-sT}}{1 + e^{-sT}} = \frac{E}{s} \tanh (\tfrac{1}{2}sT).$$

22. A function $f(t)$ is periodic with period T so that $f(t + rT) = f(t)$ where r is any integer. Show that

$$L\{f(t)\} = \int_0^\infty e^{-st}f(t)dt$$
$$= \int_0^T e^{-st}f(t)dt + e^{-sT}\int_0^T e^{-st}f(t)dt + \cdots$$
$$= (1 + e^{-sT} + e^{-2sT} + \cdots) \int_0^T e^{-st}f(t)dt$$
$$= \frac{1}{1 - e^{-sT}} \int_0^T e^{-st}f(t)dt$$

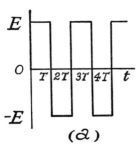

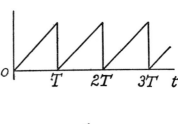

FIG. 31.

23. The " saw toothed " wave Fig. 31(b) is given by
$$y = m(t - rT) \text{ for } rT < t < (r + 1)T.$$
where $r = 0, 1, \ldots$ Using the above example show that
$$\bar{y} = \frac{m}{s^2} - \frac{mTe^{-sT}}{s(1 - e^{-sT})}.$$

24. If $y = 1, 2, 3 \ldots$ for the intervals $(0, T)$, $(T, 2T)$, $(2T, 3T,) \ldots$ etc., (the staircase function) show that
$$\bar{y} = \frac{1}{s(1 - e^{-sT})}.$$

CHAPTER 11

FOURIER SERIES

" This process [Harmonic analysis] is perhaps the most valuable contribution mathematics has made to engineering science. . . . Harmonic causes produce harmonic effects and by harmonically analysing a cause we seem to lift a veil and to visualize perhaps but dimly the way nature reacts to dynamical problems."

Sir Charles Inglis, *Applied Mechanics for Engineers.*

11.1. Introduction

We can find the equation to a curve passing through a given number of points whose (x, y) co-ordinates are given. If there are, say, four such points we usually assume an equation of the form

$$y = a_0 + a_1x + a_2x^2 + a_3x^3.$$

Substitution of the given data will give four equations from which the four unknowns a_0, a_1, a_2, a_3 can be found. If we assume a different type of equation such as

$$y = a_0 + a_1x + a_2x^2 + a_3 \sin 3x$$

we will find a curve passing through the given points but differing from the first curve. The form of the curve is therefore not fixed by the points given, but depends upon the equation assumed by the student.

Given n points, all of whose xs lie in the interval 0 to π, we can assume a series of the form

$$y = a_1 \sin x + a_2 \sin 2x + a_3 \sin 3x + \ldots + a_n \sin nx \quad (1)$$

and find as above the n unknown coefficients. We will then have a function of period 2π passing through the n points and repeating its shape in every interval of 2π. If the n given points all lie on $y = f(x)$, then the curve (1) will have n points in common with $y = f(x)$. If now we can let n tend to infinity, (1) will have an infinite number of points in common with $y = f(x)$, *i.e.*, will coincide completely with it in the interval 0 to π. We will then have a curve $y = f(x)$ represented by a sum of *periodic* functions. There is a much simpler way of finding this infinite number of coefficients than by the method indicated.

11.2. Development in Sine Series

Assume that $y = f(x)$, given in the range 0 to π, can be represented by an infinite sine series in this interval, *i.e.*,

$$f(x) = a_1 \sin x + a_2 \sin 2x + \ldots + a_n \sin nx + \ldots \quad (2)$$

Multiply both sides by $\sin nx$, where n is the suffix of the coefficient a_n that we wish to find (note that n is a positive integer) and integrate from 0 to π.

$$\int_0^\pi f(x) \sin nx\, dx = a_1 \int_0^\pi \sin x \sin nx\, dx + a_2 \int_0^\pi \sin 2x \sin nx\, dx + \ldots$$

$$\ldots + a_n \int_0^\pi \sin^2 nx\, dx + a_{n+1} \int_0^\pi \sin (n+1)x \sin nx\, dx + \ldots$$

On the right-hand side, except for the term in a_n, each integrand is of the form $(p \neq n)$

$$\sin px \sin nx = \tfrac{1}{2}[\cos (p-n)x - \cos (p+n)x]$$

$$\therefore \quad \int_0^\pi \sin px \sin nx\, dx = \tfrac{1}{2}\left[\frac{\sin (p-n)x}{p-n} - \frac{\sin (p+n)x}{p+n}\right]_0^\pi$$

$$= 0.$$

Every term on the right-hand side therefore vanishes except the integral associated with a_n, which gives

$$\int_0^\pi \sin^2 nx\, dx = \int_0^\pi \tfrac{1}{2}(1 - \cos 2nx)dx.$$

$$= \tfrac{1}{2}\pi$$

We thus find

$$\int_0^\pi f(x) \sin nx\, dx = \tfrac{1}{2}\pi a_n \quad \ldots \quad (3)$$

or a_n is given by the equation

$$a_n = \frac{2}{\pi} \int_0^\pi f(x) \sin nx\, dx \quad \ldots \quad (4)$$

When this has been evaluated we obtain the successive coefficients by putting $n = 1, 2, 3 \ldots$ in the result.

The series such as (2), where the coefficients are given by (4), is called a half-range Fourier series, the term half-range referring to the interval 0 to π of length π instead of the full-range length 2π.

Example 1. Find a sine series for $f(x) = c$ (a constant) in the range 0 to π.

If

$$f(x) = \sum_1^\infty a_n \sin nx,$$

then by (3)

$$\frac{\pi}{2} a_n = \int_0^\pi f(x) \sin nx\, dx = \int_0^\pi c \sin nx\, dx$$

$$= \frac{c}{n}[1 - \cos n\pi] = 0 \text{ if } n \text{ is even}$$

$$= \frac{2c}{n} \text{ if } n \text{ is odd}.$$

The even terms are absent from the series, and

$$f(x) \equiv c = \Sigma \frac{4c}{\pi n} \sin nx \qquad \text{(for } n \text{ odd)}$$

$$= \frac{4c}{\pi}\left[\frac{\sin x}{1} + \frac{\sin 3x}{3} + \frac{\sin 5x}{5} + \dots \frac{\sin (2n-1)x}{2n-1} + \dots\right]$$

$$= \frac{4c}{\pi} \sum_{n=1}^{\infty} \frac{\sin (2n-1)x}{2n-1}.$$

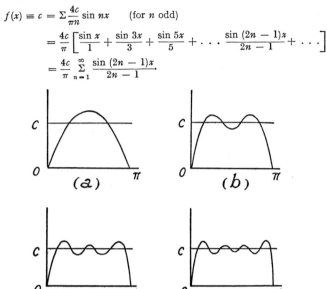

FIG. 32.

Fig. 32 shows the manner in which, as the Fourier series is added together term by term, the result approximates more and more closely to the function it represents. Thus the figures show successively

(a) $\quad y = \frac{4c}{\pi} \sin x,$

(b) $\quad y = \frac{4c}{\pi}\left[\sin x + \frac{\sin 3x}{3}\right],$

(c) $\quad y = \frac{4c}{\pi}\left[\sin x + \frac{\sin 3x}{3} + \frac{\sin 5x}{5}\right],$

(d) $\quad y = \frac{4c}{\pi}\left[\frac{\sin x}{1} + \frac{\sin 3x}{3} + \frac{\sin 5x}{5} + \frac{\sin 7x}{7}\right],$

drawn on the same diagram as $y = c$.

The student should note that so far we have found a representation for $y = c$ as the sum of an infinite series of sines *in the interval 0 to π*. As yet we will not consider what happens outside this range.

11.3. The theory of Fourier series is required in many problems. It is necessary in the solution of problems arising in mechanical and electrical vibrations, sound vibration, heat conduction, wireless waves, etc. One great advantage of Fourier series is that it can be applied to many more functions than can Taylor's (and Maclaurin's) theorem.

166

This later requires that the functions should be *continuous* and possess derivatives of all orders in order that the function should have a power series in x. Fourier series can be found for functions which have a *finite discontinuity* [*i.e.*, the value of the function makes a finite jump at some point (or points) in the interval]. The following example shows this.

Example 2. Find a sine series for

$$f(x) = x \qquad 0 < x < \frac{\pi}{2}$$

$$= 0 \qquad \frac{\pi}{2} < x < \pi.$$

Since $f(x)$ is represented by two different functions, we must split the range of integration and use the appropriate function for each part. Thus

$$\frac{\pi}{2} a_n = \int_0^\pi f(x) \sin nx dx$$

$$= \int_0^{\frac{1}{2}\pi} f(x) \sin nx dx + \int_{\frac{1}{2}\pi}^\pi f(x) \sin nx dx$$

$$= \int_0^{\frac{1}{2}\pi} x \sin nx dx + \int_{\frac{1}{2}\pi}^\pi 0 . \sin nx dx$$

$$= \int_0^{\frac{1}{2}\pi} x \sin nx dx = \left[- x \frac{\cos nx}{n} + \frac{\sin nx}{n^2} \right]_0^{\frac{1}{2}\pi}$$

$$= \frac{1}{n^2} \sin\left(\frac{n\pi}{2}\right) - \frac{\pi}{2n} \cos\left(\frac{n\pi}{2}\right).$$

It is not sufficient here to consider whether n is odd or even since if $n = 6$, $\cos(\frac{1}{2}n\pi) = -1$, whilst if $n = 8$, $\cos(\frac{1}{2}n\pi) = +1$. We are concerned with the form of n after we have divided by 2. It is therefore necessary to consider n in the four possible forms $4p + 1$, $4p + 2$, $4p + 3$, $4p + 4$, where, as p varies from 0 to ∞ (by integral values), we get all possible values of n. This gives

$$\frac{\pi}{2} a_n = \frac{1}{n^2} \qquad \text{if} \qquad n = 4p + 1$$

$$= \frac{\pi}{2n} \qquad n = 4p + 2$$

$$= -\frac{1}{n^2} \qquad n = 4p + 3$$

$$= -\frac{\pi}{2n} \qquad n = 4p + 4.$$

$$\therefore \quad f(x) = \frac{2}{\pi} \left[\frac{\sin x}{1^2} + \frac{\pi \sin 2x}{4} - \frac{\sin 3x}{3^2} - \frac{\pi \sin 4x}{8} \cdots \right]$$

The following figures (Fig. 33) show the manner in which, term by term, the Fourier series, approximates to the function it represents (drawn in dashes). (*a*) represents the first term of the Fourier series, (*b*) the sum of the first four and (*c*) of the first ten terms.

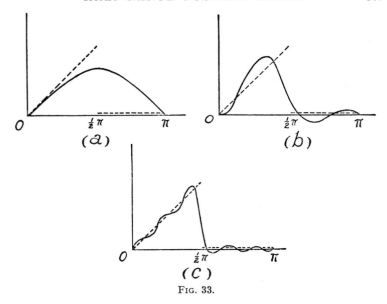

(a)   (b)

(c)

Fig. 33.

11.4. Development in Cosine Series

In an exactly similar manner we can find a half-range Fourier series of cosine terms only to represent a function given in the range 0 to π. Since $\cos(0x) = 1$, we assume

$$f(x) = b_0 + b_1 \cos x + b_2 \cos 2x + \ldots + b_n \cos nx + \ldots \quad (5)$$

If we integrate both sides of (5) between the limits 0 and π, then since

$$\int_0^\pi b_n \cos nx\,dx = b_n \left[\frac{1}{n}\sin nx\right]_0^\pi = 0 \quad \cdot \quad \cdot \quad \cdot \quad (6)$$

but

$$\int_0^\pi b_0\,dx = b_0\pi$$

each term on the right-hand side except the first vanishes, and

$$\int_0^\pi f(x)dx = b_0\pi,$$

or

$$b_0 = \frac{1}{\pi}\int_0^\pi f(x)dx \quad \cdot \quad \cdot \quad \cdot \quad \cdot \quad (7)$$

If we multiply both sides of (5) by $\cos nx$ and again integrate between 0 and π, then on the right-hand side the coefficient of b_0 vanishes as in (6). The coefficient of the one term containing b_n is

$$\int_0^\pi \cos^2 nx\,dx = \frac{\pi}{2}.$$

Every other coefficient, *e.g.*, of a_p ($p \neq n$ or 0) is

$$\int_0^\pi \cos nx \cos pxdx = \int_0^\pi \tfrac{1}{2}[\cos (n + p)x + \cos (n - p)x]dx.$$
$$= 0.$$

Finally,

$$\int_0^\pi f(x) \cos nxdx = \frac{\pi}{2} b_n \qquad\qquad (n \neq 0)$$

or $$b_n = \frac{2}{\pi} \int_0^\pi f(x) \cos nxdx \quad \cdot \quad \cdot \quad \cdot \quad \cdot \quad (8)$$

We note that the formula for b_0 in (7) is half the value that (8) gives with $n = 0$. To save an extra formula we always assume

$$f(x) = \tfrac{1}{2}b_0 + b_1 \cos x + b_2 \cos 2x + \ldots$$

and now the formula

$$b_n = \frac{2}{\pi} \int_0^\pi f(x) \cos nxdx \quad \cdot \quad \cdot \quad \cdot \quad \cdot \quad (9)$$

holds also for $n = 0$. With $\sin nx$ instead of $\cos nx$, it also holds for the coefficients in the sine expansion.

Example 3. Find a cosine series in the range 0 to π for

$$f(x) = x \qquad\qquad 0 < x < \frac{\pi}{2}$$

$$f(x) = \pi - x \qquad \frac{\pi}{2} < x < \pi.$$

If $$f(x) = \tfrac{1}{2}b_0 + \sum_1^\infty b_n \cos nx$$

by (9) $$\frac{\pi}{2} b_0 = \int_0^\pi f(x)dx$$

$$= \int_0^{\frac{1}{2}\pi} xdx + \int_{\frac{1}{2}\pi}^\pi (\pi - x)dx = \frac{\pi^2}{4}$$

$$\therefore \quad \tfrac{1}{2}b_0 = \frac{\pi}{4} \cdot \quad \cdot \quad \cdot \quad \cdot \quad \cdot \quad \cdot \quad \cdot \quad \cdot \quad \cdot \quad \cdot \quad (10)$$

Also $$\frac{\pi}{2} b_n = \int_0^{\frac{1}{2}\pi} x \cos nxdx + \int_{\frac{1}{2}\pi}^\pi (\pi - x) \cos nxdx.$$

$$= \frac{1}{n^2}\left[2 \cos \frac{n\pi}{2} - \cos n\pi - 1\right] \quad \cdot \quad \cdot \quad \cdot \quad \cdot \quad (11)$$

As in Example 2 we must consider the four possible forms for n, and find ($p = 0, 1, 2 \ldots$).

$$\frac{\pi}{2} b_n = 0 \qquad \text{when } n = 4p + 1$$

$$= -\frac{4}{n^2} \qquad n = 4p + 2$$

$$= 0 \qquad n = 4p + 3$$

$$= 0 \qquad n = 4p + 4$$

Finally, $$f(x) = \tfrac{1}{4}\pi - \frac{2}{\pi}\left[\frac{\cos 2x}{1^2} + \frac{\cos 6x}{3^2} + \frac{\cos 10x}{5^2} + \ldots\right].$$

EXERCISE 22

1. In the series obtained in Example 1 put $x = \frac{1}{2}\pi$ and show that

$$\frac{\pi}{4} = 1 - \frac{1}{3} + \frac{1}{5} - \frac{1}{7} + \cdots$$

2. Show that as a Fourier sine series in the range 0 to π the function

$$f(x) = x \qquad \text{for } 0 < x < \frac{\pi}{2}$$

$$= \pi - x \text{ for } \frac{\pi}{2} < x < \pi$$

is represented by

$$f(x) = \frac{4}{\pi} \left[\frac{\sin x}{1} - \frac{\sin 3x}{9} + \frac{\sin 5x}{25} \cdots \right].$$

Put $x = \frac{1}{2}\pi$ in this result and show that

$$\frac{\pi^2}{8} = 1 + \frac{1}{3^2} + \frac{1}{5^2} + \frac{1}{7^2} + \cdots$$

3. In Example 3 the function of Question 2 above was represented as a cosine series. Put $x = \frac{1}{2}\pi$ in this cosine series and obtain the result above for $\frac{1}{8}\pi^2$.

4. Show that as a sine series in the range 0 to π

$$f(x) = 0 \qquad 0 < x < \frac{1}{2}\pi$$

$$= c \qquad \frac{1}{2}\pi < x < \pi$$

can be represented as

$$f(x) = \frac{2c}{\pi} \left[\frac{\sin x}{1} - \frac{2 \sin 2x}{2} + \frac{\sin 3x}{3} + \frac{\sin 5x}{5} - \frac{2 \sin 6x}{6} \cdots \right],$$

but as a cosine series in the same range it is

$$f(x) = \frac{2c}{\pi} \left[\frac{\pi}{4} - \frac{\cos x}{1} + \frac{\cos 3x}{3} - \frac{\cos 5x}{5} \cdots \right].$$

5. For the function $y = x$ in the range 0 to π show that as:

(a) a sine series

$$y = 2 \left[\frac{\sin x}{1} - \frac{\sin 2x}{2} + \frac{\sin 3x}{3} \cdots \right];$$

(b) a cosine series

$$y = \frac{\pi}{2} - \frac{4}{\pi} \left[\frac{\cos x}{1^2} + \frac{\cos 3x}{3^2} + \cdots \right].$$

6. Show that in the range 0 to π, the trapezoidal wave

$$y = E_0 \frac{x}{\alpha} \qquad 0 < x < \alpha$$

$$= E_0 \qquad \alpha < x < \pi - \alpha$$

$$= \frac{E_0}{\alpha} (\pi - x) \qquad \pi - \alpha < x < \pi$$

can be represented by the sine series

$$y = \frac{4E_0}{\pi\alpha} \left[\sin \alpha \sin x + \frac{\sin 3\alpha \sin 3x}{3^2} + \frac{\sin 5\alpha \sin 5x}{5^2} + \cdots \right].$$

7. When an e.m.f. E $\sin \omega t$ is acting on a full-wave rectifier the output voltage is $e = E |\sin \omega t|$. Show that e may be represented by the half-range Fourier series

$$E \left(\frac{2}{\pi} - \frac{4}{\pi} \sum_{n=1}^{\infty} \frac{\cos 2n\omega t}{4n^2 - 1} \right).$$

11.5. The General Fourier Series

The half-range sine and cosine series obtained above are special cases of the general series which is the sum of the two. Thus for a function $f(x)$ *defined in the interval* $-\pi$ *to* $+\pi$ we assume

$$f(x) = \tfrac{1}{2}b_0 + b_1 \cos x + b_2 \cos 2x + \ldots + b_n \cos nx + \ldots$$
$$+ a_1 \sin x + a_2 \sin 2x + \ldots + a_n \sin nx + \ldots \quad . \quad (1)$$

$$= \tfrac{1}{2}b_0 + \sum_{n=1}^{\infty} (b_n \cos nx + a_n \sin nx).$$

The coefficients are obtained in much the same way as before. Thus if we integrate both sides of (1) between $-\pi$ and $+\pi$, then since

$$\int_{-\pi}^{+\pi} \cos nx\, dx = 0 = \int_{-\pi}^{+\pi} \sin nx\, dx$$

each term after the first on the right-hand side vanishes so that

$$\int_{-\pi}^{+\pi} f(x)\, dx = \int_{-\pi}^{+\pi} \tfrac{1}{2}b_0\, dx = b_0\pi,$$

or

$$b_0 = \frac{1}{\pi}\int_{-\pi}^{+\pi} f(x)\, dx \quad . \quad . \quad . \quad (2)$$

If we now multiply both sides of (1) by $\cos nx$ and integrate from $-\pi$ to $+\pi$, the right-hand side contains such terms as

$$\tfrac{1}{2}b_0 \int_{-\pi}^{+\pi} \cos nx\, dx = 0, \qquad b_n\int_{+\pi}^{-\pi} \cos^2 nx\, dx = b_n\pi$$

$$b_p \int_{-\pi}^{+\pi} \cos px \cos nx\, dx = \tfrac{1}{2}b_p \int_{-\pi}^{+\pi} [\cos (p+n)x - \cos (p-n)x]dx$$

$$= 0 \qquad (\text{for } p \neq n),$$

$$a_p \int_{-\pi}^{+\pi} \sin px \cos nx\, dx = \tfrac{1}{2}a_p \int_{-\pi}^{+\pi} [\sin (p+n)x + \sin (p-n)x]dx$$

$$= 0 \qquad (\text{whether } p = n \text{ or not}).$$

Again only one term remains on the right-hand side and

$$\int_{-\pi}^{+\pi} f(x) \cos nx\, dx = b_n\pi$$

or

$$b_n = \frac{1}{\pi}\int_{-\pi}^{+\pi} f(x) \cos nx\, dx \quad . \quad . \quad (3)$$

an equation that also holds for $n = 0$, as seen by (2).

To find a_n we multiply both sides of (1) by $\sin nx$ and again integrate from $-\pi$ to $+\pi$. The student will verify that each term except a_n on the right-hand side vanishes, giving

$$\int_{-\pi}^{+\pi} f(x) \sin nx dx = a_n \int_{-\pi}^{+\pi} \sin^2 nx dx = a_n \pi,$$

or
$$a_n = \frac{1}{\pi} \int_{-\pi}^{+\pi} f(x) \sin nx dx \quad . \quad . \quad . \quad (4)$$

Exactly the same formulae hold for a function defined in the range 0 to 2π, the limits of integration in (2), (3) and (4) now being 0 and 2π.

Example 4. Show that

$$x^2 = \frac{\pi^2}{3} + 4 \sum_{n=1}^{\infty} (-1)^n \frac{\cos nx}{n^2} \qquad (-\pi \leqslant x \leqslant +\pi).$$

If
$$f(x) = \tfrac{1}{2} b_0 + \Sigma(b_n \cos nx + a_n \sin nx)$$

$$\pi b_0 = \int_{-\pi}^{+\pi} x^2 dx = \frac{2\pi^3}{3}, \qquad b_0 = \frac{2\pi^2}{3} \quad . \quad . \quad . \quad . \quad (1)$$

$$\pi b_n = \int_{-\pi}^{+\pi} x^2 \cos nx dx$$

$$= \left[x^2 \frac{\sin nx}{n} + \frac{2x \cos nx}{n^2} - \frac{2}{n^3} \sin nx \right]_{-\pi}^{+\pi}$$

$$= \frac{4}{n^2} \pi \cos n\pi,$$

$$\left. \begin{aligned} b_n = \frac{4}{n^2} \cos n\pi &= \frac{4}{n^2} \text{ for } n \text{ even} \\ &= -\frac{4}{n^2} \text{ for } n \text{ odd} \end{aligned} \right\} \quad . \quad . \quad . \quad . \quad (2)$$

$$\pi a_n = \int_{-\pi}^{+\pi} x^2 \sin nx dx$$

$$= \left[-\frac{x^2}{n} \cos nx + \frac{2x}{n^2} \sin nx + \frac{2}{n^3} \cos nx \right]_{-\pi}^{+\pi}$$

$$= 0 \quad . \quad . \quad . \quad . \quad . \quad . \quad . \quad . \quad . \quad . \quad . \quad (3)$$

showing that there are no sine terms in this series.

$$\therefore \quad x^2 = \frac{\pi^2}{3} + 4\left[-\frac{\cos x}{1^2} + \frac{\cos 2x}{2^2} - \frac{\cos 3x}{3^2} \cdots \right]$$

$$= \frac{\pi^2}{3} + 4\sum_{1}^{\infty} (-1)^n \frac{\cos nx}{n^2}.$$

Putting $x = \pi$ in this series gives

$$\frac{\pi^2}{6} = 1 + \frac{1}{2^2} + \frac{1}{3^2} + \frac{1}{4^2} + \cdots$$

$x = 0$ gives

$$\frac{\pi^2}{12} = 1 - \frac{1}{2^2} + \frac{1}{3^2} - \frac{1}{4^2} + \cdots$$

adding:

$$\frac{\pi^2}{8} = 1 + \frac{1}{3^2} + \frac{1}{5^2} + \cdots$$

This last series which contains odd terms only will cause half the amount of calculation needed in the two previous series when used to calculate π to a given degree of accuracy.

11.6. Representation of the Given $f(x)$ Outside the Range

If a function is periodic with period 2π and we have found a Fourier series for it in the range $-\pi$ to $+\pi$ or 0 to 2π, then since the function and the series repeat in each interval of 2π the series will represent the function everywhere. If, however, the function $f(x)$ is given only in the range $-\pi$ to $+\pi$ or 0 to 2π, then the repetitions of the Fourier series are of no consequence to the function, which does not exist outside its given range. In the given range the Fourier series represents it, but outside the range the $f(x)$ does not exist—as far as our information goes—and therefore cannot be represented by a Fourier series or any other series.

11.7. Continuity

We have so far, until this chapter on Fourier series, dealt only with continuous functions, *i.e.*, with functions that could be represented

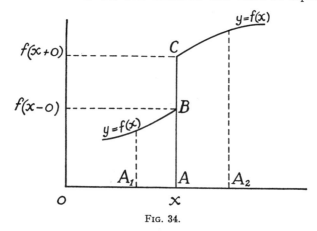

Fig. 34.

by a curve that did not show any gaps or finite jumps. We must now introduce a notation to enable us to deal with such cases. If we proceed along the curve (Fig. 34) from the left to the point whose co-ordinate is x, then the value of the function (AB in the figure) is denoted by

$$f(x -) \quad \text{or} \quad f(x - 0)$$

showing that the co-ordinate reaches x from below. If we proceed along the curve from the right the value of the function (AC in the figure) is denoted by

$$f(x +) \quad \text{or} \quad f(x + 0)$$

showing that the co-ordinate reaches the value x from above. If the function is continuous for a value x, then clearly

$$f(x -) = f(x) = f(x +)$$

as at A_1 or A_2.

In Example 2 in Section 3 of this chapter the function is discontinuous at $x = \frac{\pi}{2}$ and

$$f\left(\frac{\pi}{2}-\right) = \frac{\pi}{2}, \quad \text{but} \quad f\left(\frac{\pi}{2}+\right) = 0.$$

11.8. The Value of a Fourier Series at a Point of Discontinuity.

It can be shown that at any point of discontinuity x such as those considered, where a function is represented by a Fourier series, the value of the Fourier series is

$$\tfrac{1}{2}[f(x+) + f(x-)].$$

Thus in Example 2 in Section 3 above, where there is a discontinuity at $x = \frac{1}{2}\pi$

$$f\left(\frac{\pi}{2}-\right) = \tfrac{1}{2}\pi \quad \text{and} \quad f\left(\frac{\pi}{2}+\right) = 0,$$

$$\therefore \quad \tfrac{1}{2}\left[f\left(\frac{\pi}{2}+\right) + f\left(\frac{\pi}{2}-\right)\right] = \tfrac{1}{2}\left[0 + \tfrac{1}{2}\pi\right] = \tfrac{1}{4}\pi.$$

The Fourier series is

$$\frac{2}{\pi}\left[\frac{\sin x}{1^2} + \frac{\pi \sin 2x}{4} - \frac{\sin 3x}{3^2} - \frac{\pi \sin 4x}{8} \cdots\right].$$

At $x = \tfrac{1}{2}\pi$ the value is

$$\frac{2}{\pi}\left[\frac{1}{1^2} + \frac{1}{3^2} + \frac{1}{5^2}\cdots\right] = \frac{2}{\pi} \times \frac{\pi^2}{8} \quad \text{(by Example 4)}$$

$$= \frac{\pi}{4},$$

thus checking the above statement.

11.9. Odd and Even Functions

In Example 4 the sine terms were found to be zero. We can often tell this in advance, and thus save work, by means of the following considerations.

(a) A function $f(x)$ is said to be an odd function if $f(-x) = -f(x)$. x, $\sin x$, $y = 1 + x$ for $0 < x < \pi$ but $y = -1 + x$ for $-\pi < x < 0$ are all odd functions of x. In each case the graph is symmetrical with respect to the origin. To find a Fourier series for such an " odd " function *given in the range* $-\pi$ to $+\pi$, as usual assume

$$f(x) = \tfrac{1}{2}b_0 + \Sigma(b_n \cos nx + a_n \sin nx) \quad . \quad . \quad (1)$$

Since the function is given in the negative range $-\pi$ to 0, we may change the sign of x to find

$$f(-x) = \tfrac{1}{2}b_0 + \Sigma(b_n \cos nx - a_n \sin nx),$$

since $\quad \cos(-nx) = \cos nx \quad$ but $\quad \sin(-nx) = -\sin nx.$

$f(-x) = -f(x)$ for an odd function, therefore this equation may be written

$$-f(x) = \tfrac{1}{2}b_0 + \Sigma(b_n \cos nx - a_n \sin nx) \quad . \quad . \quad (2)$$

(1) – (2) gives

$$f(x) = \Sigma a_n \sin nx,$$

showing that the constant and cosine terms are absent in the series for $f(x)$.

(b) $f(x)$ is said to be an even function if $f(-x) = f(x)$, $y = x^2$, $y = \cos x$, $y = 1 + x$ for $0 < x < \pi$ but $y = 1 - x$ for $-\pi < x < 0$ are all even functions. In each case the graph is symmetrical with respect to the y axis. If such a function is given in the range $-\pi$ to $+\pi$ we assume as usual (1) above. For the negative range we change the sign of x, and since $f(-x) = f(x)$, equation (1) becomes

$$f(x) = \tfrac{1}{2}b_0 + \Sigma(b_n \cos nx - a_n \sin nx) \quad . \quad . \quad . \quad (3)$$

(1) + (3) gives

$$f(x) = \tfrac{1}{2}b_0 + \Sigma b_n \cos nx,$$

showing that the sine terms are absent in this case. This consideration applied to Example 4 would have shown in advance that $y = x^2$ in the range $-\pi$ to $+\pi$ has no sine terms in its Fourier series.

(c) If a function is given *in the range* 0 *to* 2π we cannot change x into $-x$ to find its value for negative values of x. We cannot therefore use the above method for showing that sine or cosine terms may be absent in the Fourier series.

(d) A function such as $y = e^x$ is neither odd nor even, and so we cannot say in advance that sine or cosine terms will not be present in its Fourier series.

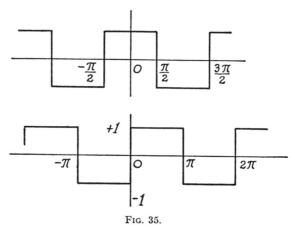

Fig. 35.

(e) For a periodic function we can often choose the origin so as to make the function symmetrical with regard to the origin when the

Fourier series will have sine terms only or symmetrical with regard to the y axis when the Fourier series will have cosine terms only. Thus Fig. 35(a) shows an even function, whilst in (b) the same function is an odd function. The student will easily obtain for the first case

$$f(x) = \frac{4}{\pi}\left[\cos x + \frac{1}{3}\cos 3x + \frac{1}{5}\cos 5x + \cdots\right],$$

whilst in the second case

$$F(x) = \frac{4}{\pi}\left[\sin x + \frac{1}{3}\sin 3x + \frac{1}{5}\sin 5x + \cdots\right],$$

showing that the values of the respective harmonics are unchanged by a change of origin.

(f) It will also be appreciated that if we are given a function in the range 0 to π, such as

$$y = x \qquad 0 < x < \pi/2,$$
$$y = \pi - x \qquad \pi/2 < x < \pi,$$

and fit a half-range cosine series to this, then the *extension of the Fourier series* thus found beyond the range 0 to π is as in Fig. 36(a), a

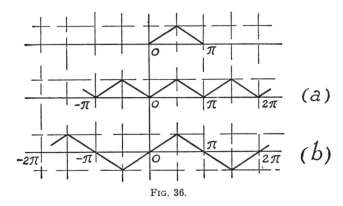

Fig. 36.

curve symmetrical with respect to the y axis. If, however, we find a half-range sine series, then its extension is as shown in Fig. 36(b), a curve symmetrical with respect to the origin. In the range 0 to π each form of series represents the function completely, but outside this range they differ.

Example 5. Find in the range $-\pi$ to $+\pi$ a Fourier series for,

(a) $y = 1 + x \qquad 0 < x < \pi$
$\quad\ y = -1 + x \qquad -\pi < x < 0$

(b) $y = 1 + x \qquad 0 < x < \pi$
$\quad\ y = 1 - x \qquad -\pi < x < 0$

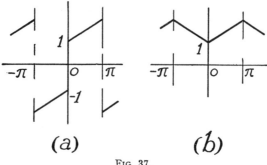

FIG. 37.

(a) The graph is symmetrical with respect to the origin, therefore the series contains sine terms only.

If $\qquad f(x) = \Sigma a_n \sin nx$

$$\pi a_n = \int_{-\pi}^{0} (-1 + x) \sin nx dx + \int_{0}^{\pi} (1 + x) \sin nx dx \quad . \quad . \quad (1)$$

If we put $x = -y$ in the first integral it changes into the second and

$$\pi a_n = 2 \int_{0}^{\pi} (1 + x) \sin nx dx \quad . \quad . \quad . \quad . \quad (2)$$

but this is the formula for a_n in a half-range sine series [11.2(4)]. We note then that when finding a Fourier series in the range $-\pi$ to $+\pi$ that will contain sine terms only we may ignore the negative half of the range and find a half-range sine series for the function given in the positive half.

From (2)

$$a_n = \frac{2}{\pi n} (1 - \cos n\pi - \pi \cos n\pi),$$

$$\therefore \quad f(x) = \frac{2}{\pi} \left[(2 + \pi) \sin x - \frac{\pi}{2} \sin 2x + \frac{1}{3} (2 + \pi) \sin 3x - \frac{\pi}{4} \sin 4x + \ldots \right].$$

(b) The graph is symmetrical with respect to the y axis, and the series contains cosine terms only.

If $\qquad f(x) = \frac{1}{2}b_0 + \Sigma b_n \cos nx,$

$$\pi b_0 = \int_{-\pi}^{0} (1 - x)dx + \int_{0}^{\pi} (1 + x)dx.$$

If we put $x = -y$ in the first integral it changes into the second integral

$$\therefore \quad \pi b_0 = 2 \int_{0}^{\pi} (1 + x)dx \quad . \quad . \quad . \quad . \quad . \quad . \quad (3)$$

also

$$\pi b_n = \int_{-\pi}^{0} (1 - x) \cos nx dx + \int_{0}^{\pi} (1 + x) \cos nx dx$$

$$= 2 \int_{0}^{\pi} (1 + x) \cos nx dx \quad . \quad . \quad . \quad . \quad . \quad . \quad . \quad (4)$$

when the substitution $x = -y$ is made.

(3) and (4) are the formulae for b_0 and b_n when a half-range cosine series is fitted to the function given for the positive half of the range, so this short method may again be used when it has been noticed that the graph is symmetrical with respect to the y axis.

By integration

$$b_0 = \pi + 2,$$

$$b_n = \frac{2}{\pi n^2} (\cos n\pi - 1) \qquad (n \neq 0),$$

$$\therefore \quad f(x) = \frac{\pi}{2} + 1 - \frac{4}{\pi} \left(\cos x + \frac{1}{3^2} \cos 3x + \frac{1}{5^2} \cos 5x + \ldots \right).$$

11.10. Odd and Even Harmonics

It can also be seen in advance when a function expressed as a Fourier series will have only odd or even harmonics. This is independent of whether or not it has sine or cosine terms only.

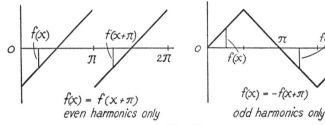

FIG. 38.

If
$$f(x) = \tfrac{1}{2}a_0 + a_1 \cos x + a_2 \cos 2x + \ldots$$
$$+ b_1 \sin x + b_2 \sin 2x + \ldots$$

then
$$f(x + \pi) = \tfrac{1}{2}a_0 - a_1 \cos x + a_2 \cos 2x + \ldots$$
$$- b_1 \sin x + b_2 \sin 2x + \ldots$$

so that if $f(x) = f(x + \pi)$, $a_1 = a_3 = \ldots = 0$ and $b_1 = b_3 = \ldots = 0$ and the series contains even harmonics only.

Similarly, if $f(x + \pi) = -f(x)$, then there are odd harmonics only in the Fourier expansion of $f(x)$.

EXERCISE 23

1. Show that a Fourier series for $y = x$ in the range $-\pi$ to $+\pi$ is
$$y = 2[\sin x - \tfrac{1}{2} \sin 2x + \tfrac{1}{3} \sin 3x \ldots].$$

Deduce that
$$\tfrac{1}{4}\pi = 1 - \tfrac{1}{3} + \tfrac{1}{5} - \tfrac{1}{7} \ldots$$

2. In the above if the range is 0 to 2π, show that the series becomes
$$y = \pi - 2[\sin x + \tfrac{1}{2} \sin 2x + \tfrac{1}{3} \sin 3x + \ldots].$$

3. Find a Fourier series for
$$y = x \qquad 0 < x < \pi,$$
$$y = 0 \qquad \pi < x < 2\pi.$$

From the value of the series at the discontinuity $x = \pi$ deduce the series for $\tfrac{1}{8}\pi^2$.

4. Find a Fourier series in the range 0 to 2π for $y = 1 - x/\pi$.

5. Express as a Fourier series the function

$$y = a \qquad 0 \leqslant x < \pi$$
$$y = -a \qquad -\pi < x < 0.$$

Check that the value of the series and of the function agree at the discontinuity $x = 0$.

6. Find a Fourier series for the lines

$$y = \tfrac{1}{2}\pi + x \qquad -\pi < x < 0,$$
$$y = \tfrac{1}{2}\pi - x \qquad 0 < x < \pi.$$

7. Find a Fourier series for the half-sine wave in Fig. 39(a).

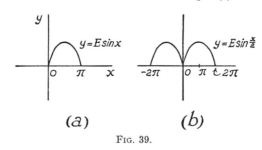

(a)   (b)

Fig. 39.

8. Find a Fourier series for the rectified sine wave in Fig. 39(b).

9. Show that in the range $-\pi$ to $+\pi$, $y = e^x$ as a Fourier series is

$$e^x = \frac{\sinh \pi}{\pi}\left[1 + 2\sum_1^\infty \frac{(-1)^n}{n^2+1}(\cos nx - n \sin nx)\right].$$

Deduce from this that

$$\frac{\pi}{\sinh \pi} = 2\sum_2^\infty \frac{(-1)^n}{n^2+1}.$$

10. In the above if the range is 0 to 2π, show that

$$e^x = \frac{e^{2\pi}-1}{\pi}\left[\tfrac{1}{2} + \sum_1^\infty \frac{\cos nx}{n^2+1} - \sum_1^\infty \frac{n \sin nx}{n^2+1}\right].$$

11. Find a Fourier series for $y = \cos ax$, $-\pi \leqslant x \leqslant +\pi$, given that a is not an integer or zero.

12. Find the sine and cosine half-range series for $y = e^x$, $0 < x < \pi$. Why do these results differ from those of Questions 9 and 10?

13. Show that as a cosine series in the half range 0 to π

$$\sin x = \frac{4}{\pi}\left[\frac{1}{2} - \frac{\cos 2x}{1 \cdot 3} - \frac{\cos 4x}{3 \cdot 5} - \frac{\cos 5x}{5 \cdot 7} \cdots\right].$$

Whilst as a sine series in the same range

$$\cos x = \frac{4}{\pi}\left[\frac{2 \sin 2x}{2^2-1} + \frac{4 \sin 4x}{4^2-1} + \frac{6 \sin 6x}{6^2-1} + \cdots\right].$$

14. If $f(x)$ and $g(x)$ are two functions expressed as Fourier series in the range 0 to 2π by

$$f(x) = \tfrac{1}{2}a_0 + \sum_1^\infty[a_n \cos nx + b_n \sin nx],$$

$$g(x) = \tfrac{1}{2}A_0 + \sum_1^\infty[A_n \cos nx + B_n \sin nx],$$

THIS IS NOT USED

show that the average value of the product of the two functions in the interval 0 to 2π, which is

$$\frac{1}{2\pi}\int_0^{2\pi} f(x)g(x)dx,$$

is given by

$$\tfrac{1}{4}a_0A_0 + \tfrac{1}{2}\sum_{n=1}^{\infty}(a_nA_n + b_nB_n) \quad \cdot \quad \cdot \quad \cdot \quad \cdot \quad \cdot \quad (1)$$

Deduce that

$$\frac{1}{2\pi}\int_0^{2\pi}[f(x)]^2 dx = \tfrac{1}{4}a_0{}^2 + \tfrac{1}{2}\sum_1^{\infty}(a_n{}^2 + b_n{}^2) \quad \cdot \quad \cdot \quad \cdot \quad (2)$$

15. Put $x = \omega t$ in Question 14 above and deduce that if a periodic voltage e expressed as a Fourier series causes a periodic current i of the same fundamental period to flow in a circuit, then the average value of the power delivered, which is

$$\frac{1}{2\pi}\int_0^{2\pi} eid(\omega t),$$

is given by expression (1).

16. If the current i mentioned in Question 15 passes through a resistance R, show that the average value of the power dissipated

$$\frac{1}{2\pi}\int_0^{2\pi} Ri^2 d(\omega t)$$

is given by expression (2), Question 14 multiplied throughout by R

17. If $f(x)$ is a function with a period of 2π, which is defined as being zero in each of the ranges $-\pi < x < -\tfrac{1}{3}\pi$ and $\tfrac{1}{3}\pi < x < \pi$ and having the value $\cos(3x/4)$ in the range $-\tfrac{1}{3}\pi < x < \tfrac{1}{3}\pi$, show that the Fourier series for $f(x)$ has cosine terms only; find the constant term and show that the coefficient of $\cos nx$ is

$$\frac{4n \sin \tfrac{1}{3}n\pi - 3\cos \tfrac{1}{3}n\pi}{2\pi\sqrt{2}\{n^2 - (\tfrac{3}{4})^2\}}.$$ [L.U.]

18. A periodic function $f(x)$ whose period is $\pi/2$ is zero in the ranges $-\dfrac{\pi}{4} < x < -\alpha$ and $\alpha < x < \pi/4$, while in the range $-\alpha < x < \alpha$, $f(x)$ has the constant value $1/\alpha$. Show that $f(x)$ can be expanded in the Fourier series

$$f(x) = \frac{8}{\pi}\left\{\frac{1}{2} + \sum_{r=1}^{\infty} \frac{\sin 4r\alpha}{4r\alpha}\cos 4rx\right\}.$$ [L.U.]

19. Find a_n, b_n so that the integral

$$\int_{-\pi}^{+\pi}\left[f(x) - (\tfrac{1}{2}a_0 + \sum_1^{\infty} a_n\cos nx + \sum_1^{\infty} b_n\sin nx)\right]^2 dx$$

is a minimum. (Note that since a_n, b_n as found are the Fourier coefficients for $f(x)$, a function given in the interval $-\pi$ to $+\pi$, a Fourier series represents the best fit to $f(x)$ in the " least squares " sense is given in Chapter 18.)

20. If the $f(x)$ can be expanded in the Fourier series

$$\tfrac{1}{2}a_0 + \sum_{n=1}^{\infty}(a_n\cos nx + b_n\sin nx) \qquad 0 < x < 2\pi,$$

show that

$$a_n + ib_n = \frac{1}{\pi}\int_0^{2\pi} f(x)e^{-inx}dx.$$

Hence or otherwise expand the function $x(x-1)(x-2)$ in a full-range Fourier series valid in the interval $(0, 2)$ giving the general term. Illustrate your result by drawing a graph of the function represented by the series for values of x outside the range. [L.U.]

21. What conditions must be satisfied by a periodic function $f(x)$ of period 2π if its Fourier series is to contain only sines of odd multiples of x? Show that if these conditions are satisfied the coefficient of $\sin(2n+1)x$ is

$$b_{2n+1} = \frac{4}{\pi}\int_0^{\pi/2} f(x)\sin(2n+1)x\,dx.$$

If such a function is zero for $0 < x < \frac{1}{3}\pi$ and unity for $\frac{1}{3}\pi < x < \frac{1}{2}\pi$, determine the first four terms of its Fourier series. [L.U.]

11.11. Change of Length of Interval

In practice we often require to find a Fourier series for an interval which is not of length π or 2π; thus we may be concerned with a rod of length $2l$.

Suppose we are given $y = f(x)$ defined in the range $-l$ to $+l$. We change this interval into one of length 2π by the substitution

$$z = \frac{\pi x}{l},$$

so that when $x = -l$, $z = -\pi$ and when $x = +l$, $z = +\pi$.

Now instead of $\qquad y = f(x), \quad (-l < x < +l)$

we have

$$y = f\left(\frac{lz}{\pi}\right), \quad (-\pi < z < +\pi).$$

This, being a function defined in the range $-\pi$ to $+\pi$, can be expanded as a Fourier series in the usual manner. If it is

$$\tfrac{1}{2}b_0 + \sum_1^\infty (b_n\cos nz + a_n\sin nz) \quad . \quad . \quad . \quad . \quad (1)$$

then as usual

$$b_0 = \frac{1}{\pi}\int_{-\pi}^{+\pi} f\left(\frac{lz}{\pi}\right)dz,$$

$$b_n = \frac{1}{\pi}\int_{-\pi}^{+\pi} f\left(\frac{lz}{\pi}\right)\cos nz\,dz,$$

$$a_n = \frac{1}{\pi}\int_{-\pi}^{+\pi} f\left(\frac{lz}{\pi}\right)\sin nz\,dz.$$

Having found the coefficients for (1), we may now revert to the original variable $x = lz/\pi$ and obtain, instead of (1),

$$f(x) = \tfrac{1}{2}b_0 + \sum_1^\infty \left(b_n\cos\frac{n\pi x}{l} + a_n\sin\frac{n\pi x}{l}\right),$$

Example 6. Find a Fourier series for
$$y = x, \qquad -l < x < +l.$$

With the substitution $z = \frac{\pi x}{l}$, this becomes

$$y = \frac{lz}{\pi}, \qquad -\pi < z < +\pi.$$

The graph of this function is symmetrical with respect to the origin, and the Fourier series will therefore consist of sine terms only.

If
$$\frac{lz}{\pi} = \sum_1^\infty a_n \sin nz,$$

$$\pi a_n = \int_{-\pi}^{+\pi} \frac{lz}{\pi} \sin nz\, dz = -\frac{2l}{n} \cos n\pi$$

$$= \frac{2l}{n} \quad (n \text{ odd})$$

$$= -\frac{2l}{n} \quad (n \text{ even}),$$

$$\therefore \quad \frac{lz}{\pi} = \frac{2l}{\pi} [\sin z - \tfrac{1}{2} \sin 2z + \tfrac{1}{3} \sin 3z \ldots],$$

which reverts to

$$x = \frac{2l}{\pi} \left[\sin \frac{\pi x}{l} - \tfrac{1}{2} \sin \frac{2\pi x}{l} + \tfrac{1}{3} \sin \frac{3\pi x}{l} - \cdots \right].$$

In practice it is more usual to work directly with the Fourier series (1) and the integrals following it expressed in terms of x. Instead of (1) and the following integrals we then have, using $z = \pi x/l$,

$$f(x) = \tfrac{1}{2} b_0 + \Sigma \left(b_n \cos \frac{n\pi x}{l} + a_n \sin \frac{n\pi x}{l} \right), \qquad (-l < x < +l),$$

where

$$b_0 = \frac{1}{l} \int_{-l}^{+l} f(x)\, dx,$$

$$b_n = \frac{1}{l} \int_{-l}^{+l} f(x) \cos \frac{n\pi x}{l}\, dx,$$

$$a_n = \frac{1}{l} \int_{-l}^{+l} f(x) \sin \frac{n\pi x}{l}\, dx.$$

In the same way for a half-range sine series
$$y = f(x) \qquad 0 < x < l$$

we assume
$$f(x) = \Sigma a_n \sin \frac{n\pi x}{l},$$

where
$$a_n = \frac{2}{l} \int_0^l f(x) \sin \frac{n\pi x}{l}\, dx.$$

For a half-range cosine series
$$y = f(x) \qquad (0 < x < l)$$

we assume
$$f(x) = \tfrac{1}{2} b_0 + \Sigma b_n \cos \frac{n\pi x}{l},$$

where

$$b_0 = \frac{2}{l} \int_0^l f(x)dx, \quad b_n = \frac{2}{l} \int_0^l f(x) \cos \frac{n\pi x}{l} dx.$$

Example 7. Find a cosine series for

$$y = 2x - 1, \qquad 0 < x < 1.$$

If

$$y = \tfrac{1}{2}b_0 + \Sigma b_n \cos n\pi x,$$

$$b_0 = 2 \int_0^1 (2x - 1)dx = 0$$

$$b_n = 2 \int_0^1 (2x - 1) \cos n\pi x dx$$

$$= \frac{4}{n^2\pi^2} (\cos n\pi - 1) = 0 \quad (n \text{ even})$$

$$= -\frac{8}{n^2\pi^2} \quad (n \text{ odd}).$$

$$\therefore \quad 2x - 1 = -\frac{8}{\pi^2} \left\{ \frac{1}{1^2} \cos \pi x + \frac{1}{3^2} \cos 3\pi x + \ldots \right\}$$

$$= -\frac{8}{\pi^2} \sum_1^\infty \frac{\cos (2n - 1)\pi x}{(2n - 1)^2}, \quad \text{where} \quad 0 < x < 1.$$

EXERCISE 24

1. Find a Fourier series in the range $(-2, 2)$ for the function
$$f(x) = 0 \text{ for } -2 < x < 0, \quad f(x) = 1 \text{ for } 0 < x < 2.$$

2. Find a Fourier series for
$$f(x) = 2x - 1, \quad -1 < x < 1.$$

3. Find a sine expansion for $y = 2x - 1$ in the range $0 < x < 1$.

4. Find (a) a sine expansion, (b) a cosine expansion for $y = x^2$ in the range $(0, 3)$.

5. Find a half-range sine series for
$$y = \frac{2}{l} x, \quad 0 < x < \tfrac{1}{2}l,$$

$$y = \frac{2}{l} (l - x), \quad \tfrac{1}{2}l < x < l.$$

6. An alternating e.m.f. e is defined as $e = E_0$ for $0 < x < \pi/w$ and $e = -E_0$ for $\pi/w < x < 2\pi/w$ is applied to a pure inductance L. Assuming that $i = 0$ when $t = 0$, show that the maximum current is $\pi E_0/wL$.

11.12. The R.M.S. Value of a Function

We have already dealt with R.M.S. values in the previous volume [Vol. I (19.9)], but here will deal specifically with current and voltages represented by Fourier series.

If

$$y = \tfrac{1}{2}b_0 + \sum_1^\infty (b_n \cos n\omega t + a_n \sin n\omega t)$$

so that $T = 2\pi/\omega$ is the fundamental period, then the R.M.S. value Y is defined as

$$Y^2 = \frac{1}{T}\int_0^T y^2 dt.$$

The square of y will contain such terms as

$$\tfrac{1}{4}b_0^2, \quad b_n^2 \cos^2 n\omega t, \quad a_n^2 \sin^2 n\omega t, \quad b_0 b_n \cos n\omega t, \quad b_0 a_n \sin n\omega t,$$
$$2b_n b_m \cos n\omega t \cos m\omega t, \quad 2b_n a_m \cos n\omega t \sin m\omega t,$$
$$2a_n a_m \sin n\omega t \sin m\omega t,$$

where $n \neq m$, but each will have any integral value from 1 to ∞.

The integral of each of the last five over the range 0 to T is zero, whilst the first three terms give :

$$\frac{1}{T}\int_0^T \tfrac{1}{4}b_0^2 dt = \tfrac{1}{4}b_0^2,$$

$$\frac{1}{T}\int_0^T b_n^2 \cos^2 n\omega t\, dt = \tfrac{1}{2}b_n^2,$$

$$\frac{1}{T}\int_0^T a_n^2 \sin^2 n\omega t\, dt = \tfrac{1}{2}a_n^2,$$

$$\therefore \quad Y^2 = \tfrac{1}{4}b_0^2 + \tfrac{1}{2}\sum_1^\infty (b_n^2 + a_n^2).$$

In the case of an alternating voltage

$$e = \sum_1^\infty E_n \sin (n\omega t + \alpha_n)$$

$$= \sum [(E_n \cos \alpha_n) \sin n\omega t + (E_n \sin \alpha_n) \cos n\omega t]$$

the R.M.S. value E is given by

$$E^2 = \tfrac{1}{2}\sum (E_n^2 \cos^2 \alpha_n + E_n^2 \sin^2 \alpha_n)$$

$$= \tfrac{1}{2}\sum E_n^2$$

and $\qquad E = \sqrt{\tfrac{1}{2}(E_1^2 + E_2^2 + E_3^2 + \ldots + E_n^2 + \ldots)}.$

Similarly, if, using cosine instead of sine to show that the result is not affected,

$$i = \sum_1^\infty I_n \cos (n\omega t + \beta_n)$$

the R.M.S. value I is given by

$$I = \sqrt{\tfrac{1}{2}(I_1^2 + I_2^2 + \ldots + I_n^2 + \ldots)}.$$

11.13. Harmonic Analysis

In practice, the function to be resolved into its harmonic constituents is seldom given by a mathematical expression. Usually we are given an irregular graph, and by careful measurement must obtain the value of the ordinate at a number of equidistant points on the x axis, the total range on the x axis corresponding to the period 2π. In the following examples we will assume that these values of the ordinates have been obtained.

There are a number of attractive methods (see *Waveform Analysis*, by R. G. Manley), but we will consider only the following, which is the same in principle however many ordinates are taken, also with it any given harmonic may be calculated without any time being spent in finding unwanted coefficients.

Suppose the range 0 to 2π is divided into N equal parts by x_0, x_1, . . . x_N, where x_0 is the point 0 and x_N the point 2π. The ordinates at these points are y_0, y_1, . . . y_N, where because of the periodicity $y_0 = y_N$.

Remembering that an integral of y is the limiting value of a sum of terms such as $y_r \Delta x_r$, we will take for Δx_r the constant value $2\pi/N$, which is the length of each of the above intervals such as $x_r x_{r+1}$ and for y_r we will take the value of the ordinate at the end of each such interval. If then

$$y = \tfrac{1}{2}a_0 + \sum_{r=1}^{\infty} (a_r \cos rx + b_r \sin rx),$$

so that each coefficient is given by the usual integral formula, then using the sum of terms such as $y_r \Delta x_r$ as an approximation to the value of the integral, the Fourier coefficients are given by

$$a_0 = \frac{1}{\pi}\int_0^{2\pi} y\,dx \simeq \frac{1}{\pi} \sum_{k=1}^{N} y_k \cdot \frac{2\pi}{N} = \frac{2}{N} \sum_{k=1}^{N} y_k,$$

$$a_r = \frac{1}{\pi}\int_0^{2\pi} y \cos rx\,dx \simeq \frac{1}{\pi} \sum_{k=1}^{N} y_k \cos (rx_k) \cdot \frac{2\pi}{N} = \frac{2}{N} \sum_{k=1}^{N} y_k \cos (rx_k),$$

$$b_r = \frac{1}{\pi}\int_0^{2\pi} y \sin rx\,dx \simeq \frac{1}{\pi} \sum_{k=1}^{N} y_k \sin (rx_k) \cdot \frac{2\pi}{N} = \frac{2}{N} \sum_{k=1}^{N} y_k \sin (rx_k).$$

We could equally well use the value of y_k at the beginning of each interval $2\pi/N$ and write each of the above sums as

$$\sum_{k=0}^{N-1}$$

The following example shows the application of the method.

Example 8. A curve of current i covers one period of 2π. The values for twelve equidistant values of t covering the range and taken from the curve are 2·340, 3·012, 3·685, 4·149, 3·685, 2·203, 0·825, 0·513, 0·875, 1·085, 1·189, 1·637.

Find a Fourier series if terms of the fourth and higher harmonics are negligible.

θ.	i.	$\cos \theta$.	$i \cos \theta$.	$\sin \theta$.	$i \sin \theta$.	$\cos 2\theta$.	$i \cos 2\theta$.	$\sin 2\theta$.	$i \sin 2\theta$.	$\sin 3\theta$.	$i \sin 3\theta$.	$\cos 3\theta$.	$i \cos 3\theta$.
30°	2·340	0·866	2·026	0·5	1·170	0·5	1·170	0·866	2·026	1	2·340	0	0
60°	3·012	0·5	1·506	0·866	2·608	−0·5	−1·506	0·866	2·608	0	0	−1	−3·012
90°	3·685	0	0	1	3·685	−1	−3·685	0	0	−1	−3·685	0	0
120°	4·149	−0·5	−2·074	0·866	3·593	−0·5	−2·074	−0·866	−3·593	0	0	1	4·149
150°	3·685	−0·866	−3·191	0·5	1·843	0·5	1·843	−0·866	−3·191	1	3·685	0	0
180°	2·203	−1	−2·203	0	0	1	2·203	0	0	0	0	−1	−2·203
210°	0·825	−0·866	−0·714	−0·5	−0·413	0·5	0·413	0·866	0·714	−1	−0·825	0	0
240°	0·513	−0·5	−0·257	−0·866	−0·444	−0·5	−0·257	0·866	0·444	0	0	1	0·513
270°	0·875	0	0	−1	−0·875	−1	−0·875	0	0	1	0·875	0	0
300°	1·085	0·5	0·542	−0·866	−0·940	−0·5	−0·542	−0·866	−0·940	0	0	−1	−1·085
330°	1·189	0·866	1·030	−0·5	−0·595	0·5	0·595	−0·866	−1·030	−1	−1·189	0	0
360°	1·637	1	1·637	0	0	1	1·637	0	0	0	0	1	1·637
	25·198		−1·698		9·632		−1·078		−2·962		1·201		−0·001

$$a_0 = \tfrac{2}{12} \times 25\cdot198 = 4\cdot199 = 4\cdot20$$
$$a_1 = \tfrac{1}{6} \times - 1\cdot698 = - 0\cdot283, \qquad b_1 = \tfrac{1}{6} \times 9\cdot632 = 1\cdot605$$
$$a_2 = \tfrac{1}{6} \times - 1\cdot078 = - 0\cdot180, \qquad b_2 = \tfrac{1}{6} \times - 2\cdot962 = - 0\cdot494,$$
$$a_3 = \tfrac{1}{6} \times - 0\cdot001 = - 0\cdot0002, \qquad b_3 = \tfrac{1}{6} \times 1\cdot201 = 0\cdot200.$$

Hence

$$i = 2\cdot10 - 0\cdot283 \cos\theta - 0\cdot180 \cos 2\theta - 0\cdot0002 \cos 3\theta$$
$$+ 1\cdot605 \sin\theta - 0\cdot494 \sin 2\theta + 0\cdot200 \sin 3\theta.$$

It will be noticed that:

(1) The values of the trigonometrical functions repeat themselves with at most a change of sign and that the arithmetic involved is not very heavy.

(2) The method is independent of the number of ordinates given, but the more ordinate values given, the greater the accuracy.

(3) We can continue and find any further harmonics required.

EXERCISE 25

1. Determine the first 3 harmonics of the Fourier series for the values:

x	.	.	.	0°	60°	120°	180°	240°	300°
y	.	.	.	0·47	1·77	2·20	−2·20	−1·64	−0·49

2. The values of y, a periodic function of t, are given below for twelve equidistant values of t covering the whole period. Express y in a Fourier series as far as the second harmonic if the first value is for $\theta = 30°$.

13·602, 18·468, 20·671, 20·182, 17·820, 14·346, 10·130, 5·612, 1·877, 0·486, 2·500, 7·606.

3. The displacement x of a sliding piece from a point in its path is measured for every 30° of rotation of the crank. Find the first five coefficients in the Fourier series if the first value is for $\theta = 30°$.

$x = 2\cdot15,\ 2\cdot70,\ 3\cdot00,\ 3\cdot00,\ 2\cdot57,\ 1\cdot60,\ 0\cdot75,\ 0\cdot32,\ 0\cdot32,\ 0\cdot60,\ 1\cdot05,\ 1\cdot54.$

4. The base of a curve covering the range 0 to 2π is divided into twelve equal parts, and the values of the ordinates at $x = 2\pi/12,\ 4\pi/12,\ \ldots\ 24\pi/12$ are 3·5, 6·09, 7·82, 8·58, 8·43, 7·73, 6·98, 6·19, 6·04, 5·55, 5·01, 3·35. Find the first seven coefficients in the Fourier series fitting this data.

CHAPTER 12

PARTIAL DIFFERENTIATION

12.1. Partial Derivatives

In Vol. I (27) partial differentiation was dealt with in an elementary fashion. It was seen that, for example, if

$$z = f(x, y, t, \ldots),$$
$$\frac{\partial z}{\partial y} = \underset{\Delta y \to 0}{\mathfrak{L}t.} \frac{f(x, y + \Delta y, t \ldots) - f(x, y, t, \ldots)}{\Delta y}$$

or to find the partial derivative of z with respect to y we differentiate as usual with respect to y, regarding the other variables as constants. In this way we can find

$$z_x, \; z_y, \; z_t, \; \ldots$$

or in other notations

$$\frac{\partial z}{\partial x}, \; \frac{\partial z}{\partial y}, \; \frac{\partial z}{\partial t} \ldots$$

and

$$\frac{\partial f}{\partial x}, \; \frac{\partial f}{\partial y}, \; \frac{\partial f}{\partial t} \ldots$$

using f or z as is convenient.

From these functions we can form second-order partial derivatives such as

$$\frac{\partial}{\partial x}\left(\frac{\partial z}{\partial x}\right), \quad \frac{\partial}{\partial x}\left(\frac{\partial z}{\partial y}\right) \ldots$$

which can be written as

$$z_{xx}, \; z_{yx}, \; \ldots$$

or

$$\frac{\partial^2 z}{\partial x^2}, \; \frac{\partial^2 z}{\partial x \partial y}, \; \ldots$$

and

$$\frac{\partial^2 f}{\partial x^2}, \; \frac{\partial^2 f}{\partial x \partial y}, \; \ldots$$

In all the examples we shall meet with

$$\frac{\partial}{\partial x}\left(\frac{\partial z}{\partial y}\right) = \frac{\partial}{\partial y}\left(\frac{\partial z}{\partial x}\right)$$

or

$$z_{yx} = z_{xy},$$

showing that we may, for example, differentiate first with respect to x and then y or vice versa, the final result being the same.

Example 1. If $z = x^3y^4t^5 + f(x) + F(y) + G(t)$ find all the first-order partial derivatives and show that

$$z_{xt} = z_{tx}.$$

By differentiation

$$z_x = 3x^2y^4t^5 + f'(x),$$
$$z_y = 4x^3y^3t^5 + F'(y),$$
$$z_t = 5x^3y^4t^4 + G'(t).$$

also from z_x and z_t

$$z_{xt} = 15x^2y^4t^4,$$
$$z_{tx} = 15x^2y^4t^4.$$

12.2. The Total Differential

(i) In the case of $y = f(x)$ a function of a single independent variable x, a new variable written dx is associated with x. This new variable is called the *differential* of x, is quite independent of x and may have any arbitrary value.

The differential of y, the dependent variable, is *defined* as

$$f'(x)dx \equiv \frac{d}{dx}(y) \cdot dx \quad \cdots \cdots \quad (1)$$

and denoted by dy. Since now

$$dy = f'(x)dx$$
$$\frac{dy}{dx} = f'(x)$$

where the left-hand side is a ratio of two expressions which can be separated when required. It is unfortunate that the usual notation does not help us to distinguish between $\frac{dy}{dx}$ meaning $\frac{d}{dx}(y)$ and $\frac{dy}{dx}$ meaning the ratio of two differentials, but in practice we do not need to do so.

From the above we may easily obtain the rules for differentiation in terms of differentials. Thus if

$$y = uv,$$

where u and v are functions of x, then by the definition (1)

$$du = \frac{d}{dx}(u)dx, \quad dv = \frac{d}{dx}(v)dx, \quad dy = \frac{d}{dx}(y)dx. \quad \cdot \quad (2)$$

and it follows from

$$dy = \frac{d}{dx}(uv)dx$$

$$= \left[u\frac{d}{dx}(v) + v\frac{d}{dx}(u) \right]dx$$

$$= u\left[\frac{d}{dx}(v)dx \right] + v\left[\frac{d}{dx}(u)dx \right]$$

that by (2) $\qquad d(uv) = udv + vdu.$

Similarly $\quad d\left(\dfrac{u}{v}\right) = \dfrac{v\,du - u\,dv}{v^2}.$

(This justifies the use of differentials as used in Vol. I.)

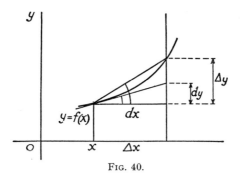

FIG. 40.

The figure helps to explain the following.
Since

$$\operatorname*{Lt.}_{\Delta x \to 0} \frac{\Delta y}{\Delta x} = f'(x)$$

$$\frac{\Delta y}{\Delta x} = f'(x) + \varepsilon$$

where $\varepsilon \longrightarrow 0$ with Δx. Therefore

$$\Delta y = f'(x)\Delta x + \varepsilon\Delta x,$$

and since if Δx is small, ε is nearly zero, $\varepsilon\Delta x$ is very much smaller than $f'(x)\Delta x$. The differential dx is arbitrarily chosen, and we can choose it to equal Δx. We now have

$$\Delta y = f'(x)dx + \varepsilon\Delta x$$
$$= dy + \varepsilon\Delta x,$$

showing that the differential dy is a first approximation to or the principal part of the increment Δy in y when $dx\ (= \Delta x)$ is small.

(ii) For a function of two or more variables we proceed in a similar fashion. Thus for $z = f(x, y)$ we associate with x and y two independent variables dx and dy, called the differentials of x and y, and *define* the *total differential* of z as

$$dz = \frac{\partial f}{\partial x}\,dx + \frac{\partial f}{\partial y}\,dy$$

so that dz is a function of four independent variables x, y, dx, dy. We relate it to the usual increment Δz of z as follows :
Let $\quad \Delta z = f(x + \Delta x, y + \Delta y) - f(x, y).$
If we add and subtract $\quad f(x, y + \Delta y),$
$$\Delta z = f(x + \Delta x, y + \Delta y) - f(x, y + \Delta y) + f(x, y + \Delta y) - f(x, y).$$

But

$$\operatorname*{Lt.}_{\Delta x \to 0} \frac{f(x + \Delta x, y + \Delta y) - f(x, y + \Delta y)}{\Delta x} = \frac{\partial f(x, y + \Delta y)}{\partial x},$$

so that

$$f(x + \Delta x, y + \Delta y) - f(x, y + \Delta y) = \left(\frac{\partial f(x, y + \Delta y)}{\partial x} + \varepsilon_1\right) \Delta x$$

where $\varepsilon_1 \longrightarrow 0$ with Δx. Also since

$$\operatorname*{Lt.}_{\Delta y \to 0} \frac{\partial f(x, y + \Delta y)}{\partial x} = \frac{\partial f(x, y)}{\partial x},$$

assuming the derivative to be continuous it follows that

$$\frac{\partial f(x, y + \Delta y)}{\partial x} = \frac{\partial f(x, y)}{\partial x} + \varepsilon_2$$

where $\varepsilon_2 \longrightarrow 0$ with Δy. Similarly,

$$f(x, y + \Delta y) - f(x, y) = \left(\frac{\partial f(x, y)}{\partial y} + \varepsilon_3\right) \Delta y$$

where $\varepsilon_3 \longrightarrow 0$ with Δy. From these

$$\Delta z = \frac{\partial f}{\partial x} \Delta x + \frac{\partial f}{\partial y} \Delta y + \varepsilon \Delta x + \varepsilon_3 \Delta y$$

where $\varepsilon = \varepsilon_1 + \varepsilon_2$.

Choosing the arbitrary values dx, dy to be Δx, Δy the expression

$$\frac{\partial f}{\partial x} \Delta x + \frac{\partial f}{\partial y} \Delta y \equiv \frac{\partial f}{\partial x} dx + \frac{\partial f}{\partial y} dy$$

is, as above, *defined* as the total differential of z and written dz.

Generally if

$$z = f(x_1, x_2, \ldots x_n)$$

then

$$dz = \sum_{r=1}^{n} \frac{\partial f}{\partial x_r} dx_r \quad \text{or} \quad \sum_{r=1}^{n} \frac{\partial z}{\partial x_r} dx_r$$

is called the *total differential* of z and it can be seen that it is the *principal part* of Δz, being a close approximation to Δz for sufficiently small values of $dx_1, dx_2 \ldots dx_n$ where each $dx_r \equiv \Delta x_r$. We have already applied this to problems on small errors, and the above method of differentials justifies the procedure adopted.

12.3. Total Derivatives

If instead of being independent variables, x and y are both functions of a variable t, we can divide the expression

$$\Delta z = \frac{\partial f}{\partial x} \Delta x + \frac{\partial f}{\partial y} \Delta y + \varepsilon \Delta x + \varepsilon_3 \Delta y$$

by Δt to give

$$\frac{\Delta z}{\Delta t} = \frac{\partial f}{\partial x} \frac{\Delta x}{\Delta t} + \frac{\partial f}{\partial y} \frac{\Delta y}{\Delta t} + \varepsilon \frac{\Delta x}{\Delta t} + \varepsilon_3 \frac{\Delta y}{\Delta t}$$

and in the limit when Δx, $\Delta y \longrightarrow 0$ with Δt,

$$\frac{dz}{dt} = \frac{\partial f}{\partial x} \frac{dx}{dt} + \frac{\partial f}{\partial y} \frac{dy}{dt} \qquad \cdots \cdots \quad (1)$$

Or since, with differentials,

$$dz = \frac{dz}{dt} dt, \quad dx = \frac{dx}{dt} dt, \quad dy = \frac{dy}{dt} dt$$

on multiplying through by dt and using differentials

$$dz = \frac{\partial f}{\partial x} dx + \frac{\partial f}{\partial y} dy \quad . \quad . \quad . \quad . \quad . \quad (2)$$

so that this form holds whether x and y are dependent on a third variable t or independent.

Generally if

$$z = f(x_1, x_2, \ldots x_n)$$

where

$$x_1 = \phi_1(t), \ldots x_n = \phi_n(t)$$

then

$$\frac{dz}{dt} = \sum_{r=1}^{n} \frac{\partial f}{\partial x_r} \frac{dx_r}{dt} \quad . \quad . \quad . \quad . \quad . \quad . \quad . \quad (3)$$

and in differentials

$$dz = \sum_{r=1}^{n} \frac{\partial f}{\partial x_r} dx_r \quad . \quad . \quad . \quad . \quad . \quad . \quad . \quad (4)$$

For the case of $t = x$ and two variables x, y, (3) gives

$$\frac{dz}{dx} = \frac{\partial f}{\partial x} + \frac{\partial f}{\partial y} \frac{dy}{dx}.$$

If a function is given by $f(x, y) = 0$, then since z is zero always, so is dz/dx, and the above gives

$$0 = \frac{\partial f}{\partial x} + \frac{\partial f}{\partial y} \frac{dy}{dx}$$

or

$$\frac{dy}{dx} = - f_x/f_y,$$

provided $f_y \neq 0$, giving a formula for the gradient at any point on the curve $f(x, y) = 0$.

Example 2.   Find the gradient for the curve

$$x^3 + y^3 - 3axy = 0.$$
$$f_x = 3x^2 - 3ay,$$
$$f_y = 3y^2 - 3ax,$$
$$\therefore \quad \frac{dy}{dx} = - \frac{3(x^2 - ay)}{3(y^2 - ax)} = - \left(\frac{x^2 - ay}{y^2 - ax}\right).$$

As a more general expression suppose x, y, z are each a function of a parameter t and are connected by the equation

$$f(x, y, z) = 0,$$

then by (3) above since $f = 0$,

$$\frac{\partial f}{\partial x} \frac{dx}{dt} + \frac{\partial f}{\partial y} \frac{dy}{dt} + \frac{\partial f}{\partial z} \frac{dz}{dt} = 0.$$

If here we put $t = x$, this becomes

$$\frac{\partial f}{\partial x} + \frac{\partial f}{\partial y}\frac{dy}{dx} + \frac{\partial f}{\partial z}\frac{dz}{dx} = 0.$$

But if we regard $f(x, y, z) = 0$ as an equation which gives z in terms of the independent variables x and y, then keeping y and x constant in turn,

$$\frac{\partial f}{\partial x} + \frac{\partial f}{\partial z}\frac{\partial z}{\partial x} = 0, \qquad \frac{\partial f}{\partial y} + \frac{\partial f}{\partial z}\frac{\partial z}{\partial y} = 0,$$

from which

$$\frac{\partial z}{\partial x} = -\frac{\partial f}{\partial x}\bigg/\frac{\partial f}{\partial z}, \qquad \frac{\partial z}{\partial y} = -\frac{\partial f}{\partial y}\bigg/\frac{\partial f}{\partial z}.$$

Example 3. Find z_x, z_y and dz when
$$x^2 + y^2 + z^2 = (2x - y - z)^2.$$

This means, as above, that z is regarded as a function of the two independent variables x and y, each of which can be varied independently of the other.

Keeping y and x constant in turn,

$$2x + 2z\frac{\partial z}{\partial x} = 2(2x - y - z)\left(2 - \frac{\partial z}{\partial x}\right),$$

giving

$$\frac{\partial z}{\partial x} = \frac{3x - 2y - 2z}{2x - y}$$

Similarly,

$$2y + 2z\frac{\partial z}{\partial y} = 2(2x - y - z)\left(-1 - \frac{\partial z}{\partial y}\right)$$

and

$$\frac{\partial z}{\partial y} = \frac{z - 2x}{2x - y}.$$

Or, taking the total differential

$$2xdx + 2ydy + 2zdz = 2(2x - y - z)(2dx - dy - dz),$$

from which

$$dz = \left(\frac{3x - 2y - 2z}{2x - y}\right)dx + \left(\frac{z - 2x}{2x - y}\right)dy$$

and taking y and x as constant in turn gives as above the values of z_x, z_y.

12.4. We have had

$$\frac{dz}{dt} = \Sigma\frac{\partial f}{\partial x_r}\frac{dx_r}{dt}$$

when each x_r was a function of t only. It can be shown that if x_r is a function of several variables so that

$$x_r = \phi_r(t_1, t_2, \ldots t_m) \qquad (r = 1, 2 \ldots n)$$

then

$$\frac{\partial z}{\partial t_p} = \sum_{r=1}^{n}\frac{\partial f}{\partial x_r}\frac{\partial x_r}{\partial t_p} \qquad \cdots \cdots \cdots \quad (5)$$

where t_p is any one of the variables $t_1 \ldots t_m$. This gives the total rate of change of z when only the t_p affecting each x is changed. From this we can again derive

$$dz = \Sigma\frac{\partial f}{\partial x_r}dx_r \qquad \cdots \cdots \cdots \quad (6)$$

showing that this holds whether the independent variables are $x_1 \ldots x_n$ or $t_1 \ldots t_m$.

Example 4. Let $z = e^{xy}$, where $x = \log (u + v)$, $y = \tan^{-1}(u/v)$. To find z_u, z_v.

By (5) above

$$\frac{\partial z}{\partial u} = \frac{\partial z}{\partial x}\frac{\partial x}{\partial u} + \frac{\partial z}{\partial y}\frac{\partial y}{\partial u}$$

$$= ye^{xy} \cdot \frac{1}{u + v} + xe^{xy} \cdot \frac{v}{v^2 + u^2}.$$

Similarly,

$$\frac{\partial z}{\partial v} = \frac{\partial z}{\partial x}\frac{\partial x}{\partial v} + \frac{\partial z}{\partial y}\frac{\partial y}{\partial v}$$

$$= ye^{xy} \cdot \frac{1}{u + v} - xe^{xy} \cdot \frac{u}{v^2 + u^2}.$$

Example 5. Find $\dfrac{\partial u}{\partial x}, \dfrac{\partial v}{\partial x}, \dfrac{\partial u}{\partial y}, \dfrac{\partial v}{\partial y}$ from the pair of equations $v + \log u = xy$, $u + \log v = x - y$.

We are expected to regard these equations as giving u and v each in terms of the independent variables x and y. Therefore differentiating each, keeping y constant,

$$\frac{\partial v}{\partial x} + \frac{1}{u}\frac{\partial u}{\partial x} = y,$$

$$\frac{1}{v}\frac{\partial v}{\partial x} + \frac{\partial u}{\partial x} = 1.$$

Solving these gives

$$\frac{\partial v}{\partial x} = \frac{v(1 - uy)}{1 - uv}, \quad \frac{\partial u}{\partial x} = \frac{u(y - v)}{1 - uv}.$$

Similarly, keeping x constant

$$\frac{\partial v}{\partial y} + \frac{1}{u}\frac{\partial u}{\partial y} = x,$$

$$\frac{1}{v}\frac{\partial v}{\partial y} + \frac{\partial u}{\partial y} = -1,$$

and the solution of these gives u_y and v_y.

12.5. Envelopes

Usually a curve is considered as the aggregate of a number of points, but if a large number of lines be drawn at unit distance from a fixed

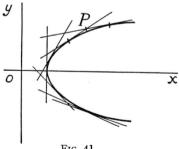

Fig. 41.

point obviously the outline shown will be a circle. Similarly, the figure shows a large number of lines $y = mx + 2/m$ drawn for different values of m. They outline a curve which soon will be shown as the parabola $y^2 = 8x$. We say that the line $y = mx + 2/m$, where m is the parameter of the family of lines, envelopes the curve and the curve is the *envelope* of the family of lines. Let P be the point of intersection of two near members of the family. It appears intuitively obvious that as the two members tend to coincidence the point P tends nearer to the envelope, and finally is a point on it. To obtain the equation to the envelope from this point of view let

$$f(x, y, m) = 0, \qquad f(x, y, m + \Delta m) = 0$$

be the two near members. Their point of intersection will also lie on

$$f(x, y, m + \Delta m) - f(x, y, m) = 0$$

and therefore on

$$\frac{f(x, y, m + \Delta m) - f(x, y, m)}{\Delta m} = 0,$$

where we have divided by Δm.

Proceeding to the limit as $\Delta m = 0$, the point of intersection also lies on

$$\frac{\partial f}{\partial m} = 0,$$

so that the pair of equations

$$f(x, y, m) = 0, \qquad \frac{\partial}{\partial m} f(x, y, m) = 0$$

provide a means of finding x, y in terms of the parameter m, (x, y) being a point on the envelope.

Example 6. Find the envelope of the line $y = mx + a/m$, where m is the parameter.

This is

$$m^2x - my + a = 0 \quad \cdots \quad \cdots \quad \cdots \quad (1)$$

$\dfrac{\partial f}{\partial m}$:

$$2mx - y = 0$$

solving for x and y:

$$x = a/m^2$$
$$y = 2a/m,$$

giving a parametric representation of the point or, on eliminating m, $y^2 = 4ax$ the equation to the envelope.

It will be seen that the result $y^2 = 4ax$ is the condition that the equation (1) a quadratic in m has equal roots, and this result should be known. Thus if m (or the parameter) occurs in a quadratic form

$$Am^2 + Bm + C = 0 \quad \cdots \quad \cdots \quad \cdots \quad (2)$$

where A, B, C are functions of x, y, then the envelope is

$$B^2 = 4AC.$$

The explanation is clear for given a point (x, y) the values substituted in (2) will give two distinct values of m or two curves through the point. If, however, the point is on the envelope the curves must coincide and the roots in m must be equal.

EXERCISE 26

1. A ship is anchored at a distance of n m from the base of a lighthouse, and the angle of elevation of the top of the lighthouse is observed to be x degrees. Later, when the tide has fallen but the ship has not otherwise moved, the elevation is $x + h$ degrees. Find how many metres the tide has fallen and show that if h is small the number is roughly equal to

$$\frac{1}{60} \pi n h \sec^2 (x°) \text{ metres}$$

2. The base b of a triangle and the base angles A and C are measured. If there are small errors p, q respectively in these base angles, show that the error in the calculated value of the height of the triangle is

$$b \left(\frac{q \sin^2 A + p \sin^2 C}{\sin^2 (A + C)} \right).$$

3. The angles of a triangle are calculated from the sides a, b, c. If small changes δa, δb, δc are made in the sides, show that approximately

$$\delta A = \frac{a}{2\Delta} (\delta a - \delta b \cos C - \delta c \cos B)$$

where Δ is the area of the triangle. Apply this to calculate in degrees and minutes the angle A of a triangle in which

$$a = 99{\cdot}5, \quad b = 100{\cdot}2, \quad c = 100{\cdot}7.$$
$$[1 \text{ degree} = 0{\cdot}0175 \text{ radians.}] \qquad \text{[L.U.]}$$

4. In a triangle ABC, A, B, c are measured and a is calculated. If there is an error of x minutes in A show that the error in a is approximately

$$\frac{\pi x c \sin B}{10\,800 \sin^2 (A + B)}.$$

5. A triangle ABC is determined by the values of a, b and C. If these measures are liable to small errors Δa, Δb, ΔC prove that the error in c is approximately

$$\cos B \, \Delta a + \cos A \, \Delta b + a \sin B\Delta C.$$

6. Two rods AC = 5m, BC = 4 m, are fastened to the points A, B on a vertical wall with A 2 m vertically above B. If B is displaced a small distance Δc vertically downwards, show that the resultant vertical displacement of C is approximately $5\Delta c/8$ downwards.

7. (a) If $f(y/x)$ is any differentiable function of y/x, find the value of $x\partial u/\partial x + y\partial u/\partial y$ when

$$u = f(y/x) + (x^2 + y^2)^{1/2}.$$

(b) The height h of an object is calculated from the angles of elevation A, B at the ends of a horizontal base line of length c in a vertical plane containing the object and on the same side of it. If δh is the greatest possible error in h due to errors δA, δB, δc in the measured quantities, show that

$$\frac{\delta h}{h} = \frac{|\delta c|}{c} + \frac{h}{c} \{\operatorname{cosec}^2 A \,|\delta A| + \operatorname{cosec}^2 B \,|\delta B|\}. \qquad \text{[L.U.]}$$

8. (a) If $u = \tan^{-1} \left(\frac{x}{y} - \frac{y}{x} \right)$, find u_x, u_y and show that $xu_x + yu_y = 0$.

(b) Verify that if $z = 3xy - y^3 + (y^2 - 2x)^{3/2}$ then $\dfrac{\partial^2 z}{\partial x^2} \dfrac{\partial^2 z}{\partial y^2} = \left(\dfrac{\partial^2 z}{\partial x \partial y} \right)^2$.

9. (a) If z is a function of u and $u = (x^2 + y^2) \tan^{-1} \left(\frac{y}{x} \right)$, prove that

$$x \frac{\partial z}{\partial y} - y \frac{\partial z}{\partial x} = (x^2 + y^2) \frac{\partial z}{\partial u}.$$

(b) If $z = xf(x + y) + yF(x + y)$, prove that $\dfrac{\partial^2 z}{\partial x^2} + \dfrac{\partial^2 z}{\partial y^2} = 2 \dfrac{\partial^2 z}{\partial x \partial y}$. [L.U.]

10. If $\dfrac{x^2}{a^2+z} + \dfrac{y^2}{b^2+z} = 1$ where a and b are constant, prove that

$$z_x{}^2 + z_y{}^2 = 2(xz_x + yz_y).$$

11. (a) If $z = x^2 + y^2 - \log(x+y)$ and x, y are both positive, find the value of z when $z_x = 0$ and $z_y = 0$.

(b) If $z = xe^{kxy}$ where k is a constant, show that

$$xz_x - yz_y = z, \qquad x\frac{\partial^2 z}{\partial x^2} - y\frac{\partial^2 z}{\partial x \partial y} = 0. \qquad \text{[L.U.]}$$

12. Find $\dfrac{du}{dt}$ when

(a) $u = xy + 1$, $\quad x = t$, $\quad y = 1/t$;

(b) $u = x^3 + y^3$, $\quad x = t/(1+t^3)$, $\quad y = t^2/(1+t^3)$;

(c) $u = xyz$, $\quad x = e^{2t}\sin t$, $\quad y = e^{-t}\sec t$, $\quad z = \cot t$.

13. If $z = uvw$ where $u = r\cos\theta$, $v = r\sin\theta$, $w = k\theta$, find $\partial z/\partial\theta$.

14. If $u = xy + yz + zx$ where $x = t$, $y = t^2$, $z = \sin t$, find du/dt.

15. The pair of equations $x^2 + y^2 + z^2 - a^2 = 0$, $x^2 - y^2 - 2z^2 - b^2 = 0$ can be regarded as equations from which x, y or z can be eliminated to give an equation in the other two. Take the total differential of each and solve for the ratios $dx : dy : dz$. Hence find dy/dx and dz/dx.

16. Find dz/dx when

(a) $z = x^3 + y^3$, $\quad x^2 + y^2 = 1$.

(b) $z = \dfrac{1}{x^2} + \dfrac{1}{y^2}$, $\quad x^2 + y^2 = a^2$.

17. If $u = xy - yz$ where $x = r + s$, $y = r - s$, $z = t$, find $\partial u/\partial r$, $\partial u/\partial s$ $\partial u/\partial t$.

18. $x^2 + y^2 + u^3 + v^3 = 0$, $x^3 + y^3 + u^2 + v^2 = 0$ can be regarded as a pair of equations from which v, for example, could be eliminated and $\partial u/\partial x$ found from the result. Find this value from the total differential of each equation.

19. Find $\dfrac{\partial u}{\partial x}, \dfrac{\partial v}{\partial x}, \dfrac{\partial u}{\partial y}, \dfrac{\partial v}{\partial y}$ if

$$u^2 + v^2 + y^2 - 2x = 0, \quad u^3 + v^3 - x^3 + 3y = 0.$$

20. (a) Four variables x, y, r, θ are connected by the relations $x = r\cos\theta$, $y = r\sin\theta$. Show that:

(i) $(x_r)_\theta(r_x)_\theta = 1$;

(ii) $(x_r)_\theta = (r_x)_y$;

(iii) $(y_x)_\theta = -(x_y)_r$;

(iv) $\dfrac{1}{r}(x_\theta)_r = r(\theta_x)_y$;

(v) $(x_r)_\theta(r_x)_y = \cos^2\theta$.

Note that $(x_r)_\theta$ means that x has been expressed as a function of r and θ and $\dfrac{\partial x}{\partial r}$ has been found.

(b) Prove that

$$\begin{vmatrix} x_r & y_r \\ x_\theta & y_\theta \end{vmatrix} \cdot \begin{vmatrix} r_x & \theta_x \\ r_y & \theta_y \end{vmatrix} = 1.$$

(c) If u is any function of x and y (and therefore of r and θ) using

$$u_r = u_x x_r + u_y y_r$$

obtain

$$u_r = \cos\theta\, u_x + \sin\theta\, u_y,$$

and similarly

$$u_\theta = -r\sin\theta\, u_x + r\cos\theta\, u_y.$$

Solve these to obtain

$$u_x = \cos\theta\, u_r - \frac{\sin\theta}{r}\, u_\theta,$$

$$u_y = \sin\theta\, u_r + \frac{\cos\theta}{r}\, u_\theta.$$

Obtain these equations directly, beginning with, for example,

$$u_x = u_r r_x + u_\theta \theta_x.$$

21. A door 2 m wide can swing open about its vertical line of hinges at 2 rad/s. A lamp of 50 candela can move towards the door at 0·5 m/s on a fixed horizontal line perpendicular to the closed door and passing through the centre of the door when it is closed. Find the rate at which the intensity of illumination is changing at the point (P) in which the line meets the door when it has opened through 45° and the lamp (L) is 3 m from the wall. It is given that the intensity of illumination at P due to L is 50 cos α/LP², where α is the angle of incidence of the ray LP.

22 (a) Find the equation of the tangent and normal at the point $(x_1,\ y_1)$ on the curve

$$x^4 + y^4 = 4a^2xy.$$

(b) If $F(x, y) = a$, $f(x, z) = b$, show that

$$\frac{dy}{dz} = \frac{F_x f_z}{F_y f_x}.$$

23. If $z = f\left(\dfrac{xy}{x+y}\right)$, show that $x^2 z_x = y^2 z_y$.

24. Given $f(v,\ p,\ T) = 0$ any relation between the three variables p, v, T, show that

$$\left(\frac{\partial v}{\partial T}\right)_p \left(\frac{\partial T}{\partial p}\right)_v \left(\frac{\partial p}{\partial v}\right)_T = -1.$$

25. If v, T, p are three variables each expressed in terms of two others α and β, show that

$$\left(\frac{\partial v}{\partial T}\right)_p = \frac{v_\alpha p_\beta - v_\beta p_\alpha}{T_\alpha p_\beta - T_\beta p_\alpha}$$

where $v_\alpha \equiv \dfrac{\partial v}{\partial \alpha}$ and similarly for the others.

26. Four variables u, t, p, v are such that any one of them can be expressed as a function of any two of the others. (a) Regard u as a function of t, p, i.e., from $u \equiv u(t,\ p)$ obtain

$$du = (u_t)_p dt + (u_p)_t dp.$$

Obtain the similar equations from $u \equiv u(t,\ v)$ and $v \equiv v(t,\ p)$. (b) Eliminate du between the first two to obtain dv in terms of dt and dp. Equate coefficients in this eliminant and the equation in dv to obtain:

$$(u_t)_p - (u_t)_v = (u_v)_t (v_t)_p,$$

$$(u_p)_t = (u_v)_t (v_p)_t.$$

27. If when v is eliminated between the equations $y = f(x, v)$ and $z = g(x, v)$ the equation $z = F(x, y)$ is obtained, show that,

(a) $(z_x)_y (y_v)_x = (y_v)_x (z_x)_v - (y_x)_v (z_v)_x$;

(b) $(z_y)_x (y_v)_x = (z_v)_x.$

28. (a) If $z = (x + y) \log \dfrac{x}{y}$, prove that $x^2 \dfrac{\partial^2 z}{\partial x^2} = y^2 \dfrac{\partial^2 z}{\partial y^2}.$

(b) If $x + y + u + v =$ constant and $x^2 + y^2 + u^2 + v^2 =$ constant, find $\left(\dfrac{\partial u}{\partial x}\right)_y$ and $\left(\dfrac{\partial v}{\partial x}\right)_y$, the suffix denoting the quantity which remains constant in the differentiation, and verify that

$$\left(\frac{\partial u}{\partial x}\right)_y \left(\frac{\partial x}{\partial u}\right)_v = \left(\frac{\partial v}{\partial y}\right)_x \left(\frac{\partial y}{\partial v}\right)_u.$$

[L.U.]

29. (a) If z is a function of x and y, and $u = e^x$, $v = e^y$, show that $\dfrac{\partial^2 z}{\partial x\,\partial y} = uv\,\dfrac{\partial^2 z}{\partial u\,\partial v}$.

(b) If A, B, C are three points and P is a variable point in their plane, show that for $PA^2 + PB^2 + PC^2$ to be a true minimum P must be the centroid of the triangle ABC. [L.U.]

30. (a) If any one of x, y, z is expressed in terms of the other two by means of the relation $2xyz - 5x + 2y - 3z = 1$, prove that

$$\left(\frac{\partial y}{\partial z}\right)_x \left(\frac{\partial z}{\partial x}\right)_y \left(\frac{\partial x}{\partial y}\right)_z = -1.$$

(b) If $(x - y)\tan u = x^3 + y^3$, prove that

$$x\,\frac{\partial u}{\partial x} + y\,\frac{\partial u}{\partial y} = \sin 2u.$$

(c) If $z = \dfrac{x}{x^2 + y^2}$, verify that $\dfrac{\partial^2 z}{\partial x\,\partial y} = \dfrac{\partial^2 z}{\partial y\,\partial x}$ [L.U.]

31. (a) If $u = xy/(x^2 + y^2)^2$, verify that

$$\frac{\partial^2 u}{\partial x^2} + \frac{\partial^2 u}{\partial y^2} = 0.$$

(b) If ϕ, u, p are functions of v and T and $d\phi = \dfrac{1}{T}(du + p\,dv)$, express $d\phi$ in the form $P\,dv + Q\,dT$ and prove that

$$\left(\frac{\partial \phi}{\partial v}\right)_T = \frac{1}{T}\left\{p + \left(\frac{\partial u}{\partial v}\right)_T\right\},$$

and, since du is an exact differential

$$\left(\frac{\partial T}{\partial v}\right)_\phi = -\left(\frac{\partial p}{\partial \phi}\right)_v.$$

[L.U.]

32. The variables P, V, T, ϕ are connected by the relations

$$PV = RT, \quad \phi = C_v \log P + C_p \log V$$

where C_p, C_v and R are constants and $C_p - C_v = R$. Prove that

$$\frac{\partial T}{\partial P}\frac{\partial \phi}{\partial V} - \frac{\partial T}{\partial V}\frac{\partial \phi}{\partial P} = \frac{\partial P}{\partial T}\frac{\partial V}{\partial \phi} - \frac{\partial P}{\partial \phi}\frac{\partial V}{\partial T} = 1.$$

Explain briefly the geometrical significance of the result. [L.U.]

33. If $\tan^2 \theta = y/x$ and $\sec^2 \phi = x + y$ and V is any function of θ and ϕ, prove that:

(a) $\left(\dfrac{\partial \theta}{\partial x}\right)_y \left(\dfrac{\partial \theta}{\partial y}\right)_x = -\dfrac{1}{4(x + y)^2}$;

(b) $\dfrac{1}{2}\cot \phi \left(\dfrac{\partial V}{\partial \phi}\right)_\theta = x\left(\dfrac{\partial V}{\partial x}\right)_y + y\left(\dfrac{\partial V}{\partial y}\right)_x.$ [L.U.]

34. (a) If the four variables x, y, z, u are connected by the two relations

$$u = \sin x \cosh y, \quad \log(x + y) + 2y - 3\log z = 4,$$

find $\partial u/\partial x$, when x and z are the independent variables and $\dfrac{\partial u}{\partial z}$.

(b) If $x = u + v$, $y = uv$, and Z is any function of x and y, show that

$$u\,\partial Z/\partial u + v\,\partial Z/\partial v = x\,\partial Z/\partial x + 2y\,\partial Z/\partial y.$$

[L.U.]

35. If z is a function of x and y and these are functions of u and v defined by

$$x = uv \qquad y = (u + v)/(u - v),$$

show that

$$2x \frac{\partial z}{\partial x} = u \frac{\partial z}{\partial u} + v \frac{\partial z}{\partial v},$$

$$2y \frac{\partial z}{\partial y} = \tfrac{1}{2}(u^2 - v^2) \left\{ \frac{1}{u} \frac{\partial z}{\partial v} - \frac{1}{v} \frac{\partial z}{\partial u} \right\}. \qquad \text{[L.U.]}$$

36. If u, v are given as functions of x and y by means of the equations

$$f(x, y, u) = 0 \qquad F(u, v, x) = 0,$$

show that

$$\frac{\partial v}{\partial x} = \frac{f_x F_u - f_u F_x}{f_u F_v}, \quad \frac{\partial v}{\partial y} = \frac{f_y F_u}{f_u F_v}.$$

37. Find the envelope of the family of lines

$$x \cos \theta + y \sin \theta = 1.$$

38. Find the envelope of the family of curves

$$x^2 \cos \theta + y^2 \sin \theta = a^2.$$

39. Find the normal to the parabola $y^2 = 4ax$ at the point $(at^2, 2at)$ and the envelope of normals.

40. A straight line moves so that the sum of the intercepts on the axes is constant, *i.e.*, the line is $x/a + y/b = 1$ where $a + b = c$ a constant. Express the line in terms of one variable, and hence find its envelope.

41. Particles are projected from a point in all directions in a vertical plane. Using that the position of a particle at time t can be written for the particle projected with velocity V at angle θ

$$x = \text{V} \cos \theta \, t, \quad y = \text{V} \sin \theta \, t - \tfrac{1}{2}gt^2,$$

obtain the equation to the path as

$$y = x \tan \theta - \frac{gx^2}{2\text{V}^2} (1 + \tan^2 \theta).$$

Put $\tan \theta = m$ and find the envelope of the paths.

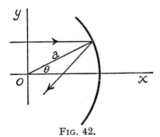

FIG. 42.

42. The figure represents a ray of light reflected from a circular mirror centre O. Show that the envelope of the reflected ray (called a caustic curve), is given by

$$x = \tfrac{1}{2}a \cos \theta (3 - 2 \cos^2 \theta), \quad y = a \sin^3 \theta.$$

12.6. Laplace's Equation

Suppose water of unit depth streaming over the (x, y) plane. With axes anywhere in the plane let the velocity of the water at the point (x, y) have components (u, v) parallel to the respective axes. Suppose further a rectangle of sides dx by dy surrounding the point and with its centre at (x, y).

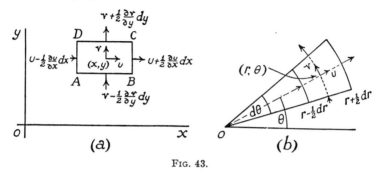

Fig. 43.

Since u is the velocity at x, at the centre of the side BC, where x has changed to $x + \frac{1}{2}dx$ but y remains unchanged, the u velocity component is

$$u + \frac{\partial u}{\partial x}(\tfrac{1}{2}dx),$$

since $\frac{\partial u}{\partial x}$ is the rate of change of u with respect to x, and the small change in x is $\frac{1}{2}dx$. Similarly, at the centre of AD, the u velocity component is

$$u + \frac{\partial u}{\partial x}(-\tfrac{1}{2}dx).$$

The volume of water leaving this imaginary rectangle due to the flow across the sides AD and BC is then

$$\left(u + \tfrac{1}{2}\frac{\partial u}{\partial x}dx\right)dy - \left(u - \tfrac{1}{2}\frac{\partial u}{\partial x}dx\right)dy = \frac{\partial u}{\partial x}dxdy.$$

For the sides AB and CD this is similarly

$$\frac{\partial v}{\partial y}dxdy.$$

The total amount is therefore

$$\left(\frac{\partial u}{\partial x} + \frac{\partial v}{\partial y}\right)dxdy.$$

If there is no source or sink within the rectangle and the water is assumed to be of constant density, this is zero, and therefore

$$\frac{\partial u}{\partial x} + \frac{\partial v}{\partial y} = 0.$$

In the cases usually considered, both u and v are derived from a single function V, the velocity potential, and such that

$$u = -\frac{\partial V}{\partial x}, \quad v = -\frac{\partial V}{\partial y}.$$

Substitution of these values gives

$$\frac{\partial^2 V}{\partial x^2} + \frac{\partial^2 V}{\partial y^2} = 0.$$

An equation of this form is obeyed in all cases of the continuous movement, subject to the conditions assumed, of liquids, heat, magnetic or electric flux, etc. This equation, called *Laplace's equation*, is therefore the starting point in many cases and its different solutions to satisfy given boundary conditions, an important matter in Mathematical Physics.

12.7. The Polar Form

In many problems it is advantageous to use the above equation in the polar form.

Example 8. Transform into the polar form the expression

$$\frac{\partial^2 V}{\partial x^2} + \frac{\partial^2 V}{\partial y^2}$$

The relations to be used are:

$$x = r \cos \theta, \quad y = r \sin \theta,$$
$$r^2 = x^2 + y^2, \quad \tan \theta = y/x.$$

Since

$$2r \frac{\partial r}{\partial x} = 2x, \quad \frac{\partial r}{\partial x} = \frac{x}{r} = \cos \theta,$$

and

$$\sec^2 \theta \frac{\partial \theta}{\partial x} = -\frac{y}{x^2}, \quad \frac{\partial \theta}{\partial x} = -\frac{y}{r^2} = -\frac{\sin \theta}{r}.$$

Using these, since

$$\frac{\partial V}{\partial x} = \frac{\partial V}{\partial r} \frac{\partial r}{\partial x} + \frac{\partial V}{\partial \theta} \frac{\partial \theta}{\partial x},$$

$$\therefore \quad \frac{\partial V}{\partial x} = \cos \theta \frac{\partial V}{\partial r} - \frac{\sin \theta}{r} \frac{\partial V}{\partial \theta} \quad \cdot \quad \cdot \quad \cdot \quad \cdot \quad (1)$$

so that, as equivalent operators, we may use

$$\frac{\partial}{\partial x} \equiv \cos \theta \frac{\partial}{\partial r} - \frac{\sin \theta}{r} \frac{\partial}{\partial \theta}.$$

Operating with these on (1)

$$\frac{\partial}{\partial x}\left(\frac{\partial V}{\partial x}\right) = \left(\cos \theta \frac{\partial}{\partial r} - \frac{\sin \theta}{r} \frac{\partial}{\partial \theta}\right)\left(\cos \theta \frac{\partial V}{\partial r} - \frac{\sin \theta}{r} \frac{\partial V}{\partial \theta}\right)$$

$$= \cos \theta \left(\cos \theta \frac{\partial^2 V}{\partial r^2} + \frac{\sin \theta}{r^2} \frac{\partial V}{\partial \theta} - \frac{\sin \theta}{r} \frac{\partial^2 V}{\partial r \partial \theta}\right)$$

$$- \frac{\sin \theta}{r}\left(-\sin \theta \frac{\partial V}{\partial r} + \cos \theta \frac{\partial^2 V}{\partial \theta \partial r} - \frac{\cos \theta}{r} \frac{\partial V}{\partial \theta} - \frac{\sin \theta}{r} \frac{\partial^2 V}{\partial \theta^2}\right)$$

$$= \cos^2 \theta \frac{\partial^2 V}{\partial r^2} - \frac{2 \sin \theta \cos \theta}{r} \frac{\partial^2 V}{\partial r \partial \theta} + \frac{\sin^2 \theta}{r^2} \frac{\partial^2 V}{\partial \theta^2}$$

$$+ \frac{\sin^2 \theta}{r} \frac{\partial V}{\partial r} + \frac{2 \sin \theta \cos \theta}{r^2} \frac{\partial V}{\partial \theta} \quad \cdot \quad \cdot \quad \cdot \quad (2)$$

Similarly, as equivalent operators, we can obtain

$$\frac{\partial}{\partial y} \equiv \sin \theta \frac{\partial}{\partial r} + \frac{\cos \theta}{r} \frac{\partial}{\partial \theta}$$

If we put $\theta_1 = \theta - \frac{1}{2}\pi$ so that $d\theta_1 = d\theta$, this becomes

$$\frac{\partial}{\partial y} \equiv \cos \theta_1 \frac{\partial}{\partial r} - \frac{\sin \theta_1}{r} \frac{\partial}{\partial \theta_1},$$

which is of the same form as (1). We may therefore obtain $\partial^2 V/\partial y^2$ from $\partial^2 V/\partial x^2$ by this substitution. Using (2), we find

$$\frac{\partial^2 V}{\partial y^2} = \sin^2 \theta \frac{\partial^2 V}{\partial r^2} + \frac{2 \sin \theta \cos \theta}{r} \frac{\partial^2 V}{\partial r \partial \theta} + \frac{\cos^2 \theta}{r^2} \frac{\partial^2 V}{\partial \theta^2}$$

$$+ \frac{\cos^2 \theta}{r} \frac{\partial V}{\partial r} - \frac{2 \sin \theta \cos \theta}{r} \frac{\partial V}{\partial \theta} \quad . \quad . \quad . \quad . \quad (3)$$

(2) and (3) give

$$\frac{\partial^2 V}{\partial x^2} + \frac{\partial^2 V}{\partial y^2} = \frac{\partial^2 V}{\partial r^2} + \frac{1}{r^2} \frac{\partial^2 V}{\partial \theta^2} + \frac{1}{r} \frac{\partial V}{\partial r},$$

the right-hand side being the required polar form. The student should also obtain this by assuming water flowing over the plane (in polar co-ordinates) with values (u, v) at the point (r, θ) inside an element of area $r d\theta$ by dr. The method is exactly similar to that used above in (x, y) co-ordinates (see Fig. 43 (b)).

In three dimensions Laplace's equation is

$$\frac{\partial^2 V}{\partial x^2} + \frac{\partial^2 V}{\partial y^2} + \frac{\partial^2 V}{\partial z^2} = 0$$

and can be obtained similarly to the method above by imagining water flowing through a box of sides dx, dy, dz referred to the x, y, z axes. It occurs often, and is written $\nabla^2 V = 0$ where ∇^2 is

$$\frac{\partial^2}{\partial x^2} + \frac{\partial^2}{\partial y^2} \quad \text{or} \quad \frac{\partial^2}{\partial x^2} + \frac{\partial^2}{\partial y^2} + \frac{\partial^2}{\partial z^2}$$

according as we are dealing with two or three dimensions.

It can be shown that in polar form (three dimensions) Laplace's equation is

$$\frac{\partial^2 V}{\partial r^2} + \frac{2}{r} \frac{\partial V}{\partial r} + \frac{1}{r^2} \frac{\partial^2 V}{\partial \theta^2} + \frac{\cot \theta}{r^2} \frac{\partial V}{\partial \theta} + \frac{1}{r^2 \sin^2 \theta} \frac{\partial^2 V}{\partial \phi^2}.$$

Example 9. (a) If $r = \sqrt{(x^2 + y^2)}$ and $u = f(r)$, show that

$$\frac{\partial^2 u}{\partial x^2} + \frac{\partial^2 u}{\partial y^2} = f''(r) + \frac{1}{r} f'(r)$$

and find u in terms of r if $\dfrac{\partial^2 u}{\partial x^2} + \dfrac{\partial^2 u}{\partial y^2} = 0$.

(b) If $x = r \cos \theta$, $y = r \sin \theta$, $u = r^a \cos \theta$ and $\dfrac{\partial^2 u}{\partial x^2} + \dfrac{\partial^2 u}{\partial y^2} = 0$, find the possible values of the constant a. [L.U.]

(a) (i)
$$\frac{\partial u}{\partial x} = f'(r) \frac{\partial r}{\partial x} = \frac{x}{\sqrt{(x^2 + y^2)}} f'(r),$$

$$\frac{\partial^2 u}{\partial x^2} = \frac{x}{\sqrt{(x^2 + y^2)}} f''(r) \frac{\partial r}{\partial x} + f'(r) \frac{\partial}{\partial x} \left\{ \frac{x}{\sqrt{(x^2 + y^2)}} \right\}$$

$$= \frac{x^2}{x^2 + y^2} f''(r) + \frac{y^2}{(x^2 + y^2)^{3/2}} f'(r).$$

Similarly,

$$\frac{\partial^2 u}{\partial y^2} = \frac{y^2}{x^2 + y^2} f''(r) + \frac{x^2}{(x^2 + y^2)^{3/2}} f'(r),$$

$$\therefore \quad \frac{\partial^2 u}{\partial x^2} + \frac{\partial^2 u}{\partial y^2} = f''(r) + \frac{1}{r} f'(r).$$

(ii) The shorter way is

$$\left(\frac{\partial^2}{\partial x^2} + \frac{\partial^2}{\partial y^2}\right) u = \left(\frac{\partial^2}{\partial r^2} + \frac{1}{r^2}\frac{\partial^2}{\partial \theta^2} + \frac{1}{r}\frac{\partial}{\partial r}\right) f(r)$$

$$= f''(r) + \frac{1}{r}f'(r),$$

since $$\qquad \frac{\partial}{\partial \theta}\{f(r)\} = 0.$$

When the L.H.S. is zero

$$\frac{f''(r)}{f'(r)} = -\frac{1}{r},$$

$$\therefore \quad \log f'(r) = -\log r + \log A$$

$$= \log \frac{A}{r},$$

$$f'(r) = \frac{A}{r},$$

$$\underline{f(r) = A \log r + B.}$$

(b) In polar co-ordinates

$$\left(\frac{\partial^2}{\partial r^2} + \frac{1}{r^2}\frac{\partial^2}{\partial \theta^2} + \frac{1}{r}\frac{\partial}{\partial r}\right)(r^a \cos \theta) = 0,$$

$$\therefore \quad a(a - 1)r^{a-2}\cos\theta + \frac{1}{r}ar^{a-1}\cos\theta - \frac{1}{r^2}r^a\cos\theta = 0,$$

$$\therefore \quad a^2 = 1,$$

$$\underline{a = \pm 1.}$$

EXERCISE 27

1. If $u = f(x + at) + F(x - at)$, prove that

$$\frac{\partial^2 u}{\partial t^2} = a^2 \frac{\partial^2 u}{\partial x^2}.$$

2. If $u = x^n F(x/y)$ where F denotes an arbitrary function, show that

$$x\frac{\partial u}{\partial x} + y\frac{\partial u}{\partial y} = nu$$

and hence that

$$x^2\frac{\partial^2 u}{\partial x^2} + 2xy\frac{\partial^2 u}{\partial x\partial y} + y^2\frac{\partial^2 u}{\partial y^2} = n(n-1)u. \qquad \text{[L.U.]}$$

3. (i) If $u = x/r^3$ and $r^2 = x^2 + y^2 + z^2$, show that

$$\frac{\partial^2 u}{\partial x^2} + \frac{\partial^2 u}{\partial y^2} + \frac{\partial^2 u}{\partial z^2} = 0.$$

(ii) If x increases at the rate of 2 m/s at the instant when $x = 3$ m and $y = 1$ m, at what rate must y be changing in order that the function $2xy - 3x^2y$ shall be neither increasing nor decreasing when x and y have these values.

[L.U.]

4. (i) If $u = x^2 \tan^{-1}\frac{y}{x} - y^2\tan^{-1}\frac{x}{y}$, evaluate $x\frac{\partial u}{\partial x} + y\frac{\partial u}{\partial y}$.

(ii) If $u = x^3 - 3xy^2$, prove that

$$\frac{\partial^2 u}{\partial x^2} + \frac{\partial^2 u}{\partial y^2} = 0.$$

Prove also that if $z = r^n u$ where $r^2 = x^2 + y^2$, then

$$r^2\left\{\frac{\partial^2 z}{\partial x^2} + \frac{\partial^2 z}{\partial y^2}\right\} = (n^2 + 6n)z. \qquad \text{[L.U.]}$$

5. If $z = (x + y)\phi(y/x)$
where ϕ is an arbitrary function, prove that

$$x \frac{\partial z}{\partial x} + y \frac{\partial z}{\partial y} = z$$

and that

$$x^2 \frac{\partial^2 z}{\partial x^2} + 2xy \frac{\partial^2 z}{\partial x \partial y} + y^2 \frac{\partial^2 z}{\partial y^2} = 0.$$

[L.U.]

6. (a) If $V = f(x^2 + y^2)$ where f is any function, show that:

(i) $y \dfrac{\partial V}{\partial x} - x \dfrac{\partial V}{\partial y} = 0$;

(ii) $y^2 \dfrac{\partial^2 V}{\partial x^2} - 2xy \dfrac{\partial^2 V}{\partial x \partial y} + x^2 \dfrac{\partial^2 V}{\partial y^2} = x \dfrac{\partial V}{\partial x} + y \dfrac{\partial V}{\partial y}.$

(b) Show that if $u = \log r$ where $r^2 = x^2 + y^2$

$$\frac{\partial^2 u}{\partial x^2} + \frac{\partial^2 u}{\partial y^2} = 0.$$

[L.U.]

7. If $u = \log r$ where $r^2 = x^2 + y^2 + z^2$, show that

$$\frac{\partial^2 u}{\partial x^2} + \frac{\partial^2 u}{\partial y^2} + \frac{\partial^2 u}{\partial z^2} = \frac{1}{r^2}.$$

8. (i) If $V = f(r)$ where $r = \sqrt{(x^2 + y^2 + z^2)}$, prove that

$$\frac{\partial^2 V}{\partial x^2} + \frac{\partial^2 V}{\partial y^2} + \frac{\partial^2 V}{\partial z^2} = \frac{1}{r} \frac{d^2}{dr^2} (rV).$$

(ii) If $u^2 = x^2 + y^2$ and $v^2 = 2xy$, prove that

$$\left(\frac{\partial y}{\partial x}\right)_u \left(\frac{\partial y}{\partial x}\right)_v = 1 \quad \text{and} \quad \left(\frac{\partial v}{\partial u}\right)_x \left(\frac{\partial v}{\partial u}\right)_y = \frac{u^2}{v^2}.$$

[L.U.]

9. Find a if $V = x^3 + axy^2$ satisfies $\dfrac{\partial^2 V}{\partial x^2} + \dfrac{\partial^2 V}{\partial y^2} = 0$. Taking this value of V, show that if $u = r^n V$ where $r^2 = x^2 + y^2$, then

$$\frac{\partial^2 u}{\partial x^2} + \frac{\partial^2 u}{\partial y^2} = n(n + 6)r^{n-2}V.$$

[Put V into the polar form $r^3 \cos 3\theta$ and use this form of $\nabla^2 u$.]　　　[L.U.]

10 If $V = f(z)$ and $z^2 = x^2/4t$, find $\dfrac{\partial V}{\partial t}$ and $\dfrac{\partial^2 V}{\partial x^2}$. Show that if V is to satisfy the equation

$$\frac{\partial V}{\partial t} = \frac{\partial^2 V}{\partial x^2}, \text{ then } f(z) = A \int e^{-z^2} dz + B.$$

[L.U.]

11. If $V = (x^2 - y^2)f(xy)$, show $\dfrac{\partial^2 V}{\partial x^2} + \dfrac{\partial^2 V}{\partial y^2} = (x^4 - y^4)f''(xy)$

and

$$\frac{\partial^2 V}{\partial x \partial y} = (x^2 - y^2)\{3f'(xy) + xyf''(xy)\}.$$

Find the form of $f(xy)$ for which $\dfrac{\partial^2 V}{\partial x \partial y} = 0.$

[L.U.]

12. Show that, a, b and λ being constants,

$$V = \frac{1}{2a\sqrt{(\pi t)}} \exp\left\{-b^2 t - \frac{(\lambda - x)^2}{4a^2 t}\right\}$$

satisfies the equation

$$\frac{\partial V}{\partial t} = a^2 \frac{\partial^2 V}{\partial x^2} - b^2 V.$$

Hence or otherwise prove that

$$u = \frac{\lambda - x}{2a^2 t} V$$

is also a solution.

[L.U.]

13. (i) If V is a function of r alone where $r^2 = x^2 + y^2 + z^2$, show that

$$\nabla^2 V = \frac{d^2 V}{dr^2} + \frac{2}{r}\frac{dV}{dr}$$

(ii) If in the above $V = (x + y + z)/r^2$, prove that

$$\nabla^2 V = -\frac{2V}{r^2}$$

(iii) If $V = f(ax + by + cz)$ and $\nabla^2 V = 0$ for all forms of f, find the relation between a, b, c.

14. (a) If z is a function of x and y and u, v be two other variables such that $u = lx + my$, $v = ly - mx$, show that

$$\frac{\partial^2 z}{\partial x^2} + \frac{\partial^2 z}{\partial y^2} = (l^2 + m^2)\left(\frac{\partial^2 z}{\partial u^2} + \frac{\partial^2 z}{\partial v^2}\right).$$

(b) If $u = e^{x^2-y^2}\cos 2xy$, $v = e^{x^2-y^2}\sin 2xy$, prove that

(i) $\dfrac{\partial u}{\partial x} = \dfrac{\partial v}{\partial y},\quad \dfrac{\partial u}{\partial y} = -\dfrac{\partial v}{\partial x};$

(ii) $\dfrac{\partial^2 u}{\partial x^2} + \dfrac{\partial^2 u}{\partial y^2} = 0,\quad \dfrac{\partial^2 v}{\partial x^2} + \dfrac{\partial^2 v}{\partial y^2} = 0.$

15. If $u = \cos\theta\cos\alpha + \sin\theta\sin\alpha\cos(\phi - \beta)$, where θ, ϕ, α and β are independent of each other, prove that:

(i) $\left(\dfrac{\partial u}{\partial \theta}\right)^2 + \dfrac{1}{\sin^2\theta}\left(\dfrac{\partial u}{\partial \phi}\right)^2 = 1 - u^2;$

(ii) $\dfrac{\partial^2 u}{\partial\theta^2} + \cot\theta\,\dfrac{\partial u}{\partial\theta} + \dfrac{1}{\sin^2\theta}\dfrac{\partial^2 u}{\partial\phi^2} = -2u.$ [L.U.]

16. A function $f(x, y)$ is transformed into a function of u and v by the substitution $u = x^2 - y^2$, $v = xy$. Prove that

$$\frac{\partial f}{\partial u} = \frac{1}{2(x^2 + y^2)}\left[x\,\frac{\partial f}{\partial x} - y\,\frac{\partial f}{\partial y}\right]$$

$$\frac{\partial f}{\partial v} = \frac{1}{x^2 + y^2}\left[y\,\frac{\partial f}{\partial x} + x\,\frac{\partial f}{\partial y}\right]$$ [L.U.]

17. (i) If $z = f(x + y)g(x - y)$ where f and g are arbitrary functions, prove that

$$z\,\frac{\partial^2 z}{\partial x^2} - z\,\frac{\partial^2 z}{\partial y^2} = \left(\frac{\partial z}{\partial x}\right)^2 - \left(\frac{\partial z}{\partial y}\right)^2$$

(ii) If the variables x, y, z are connected by the equations

$$f(x, y, z) = 0, \quad x^2 + y^2 + z^2 = \text{constant},$$

prove that

$$\frac{dy}{dx} = -\left(z\,\frac{\partial f}{\partial x} - x\,\frac{\partial f}{\partial z}\right)\Big/\left(z\,\frac{\partial f}{\partial y} - y\,\frac{\partial f}{\partial z}\right).$$ [L.U.]

18. If $x = u^2 + v^2$ and $y = u^2 - v^2$, prove that:

(i) $2\left(\dfrac{\partial x}{\partial u}\right)_v \left(\dfrac{\partial u}{\partial x}\right)_y = 1.$

(ii) If f is a function of x and y and therefore of u and v,

$$v^2\,\frac{\partial^2 f}{\partial u^2} - u^2\,\frac{\partial^2 f}{\partial v^2} = 4(x^2 - y^2)\frac{\partial^2 f}{\partial x\,\partial y} - 2y\,\frac{\partial f}{\partial x} + 2x\,\frac{\partial f}{\partial y}.$$ [L.U.]

19. If $x = e^r\cos\theta$, $y = e^r\sin\theta$, show that, V being a function of x and y and hence of r and θ,

$$\frac{\partial^2 V}{\partial x^2} + \frac{\partial^2 V}{\partial y^2} = e^{-2r}\left[\frac{\partial^2 V}{\partial r^2} + \frac{\partial^2 V}{\partial \theta^2}\right].$$

20. If $x = \operatorname{ch} u \cos v,\ y = \operatorname{sh} u \sin v$, prove that:

(i) $\coth u \dfrac{\partial \phi}{\partial u} + \cot v \dfrac{\partial \phi}{\partial v} = (\operatorname{ch}^2 u - \cos^2 v)\dfrac{1}{y}\dfrac{\partial \phi}{\partial y};$

(ii) $\dfrac{\partial^2 \phi}{\partial u^2} + \dfrac{\partial^2 \phi}{\partial v^2} = (\operatorname{ch}^2 u - \cos^2 v)\left(\dfrac{\partial^2 \phi}{\partial x^2} + \dfrac{\partial^2 \phi}{\partial y^2}\right).$

Deduce that $\phi = \log\{\operatorname{th} \tfrac{1}{2}u\}$ is a solution of

$$\frac{\partial^2 \phi}{\partial x^2} + \frac{\partial^2 \phi}{\partial y^2} + \frac{1}{y}\frac{\partial \phi}{\partial y} = 0.$$

21. If $x = r \cos \theta,\ y = r \sin \theta$, show that

$$(x^2 - y^2)\left(\frac{\partial^2 u}{\partial x^2} - \frac{\partial^2 u}{\partial y^2}\right) + 4xy\frac{\partial^2 u}{\partial x\,\partial y} = r^2\frac{\partial^2 u}{\partial r^2} - r\frac{\partial u}{\partial r} - \frac{\partial^2 u}{\partial \theta^2}.$$

22. If $x = \operatorname{ch} u \operatorname{ch} v,\ y = \operatorname{sh} u \operatorname{sh} v$, prove that

$$\frac{\partial^2 z}{\partial u^2} - \frac{\partial^2 z}{\partial v^2} = (\operatorname{sh}^2 u - \operatorname{sh}^2 v)\left(\frac{\partial^2 z}{\partial x^2} - \frac{\partial^2 z}{\partial y^2}\right).$$

23. Given that $x = u/(u^2 + v^2),\ y = v/(u^2 + v^2)$ and f is any function of x and y, show that

$$(x^2 + y^2)\left(\frac{\partial^2 f}{\partial x^2} + \frac{\partial^2 f}{\partial y^2}\right) = (u^2 + v^2)\left(\frac{\partial^2 f}{\partial u^2} + \frac{\partial^2 f}{\partial v^2}\right).$$

CHAPTER 13

THE COMPLEX VARIABLE AND TRANSFORMATIONS

" The possibility that an algebraical letter might stand for something other than a real number was first envisaged more than two hundred years ago, when the so-called imaginary unit $\sqrt{-1}$ was introduced and its properties investigated. Such a formula as $\pi = \log(-1)/\sqrt{-1}$ in which the ratio of the circumference of a circle to its diameter is expressed in terms of this new quantity, stirred the imagination of the eighteenth-century mathematicians. But $\sqrt{-1}$, though not a number in the old sense, obeyed the commutative law of multiplication and the other classical rules and was essentially a new arithmetical entity, not an element of a new algebra."

Sir E. Whittaker, *From Euclid to Eddington.*

13.1. Complex Numbers

In Vol. I (28) we showed that a complex number $a + jb$ can be put into the form $r(\cos\theta + j\sin\theta)$ or $re^{j\theta}$. A demonstration was given of Demoivre's Theorem that for all values of n

$$(\cos\theta + j\sin\theta)^n = \cos n\theta + j\sin n\theta$$

and its use in finding the roots of a number.

13.2. Relation between Circular and Hyperbolic Functions

By the expansion of $e^{j\theta}$, it was shown that

$$e^{j\theta} = \cos\theta + j\sin\theta \quad \cdots \cdots \quad (1)$$

Similarly, $\qquad e^{-j\theta} = \cos\theta - j\sin\theta.$

From this pair of relations

$$\cos\theta = \tfrac{1}{2}(e^{j\theta} + e^{-j\theta}), \quad \sin\theta = \frac{1}{2j}(e^{j\theta} - e^{-j\theta}) \quad \cdot \quad \cdot \quad (2)$$

Since $\qquad \cosh x = \tfrac{1}{2}(e^x + e^{-x}), \quad \sinh x = \tfrac{1}{2}(e^x - e^{-x}),$

changing x into jx,

$$\left.\begin{array}{l} \cosh jx = \tfrac{1}{2}(e^{jx} + e^{-jx}) = \cos x, \\ \sinh jx = \tfrac{1}{2}(e^{jx} - e^{-jx}) = j\sin x \end{array}\right\} \quad \cdots \quad (3)$$

Similarly, putting jx for θ in (2),

$$\left.\begin{array}{l} \cos jx = \tfrac{1}{2}(e^{j\cdot jx} + e^{-j\cdot jx}) = \cosh x \\ \sin jx = \dfrac{1}{2j}(e^{j\cdot jx} - e^{-j\cdot jx}) = j\sinh x \end{array}\right\} \quad \cdots \quad (4)$$

By division,

$$\tanh jx = j\tan x \text{ and } \tan jx = j\tanh x \quad \cdot \quad \cdot \quad (5)$$

Using these, it is easy to obtain hyperbolic relations from the usual circular relations. Thus, since

$$\sin (\theta - \phi) = \sin \theta \cos \phi - \cos \theta \sin \phi,$$

replacing θ by $j\theta$ and ϕ by $j\phi$,

$$\sin j(\theta - \phi) = \sin j\theta \cos j\phi - \cos j\theta \sin j\phi$$

or by (4), $\quad\sinh (\theta - \phi) = \sinh \theta \cosh \phi - \cosh \theta \sinh \phi$

Similarly, since

$$\tan (A + B) = \frac{\tan A + \tan B}{1 - \tan A \tan B}$$

$$\tan j(A + B) = \frac{\tan jA + \tan jB}{1 - \tan jA \tan jB}$$

which becomes by (5)

$$\tanh (A + B) = \frac{\tanh A + \tanh B}{1 + \tanh A \tanh B}.$$

13.3. Separation into Real and Imaginary Parts

For certain purposes it is necessary to separate a complex expression into the form $A + jB$, where A and B are real.

Example 1. Separate $\cos (A + jB)$ into real and imaginary components, and hence evaluate $\cos (1 + j2)$.

$$\cos (A + jB) = \cos A \cos jB - \sin A \sin jB$$
$$= \cos A \cosh B - \sin A \cdot j \sinh B$$

showing the two components. From this

$$\cos (1 + j2) = \cos 1 \cosh 2 - j \sin 1 \sinh 2$$
$$= 0 \cdot 540 \times 3 \cdot 762 - j \times 0 \cdot 842 \times 3 \cdot 627$$
$$= 2 \cdot 031 - j3 \cdot 054.$$

Similarly, if the expansion of $\cosh (x + jy)$ is not known,

$$\cosh (A + jB) = \cos j(A + jB) = \cos (jA - B)$$
$$= \cos jA \cos B + \sin jA \sin B$$
$$= \cosh A \cos B + j \sinh A \sin B$$

Example 2. Separate $\cot (A + jB)$ into its real and j components.

$$\cot (A + jB) = \frac{\cos (A + jB)}{\sin (A + jB)}$$

If we can obtain cosine functions only in the denominator, then, since $\cos jx = \cosh x$, the denominator will be cleared of j components and no further rationalisation is needed. To this end multiply above and below by $2 \sin (A - jB)$ to give

$$\frac{\cos (A + jB)}{\sin (A + jB)} \times \frac{2 \sin (A - jB)}{2 \sin (A - jB)} = \frac{\sin 2A - \sin 2jB}{\cos 2jB - \cos 2A}$$

$$= \frac{\sin 2A - j \sinh 2B}{\cosh 2B - \cos 2A},$$

from which the two components can be seen.

Example 3. If $\sin (\theta + j\phi) = \rho (\cos \alpha + j \sin \alpha)$, show that

$$\rho^2 = \tfrac{1}{2}(\text{ch } 2\phi - \cos 2\theta), \quad \tan \alpha = \text{th } \phi \cot \theta.$$

Since $\qquad \rho (\cos \alpha + j \sin \alpha) = \sin (\theta + j \phi)$
$$= \sin \theta \cosh \phi + j \cos \theta \sinh \phi,$$

equating real and j parts,

$$\rho \cos \alpha = \sin \theta \cosh \phi$$
$$\rho \sin \alpha = \cos \theta \sinh \phi$$

Square and add:

$$\begin{aligned}
\rho^2 &= \sin^2 \theta \, \mathrm{ch}^2 \, \phi + \cos^2 \theta \, \mathrm{sh}^2 \, \phi \\
&= \tfrac{1}{4}\{(1 - \cos 2\theta)(1 + \mathrm{ch} \, 2\phi) \\
&\qquad + (1 + \cos 2\theta)(\mathrm{ch} \, 2\phi - 1)\} \\
&= \tfrac{1}{2}(\mathrm{ch} \, 2\phi - \cos 2\theta).
\end{aligned}$$

Divide: $\qquad\qquad\qquad \tan \alpha = \cot \theta \tanh \phi$

Note: If $\qquad\qquad u + jv = f(x + jy),$

then we can change the sign of j to obtain

$$u - jv = f(x - jy),$$

applying this to the above,

$$\rho (\cos \alpha + j \sin \alpha) = \sin (\theta + j\phi),$$

therefore $\qquad \rho (\cos \alpha - j \sin \alpha) = \sin (\theta - j\phi).$

Multiplication gives

$$\begin{aligned}
\rho^2 (\cos^2 \alpha + \sin^2 \alpha) &= \sin (\theta + j\phi) \sin (\theta - j\phi) \\
&= \tfrac{1}{2}[\cos j2\phi - \cos 2\theta]
\end{aligned}$$

or $\qquad\qquad\qquad\qquad \rho^2 = \tfrac{1}{2}[\mathrm{ch} \, 2\phi - \cos 2\theta].$

Example 4. ABCD is a parallelogram whose diagonals AC and BD intersect at E. The angle AED is 45°, the lengths AC : BD are as 3 : 2 and the sense of description of ABCD is counter-clockwise. If A is the point $-2 - i3$ and C is $4 + i$ find the numbers representing B and D. [L.U.]

$$\overline{OE} = \tfrac{1}{2}(\overline{OA} + \overline{OC}) = \tfrac{1}{2}(-2 - i3 + 4 + i) = 1 - i,$$

therefore $\qquad\qquad \overline{EC} = (4 + i) - (1 - i) = 3 + i2.$

Since $\overline{ED}$ is $\overline{EC}$ rotated through 135° and decreased in the ratio 2 : 3,

$$\begin{aligned}
\overline{ED} &= \tfrac{2}{3}(3 + i2)e^{i3\pi/4} \\
&= \left(2 + i\tfrac{4}{3}\right)\left(-\frac{1}{\sqrt{2}} + i\frac{1}{\sqrt{2}}\right) \\
&= -\frac{5\sqrt{2}}{3} + i\frac{\sqrt{2}}{3},
\end{aligned}$$

therefore $\qquad\qquad \overline{OD} = \overline{OE} + \overline{ED}$

$$= (1 - i) + \left(-\frac{5\sqrt{2}}{3} + i\frac{\sqrt{2}}{3}\right)$$

$$D = \left(1 - \frac{5\sqrt{2}}{3}\right) + i\left(\frac{\sqrt{2}}{3} - 1\right).$$

B can be obtained similarly, or since E is the mid point of BD, if B is $a + ib$,

$$2(1 - i) = (a + ib) + \left(1 - \frac{5\sqrt{2}}{3}\right) + i\left(\frac{\sqrt{2}}{3} - 1\right),$$

giving $\qquad\qquad a = 1 + \frac{5\sqrt{2}}{3}, \qquad b = -1 - \frac{\sqrt{2}}{3}$

and $\qquad\qquad B = \left(1 + \frac{5\sqrt{2}}{3}\right) - i\left(1 + \frac{\sqrt{2}}{3}\right).$

EXERCISE 28

1. Find the conjugate of $(2 + j3)/(1 - j2)^2$. State the modulus and amplitude.

2. If $w = z/(z + 3)$, where $w = u + jv$, $z = x + jy$ and if z moves on the circle $(x + 3)^2 + y^2 = 1$, find the locus of the point w.

3. Prove that the four points 1, -1, $3 + j4$, $1/(3 + j4)$ lie on a circle of radius $\sqrt{10}$ on the Argand diagram and find its centre.

4. Given that $3 - j1$ is one root of the equation

$$x^4 - 8x^3 + 39x^2 - 122x + 170 = 0,$$

find the other roots. [Note that complex roots occur in conjugate pairs.]

5. The numbers $2 + j3$, $8 + j11$, $j17$ are the points A, B, C on the Argand diagram. Show that A, B, C are three vertices of a square and find the fourth vertex D.

6. If $x = \cos\alpha + j\sin\alpha$, $y = \cos\beta + j\sin\beta$, prove that:

(i) $\dfrac{x - y}{x + y} = j\tan\tfrac{1}{2}(\alpha - \beta)$;

(ii) $\dfrac{(x + y)(xy - 1)}{(x - y)(xy + 1)} = \dfrac{\sin\alpha + \sin\beta}{\sin\alpha - \sin\beta}$.

7. (a) Find the complex number with modulus twice that of $4 + j3$ and amplitude less than that of $4 + j3$ by $\pi/4$.

(b) If ω is one of the complex cube roots of unity show that

$$\frac{3}{x^3 - 1} = \frac{1}{x - 1} + \frac{1}{\omega x - 1} + \frac{1}{\omega^2 x - 1}.$$

Deduce the identity

$$3\cot 3\theta = \cot\theta + \cot(\theta + \tfrac{1}{3}\pi) + \cot(\theta + \tfrac{2}{3}\pi)$$

by putting $x = \cos 2\theta + j\sin 2\theta$.

8. (a) If $x = j2/(3 - j4)$, find $|z|$ and arg z.

(b) If arg $(z + 1) = \tfrac{1}{6}\pi$ and arg $(z - 1) = \tfrac{2}{3}\pi$ find z.

(c) The complex quantities $e^{j\alpha}$, $e^{j\beta}$ are represented in an Argand diagram by vectors $\overrightarrow{OA}$, $\overrightarrow{OB}$ respectively. If $\overrightarrow{OP} = \overrightarrow{OA} + \overrightarrow{OB}$ prove that $\overrightarrow{OP}$ represents the complex quantity

$$2\cos\frac{\alpha - \beta}{2}\, e^{j\left(\frac{\alpha + \beta}{2}\right)}$$

If $\overrightarrow{OA}$, $\overrightarrow{OB}$ rotate in the anti-clockwise direction with constant angular speeds ω and 2ω respectively starting from the line $\theta = 0$ at time $t = 0$, show that P describes a curve whose polar equation is $r = 2\cos(\theta/3)$ and sketch the curve.
[L.U.]

9. Show that the joins of the numbers a to b and c to d are parallel if $(a - b)/(c - d)$ is real and perpendicular if this expression is imaginary.

10. If $x + jy = \text{sh}\,(3 + j4)$, show that $x = -6.5$, $y = -7.6$.

11. (i) If $z = (2 + i)/(1 - i)$, find the real and imaginary parts of $z + z^{-1}$.

(ii) When the vertices of a square A, B, C, D are taken anti-clockwise in that order the points A, B represent the complex numbers $-1 + 4i$, $-3 + 0i$ in the Argand diagram. Find the complex numbers represented by the other vertices and by the centre of the square. [L.U.]

12. If $(x + jy - c)/(x + jy + c) = e^{u+jv}$ where x, y, u, v, c are all real and $j = \sqrt{-1}$, prove that

$$x = -\frac{c\sinh u}{\cosh u - \cos v}, \quad y = \frac{c\sin v}{\cosh u - \cos v}.$$

If also $x^2 + y^2 = c^2$, prove that $v = (2n + 1)\dfrac{\pi}{2}$ where n is any integer. [L.U.]

13. (i) If $z = \sqrt{3}e^{j\pi/3} - 2e^{j\pi/6}$ where $j = \sqrt{-1}$, express z in the form $re^{j\theta}$.

(ii) If $u + jv = (2 + j)/(z + 3)$ where $z = x + jy$, find u and v as functions of x and y.   [L.U.]

14. (i) On an Argand diagram a variable point is represented by z. If the real part of $(z - 4)/(z - j2)$ is zero, prove that the locus of the point is a circle of radius $\sqrt{5}$.

(ii) If $\tan^{-1}(x + jy) = p + jq$, prove that

$$x = \frac{\tan p(1 - \tanh^2 q)}{1 + \tan^2 p \tanh^2 q}.$$ [L.U.]

15. (a) Write down the modulus and amplitude of each of

$$1 + i, \quad -1 + i, \quad 1 - i.$$

(b) The points ABCD in the Argand diagram correspond to the complex numbers $9 + i$, $4 + 13i$, $-8 + 8i$, $-3 - 4i$. Prove that ABCD is a square.   [L.U.]

16. Find the real part of

$$\frac{p}{ap^2 + bp + c}\, e^{pt},$$

where $p = j\omega, j = \sqrt{-1}$, and a, b, c, ω, t are real.

17. State the expression for $\sin \theta$ in terms of $e^{j\theta}$. Prove that

$$32 \sin^6 \theta = 10 - 15 \cos 2\theta + 6 \cos 4\theta - \cos 6\theta,$$

and evaluate

$$\int_0^{\pi/3} \sin^6 \theta\, d\theta.$$ [L.U.]

18. Express in the form $A + jB$:   (a) $\sin(x - jy)$;   (b) $\cos^2(x + jy)$; (c) ch $(x + jy)$;   (d) cosec $(x + jy)$;   (e) exp $\{\sin(x + jy)\}$.

19. Simplify

$$\frac{\cos(x + jy)}{\cos(x - jy)} + \frac{\cos(x - jy)}{\cos(x + jy)}.$$

20. Prove that

$$\tan \tfrac{1}{2}(u + jv) = \frac{\sin u + j \operatorname{sh} v}{\cos u + \operatorname{ch} v}.$$

21. If $\cos(A + jB) = \cos x + j \sin x$, show that $\cos 2A + \operatorname{ch} 2B = 2$.

22. Express $3 + j4$ in the form $re^{j\theta}$ and hence put $\log(3 + j4)$ into the form $a + jb$.

23. If $x + jy = u + a^2/u$ and $u = \tfrac{1}{2}a(3 + e^{j\theta})$ where a and θ are real, find the explicit expressions for x and y in terms of a and θ.   [L.U.]

24. If $x + jy = \cos(u + jv)$, show that $\cos^2 u$ and $\operatorname{ch}^2 v$ are the roots of the equation $\lambda^2 - (x^2 + y^2 + 1)\lambda + x^2 = 0$. If $\cos(u + jv) = \tfrac{1}{6}(5 + j4\sqrt{3})$, find u and v.   [L.U.]

13.4. Transformations

Consider a simple relation

$$u + jv = (x + jy)^2$$

in which u, v, x, y are real. We may equate the real and imaginary parts to obtain

$$u = x^2 - y^2, \quad v = 2xy \quad . \quad . \quad . \quad . \quad . \quad (1)$$

If, as is usual in the simpler case of $y = f(x)$, we wish to graph the relation, we find here four variables u, v, x and y instead of the usual two, and we cannot draw four-dimensional pictures. A very ingenious way out of the difficulty has been invented. Let (u, v) be a point in one plane — the w plane and (x, y) a point in another —the z plane. We may then regard the general relation

i.e., $$w = f(z)$$
or $$u + jv = f(x + jy)$$

which in our case is

$$u + jv = (x + jy)^2,$$

as relating the point $w(= u + jv)$ to the point $z(= x + jy)$. As w describes a given curve on its plane (or Argand diagram) we require

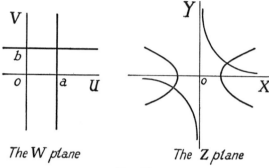

FIG. 44.

the curve described by the corresponding point z in its plane (or Argand diagram) and vice versa.

We can further simplify matters by assuming that u describes a line $u = a$ and v a line $v = b$. Equation (1) then shows that the point in the z plan describes for $u = a$ the curve

$$x^2 - y^2 = a,$$

and for $v = b$ the curve

$$2xy = b.$$

So that we may regard the line $u = a$ as *transformed* into the hyperbola $x^2 - y^2 = a$ and $v = b$ into the hyperbola $2xy = b$. As a varies, the series of parallel lines $u = a$ will become the family of conics $x^2 - y^2 = a$, and similarly for $v = b$.

The next section will give some of the general properties of relations of the form $w = f(z)$. Here we will mention only that $u = a$, $v = b$, clearly cut at right angles, and it can be shown that the curves $x^2 - y^2 = a$, $2xy = b$ also cut at the same angle, which in this case is a right angle.

Example 5. For the transformation $w = \cosh z$, *i.e.,* $u + jv = \cosh(x + jy)$, what are the loci (i) $x = a$, (ii) $y = b$.

$$u + jv = \cosh(x + jy)$$
$$= \cosh x \cosh jy + \sinh x \sinh jy$$
$$= \cosh x \cos y + j \sinh x \sin y,$$
$$\therefore \quad u = \cosh x \cos y, \qquad v = \sinh x \sin y.$$

(i) If $x = a$, a constant

$$u = \cosh a \cos y$$
$$v = \sinh a \sin y$$

eliminating y, which is variable,

$$\frac{u^2}{\operatorname{ch}^2 a} + \frac{v^2}{\operatorname{sh}^2 a} = \cos^2 y + \sin^2 y,$$
$$= 1$$

showing that the line $x = a$ in the (x, y) plane is transformed into an ellipse in the (u, v) plane.

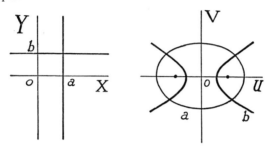

FIG. 45.

(ii) For $y = b$

$$u = \operatorname{ch} x \cos b,$$
$$v = \operatorname{sh} x \sin b,$$

eliminating x, the variable,

$$\frac{u^2}{\cos^2 b} - \frac{v^2}{\sin^2 b} = \operatorname{ch}^2 x - \operatorname{sh}^2 x,$$
$$= 1$$

showing that the line $y = b$ in the (x, y) plane becomes an hyperbola in the (u, v) plane.

In the case of the ellipse [Vol. I (14.2)]

$$\frac{x^2}{A^2} + \frac{y^2}{B^2} = 1$$

the distance of a focus from the centre is $\sqrt{(A^2 - B^2)}$, whilst for an hyperbola

$$\frac{x^2}{p^2} - \frac{y^2}{q^2} = 1$$

this distance is $\sqrt{(p^2 + q^2)}$. Applying these to the above conics, both of which clearly have the same axes $u = 0$, $v = 0$, and the same centre—the origin in the (u, v) plane—we find that for the ellipse

$$\sqrt{(A^2 - B^2)} = \sqrt{(\operatorname{ch}^2 a - \operatorname{sh}^2 a)} = 1,$$

whilst for the hyperbola

$$\sqrt{(p^2 + q^2)} = \sqrt{(\cos^2 b + \sin^2 b)} = 1,$$

so that both conics have the same foci for all values of a and b. This show that the family of lines $x = a$ (for all a) and the family of lines $y = b$ (for all b) are transformed into families of *confocal conics*.

Example 6. If $w = z^2$, sketch in an Argand diagram the path traced out by the point w as the point z describes the rectangle whose vertices are at the points $\pm a$, $\pm a + ja$ where a is real.　　　　　　　　　　　　　　　　　[L.U.]

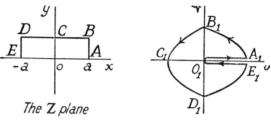

The Z plane

FIG. 46.

Given

$$w = z^2$$

or

$$u + jv = (x + jy)^2$$

therefore

$$u = x^2 - y^2, \qquad v = 2xy \qquad \cdot \quad \cdot \quad \cdot \quad \cdot \quad \cdot \quad (1)$$

The x axis, *i.e.*, $y = 0$ gives

$$u = x^2, \qquad v = 0$$

so that points on the x axis transform into points on the u axis, the point $E (- a, 0)$ giving $E_1 (a^2, 0)$, $O (0, 0)$ giving $O_1 (0, 0)$, $A (a, 0)$ giving $A_1 (a^2, 0)$. It appears then that as z moves from E to A, the point w moves from E_1 to O_1 and back again to A_1.

From (1) the line AB $(x = a)$ gives

$$u = a^2 - y^2, \qquad v = 2ay \qquad \cdot \quad \cdot \quad \cdot \quad \cdot \quad \cdot \quad (2)$$

eliminating y, the variable to obtain the (u, v) locus for all y

$$u = a^2 - v^2/4a^2$$

or

$$v^2 = 4a^2(a^2 - u)$$

the parabola shown. B (a, a) by (2) becomes the point B_1 $(u = 0, v = 2a^2)$, so that as z moves from A to B the point w moves from A_1 to B_1.

On BD, $y = a$, so that (1) gives

$$u = x^2 - a^2, \qquad v = 2ax$$

eliminating the variable x gives

$$u = \frac{v^2}{4a^2} - a^2$$

or

$$v^2 = 4a^2(u + a^2).$$

Continuing as above, we find that as z moves from B to D, w moves along this parabola from B_1 to C_1 to D_1 and then the path D_1E_1 to correspond to DE.

13.5. The Bilinear Transformation

The relation

$$w = \frac{az + b}{cz + d}$$

where in general a, b, c, d are complex numbers, considered as a method of transforming points in the z plane into points in the w plane is called *bilinear*, since to each value of z corresponds only one value of w and, by solving for z, it can be seen that to each value of w corresponds only one value of z.

For general values this may be written

$$w = \frac{(bc - ad)/c^2}{z + \dfrac{d}{c}} + \frac{a}{c}$$

and we may therefore consider the transformation as consisting of four consecutive transformations :

(i) $z_1 = z + \dfrac{d}{c}$;

(ii) $z_2 = \dfrac{1}{z_1}$;

(iii) $z_3 = mz_2 \quad \left[m \equiv \dfrac{bc - ad}{c^2} \right]$;

(iv) $w = z_3 + \dfrac{a}{c}$.

(i) and (iv) are clearly merely translations of the point, so that a figure subject to these is translated without rotation or change in size.

(ii) is an example of the transformation given the obvious name of *inversion*.

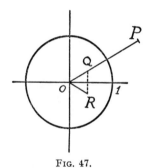

Fig. 47.

In polar co-ordinates it may be written

$$\rho e^{j\phi} = \frac{1}{r e^{j\theta}} = \frac{1}{r} e^{-j\theta}$$

so that
$$\rho = \frac{1}{r}, \quad \phi = -\theta.$$

If P is the point z_1 and Q for convenience drawn on the same diagram the point such that $OP \cdot OQ = 1$, where $OQ = \rho$, then z_2 is the point R where R is the image of Q in the x axis. Thus any figure subject to this transformation is inverted in the circle of unit radius centre at O and then reflected in the x axis. In cartesian co-ordinates

$$u + jv = \frac{1}{x + jy}$$

giving

$$u = \frac{x}{x^2 + y^2}, \qquad v = \frac{-y}{x^2 + y^2}$$

and

$$x = \frac{u}{u^2 + v^2}, \qquad y = \frac{-v}{u^2 + v^2}.$$

Any circle in the (x, y) plane may be written (A, B, C, D being real numbers)

$$A(x^2 + y^2) + Bx + Cy + D = 0 \quad . \quad . \quad . \quad (1)$$

which by the above equations transforms into

$$\frac{A(u^2 + v^2)}{(u^2 + v^2)^2} + \frac{Bu - Cv}{u^2 + v^2} + D = 0$$

or

$$D(u^2 + v^2) + Bu - Cv + A = 0 \quad . \quad . \quad . \quad (2)$$

so that generally the circle (1) transforms into a circle as shown by (2). If, however, $D = 0$ so that (1) is a circle through the origin then (2) is a line. Also if $A = 0$ so that (1) is a line then (2) is a circle passing through the origin of inversion. If A and D are both zero the line of (1) gives a line in (2) which is its image in the x axis. Therefore, considering lines as circles of infinite radius, the process of inversion changes circles into circles.

Finally, for (iii) using polars, if $m = ae^{j\alpha}$, and (R, β) denotes z_3.

$$Re^{j\beta} = ae^{j\alpha} . re^{j\theta}$$
$$= are^{j(\alpha + \theta)}$$

Therefore this transformation magnifies a distance r from the origin by $a = |m|$ and rotates it by $\alpha = \arg m$. Any figure subject to this transformation is therefore magnified and rotated.

The sum total effect of this bilinear transformation is then that circles and lines in the z plane are transformed into circles or lines in the w plane, since the processes (i), (iii), (iv) only translate, magnify and rotate and such processes would not change the nature of the circle or line resulting from (ii).

Before considering the applications of this to electrical examples we must obtain the equation to a circle in complex co-ordinates.

Since if $\qquad z = x + jy = r(\cos\theta + j\sin\theta)$

the conjugate $\bar{z}$ is given by

$$\bar{z} = x - jy = r(\cos\theta - j\sin\theta)$$

therefore

$$z\bar{z} = x^2 + y^2 = r^2 = |z|^2 = |\bar{z}|^2$$

The equation to a circle centre at 0 and radius r which is obviously

$$|z| = r$$

can therefore be written

$$z\bar{z} = r^2$$

or if the centre is at α

$$(z - \alpha)(\bar{z} - \bar{\alpha}) = r^2$$

where $\bar{z} - \bar{\alpha}$ is the conjugate of $z - \alpha$. This gives

$$z\bar{z} - \bar{\alpha}z - \alpha\bar{z} + \alpha\bar{\alpha} - r^2 = 0 \quad \ldots \quad (3)$$

as the general equation to a circle centre α and radius r.

We can apply this to find the centre and radius of a circle obtained by applying the above bilinear transformation in certain special electrical cases. Thus although the complex impedance is say $j\omega C$, it is only the C that varies by real values from zero to infinity. We can therefore consider that we have applied

$$w = \frac{az + b}{cz + d}$$

to the real axis in the z plane where a, b, c, d may be complex. In a given problem z will be C and the locus a circle.

The real axis has as its equation in the z plane

$$y = 0$$

or

$$z = \bar{z} \ldots \ldots \ldots \ldots (4)$$

which is

$$\frac{b - dw}{-a + cw} = \frac{\bar{b} - \bar{d}\bar{w}}{-\bar{a} + \bar{c}\bar{w}}$$

or

$$w\bar{w} + \frac{\bar{a}d - \bar{b}c}{\bar{d}c - d\bar{c}} w + \frac{b\bar{c} - a\bar{d}}{\bar{d}c - d\bar{c}} \bar{w} + \frac{a\bar{b} - \bar{a}b}{\bar{d}c - d\bar{c}} = 0. \quad . \quad (5)$$

comparing this with (3) will give α the centre and r the radius of the circle, *i.e.* the locus.

Consider as an example a circuit consisting of a resistance R and a condenser C in parallel where we require the variation in the impedance z as C varies

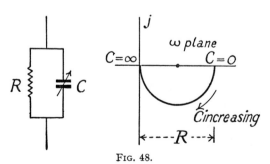

FIG. 48.

0 to ∞. In the usual notation

$$\frac{1}{z} = \frac{1}{R} + j\omega C$$

or

$$z = \frac{-jR}{R\omega C - j}$$

in the above notation this is written

$$w = \frac{-jR}{R\omega z - j} \quad \cdot \quad \cdot \quad \cdot \quad \cdot \quad \cdot \quad \cdot \quad \cdot \quad \cdot \quad (6)$$

so that by comparison with the standard form

$$a = 0, \qquad b = -jR, \qquad c = R\omega, \qquad d = -j$$

and therefore taking the conjugates

$$\bar{a} = 0, \qquad \bar{b} = jR, \qquad \bar{c} = R\omega, \qquad \bar{d} = j.$$

By (5) α, the coefficient of $-\bar{z}$ in (3), is given by

$$\alpha = \frac{a\bar{d} - b\bar{c}}{\bar{d}c - d\bar{c}} = \frac{-(-jR)(R\omega)}{j(R\omega) - (-j)(R\omega)}$$

$$= \tfrac{1}{2}R + j \cdot 0$$

Therefore

$$\alpha\bar{\alpha} - r^2 = (\tfrac{1}{2}R)(\tfrac{1}{2}R) - r^2$$

$$= \frac{ab - \bar{a}b}{\bar{d}c - d\bar{c}} = 0,$$

or

$$r^2 = (\tfrac{1}{2}R)^2, \quad r = \tfrac{1}{2}R.$$

We now have the centre of the circle a and its radius r. The circle is therefore as shown giving the variation in z as C varies.

It is often simpler to solve (6) for z

$$z = \frac{-jR + jw}{\omega Rw},$$

and by (4) equating z to $\bar{z}$

$$\frac{j(w - R)}{\omega Rw} = \frac{-j(\bar{w} - R)}{\omega R\bar{w}}$$

or

$$jw\bar{w} - jR\bar{w} = -jw\bar{w} + jwR$$

$$2jw\bar{w} - jR\bar{w} - jRw = 0,$$

$$w\bar{w} - \frac{R}{2}\bar{w} - \frac{R}{2}w = 0.$$

Comparing with (3),

$$\alpha = \frac{R}{2}, \quad \alpha\bar{\alpha} - r^2 = 0$$

or

$$r^2 = \alpha\bar{\alpha} = \left(\frac{R}{2}\right)^2,$$

Therefore

$$a = \frac{R}{2} + j0, \quad r = \tfrac{1}{2}R.$$

As a further example consider the circuit shown where C is to be varied and we require the locus of the total impedance. Using z instead of w

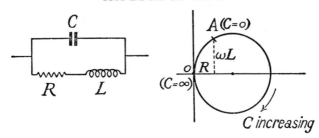

FIG. 49.

$$\frac{1}{z} = \frac{1}{R + j\omega L} + j\omega C$$

or

$$z = \frac{R + j\omega L}{1 + (jR\omega - \omega^2 L)C},$$

giving

$$C = \frac{R + j\omega L - z}{z(jR\omega - \omega^2 L)},$$

so that, again equating the conjugates,

$$\frac{(R + j\omega L) - z}{z(jR\omega - \omega^2 L)} = \frac{(R - j\omega L) - \bar{z}}{\bar{z}(-jR\omega - \omega^2 L)}.$$

$$z\bar{z}(jR + \omega L) - \bar{z}(jR + \omega L)(R + j\omega L)$$
$$= -z\bar{z}(jR - \omega L) + z(R - j\omega L)(jR - \omega L).$$

$$z\bar{z} - \bar{z}\frac{(jR + \omega L)(R + j\omega L)}{2jR} - zj\frac{(R - j\omega L)(R + j\omega L)}{2jR} = 0.$$

or

$$z\bar{z} - \bar{z}\frac{(R^2 + \omega^2 L^2)}{2R} - z\frac{(R^2 + \omega^2 L^2)}{2R} = 0.$$

By comparison with (3)

$$\alpha = \frac{1}{2R}(R^2 + \omega^2 L^2),$$

$$r^2 - \alpha\bar{\alpha} = 0, \quad r^2 = \alpha\bar{\alpha}.$$

Therefore

$$r = \frac{1}{2R}(R^2 + \omega^2 L^2).$$

The circle is as shown. When $C = 0$, $z = R + j\omega L$, giving the point A on the circle. When $C = \infty$, $z = 0$, the origin. The (minor) arc of the circle OA therefore applies to negative values of C and is inadmissible.

13.6. Functions of a Complex Variable

We will consider further the general relation $w = f(z)$. First we are concerned with discovering the conditions under which a differential coefficient exists where it is defined [by analogy with $y = f(x)$] as

$$\frac{dw}{dz} = f'(z) = \lim_{\Delta z \to 0} \frac{f(z + \Delta z) - f(z)}{\Delta z}.$$

Since $\Delta z = \Delta x + j\Delta y$ the value of the derivative might depend upon the way in which the increment tends to zero. Suppose that $f(z)$ has been separated into its real and j components, so that

$$f(z) = u(x, y) + jv(x, y)$$

and $f(z)$ has a unique derivative at the point z. Then let Δy approach

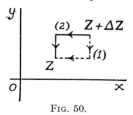

FIG. 50.

zero first and then Δx as in the first path. In this case

$$f'(z) = \lim_{\Delta x \to 0} \frac{u(x + \Delta x, y) - u(x, y)}{\Delta x} + j \lim_{\Delta x \to 0} \frac{v(x + \Delta x, y) - v(x, y)}{\Delta x}$$

$$= \frac{\partial u}{\partial x} + j \frac{\partial v}{\partial x}.$$

Using the path marked (2) where Δx approaches zero first and then Δy, since after $\Delta x = 0$, $\Delta z = j\Delta y$,

$$f'(z) = \lim_{\Delta y \to 0} \frac{u(x, y + \Delta y) - u(x, y)}{j\Delta y} + j \lim_{\Delta y \to 0} \frac{v(x, y + \Delta y) - v(x, y)}{j\Delta y}$$

$$= \frac{1}{j} \frac{\partial u}{\partial y} + \frac{\partial v}{\partial y}$$

If a unique derivative exists at the point (x, y) these values are equal or

$$\frac{\partial u}{\partial x} + j \frac{\partial v}{\partial x} = \frac{1}{j} \frac{\partial u}{\partial y} + \frac{\partial v}{\partial y},$$

from which

$$\frac{\partial u}{\partial x} = \frac{\partial v}{\partial y}, \quad \frac{\partial u}{\partial y} = -\frac{\partial v}{\partial x}.$$

These are known as the *Cauchy–Riemann* conditions.

From these by partial differentiation

$$\frac{\partial}{\partial x}(u_x) = \frac{\partial}{\partial x}(v_y)$$

and

$$\frac{\partial}{\partial y}(v_x) = -\frac{\partial}{\partial y}(u_y),$$

giving

$$u_{xx} = v_{xy} = -u_{yy}$$

or

$$u_{xx} + u_{yy} = 0$$

Similarly,

$$v_{xx} + v_{yy} = 0,$$

showing that if from the relation

$$u + jv = f(x + jy)$$

we sort out the two functions u and v, then each separately will satisfy Laplace's equation, provided we deal only with functions having generally a unique derivative. But from

$$u + jv = f(x + jy)$$

since u, v are functions of x and y

$$u_x + jv_x = f'(x + jy)$$
$$u_y + jv_y = jf'(x + jy)$$
$$\therefore \quad j(u_x + jv_x) = u_y + jv_y$$

so that

$$u_x = v_y, \qquad v_x = -u_y$$

which are the required conditions for a definite derivative to exist. This shows that provided we keep to functions of the variable $x + jy$ (and not say $2x + j3y$, etc.) the derivative exists except perhaps at certain points, *e.g.*, the origin for $w = 1/z$. Since the derivative exists, Laplace's equation is satisfied everywhere in the plane except perhaps at these certain points.

As a simple example, consider as above

$$u + jv = (x + jy)^2,$$

giving

$$u = x^2 - y^2, \qquad v = 2xy$$

Since we have used $x + jy$, we know that

$$u_x = 2x = v_y, \quad v_x = 2y = -u_y$$

as here is easily verified. From this each function u and v will satisfy Laplace's equation as here again is easily verified.

There is another important property. If we take any two curves of the systems

$$u(x, y) = a \qquad v(x, y) = b$$

where u is written as $u(x, y)$ as a reminder that it is a function of x and y, then by differentiation

$$u_x + u_y \frac{dy}{dx} = 0, \qquad v_x + v_y \frac{dy}{dx} = 0.$$

The product of these two gradients is

$$-\frac{u_x}{u_y} \times -\frac{v_x}{v_y},$$

which, by the Cauchy–Riemann Equations,

$$= \frac{u_x}{u_y} \times -\frac{u_y}{u_x} = -1.$$

This shows that two curves, one of each family, will always cut at right angles. This is a characteristic of lines of force and equipotential lines or of lines of flow and velocity potential lines. The relation

$$w = f(z)$$

for all regular forms of f then provides us with a large number of forms of curves for dealing with important problems of flow in Engineering and Science. The forms u and v obtained from such a relation *automatically* satisfy Laplace's equation, *i.e.*, the requirements of, for example, continuity in the medium or the constancy of the quantity of matter involved. Also the curves obtained automatically cut at right angles. They therefore satisfy two of the conditions always arising in these problems. (*Conjugate Functions for Engineers*, by Miles Walker, is one of the many excellent books for students wishing to learn more about this important and fascinating branch of Mathematics.)

EXERCISE 29

1. If the ratio $(z - j)/(z - 1)$ is purely imaginary, show that the point z lies on a circle whose centre is $\frac{1}{2}(1 + j)$ and radius $1/\sqrt{2}$. [L.U.]

2. If the amplitude of $(z - 1)/(z + 1)$ is $\frac{1}{4}\pi$, show that z lies on a fixed circle of radius $\sqrt{2}$ and centre $(0, 1)$. [L.U.]

3. If $w = \log z$ is a transformation, what are the loci: (i) $u = a$; (ii) $v = b$.

4. If $x + jy = \text{th}\ (u + \frac{1}{4}j\pi)$, show that $x^2 + y^2 = 1$. [L.U.]

5. If $u + jv = a/z$, where $z = x + jy$ and a is real, show that the curves in the (x, y) plane along which u and v are respectively constant are circles and that they intersect orthogonally.

6. If $w^2 = z$ transforms points in the (u, v) plane into points in the (x, y) plane, what are the loci: (i) $u = a$; (ii) $v = b$.

7. P is the point on an Argand diagram representing z and Q is the point representing $1/(z - 3) + 17/3$. Find the locus of Q as P describes the circle $|z - 3| = 3$.

8. Apply the transformation $w = e^z$ to the lines: (i) $x = a$; (ii) $y = b$.

9. If $A + jB = c \tan (x + jy)$, show that $\tan 2x = 2cA/(c^2 - A^2 - B^2)$. [From the given relation we deduce that $A - jB = c \tan (x - jy)$. We now have

$$\tan 2x = \tan \overline{(x + jy} + \overline{x - jy)},$$

and the result follows by expanding the right-hand side as $\tan (C + D)$ in terms of $\tan C$ and $\tan D$ and substituting for the given tangents in terms of $A \pm jB$.]

10. If $\tan (A + jB) = x + jy$, prove that

$$x^2 + y^2 + 2x \cot 2A - 1 = 0,$$
$$x^2 + y^2 - 2y \coth 2B + 1 = 0.$$ [L.U.]

11. Prove that

$$2(\text{sh}^2 x \cos^2 y + \text{ch}^2 x \sin^2 y) = \text{ch}\ 2x - \cos 2y.$$

The current I at the receiving end of a long underground cable of length l is the real part of the complex quantity $-jKe^{j(\omega t + \phi)}/\text{sh}\ Pl$, where $P = \alpha + j\beta$, $j = \sqrt{(-1)}$ and K, ω, ϕ, α and β are real constants of the problem. Show that

$$I = \sqrt{2}K \sin (\omega t + \phi - \theta)/(\text{ch}\ 2\alpha l - \cos 2\beta l)^{\frac{1}{2}},$$

where $\tan \theta = \tan \beta l/\text{th}\ \alpha l$. [L.U.]

12. If $u + iv = e^{x + iy} \cos (x + iy)$, find u and v in terms of x and y.

By differentiating the above equation partially with respect to x, show that

$$\frac{\partial^2 u}{\partial x^2} + i \frac{\partial^2 v}{\partial x^2} = -2e^{x + iy} \sin (x + iy).$$

Find the corresponding result for differentiation with respect to y, and deduce that

$$\frac{\partial^2 u}{\partial x^2} + \frac{\partial^2 u}{\partial y^2} = 0 = \frac{\partial^2 v}{\partial x^2} + \frac{\partial^2 v}{\partial y^2}.$$

13. ABCDEF is a regular hexagon inscribed in the circle $|z| = a$ in the Argand Diagram, A being the point $(a, 0)$. If P representing the complex number z is

any point on the circle write down the complex numbers represented by the six points obtained by drawing lines from the origin equal and parallel to the directed lines $\overrightarrow{AP}$, $\overrightarrow{BP}$, $\overrightarrow{CP}$, $\overrightarrow{DP}$, $\overrightarrow{EP}$, $\overrightarrow{FP}$ and prove that their product is $z^6 - a^6$. Hence prove that

$$AP \cdot BP \cdot CP \cdot DP \cdot EP \cdot FP \leqslant 2a^6. \qquad \text{[L.U.]}$$

14. The point P represented in the Argand diagram by $z(= x + iy)$ lies on the line $6x + 8y = R$ where R is real. Q is the point represented by R^2/z. Show that the locus of Q is a circle and find its centre and radius. [L.U.]

15. If $u + iv = (x + iy)^{1/2}$, where u, v are real and the point (x, y) describes the circle $(x - 1)^2 + y^2 = 1$, find the polar equation of the locus of the point (u, v). [L.U.]

16. (i) Find the curve into which the circle $x^2 + y^2 = 1$ is transformed by the relation $w = (a + b)z + (a - b)z^{-1}$, where a, b are real constants x, y, u, v are real variables and $z = x + iy$, $w = u + iv$.

(ii) Show that the transformation $w = z + z^{-1}$ converts the straight line $\arg z = \alpha$ ($|\alpha| < \tfrac{1}{2}\pi$) into one branch of a hyperbola of eccentricity sec α. [L.U.]

17. If $x + jy = 2j/(u + jv)$, where x, y, u, v are all real and $j = \sqrt{-1}$, express x, y in terms of u and v.

If the point (x, y) moves round a trapezium ABCD whose vertices are the points $(1, 1)$ $(1, -1)$ $(2, -2)$ $(2, 2)$ in that order, find the equations of the path traced out by the point (u, v) and show it in a separate diagram. [L.U.]

18. Complex variables z, w are connected by the relation $z = (w - 1)/(w + 1)$. Prove that as the point corresponding to z describes the imaginary axis, the point corresponding to w describes a circle of radius unity and centre the origin. [L.U.]

19. (a) Express in the form $A + iB$ where $i^2 = -1$, $\cos\left(\dfrac{\pi}{4} + \dfrac{i}{2}\right)$.

(b) If $x + iy = t + 1/t$ and $t = re^{-i\theta}$, show that the locus in the (x, y) plane corresponding to $r = $ constant is an ellipse stating the lengths of its semi-axes and determine the locus corresponding to $\theta = \pi/4$ when r varies. [L.U.]

20. If $u + iv = a/z$ where $z = x + iy$, $i^2 = -1$ and a is real, show that the curves in the (x, y) plane along which u and v are respectively constant are circles and that they intersect orthogonally. [L.U.]

21. (a) Find in the form $a + ib$ all the roots of the equation

$$x^4 + 1 = 0.$$

(b) If $Z = z + a^2/z$ prove that when z describes the circle $x^2 + y^2 = a^2$, Z describes a straight line and find its length.

Prove that if z describes the circle $x^2 + y^2 = b^2$ where $b > a$, Z describes an ellipse whose foci are the ends of the above line. [L.U.]

22. Show that $w = z^2$ in the polar form $re^{i\theta} = R^2e^{2i\phi}$ will transform a quarter circle in the first quadrant of the z plane into a half-circle in the upper half of the w plane. Find the ratio of the areas between the quarter-circles of $|z| = a$, $|z| = b$ and between their semicircular transforms.

23. Show that $w = z^{\frac{1}{2}}$ transforms the area between two lines $u = a$, $u = b$ in the w plane into an area between two parabolas.

24. Show that $w = e^{\pi z/b}$ applied to the rectangle bounded by $y = 0$, $y = b$, $x = \pm a$ will transform it into two semicircles in the upper half w plane closed by the relevant parts of the u axis.

25. O, A, B are the points 0, 1, $1 + i$. If $w = z^2 + 1$, find the path traced by w as z traces the path OABO.

26. Prove that $\cosh i\theta = \cos\theta$, $\sinh i\theta = i\sin\theta$. If $x + iy = c\cosh(\alpha + i\beta)$, prove that the curve for which α has a constant value is an ellipse and that β is the eccentric angle and that all such ellipses are confocal.

Find the value of α if the semi-axes of the ellipses are a and b. [L.U.]

27. Find u and v the real and imaginary parts of

$$u + iv = (z - 1)e^{-i\alpha} + \frac{e^{i\alpha}}{z - 1},$$

where $z = x + iy$ and α is real.

Prove that the locus of the points on the Argand diagram representing the complex number z such that $v = 0$ is a circle of unit radius with centre at the point $(1, 0)$ and a straight line through the centre of the circle. [L.U.]

28. The complex numbers z_1 and z_2 are represented by the points P_1 and P_2 respectively. If $z_1(1 - z_2) = z_2$ and P_1 describes the line $2x + 1 = 0$, prove that P_2 describes a circle whose centre is at the origin. Find the radius of this circle and the sense in which it is described as P_1 moves along the line in the direction which makes its ordinate increase. [L.U.]

29. If $x + iy = c \tanh (u + iv)$, where x, y, c, u and v are all real, determine x and y in terms of real functions of u, v and c. Prove that this relationship implies both

$$x^2 + y^2 + c^2 - 2cx \coth 2u = 0$$

and

$$x^2 + y^2 - c^2 + 2cy \cot 2v = 0. \qquad \text{[L.U.]}$$

30. If $\quad \tan^{-1}(x + iy) = u + iv,\quad$ show that

$$u = \tfrac{1}{2} \tan^{-1}\left(\frac{2x}{1 - x^2 - y^2}\right)$$

$$v = \tfrac{1}{2} \tanh^{-1}\left(\frac{2y}{1 + x^2 + y^2}\right).$$

Show that if the real part of $\tan^{-1}(x + iy)$ is $\pi/8$ the representative point of the complex number $x + iy$ on an Argand diagram lies on a circle of radius $\sqrt{2}$ with its centre at the point $-1 + i0$. [L.U.]

31. Write down the modulus and amplitude (argument) of $w = e^z$, where $z = x + iy$ and x, y are real numbers.

The point representing z in the Argand diagram describes the triangle with vertices $z = 0$, 1, $1 + i$. Show that the point representing $w = e^z$ describes a curvilinear triangle which is equiangular with the triangle described by z. [L.U.]

32. Prove that if z and $\bar{z}$ be conjugate points in the Argand diagram, then $z\bar{z} - \alpha\bar{z} - \bar{\alpha}z + c = 0$, c being real and $\bar{\alpha}$ the complex conjugate of α, represents a circle.

Prove also that $\bar{\alpha}(\bar{z} - z) + \alpha(\bar{z} + z) + c = 0$ represents two straight lines and find the angle between them. [L.U.]

33. If $z = x + iy$ and a bar denotes the conjugate so that $\bar{z} = x - iy$, show that the equation of any circle which passes through the points z_1, z_2 can be written

$$(z - z_1)(\bar{z} - \bar{z}_2) + \lambda(\bar{z} - \bar{z}_1)(z - z_2) = 0.$$

Deduce the equation of the circle which circumscribes the triangle whose vertices are the points z_1, z_2, z_3. [L.U.]

[(i) Replace each z by $x + iy$ and $\bar{z}$ by $x - iy$. Simplify and show that the real part is a circle and the imaginary part a line through $(x_1 y_1)$, $(x_2 y_2)$; or

(ii) Show that $(z - z_1)/(z - z_2) = Re^{j\phi}$, where ϕ is the angle between the joins z_1 to z and z_2 to z. $(\bar{z} - \bar{z}_2)/(\bar{z} - \bar{z}_1)$ is therefore $\frac{1}{R}e^{-j\phi}$.]

34. A circuit consists of a resistance R_1 in series with a circuit containing a resistance R_2 in parallel with a circuit of impedance z where z contains a combination in parallel of R and C. Show that

$$w = R_1 + \frac{zR_2}{z + R_2}$$

where

$$z = (-jR)/(R\omega C - j).$$

If C now varies from 0 to ∞, show that the locus in the w plane is the lower half of a circle centre and radius

$$\left\{ R_1 + \frac{RR_2}{2(R + R_2)}, \ 0 \right\}, \qquad \frac{RR_2}{2(R + R_2)}$$

respectively.

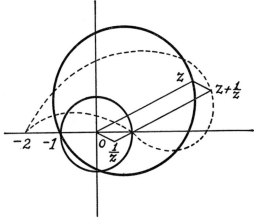

FIG. 51.

35. Draw a circle through the point $(-1, 0)$ to enclose the point $(1, 0)$. Take a series of points (z) on this circle, find $1/z$ by inverting in the circle $|z| = 1$ and add on z, i.e., find a point w corresponding to each z by means of

$$w = z + \frac{1}{z}.$$

Obtain the profile shown (known as a Joukowski aerofoil).

36. If V is a function of (x, y) or (u, v), where

$$u + iv = f(x + iy)$$

obtain

$$V_x = V_u u_x + V_v v_x.$$

and a similar expression for V_y. Obtain also expressions for V_{xx}, V_{yy}. Add these to obtain

$$V_{xx} + V_{yy} = (u_x^2 + v_x^2)(V_{uu} + V_{vv})$$
$$= |f'(x + iy)|(V_{uu} + V_{vv})$$

37. Given $x = e^r \cos \theta, y = e^r \sin \theta$ show that

$$\text{(a)} \quad r + i\theta = \log (x + iy)$$

$$\text{(b)} \quad \frac{\partial^2 V}{\partial x^2} + \frac{\partial^2 V}{\partial y^2} = e^{-2r} \left[\frac{\partial^2 V}{\partial r^2} + \frac{\partial^2 V}{\partial \theta^2} \right]$$

where V is a function of x, y and hence r, θ.

CHAPTER 14

THREE-DIMENSIONAL GEOMETRY. THE PLANE AND STRAIGHT LINE

14.1. General Considerations

In three-dimensional geometry we fix the position of a point in the (x, y) plane in the usual way by means of its co-ordinates x, y referred to two perpendicular axes Ox, Oy in the plane. For the third dimension we also need its height z above the plane and adopt a third axis Oz perpendicular to the plane xOy and therefore perpendicular to the axes Ox, Oy. We could measure the positive direction Oz up or down, but will adopt the direction that gives a right-handed set of axes. This means that we will choose the positive direction of z so that in screwing along the positive z axis with the right hand, the thumb rotates from the positive x axis to the positive y axis.

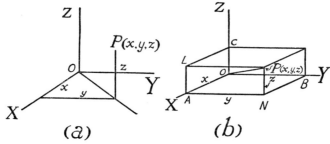

FIG. 52.

In Fig. 52(a) P(x, y, z) has x, y and z positive, but it will be evident that there are eight combinations to be obtained from $\pm x$, $\pm y$, $\pm z$. The planes xOy, xOz, yOz, extended to include negative values of the variables, will divide space into eight compartments (or *octants*). A point with given co-ordinates such as $(1, -2, -3)$ can only be in one of these.

14.2. Loci

The difference between two-dimensional and three-dimensional geometry is best appreciated by considering simple loci.

(i) With the x, y plane horizontal and on the ground, an equation $z = 3$ means that whatever the x and y of the point it is 3 units away from the ground everywhere. The locus is therefore a horizontal plane parallel to the xy plane (or ground).

226

(ii) In two dimensions the equation $y = x$ represents a line bisecting the angle xOy. In three dimensions we note that whatever the z of a point on the locus this is true and therefore the locus is a vertical plane through the Oz axis.

(iii) $x^2 + y^2 = 4$ represents a circle centre O and radius 2. In three dimensions, however, it is a cylinder, the vertical cylinder through the circle and with Oz as its axis. We further note that the circle is obtained by cutting the cylinder with the ground plane xOy, which is $z = 0$ so that the equation to the circle in three dimensions is the *pair* of equations $x^2 + y^2 = 4$, $z = 0$.

(iv) Two planes intersect generally in a line, therefore the *pair* of equations $z = 3$, $y = x$ represents the line in which these two planes intersect.

(v) It is evident that since $(x^2 + y^2 + z^2)^{\frac{1}{2}}$ is the distance of the point $P(x, y, z)$ from the origin (see Fig. 52(b)), the equation $x^2 + y^2 + z^2 = r^2$ is a sphere of radius r. Where this cuts the xOy plane ($z = 0$) we have $x^2 + y^2 = r^2$, verifying that the circle of intersection with the plane is of radius r. The equation $x^2 + y^2 + z^2 = r^2$ represents the *surface* of the sphere and to obtain a curve such as a circle we must pair off the equation with another surface such as a plane.

It appears, then, that a striking difference between three- and two-dimensional geometry is that in the first case a *single* equation generally represents a *surface*, and it needs *two* such equations treated as a pair of *simultaneous equations* to determine a line (curved or straight) by their intersection. In the second case (of two-dimensional geometry) a single equation generally represents a curve, and a pair of simultaneous equations generally represents the point or points in which these curves intersect. As a special instance of this we may regard all two-dimensional geometry as the section of three-dimensional geometry by the plane $z = 0$, and since the section of a surface will be a curve, we deal in two dimensions with curves where now in three dimensions we must deal with surfaces.

EXERCISE 30

Interpret the following equations:

1. $z = 4$.   2. $x = 3$.   3. $y = a$.   4. $x = z$.
5. $x^2 + y^2 + z^2 = 4$.   6. $y^2 + z^2 = 9$.   7. $x^2 = a^2$.
8. $y^2 = 4ax$.   9. $\dfrac{x^2}{a^2} + \dfrac{y^2}{b^2} = 1$.   10. $x^2 - y^2 = a^2$.

Interpret the following pairs of equations:

11. $z = 3$, $x = y$.   12. $x^2 + y^2 = 4$, $z = 2$.   13. $x^2 + y^2 + z^2 = a^2$, $z = 3$.
14. $y^2 = 4ax$, $x = z$.   15. $x^2 - y^2 = a^2$, $z = 3$.
16. In Fig. 52(b) if P is the point (a, b, c) write down the equation to the plane: (i) PNB, (ii) PAN, (iii) PLC.
17. With the data of the above question write down the equations to: (i) the line PN; (ii) the line PL; (iii) the line PC.

14.3. Co-ordinates of the Point Dividing a Join in a Given Ratio

It can be shown that if A is $(x_1y_1z_1)$ and B is $(x_2y_2z_2)$, the co-ordinates of a point C dividing the join AB in the ratio $m : n$ such that $AC/CB = m/n$ are :

$$x = \frac{mx_2 + nx_1}{m + n}, \qquad y = \frac{my_2 + ny_1}{m + n}, \qquad z = \frac{mz_2 + nz_1}{m + n}.$$

If the line is divided *externally*, then we must change the sign of m or n and obtain

$$x = \frac{mx_2 - nx_1}{m - n}$$

and two similar expressions for y and z. The proof depends upon the obvious fact that if from A, C, B we draw perpendiculars AN, CL, BM

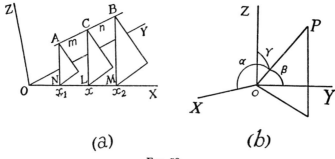

(a) (b)

Fig. 53.

to the x axis, then the feet of these lines N, L, M have co-ordinates x_1, x, and x_2 respectively. The three planes shown in Fig. 53(a) are all parallel and perpendicular to the x axis. Since transversals of parallel planes are divided in the same proportion, the proof follows.

Example 1. Find where the line joining the points (1, 2, 3), $(-1, 4, -7)$ cuts the xy plane.

If the xy plane divides the join in the ratio $m : n$, then its co-ordinates are

$$\frac{m(-1) + n(1)}{m + n}, \frac{m(4) + n(2)}{m + n}, \frac{m(-7) + n(3)}{m + n}.$$

Since the z co-ordinate is zero

$$-7m + 3n = 0, \quad \frac{n}{m} = \frac{7}{3}.$$

Therefore

$$x = \frac{-1 + \dfrac{n}{m}}{1 + \dfrac{n}{m}} = \frac{-1 + \dfrac{7}{3}}{1 + \dfrac{7}{3}} = \frac{2}{5}$$

$$y = \frac{4 + \dfrac{n}{m}2}{1 + \dfrac{n}{m}} = \frac{4 + \dfrac{7}{3}2}{1 + \dfrac{7}{3}} = \frac{13}{5}.$$

The required point is (2/5, 13/5, 0).

14.4. The Distance between Two Points

If P is the point (x, y, z) (Fig. 52(b)) and PN the perpendicular on to the xy plane so that N is the point (x, y), then

$$OP^2 = ON^2 + PN^2$$
$$= x^2 + y^2 + z^2 . \quad \ldots \quad \ldots \quad (1)$$

Similarly, if A is a point $(x_1 y_1 z_1)$ and B the point $(x_2 y_2 z_2)$, then by drawing axes through A parallel to the co-ordinate axes, we see that A may be regarded as an origin for the point B, whose co-ordinates relative to A are now

$$x_2 - x_1, \qquad y_2 - y_1, \qquad z_2 - z_1.$$

From this by using (1)

$$AB^2 = (x_2 - x_1)^2 + (y_2 - y_1)^2 + (z_2 - z_1)^2.$$

If A is a fixed point but B a variable point which will therefore be denoted by (x, y, z), then if B moves so that AB^2 is a fixed quantity a^2,

$$(x - x_1)^2 + (y - y_1)^2 + (z - z_1)^2 = a^2,$$

giving the equation to a locus which we know to be the surface of a sphere centre $(x_1 y_1 z_1)$ and radius a.

14.5. Direction Cosines of a Line

The direction of a line such as OP (Fig. 53(b)) is determined by the three angles α, β, γ which OP makes with Ox, Oy, Oz, respectively. We are usually concerned with the cosines of these angles, $\cos \alpha$, $\cos \beta$, $\cos \gamma$; these are called the *direction cosines* of OP, and often denoted by l, m, n, respectively.

If a line LM does not pass through O it will clearly have the same direction cosines as a parallel to it drawn through O and in the same sense as $\overline{\text{LM}}$.

We can find a relation between these three apparently independent quantities $\cos \alpha$, $\cos \beta$, $\cos \gamma$ in the following way. Through P (Fig. 52(b)) draw planes parallel to the co-ordinate planes so that finally we have a box with O and P diagonally opposite and the sides parallel or perpendicular to each other. In particular, the line OA is perpendicular to the side LANP, and therefore to the line AP. From this and similar relations we have

$$\cos \alpha = \frac{OA}{OP}, \qquad \cos \beta = \frac{OB}{OP}, \qquad \cos \gamma = \frac{OC}{OP}$$

$$\therefore \quad \cos^2 \alpha + \cos^2 \beta + \cos^2 \gamma = \frac{1}{OP^2}(OA^2 + OB^2 + OC^2)$$

$$= 1,$$

or
$$l^2 + m^2 + n^2 = 1,$$

an important relation showing that the direction cosines are not independent. A trio of numbers proportional to these is called the *direction ratios* of the line.

Example 2. (a) Two of the direction cosines of a line are $\frac{1}{2}$ and $\frac{1}{3}$. Find the third if it is known to be positive.

(b) The direction cosines of a line are proportional to 6, 5, -2, (*i.e.*, the direction ratios of the line are 6, 5, -2). Find them.

(a) If we are given $\cos \alpha$, and $\cos \beta$ then

$$(\tfrac{1}{2})^2 + (\tfrac{1}{3})^2 + \cos^2 \gamma = 1,$$

$$\cos^2 \gamma = \tfrac{23}{36}$$

$$\cos \gamma = \tfrac{1}{6}\sqrt{23},$$

since it is positive.

(b) Let the direction cosines be $6k$, $5k$, $-2k$, where k is the constant of proportionality. Therefore

$$(6k)^2 + (5k)^2 + (-2k)^2 = 1,$$

$$k^2 = \frac{1}{65}$$

$$k = \pm \frac{1}{\sqrt{65}}.$$

Choosing the positive value, the direction cosines are

$$\frac{6}{\sqrt{65}}, \quad \frac{5}{\sqrt{65}}, \quad \frac{-2}{\sqrt{65}}.$$

The negative value of k will give the reverse direction.

EXERCISE 31

1. Find the distance between the points (1, 2, 3) and $(-1, 3, -4)$.

2. Find the locus of a point which moves so that its distance from (0, 1, 4) is equal to its distance from $(1, -3, 2)$.

3. Find the above locus if the first distance is twice the second.

4. Prove that the values $\alpha = 30°$, $\beta = 30°$ for the direction angles are impossible.

5. The direction cosines of a line are $\frac{2}{3}$, $-\frac{2}{3}$, $\frac{1}{3}$. Find the angles the line makes with the axes and illustrate with a sketch.

6. If the direction cosines of a line are proportional to l, m, n prove that they are

$$l/\sqrt{(l^2 + m^2 + n^2)}, \qquad m/\sqrt{(l^2 + m^2 + n^2)}, \qquad n/\sqrt{(l^2 + m^2 + n^2)}.$$

7. The lengths of the projections of a line upon the axes are respectively 2, 3 and 6. Find its length.

8. Show that the direction cosines of the join (x_1, y_1, z_1) to $(x_2 y_2 z_2)$ are $(x_2 - x_1)/r$ and two similar expressions in y and z where r is the length of the join.

9. If P is the point $(x_1 y_1 z_1)$, show that the projection of OP upon a line whose direction cosines are (l, m, n) is $lx_1 + my_1 + nz_1$. If Q is the point $(x_2 y_2 z_2)$, show that the projection of PQ upon this line is $l(x_2 - x_1) + m(y_2 - y_1) + n(z_2 - z_1)$.

10. Find the co-ordinates of the point between the join $(2, -3, 6)$ to $(5, 4, -2)$ which is twice as far from the first point as from the second.

11. Show that the centre of gravity of a triangle whose vertices are (x_i, y_i, z_i) $(i = 1, 2, 3)$ is the point

$$\tfrac{1}{3}\Sigma x_i, \qquad \tfrac{1}{3}\Sigma y_i, \qquad \tfrac{1}{3}\Sigma z_i.$$

12. Show that the points (2, 4, 3), (4, 1, 9), $(10, -1, 6)$ are the vertices of an isosceles right-angled triangle. Find the co-ordinates of the centre of gravity of the triangle.

13. Find the point in which the join of $(2, -3, 1)$ to $(5, 4, 6)$ meets the zx plane.

14. Find where the join of the points $(1, 2, 1)$ to $(10, 3, 7)$ meets the cylinder $x^2 + y^2 = 16$.

14.6. The Equation to a Plane

It has already been seen that special planes are represented by first-degree equations. This will now be dealt with generally.

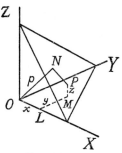

Fig. 54.

Let $ON = p$, a positive quantity be the perpendicular from the origin upon a plane and (l, m, n) the direction cosines of this normal to the plane. If $P(x, y, z)$ is any point on the plane and $OL = x$, $LM = y$, $MP = z$ is a three-step method of moving from O to P by means of steps each parallel to an axis, then the sum of the projections of each of these upon ON is equal to the projection of OP upon ON. For the point P (x, y, z) on the plane this gives

$$p = lx + my + nz,$$

giving the form of the equation to the plane in terms of the direction cosines of its normal and the length of the perpendicular upon it from the origin.

(i) Any first-degree equation represents a plane. In the general equation of the first degree $Ax + By + Cz + D = 0$ there are only three arbitrary constants, the ratios such as A/D, B/D, C/D, so that a plane can be made to satisfy three conditions.

(ii) Comparing $\qquad Ax + By + Cz + D = 0$

with $\qquad\qquad lx + my + nz - p = 0,$

in which $-p$ is a negative quantity gives

$$\frac{A}{l} = \frac{B}{m} = \frac{C}{n} = \frac{D}{-p},$$

showing that the coefficients A, B, C are proportional to the direction cosines of the normal to the plane the equation represents. With D negative the actual direction cosines of the line from O normal to the plane are

$$(A, B, C)/\sqrt{(A^2 + B^2 + C^2)}.$$

(iii) If the equation is written as

$$\frac{x}{a} + \frac{y}{b} + \frac{z}{c} = 1,$$

this meets the x axis where $y = 0$, $z = 0$, giving $x = a$. In this form, then, the intercepts on the axes are a, b, c, respectively.

Example 3. Find the direction cosines of the normal to the plane

$$3x + 2y + 4z + 7 = 0.$$

This may be written

$$- 3x - 2y - 4z - 7 = 0,$$

and further,

$$- \frac{3}{\sqrt{29}} x - \frac{2}{\sqrt{29}} y - \frac{4}{\sqrt{29}} z - \frac{7}{\sqrt{29}} = 0,$$

where $29 = (- 3)^2 + (- 2)^2 + (- 4)^2$. Therefore the direction cosines of the line from the origin normal to the plane are

$$- \frac{3}{\sqrt{29}}, \quad - \frac{2}{\sqrt{29}}, \quad - \frac{4}{\sqrt{29}}.$$

The length of the perpendicular from the origin to the plane is $7/\sqrt{29}$.

Example 4. Find the equation of the plane determined by the points $(1, 1, 0)$, $(0, 1, 1)$, $(1, 0, 1)$.

If the plane is

$$Ax + By + Cz + D = 0,$$

$$\begin{aligned}
\text{It contains } (1, 1, 0): \quad & A + B &&+ D = 0, \\
\text{,,} \quad\quad (0, 1, 1): \quad & B + C &&+ D = 0, \\
\text{,,} \quad\quad (1, 0, 1): \quad & A \quad\quad + C &&+ D = 0.
\end{aligned}$$

Eliminate the three arbitrary ratios $A : B : C : D$ from these four equations to obtain

$$\begin{vmatrix} x & y & z & 1 \\ 1 & 1 & 0 & 1 \\ 0 & 1 & 1 & 1 \\ 1 & 0 & 1 & 1 \end{vmatrix} = 0$$

which, in this case is easily reduced to $x + y + z = 2$. (The student who does not wish to use determinants can, in many cases, as easily solve the equations in A, B, C and D by the usual elementary method, obtaining A, B and C in terms of D. If D is zero the values will be found in terms of one of the non-zero coefficients.)

Example 5. Find the equation of a plane through the point $(1, 2, 3)$ parallel to the plane $2x + 3y - 7z = 4$. Find the perpendicular distance of the point $(1, 2, 3)$ from the given plane.

The equation $a(x - 1) + b(y - 2) + c(z - 3) = 0$ is of the first degree and is therefore a plane. It is clearly satisfied, from its construction, by the point $(1, 2, 3)$ and is therefore a plane through the point. Since it is to be parallel to the plane $2x + 3y - 7z = 4$, the direction cosines of the normal must be the same or a, b, c must be proportional respectively to 2, 3, $- 7$. The required equation is therefore

$$2(x - 1) + 3(y - 2) - 7(z - 3) = 0,$$

or

$$2x + 3y - 7z = - 13.$$

Using the form $lx + my + nz = p$, where p is necessarily positive, we write both equations as

$$\frac{2}{\sqrt{62}}x + \frac{3}{\sqrt{62}}y - \frac{7}{\sqrt{62}}z = \frac{4}{\sqrt{62}},$$

where $62 = 2^2 + 3^2 + 7^2$, and

$$-\frac{2}{\sqrt{62}}x - \frac{3}{\sqrt{62}}y + \frac{7}{\sqrt{62}}z = \frac{13}{\sqrt{62}}.$$

This second plane has as its normal the normal to the first plane continued backwards through O, *i.e.*, it is on the opposite side of O to the first plane. The distance between the planes is therefore the *sum* of the respective p's and is $(4 + 13)/\sqrt{62} = 17/\sqrt{62}$.

Note that if we substitute the co-ordinates of the point $(1, 2, 3)$ in the equation of the plane $2x + 3y - 7z = 4$ turned into the normal form

$$\frac{4}{\sqrt{62}} - \frac{2}{\sqrt{62}}x - \frac{3}{\sqrt{62}}y + \frac{7}{\sqrt{62}}z = 0$$

we obtain

$$\frac{4 - 2(1) - 3(2) + 7(3)}{\sqrt{62}} = \frac{17}{\sqrt{62}},$$

which is the required perpendicular distance.

This shows (compare two-dimensional geometry) that the perpendicular distance of $(x_1 y_1 z_1)$ from $Ax + By + Cz + D = 0$ is

$$\frac{Ax_1 + By_1 + Cz_1 + D}{\pm \sqrt{(A^2 + B^2 + C^2)}},$$

where we substitute the co-ordinates $(x_1 y_1 z_1)$ in the expression $Ax + By + Cz + D$ and divide by the square root of the sum of the squares of the coefficients of x, y, z, finally taking the positive value of the result.

EXERCISE 32

1. Find the intercepts of the plane $x + 2y - 3z = 7$ on the axes.

2. Find the intercepts made by the plane $2x + 3y + 4z = 30$ on the axes, and hence the volume of the tetrahedron made by this plane and the co-ordinate planes.

3. Find the equations of the planes determined by:

    (i) Intercepts on axes are $1, 2, -3$, respectively.

    (ii) Intercepts on axes are $2a, -3a, a$, respectively.

    (iii) Perpendicular from the origin upon the plane is of length 3 and makes equal angles with the axes.

    (iv) Perpendicular from the origin upon the plane is of length 4 and has direction ratios $1, 2, -3$.

4. Find the equation of the plane through the origin and the points $(-2, -3, 4)$, $(1, 3, -5)$.

5. Find the equation of the plane passing through the points $(0, 1, 2)$, $(1, 0, 4)$, $(2, -1, 5)$.

6. Find the equation of a plane if a line whose direction ratios are $(1, 2, 3)$ is perpendicular to it and the plane passes through the point $(3, -2, 1)$.

7. Find the equation of the plane which makes equal intercepts on the axes and passes through the point $(1, 0, 2)$.

8. What is the equation of the plane which passes through the z axis and contains the point $(1, -3, 7)$?

9. Find the point of intersection of the planes:

    (i) $4x - 3y + 2z = 4$ (ii) $3x - 4y + 5z = 10$
         $x + y + 3z = 1$ $2x - y + z = 3$
         $2x - y - z = 2$ $x - 3y + 2z = 1$

10. Find the equation of the plane bisecting at right angles the join of $(1, 2, 3)$ to $(-3, 2, 5)$.

11. Find the perpendicular distance between the parallel planes

$3x + 2y + 4z = 1$, and (a) $3x + 2y + 4z = 12$, (b) $3x + 2y + 4z = -12$

12. Find the perpendicular distance of the point $(1, 3, 5)$ from the plane $x - y + z = 2$.

13. A perpendicular from the origin on to a plane meets it at the point $(1, 4, 7)$. What is the equation to the plane?

14. If an area S lies on a plane making angle θ with the horizontal, its projection on to the horizontal ground has area S cos θ. A plane makes intercepts a, b, c on the axes. By projecting on to each plane xy, yz, zx in turn, show that the triangular area bounded by the co-ordinate planes is $\frac{1}{2}\sqrt{(b^2c^2 + c^2a^2 + a^2b^2)}$.

15. Find the equation of the plane through the points $(1, 2, 3)$ $(3, 2, 1)$ $(2, 3, 1)$. Noting the direction cosines of the normal to this plane, project the triangular area formed by the three points on to a co-ordinate plane, and hence find the area of the triangle which has the three points as vertices.

16. (i) Find the equation of the plane through the point $(3, 5, 4)$ which makes equal positive intercepts on the x and y axes, and a positive intercept of twice the length on the z axis. Find also the plane through the y axis perpendicular to this plane.

(ii) Obtain the distance from the origin of the plane passing through the point $(1, 0, 0)$ and through the line of intersection of the planes $2x - y - z = 5$, $x + y + z = 4$. [L.U.]

17. Find the volume of the tetrahedron formed by the three points in Question 15 and the origin.

18. A pyramid is formed by joining the point $(1, 2, 3)$ to the four points $(2, 3, 1)$ $(3, 1, 2)$ $(1, 4, 0)$ $(0, 4, -1)$. Find the volume of the pyramid, verifying that the latter four points are co-planar.

19. A formula for the volume of a tetrahedron whose vertices are (x_i, y_i, z_i), $(i = 1, 2, 3, 4)$ is

$$V = \pm \frac{1}{6} \begin{vmatrix} x_1 & y_1 & z_1 & 1 \\ x_2 & y_2 & z_2 & 1 \\ x_3 & y_3 & z_3 & 1 \\ x_4 & y_4 & z_4 & 1 \end{vmatrix}$$

where we take the positive value of the result. Use this to check the answer to Question 18.

14.7. The Angle between Two Lines

OP, OQ are two lines through O, the angle between them being θ. If OP is of length r the projection of OP on to OQ is OS $= r \cos \theta$.

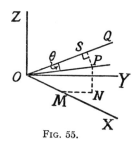

Fig. 55.

But since

$$\overline{OP} = \overline{OM} + \overline{MN} + \overline{NP}$$

(where OM, MN, NP are respectively parallel to the x, y, z axes), its projection is the sum of the projections of $\overline{OM}$, $\overline{MN}$, $\overline{NP}$ upon OQ. If now the direction cosines of OP are (l, m, n), then $OM = rl$, $MN = rm$, $NP = rn$, and if the direction cosines of OQ are $(l_1 m_1 n_1)$ the sum of the projections as above is

$$OMl_1 + MNm_1 + NPn_1$$

or

$$rll_1 + rmm_1 + rnn_1.$$

Equating the two values of the projection

$$r \cos \theta = rll_1 + rmm_1 + rnn_1,$$
$$\cos \theta = ll_1 + mm_1 + nn_1, \quad . \quad . \quad . \quad . \quad . \quad (1)$$

giving the cosine of the angle in terms of the direction cosines of each line.

(i) If the lines are perpendicular,

$$ll_1 + mm_1 + nn_1 = 0 \quad . \quad . \quad . \quad . \quad . \quad (2)$$

(ii) If we are given numbers proportional to the direction cosines such as (abc), $(a_1 b_1 c_1)$ for each respective line,

$$\cos \theta = \frac{aa_1 + bb_1 + cc_1}{\sqrt{(a^2 + b^2 + c^2)}\sqrt{(a_1^2 + b_1^2 + c_1^2)}},$$

and if the lines are perpendicular

$$aa_1 + bb_1 + cc_1 = 0.$$

Example 6. Lines OA, OB have direction cosines proportional respectively to $(1, 2, 3)$ $(-1, 3, -4)$. Find the direction cosines of the normal to the plane OAB.

If the normal has direction ratios (a, b, c), since it is perpendicular to every line in the plane, it is perpendicular to OA and OB. Therefore

$$1a + 2b + 3c = 0,$$
$$-1a + 3b - 4c = 0.$$

These give

$$\frac{a}{-17} = \frac{b}{1} = \frac{c}{5}.$$

The actual direction cosines are

$$\frac{-17, 1, 5}{\sqrt{(17^2 + 1^2 + 5^2)}} = \frac{-17, 1, 5}{\sqrt{315}}.$$

14.8. The Equations of a Straight Line

A line can be defined by means of a point $P(x_1 y_1 z_1)$ on it and its direction cosines (l, m, n).

If Q (x, y, z) is any other point on it such that $PQ = r$ then, with $\cos \alpha = l$, $\cos \beta = m$, $\cos \gamma = n$

$$l = \frac{x - x_1}{r}, \qquad m = \frac{y - y_1}{r}, \qquad n = \frac{z - z_1}{r} \quad . \quad . \quad (1)$$

From these equations we can derive a parametric form of equation of the line in terms of the parameter r :

$$x = x_1 + rl, \qquad y = y_1 + rm, \qquad z = z_1 + rn$$

where as r varies from minus infinity to plus infinity every point on the line is obtained.

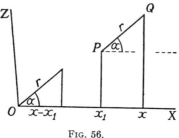

FIG. 56.

We can also obtain the *symmetrical* form of the equations to a line as

$$\frac{x - x_1}{l} = \frac{y - y_1}{m} = \frac{z - z_1}{n} = r \quad . \quad . \quad . \quad . \quad (2)$$

(i) The student will note that the first two expressions give a plane

$$mx - ly = mx_1 - ly_1$$

which is parallel to the z axis. The first and third expressions give

$$nx - lz = nx_1 - lz_1,$$

a plane parallel to the y axis. The form (2) can therefore be taken to express the line, by means of simultaneous equations, as the intersection of planes.

(ii) We can have as a special case of (2) such equations as

$$\frac{x - x_1}{l} = \frac{y - y_1}{m} = \frac{z - z_1}{0} = r.$$

This means that if $n = \cos \gamma$, then $\gamma = \frac{1}{2}\pi$, so that $\cos \gamma = 0$, and the line is perpendicular to the z axis. The above expression does not mean that we will divide by zero but multiply only. Thus if we equate to any other of the expressions and multiply across by the zero we obtain $z - z_1 = 0$, verifying the obvious fact that every point on this line parallel to the xy plane has its z co-ordinate constant at z_1.

(iii) In many cases the form

$$\frac{x - x_1}{a} = \frac{y - y_1}{b} = \frac{z - z_1}{c} = k$$

is used where a, b, c are direction ratios and not direction cosines.

(iv) The line through the points $(x_1 y_1 z_1)$, $(x_2 y_2 z_2)$ is

$$\frac{x - x_1}{x_1 - x_2} = \frac{y - y_1}{y_1 - y_2} = \frac{z - z_1}{z_1 - z_2} \quad . \quad . \quad . \quad . \quad (3)$$

since $x_1 - x_2$, $y_1 - y_2$, $z_1 - z_2$ are proportional to the direction cosines [$(x_1 - x_2)/r$ is an actual direction cosine], and so can be used as direction ratios. [We could, of course, in (3) use x_2, y_2, z_2 in the numerator instead of x_1, y_1, z_1.]

Example 7. Find where the line joining the points $(1, 2, 3)$, $(-2, 4, 3)$ meets the plane $x + y + z = 3$.

The equation of the line joining the points is

$$\frac{x - 1}{1 + 2} = \frac{y - 2}{2 - 4} = \frac{z - 3}{3 - 3} = k.$$

Therefore any point on the line may be taken as

$$x = 3k + 1, \qquad y = -2k + 2, \qquad z = 3.$$

If this point lies in the given plane

$$(3k + 1) + (-2k + 2) + 3 = 3,$$
$$k = -3.$$

The required point is $x = 3k + 1 = 3(-3) + 1 = -8$, $v = 8$, $z = 3$.

Example 8. Find the symmetrical form of the line given by the pair of equations

$$3x - 2y + z = 1, \qquad x + y + z = 4.$$

Choose any value for x, say $x = 1$. For this value of x the two equations become

$$-2y + z = -2, \qquad y + z = 3$$

which intersect at $y = 5/3$, $z = 4/3$ so that the point $(1, 5/3, 4/3)$ is on the line.

A pair of parallel planes through the origin is

$$3x - 2y + z = 0,$$
$$x + y + z = 0.$$

Solving these for the ratio $x : y : z$,

$$\frac{x}{-3} = \frac{y}{-2} = \frac{z}{5},$$

giving the line of intersection of the planes. The line of intersection of the original planes is parallel to this but through the point $(1, 5/3, 4/3)$. Its equations are therefore

$$\frac{x - 1}{-3} = \frac{y - 5/3}{-2} = \frac{z - 4/3}{5}.$$

[An alternative method is to find another point on the line in the way that the point $(1, 5/3, 4/3)$ was found. The line is then the join of the two points.]

Example 9. Find the distance of the point $(-1, 3, -5)$ from the line given by the pair of planes

$$3x + y - 2z = 1, \qquad 5x + 2y - 2z = 13.$$

If we choose $x = 3$ and repeat the above method we find the symmetrical form to be

$$\frac{x - 3}{2} = \frac{y - 6}{-4} = \frac{z - 7}{1} = \lambda \quad . \quad . \quad . \quad . \quad . \quad (1)$$

Therefore any point on this line can be taken as

$$2\lambda + 3, \qquad -4\lambda + 6, \qquad \lambda + 7 \quad . \quad . \quad . \quad . \quad (2)$$

The direction ratios of the join of this point to the point $(-1, 3, -5)$ is

$$2\lambda + 4, \qquad -4\lambda + 3, \qquad \lambda + 12.$$

If the join is perpendicular to line (1)

$$2(2\lambda + 4) - 4(-4\lambda + 3) + 1(\lambda + 12) = 0,$$
$$\lambda = -8/21.$$

Therefore using (2) the point

$$47/21, \qquad 158/21, \qquad 139/21$$

is a point on (1) whose join to $(-1, 3, -5)$ is perpendicular to the line. The distance between these points is the required perpendicular distance and is given by

$$d^2 = \left(\frac{47}{21} + 1\right)^2 + \left(\frac{158}{21} - 3\right)^2 + \left(\frac{139}{21} + 5\right)^2$$
$$= \frac{1}{21^2}[68^2 + 95^2 + 244^2], \qquad \underline{d = 12 \cdot 9.}$$

14.9. Line of Greatest Slope

The line of greatest slope in a plane is perpendicular to the line of no slope, *i.e.* the line in which the plane meets the horizontal "ground" plane.

Example 10. Find the equation of the line of greatest slope through $(1, 1, 1)$ on the plane $2x + 3y - 4z = 1$.

Let the line be

$$\frac{x-1}{l} = \frac{y-1}{m} = \frac{z-1}{n}.$$

It lies in the plane and is therefore perpendicular to its normal. From this

$$2l + 3m - 4n = 0 \quad \ldots \ldots \ldots \quad (1)$$

Where the plane meets $z = 0$

$$y = -\tfrac{2}{3}(x - \tfrac{1}{2}),$$

so that a line of no slope is

$$\frac{x - \tfrac{1}{2}}{3} = \frac{y}{-2} = \frac{z}{0}.$$

Since the line of greatest slope is perpendicular to this

$$3l - 2m + 0n = 0 \quad \ldots \ldots \ldots \quad (2)$$

(1) and (2) give

$$\frac{l}{8} = \frac{m}{12} = \frac{n}{13}.$$

The equation to the line is

$$\frac{x-1}{8} = \frac{y-1}{12} = \frac{z-1}{13}.$$

EXERCISE 33

1. If the angle between two planes is taken as the angle between their normals, find the angle between the planes $2x + 3y - 4z = 0$, $x + y + z = 1$. Find also the angle between the first plane and the line

$$\frac{x-1}{2} = \frac{y-2}{3} = \frac{z+4}{5}.$$

2. Find the angle between the planes $3x + 4y - 5z = 9$, $3x + 4y + 5z = 1$.

3. Find the angle between the lines

$$\frac{x}{1} = \frac{y}{1} = \frac{z}{1}, \qquad \frac{x-1}{2} = \frac{y+1}{4} = \frac{z-2}{-3}.$$

4. Show that the lines given by the pairs of planes

$$3x + 2y + z = 5, \quad x + y - 2z = 3 \quad \text{and} \quad 8x - 4y - 4z = 0, \quad 7x + 10y - 8z = 0$$

are perpendicular.

5. Show that the equations

$$\frac{x}{2} = \frac{y+3}{5} = \frac{z+2}{3} \quad \text{and} \quad \frac{x-2}{2} = \frac{y-2}{5} = \frac{z-1}{3}$$

represent the same line.

6. Find where the line

$$\frac{x-3}{4} = \frac{y-2}{3} = \frac{z+3}{1}$$

meets the plane $x + y + z = 1$.

7. Find in the symmetrical form the equations of the line given by the planes

$$2x + y = 21, \qquad 2z + y = 4,$$

choosing the value $y = 0$ for a point on the line.

8. Find the equation of the line through the origin parallel to the line determined by $2x + 3y - 2 = 0$, $2y - z + 9 = 0$.

9. Find the equation of the line joining the points $(1, 0, -2)$, $(2, 3, -5)$. What angle does it make with the plane $2x + 3y - z = 4$.

10. Find the point in which the join of $(1, -2, 3)$ to $(-3, 2, 1)$ is met by a perpendicular line through the point $(4, 6, -5)$. [Express any point on the join in terms of a parameter λ. Since the join of this point to $(4, 6, -5)$ is perpendicular to the line, λ can be found.]

11. Prove that the lines

$$3x + 2y + z - 5 = x + y - 2z - 3 = 0,$$
$$8x - 4y - 4z \quad = 7x + 10y - 8z = 0$$

are perpendicular to each other.

12. Find the plane through the three points $(2, 0, 6)$, $(10, 12, 0)$, $(-2, 3, 6)$. If ON is the perpendicular from the origin on to this plane, find the length ON and the co-ordinates of the point N.

13. Find the direction cosines of the line of intersection of the planes

$$x + y + z = 4, \qquad x - 2y - z = 4.$$

Find also the perpendicular distance of the point $(4, 1, 1)$ from this line.

14. Find the equation of the plane through the line $3(x - 1) = 6(y - 3) = 2z$ and the point $(2, 1, 7)$.

15. Find the direction cosines of the line of intersection of the planes

$$3x - y + z + 1 = 0 \quad \text{and} \quad 5x + y + 3z = 0.$$

Obtain the equation of the plane perpendicular to this line and passing through the point $(2, 1, 4)$.

16. Find the plane passing through the origin and the points $(1, 6, 4)$, $(6, 15, -4)$. Assuming the axis of z to be vertical, obtain the equations of the steepest line on this plane passing through the origin.

17. If Ox, Oy are horizontal and Oz is vertically upwards, show that the line

$$(x - 1)/8 = (y - 1)/12 = (z - 1)/13$$

lies entirely in the plane $2x + 3y - 4z = 1$ and that it is the line of steepest slope through the point $(1, 1, 1)$.

18. Prove that the line

$$\frac{x-3}{2} = \frac{y-4}{3} = \frac{z-5}{4}$$

is parallel to the plane $4x + 4y - 5z = 14$.

19. Show that the following lines intersect and find the point of intersection.

$$\frac{x+1}{-3} = \frac{y-3}{2} = \frac{z+2}{1}, \qquad \frac{x}{1} = \frac{y-7}{-3} = \frac{z+7}{2}.$$

20. Show that the line

$$\frac{x}{1} = \frac{y-11}{-3} = \frac{z+8}{2}$$

lies in the plane $x - y - 2z = 5$. [A line will lie in a plane if it has a point on it in the plane and is perpendicular to the normal to the plane.]

21. Show that the condition that the two lines

$$\frac{x - x_1}{l_1} = \frac{y - y_1}{m_1} = \frac{z - z_1}{n_1}, \quad \frac{x - x_2}{l_2} = \frac{y - y_2}{m_2} = \frac{z - z_2}{n_2}$$

are coplanar may be written

$$\begin{vmatrix} x_1 - x_2 & y_1 - y_2 & z_1 - z_2 \\ l_1 & m_1 & n_1 \\ l_2 & m_2 & n_2 \end{vmatrix} = 0.$$

[Assume the plane to be $ax + by + cz + d = 0$. Express that $(x_1y_1z_1)$, $(x_2y_2z_2)$ lie in it and that the lines are perpendicular to the normal to the plane.]

22. Show that the two lines following are coplanar:

$$\frac{x - 9/2}{7} = \frac{y}{2} = \frac{z + 5/2}{1}, \quad \frac{x + 3}{4} = \frac{y - 5}{-6} = \frac{z + 3}{0}.$$

(a) Use the above determinant. (b) Find a point on each line in terms of a parameter r_1 for the first, r_2 for the second. If the lines are coplanar, i.e., meet in a point, it must be possible to choose r_1 and r_2 so that the two lines have a point in common.

23. Prove that the three lines drawn through the origin with direction cosines $(l_im_in_i)$ $(i = 1, 2, 3)$ are coplanar if

$$\begin{vmatrix} l_1 & m_1 & n_1 \\ l_2 & m_2 & n_2 \\ l_3 & m_3 & n_3 \end{vmatrix} = 0.$$

24. If $u = 0$, $v = 0$ are the equations of two planes, examine the nature of the locus defined by $\lambda u + \mu v = 0$ where λ and μ are constants.

Find a and b so that the planes

$$x + y - z = 0, \quad 3x + by - 2z = 1, \quad x - 3y + az = 2$$

shall pass through one line and determine the symmetrical form of the equations to the line. [L.U.]

25. Find the direction cosines of the direction perpendicular to both the lines

(i) $x + 1 = -y = z - 2$; (ii) $\frac{x - 3}{2} = y = \frac{z - 3}{-4}.$

Find the co-ordinates of the point where the plane through the line (i) parallel to this direction meets the line (ii). [L.U.]

26. Show that the line whose equations are

$$6x + 4y - 5z = 4, \quad x - 5y + 2z = 12$$

intersects the line

$$\frac{x - 9}{2} = \frac{y + 4}{-1} = \frac{z - 5}{1}.$$

Find the equation of the plane containing these lines and also the angle between the lines. [L.U.]

27. Show that the line

$$\frac{x - 1}{2} = \frac{y - 2}{3} = \frac{z - 3}{4}$$

lies in the plane $x + 2y - 2z + 1 = 0$ and that

$$\frac{x - 3}{3} = \frac{y - 2}{2} = \frac{z - 1}{4}$$

lies in the plane $2x + y - 2z - 6 = 0$. If in addition these lines are the lines of greatest slope to the horizontal for the planes in which they lie, find the direction cosines of the vertical. [L.U.]

28. Find the equations in standard form of the projection of the line
$$\frac{x+1}{3} = \frac{y-2}{2} = \frac{z-3}{-1}$$
on the plane $x + y + 2z - 4 = 0$.
Find the projection of the point $(-1, 2, 3)$ on this plane. [L.U.]

14.10. The Shortest Distance between Two Skew Lines

We will state without proof the following theorem : Between two skew lines there is only one shortest distance, a line which is perpendicular to each of the skew lines.

Example 11. Find the direction cosines of a line which is perpendicular to both the lines
$$\frac{x-4}{2} = \frac{y}{6} = \frac{z-1}{3}, \qquad \frac{x-3}{2} = \frac{y+2}{-5} = \frac{z-2}{4}.$$

If the direction ratios of this line are a, b, c, then using the condition for perpendicularity $l_1 l_2 + m_1 m_2 + n_1 n_2 = 0$,
$$2a + 6b + 3c = 0,$$
$$2a - 5b + 4c = 0,$$
giving
$$\frac{a}{39} = \frac{b}{-2} = \frac{c}{-22}.$$

The direction cosines are
$$\frac{39, -2, -22}{\sqrt{(39^2 + 2^2 + 22^2)}} = \frac{39, -2, -22}{\sqrt{(2009)}}.$$

Example 12. Find the shortest distance between the above lines. (Fig. 57).

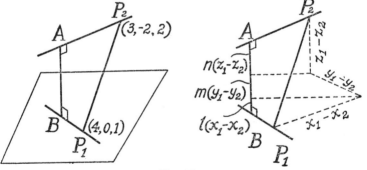

FIG. 57.

We have already used the fact that the projection of the join of the points $(x_1 y_1 z_1)$, $(x_2 y_2 z_2)$ upon a line whose direction cosines are (l, m, n) is of length
$$(x_1 - x_2)l + (y_1 - y_2)m + (z_1 - z_2)n.$$
If we choose the points $P_1 (4, 0, 1)$, $P_2 (3, -2, 2)$ which are given as being one on each line and project the join $P_1 P_2$ onto the shortest distance whose direction cosines are known, its length is found. This is
$$(4-3)\frac{39}{\sqrt{2009}} + (0+2)\frac{(-2)}{\sqrt{2009}} + (1-2)\frac{(-22)}{\sqrt{2009}}$$
$$= \frac{57}{\sqrt{2009}} = \underline{1 \cdot 27 \text{ units.}}$$

EXERCISE 34

1. Choose axes through the end of a long diagonal of a cube and find the equation of the diagonal and of an edge skew to it. If the cube is of side a, find the shortest distance between the diagonal and the edge.

2. Find the distance of the point $(2, 3, 4)$ from the line
$$\frac{x-4}{2} = \frac{y}{-6} = \frac{z-1}{3}$$

3. Find the shortest distance between the lines
$$\frac{x}{6} = \frac{y+2a}{6} = \frac{z-a}{1}, \qquad \frac{x+a}{12} = \frac{y}{6} = \frac{z}{-1}.$$

4. Find the shortest distance between the lines
$$\frac{x}{5} = \frac{y}{3} = \frac{z-2}{1}, \qquad \frac{x+3}{1} = \frac{y}{2} = \frac{z}{3}.$$

5. Find the shortest distance between the lines
$$2x = 4y = 3(z-3), \qquad 3(x+4) = 2y = 3z.$$

6. A uniform rod is of length $2a$ and mass M. Find its moment of inertia about an axis making angle θ with it and such that the shortest distance between the two is of length d and passes through the mid point of the rod.

7. Express any point on the lines
$$\frac{x-4}{2} = \frac{y}{6} = \frac{z-1}{3} = r_1, \qquad \frac{x-3}{2} = \frac{y+2}{-5} = \frac{z-2}{4} = r_2$$

in terms of r_1 and r_2 respectively. If the join of these points is the shortest distance, it will be perpendicular to each line. Hence find: (i) r_1, r_2; (ii) the special points whose join gives the shortest distance; (iii) the length of this join; (iv) the equations to the join.

8. If $u = 0$, $v = 0$ are first-degree equations denoting planes, the pair of equations together denote their line of intersection and $u + kv = 0$ where k is an arbitrary constant denotes any plane through this line. Similarly, $u_1 + k_1 v_1 = 0$ will denote any plane through the line defined by the pair of equations $u_1 = 0$, $v_1 = 0$. If we can choose k, k_1 so that the planes are parallel, then the perpendicular distance between these planes will be the shortest distance between the given lines. Apply this to the pairs of planes

(i) $\begin{aligned} 3x - 2y - 7 &= 0 \\ 4y - 3z + 11 &= 0 \end{aligned}$ (ii) $\begin{aligned} 4x - 3y + 2 &= 0 \\ 5y - 4z - 2 &= 0, \end{aligned}$

to find the shortest distance between the lines defined by each pair.

The lines in the symmetrical form are
$$\frac{x-1}{2} = \frac{y+2}{3} = \frac{z-1}{4}, \qquad \frac{x-1}{3} = \frac{y-2}{4} = \frac{z-2}{5}.$$

Apply some other method to these equations to check the result.

9. Show that the planes $x + 2y - z = 0$, $3x - 4y + z = 4$, and $4x + 3y - 2z = 24$ form a triangular prism and that the lengths of the sides of normal cross-section are in the ratio $\sqrt{13} : \sqrt{58} : 5\sqrt{3}$.   [L.U.]

10. Prove that the equation of the plane which passes through the point (α, β, γ) and the line $x = ay + b = cz + d$ is
$$\begin{vmatrix} x & ay+b & cz+d \\ \alpha & a\beta+b & c\gamma+d \\ 1 & 1 & 1 \end{vmatrix} = 0.$$

Hence find in its simplest form the equation of the plane which passes through the point $(1, 1, 1)$ and the line
$$\frac{x-1}{2} = \frac{y-2}{3} = \frac{z-3}{4}.$$

Find also the perpendicular distance of the point $(1, 1, 1)$ from the given line.

11. Find the equations of the straight line which passes through the point $(2, -1, -1)$, is parallel to the plane $4x + y + z + 2 = 0$ and is perpendicular to the line $2x + y = 0 = x - z + 5$.   [L.U.]

12. Show that the line of intersection of the planes

$$3x + 4y - 5z = 6, \qquad 2x + 3y + z = 2$$

is parallel to the plane $13x + 21y + 26z + 1 = 0$. Find the shortest distance between the line and the plane.   [L.U.]

13. Find the length and direction cosines of the shortest straight line joining the two straight lines whose equations are

$$\frac{x - 6}{3} = \frac{y - 7}{-1} = \frac{z - 4}{1}, \qquad \frac{x - 3}{-3} = \frac{y + 11}{2} = \frac{z + 2}{4}.$$ [L.U.]

14. Find the equation of the plane through the line $2x - y - 1 = 0$, $x + y - z = 1$ and parallel to the line $x - y + z + 3 = 0$, $3x + y + z + 7 = 0$ and hence or otherwise find the shortest distance between the lines.   [L.U.]

15. Find the angle between the lines

$$x + y + z = 6, \qquad 3x - y + 4z = 13$$

and

$$2x - y + 2z = 0, \qquad x - 2z = 0.$$

Find also the equation of the plane through the first line parallel to the second.   [L.U.]

16. Find the equation of the plane passing through the point $(3, 4, 5)$ and containing the line $\frac{x - 1}{2} = y + 1 = \frac{z - 2}{3}$. Show that it is parallel to the y axis and find the volume of the prism enclosed by it and the planes $x = 0$, $z = 0$, $y = 0$ and $y = 3$.   [L.U.]

17. Show that the co-ordinates of the foot of the perpendicular from a point (x_1, y_1, z_1) on to the line $x = y = -z$ are $\{\frac{1}{3}(x_1 + y_1 - z_1), \frac{1}{3}(x_1 + y_1 - z_1), -\frac{1}{3}(x_1 + y_1 - z_1)\}$. Deduce the equation of the right circular cylinder of radius a having its axis along the line $x = y = -z$ and find the area of the section of this cylinder by the plane $z = 0$.   [L.U.]

18. Find the equations to the line through the origin perpendicular to and intersecting the line

$$x + 2y + 3z + 4 = 0, \qquad 2x + 3y + 4z + 5 = 0$$

and determine the co-ordinates of the point in which these lines intersect.   [L.U.]

19. A plane passes through the points $(1, 2, 3)$, $(-1, 2, 0)$ and $(2, -1, -1)$. Find its equation, the area of the triangle formed by the lines in which it intersects the co-ordinate planes and the co-ordinates of the foot of the perpendicular from the origin to the plane.   [L.U.]

20. Find the equation of the plane through the line

$$\frac{x - 1}{3} = \frac{y - 4}{2} = \frac{z - 4}{-2}$$

and parallel to the line

$$\frac{x + 1}{3} = \frac{y - 1}{-4} = \frac{z + 2}{1}.$$

Hence or otherwise find the shortest distance between the lines.   [L.U.]

21. The points $A(2a, 0, 0)$, $B(0, 2b, 0)$, $C(0, 0, 0)$ and $D(a, b, c)$ are the corners of a tetrahedron. Find the equations to the line which intersects AC and BD and is perpendicular to AC and BD.

Hence or otherwise determine the shortest distance between the two lines.   [L.U.]

CHAPTER 15

SECOND-DEGREE SURFACES

15.1. The Sphere

The equation of the sphere with centre at the origin and radius r is clearly

$$x^2 + y^2 + z^2 = r^2$$

an equation which states that any point (x, y, z) on the surface is at a constant distance r from the origin. In the same way

$$(x - x_1)^2 + (y - y_1)^2 + (z - z_1)^2 = r^2$$

is the equation to the surface of a sphere centre at $(x_1 y_1 z_1)$ and radius r.

It will be noted that in this equation : (1) the coefficients of x^2, y^2, z^2 are equal, (2) there are no product terms such as xy, yz, zx.

The general equation

$$a(x^2 + y^2 + z^2) + bx + cy + dz + e = 0$$

where $a \neq 0$, therefore represents a sphere. We will divide by a and write the equation

$$x^2 + y^2 + z^2 + 2Ax + 2By + 2Cz + D = 0,$$

which can be expressed as

$$(x + A)^2 + (y + B)^2 + (z + C)^2 = A^2 + B^2 + C^2 - D$$

showing that the centre is $(- A, - B, - C)$ and the radius

$$\sqrt{(A^2 + B^2 + C^2 - D)}.$$

Example 1. Find the equation of the sphere passing through the four points $(0, 0, 0)$, $(1, 1, 1)$, $(2, 2, - 2)$, $(3, - 1, - 1)$.

It is clear that these four points must not be coplanar. Assume as the equation of the sphere

$$x^2 + y^2 + z^2 + ux + vy + wz + d = 0.$$

Substituting the co-ordinates of the points in turn,

$$\begin{aligned}
(0, 0, 0): &\qquad\qquad d = 0, \\
(1, 1, 1): &\qquad 3 + u + v + w + d = 0, \\
(2, 2, - 2): &\qquad 12 + 2u + 2v - 2w + d = 0, \\
(3, - 1, - 1): &\qquad 11 + 3u - v - w + d = 0.
\end{aligned}$$

We solve these and find

$$u = - 3\tfrac{1}{2}, \qquad v = - 1, \qquad w = 1\tfrac{1}{2}, \qquad d = 0.$$

The equation is

$$x^2 + y^2 + z^2 - 3\tfrac{1}{2}x - y + 1\tfrac{1}{2}z = 0,$$

which can be written as

$$(x - \tfrac{7}{4})^2 + (y - \tfrac{1}{2})^2 + (z + \tfrac{3}{4})^2 = \tfrac{62}{16},$$

showing that the centre is at the point $(\tfrac{7}{4}, \tfrac{1}{2}, - \tfrac{3}{4})$ and the radius $\tfrac{1}{4}\sqrt{62}$.

Example 2. Find the area of the circle in which the plane $x - 2y - 2z + 7 = 0$ cuts the sphere

$$x^2 + y^2 + z^2 - 2x + 6y + 4z - 35 = 0.$$

The equation to the sphere may be written

$$(x - 1)^2 + (y + 3)^2 + (z + 2)^2 = 49.$$

The length of the perpendicular from its centre $(1, -3, -2)$ to the plane is

$$\frac{1 - 2(-3) - 2(-2) + 7}{\sqrt{(1^2 + 2^2 + 2^2)}} = 6.$$

Since the radius of the sphere is 7, the radius of the circle is $\sqrt{(49 - 36)} = \sqrt{13}$. The area of the circle is therefore 13π square units.

The centre of the circle may be found as the intersection of the line through $(1, -3, -2)$ perpendicular to the plane. It is $(-1, 1, 2)$.

Example 3. Find the equation to the sphere passing through the origin and the circle defined by the sphere $x^2 + y^2 + z^2 = 4$ and the plane $x + y + z = 3$. Consider the equation

$$x^2 + y^2 + z^2 - 4 + k(x + y + z - 3) = 0.$$

This is clearly a sphere, and it is satisfied by the co-ordinates of any point that lies on both the sphere and the plane, *i.e.*, on the circle. This sphere therefore passes through the circle. We have one unknown constant k in the equation, and so can satisfy one other condition. If the sphere is to pass through $(0, 0, 0)$

$$- 4 + k(- 3) = 0.$$

The equation to the sphere is therefore

$$3(x^2 + y^2 + z^2) - 4(x + y + z) = 0.$$

EXERCISE 35

1. Find the equation to the sphere whose centre is the point $(2, -2, 1)$ and whose radius is 3. Show that it passes through the origin.

2. Find the equation to the sphere whose centre is the point $(1, 2, 3)$ and which passes through the point $(-1, -2, 4)$.

3. Find the centre and radius of each of the following spheres:

    (a) $x^2 + y^2 + z^2 - 4x + 8y + 10z = 0$;

    (b) $2(x^2 + y^2 + z^2) - 2x - 4y + 5z + 1 = 0$.

4. If a point (x, y, z) on a sphere is joined to two points on the sphere which are at opposite ends of a diameter, the joins will be at right angles. Use this to find the equation of a sphere which has the points $(1, 2, 3)$, $(4, 7, 5)$ at opposite ends of a diameter.

5. Obtain the equation required in the exercise above by using the facts that the centre of the sphere is the mid point of the join and its radius is half the distance between the points.

6. Find the equation to the sphere passing through the points $(0, 1, 3)$, $(1, 2, 4)$, $(2, 3, 1)$, $(3, 0, 2)$.

7. Find the intersections of the sphere $x^2 + y^2 + z^2 = 16$ with the line

$$\frac{x - 1}{1} = \frac{y - 1}{2} = \frac{z + 1}{1}.$$

8. Find two points diametrically opposite on the sphere

$$x^2 + y^2 + z^2 - 2x - 4y - 5 = 0$$

whose join is parallel to the line

$$\frac{x}{1} = \frac{y}{2} = \frac{z}{3}.$$

9. The tangent plane to a sphere at $(x_1y_1z_1)$, a point on its surface will be a plane through the point $(x_1y_1z_1)$ and will have the join of the point to the sphere's centre as a normal. Its equation is therefore

$$a(x - x_1) + b(y - y_1) + c(z - z_1) = 0,$$

where a, b, c are numbers proportional to the direction cosines of the normal. Use this to find the equations of the tangent planes to the following spheres at the points shown.

(i) $\qquad x^2 + y^2 + z^2 + 6x - 2y - 4z = 35$, at (3, 4, 4);

(ii) $x^2 + y^2 + z^2 - 6x + 4y + 10z - 11 = 0$ at (1, 1, 1).

10. If any tangent plane to the sphere $x^2 + y^2 + z^2 = r^2$ makes intercepts a, b and c on the axes prove that

$$\frac{1}{a^2} + \frac{1}{b^2} + \frac{1}{c^2} = \frac{1}{r^2}.$$

11. Show that the radius of the circular section of the sphere $x^2 + y^2 + z^2 = r^2$ by the plane $lx + my + nz = p$ is $\sqrt{(r^2 - p^2)}$.

12. Find the radius and centre of the section of the sphere

$$x^2 + y^2 + z^2 - 2x - 4y + 2z = 19,$$

by the plane

$$2x + 2y - z + 7 = 0.$$

13. Find the volume of the tetrahedron cut off from the first octant by the tangent plane at (1, 2, 3) to the sphere

$$x^2 + y^2 + z^2 + x - \tfrac{3}{2}y - 2z = 6.$$

14. A point moves so that the sum of the squares of its distances from the six faces of a cube is constant. Show that its locus is a sphere. (Take the origin at a corner and axes as the three edges that meet there. Assume the cube of side a.)

15. Find the area of the circle in which the plane $x + y + z = 1$ cuts the sphere $x^2 + y^2 + z^2 - 2x - 4y + 6z = 5$.

16. Find the equations to the spheres passing through the circle defined by $x^2 + y^2 + z^2 = 5$ and $x + 2y + 3z = 3$ and touching the plane $4x + 3y - 15 = 0$.

17. Show that there are two spheres passing through the points (4, 0, 3), (5, 4, 0), (5, 1, 3), and with radius 3. Find their equations.

18. Show that the spheres:

(a) $x^2 + y^2 + z^2 = 9$, $\qquad x^2 + y^2 + z^2 - 6x + 8y + 9 = 0$;

(b) $x^2 + y^2 + z^2 + 12x + 10z + 36 = 0$,
$\qquad x^2 + y^2 + z^2 - 12x - 8y + 4z - 88 = 0$

cut orthogonally. (Two spheres cut orthogonally if $d^2 = r_1{}^2 + r_2{}^2$, where d is the distance between their centres, r_1, r_2 the respective radii.)

19. Show that the two circles, given respectively by the following pairs of equations, lie on a sphere and find its equation:

$$x^2 + y^2 + z^2 - 9x + 4y + 5z - 1 = 0,$$
$$7x - 2y + z - 4 = 0.$$
$$x^2 + y^2 + z^2 + 6x - 10y + 6z - 7 = 0,$$
$$4x - 6y - 1 = 0. \qquad\qquad \text{[L.U.]}$$

[If $S_r = 0$ denotes a sphere and $L_r = 0$ a plane $(r = 1, 2)$, then $S_1 + k_1L_1 = 0$ denotes a sphere through the first circle and $S_2 + k_2L_2 = 0$, a sphere through the second circle. It must be possible to choose k_1 and k_2 so that the equations become identical and therefore denote the same sphere.]

20. Find the equation of the sphere containing the circles given by the following pairs of equations:

$$x^2 + y^2 + z^2 - 2x + 3y + 4z - 5 = 0,$$
$$5y + 6z + 1 = 0.$$
$$x^2 + y^2 + z^2 - 3x - 4y + 5z - 6 = 0,$$
$$x + 2y - 7z = 0.$$

21. Show that the curve
$$x = a \sin^2 t, \qquad y = a \sin t \cos t, \qquad z = a \cos t$$
lies on the sphere $\qquad x^2 + y^2 + z^2 = a^2.$

22. Show that the equation
$$x^2 + y^2 + z^2 + 2ax + 2by + 2cz + d + 2\lambda(\alpha x + \beta y + \gamma z + \delta) = 0$$
represents a sphere through the circle common to the sphere
$$x^2 + y^2 + z^2 + 2ax + 2by + 2cz + d = 0$$
and the plane
$$\alpha x + \beta y + \gamma z + \delta = 0.$$
Find the equations of the two spheres through the common circle of the sphere $x^2 + y^2 + z^2 + 2x + 2y = 0$ and the plane $x + y + z + 4 = 0$ and which intersect the plane $x + y = 0$ in circles of radius 3. [L.U.]

23. Find the equation to the diameter of the sphere $x^2 + y^2 + z^2 = 9$ such that a rotation about it will transfer the point $(2, 1, 2)$ to the point $(1, 2, 2)$ along a great circle of the sphere.

Find the angle through which the sphere must be rotated and the co-ordinates of the point into which $(1, 2, 2)$ is transferred. [L.U.]

15.2. The Cylinder

If through every point on a curve lying in a plane, normals to the plane are erected, the surface formed by these normals (or *generators*) is called a cylinder. We have already seen that if the equation to a surface does not contain, for example, z, the cylinder has its generators parallel to the z axis. We may therefore note that any curved surface is a cylinder with its generators parallel to a co-ordinate axis when the equation to the surface does not contain the variable corresponding to that axis.

From this point of view we may regard every curve in the x, y plane such as

$$\frac{x^2}{a^2} + \frac{y^2}{b^2} = 1, z = 0$$

as a section of a cylinder by the plane $z = 0$, in this case the cylinder having an elliptical cross-section parallel to the xy plane.

(We can, of course, have cylinders whose generators are not parallel to an axis, and for these the above test does not apply.)

15.3. The Cone

If a closed plane curve has every point on it joined to another point O not in the plane by lines continued indefinitely in either direction, the surface formed is called a *cone*. The given point O is the vertex, and the joining lines are called *generators*. If the given closed curve is a circle and O, the vertex, lies on its axis the resulting surface is a right circular cone.

Consider a surface as

$$3x^2 + 2y^2 - 4z^2 = 0 \qquad \cdots \cdots \quad (1)$$

If we substitute kx for x, ky for y and kz for z this becomes

$$k^2(3x^2 + 2y^2 - 4z^2),$$

which, of course, is zero. This shows that if (x, y, z) is any point on the surface, so is (kx, ky, kz) for all values of k. This is a feature of every point on the generator of a cone vertex at the origin. The above substitution will hold for any homogeneous equation, *i.e.*, an equation such that if n is the degree of the equation $F(x, y, z) = 0$

$$F(kx, ky, kz) = k^n F(x, y, z).$$

We conclude then that every *homogeneous* equation represents a *cone*, vertex at the origin.

The above cone in (1) can easily be traced, since the plane $z = a$ cuts it in the ellipse

$$3x^2 + 2y^2 = 4a^2.$$

The equation (1) therefore represents a cone whose axis is the z axis and whose cross-section, parallel to the xy plane is an ellipse.

Note that if

$$\frac{x}{l} = \frac{y}{m} = \frac{z}{n} = k$$

is a generator of the cone $F(x, y, z) = 0$ [a homogeneous equation] then

$$F(kl, km, kn) = 0$$

or since this equation is homogeneous

$$k^n F(l, m, n) = 0$$

and

$$F(l, m, n) = 0$$

is still a homogeneous equation with (l, m, n) substituted for (x, y, z). We deduce that if the direction ratios of a line through a fixed point satisfy a homogeneous equation, the line is a generator of a cone with vertex at the point.

Example 4. The direction ratios of a line through the fixed point (a, b, c) satisfy the equation $Al^2 + Bm^2 + Cn^2 = 0$. Find the cone all such lines generate.

If the line is

$$\frac{x - a}{l} = \frac{y - b}{m} = \frac{z - c}{n} = k$$

then $\qquad l = (x - a)/k, \qquad m = (y - b)/k, \qquad n = (z - c)/k.$

Substitute in the given relation and we have

$$A(x - a)^2 + B(y - b)^2 + C(z - c)^2 = 0$$

a cone with vertex at the point (a, b, c).

Example 5. Find the equation to the cone joining the origin to the curve of intersection of the sphere $x^2 + y^2 + z^2 + ax + by + cz + d = 0$ and the plane $lx + my + nz = p$.

The equation required will be homogeneous. Consider the equation

$$x^2 + y^2 + z^2 + (ax + by + cz)\frac{(lx + my + nz)}{p} + d\frac{(lx + my + nz)^2}{p^2} = 0.$$

This is homogeneous in x, y, z, and therefore represents a cone. At any point on the given plane $(lx + my + nz)/p = 1$, and therefore for such a point the equation to the cone that we have put down reduces to that of the original sphere. If this point on the plane is also on the sphere this equation (of the sphere) is satisfied. The equation to the cone is therefore satisfied by any point on the curve of intersection of the sphere and plane, *i.e.*, it is the equation required.

Example 6. Find the equation to the sphere determined by the four points $(0, 1, 3)$, $(1, 2, 4)$, $(2, 3, 1)$, $(3, 0, 2)$. Find also the equation to the shadow of this sphere cast upon the plane $z = 0$ by a point source of light at $(0, 0, 6)$.

[L.U.]

Assume the equation of the sphere to be

$$x^2 + y^2 + z^2 + ax + by + cz + d = 0,$$

and substitute the co-ordinates of the points in turn. Solve the equations in a, b, c and d to find the sphere as

$$x^2 + y^2 + z^2 - 3\tfrac{1}{2}x - 3y - 4\tfrac{1}{2}z + 6\tfrac{1}{2} = 0.$$

Any line through $(0, 0, 6)$ is

$$\frac{x}{l} = \frac{y}{m} = \frac{z - 6}{n} = r \quad . \quad . \quad . \quad . \quad . \quad . \quad (1)$$

This meets the sphere in points given by

$$r^2(l^2 + m^2) + (nr + 6)^2 - r(3\tfrac{1}{2}l + 3m) - 4\tfrac{1}{2}(nr + 6) + 6\tfrac{1}{2} = 0$$

or

$$r^2(l^2 + m^2 + n^2) + r(7\tfrac{1}{2}n - 3\tfrac{1}{2}l - 3m) + 15\tfrac{1}{2} = 0.$$

This line will be a tangent if

$$(7\tfrac{1}{2}n - 3\tfrac{1}{2}l - 3m)^2 = 4 \times 15\tfrac{1}{2} \times (l^2 + m^2 + n^2),$$

a homogeneous equation in l, m, n showing that the line (1) is a generator of the cone vertex at $(0, 0, 6)$. Its Cartesian equation is

$$[3\tfrac{1}{2}x + 3y - 7\tfrac{1}{2}(z - 6)]^2 = 62[x^2 + y^2 + (z - 6)^2].$$

This meets $z = 0$ in the curve

$$[7x + 6y + 90]^2 = 248[x^2 + y^2 + 36]$$

which must clearly reduce to the equation of an ellipse (since the shadow is elliptical). It is

$$199x^2 - 84xy + 212y^2 - 1260x - 1080y + 828 = 0.$$

15.4. Other Conicoids

Certain second-degree surfaces such as the sphere, cone and cylinder have been considered. We will now consider the remaining second-degree surfaces. All such surfaces of the second degree are called *conicoids*, since plane sections are *conics*.

(a) The Equation $\dfrac{x^2}{a^2} + \dfrac{y^2}{b^2} + \dfrac{z^2}{c^2} = 1$

(1) If the point (x, y, z) is on the surface, so is the point $(-x, -y, -z)$. The origin is therefore the mid-point of the join, and is the *centre* of the surface.

(2) The surface cuts the axes which are its axes of symmetry at $(\pm a, 0, 0)$ $(0, \pm b, 0)$ $(0, 0, \pm c)$. The section by the x, y plane is the ellipse

$$\frac{x^2}{a^2} + \frac{y^2}{b^2} = 1,$$

and similarly for the sections by the other co-ordinate planes.

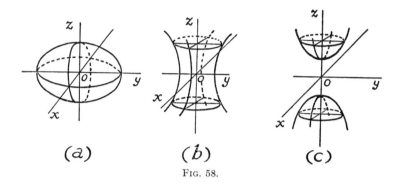

(a)   (b)   (c)

FIG. 58.

The plane $z = k$ cuts the surface in the ellipse

$$\frac{x^2}{a^2} + \frac{y^2}{b^2} = 1 - \frac{k^2}{c^2}.$$

The section is therefore a point section for $k = c$ and imaginary for $k > c$. Similar considerations show that the surface is a bounded surface lying entirely within the box formed by the planes

$$x = \pm a, \qquad y = \pm b, \qquad z = \pm c.$$

Such a surface is called an *ellipsoid*.

(b) The Equation $\dfrac{x^2}{a^2} + \dfrac{y^2}{b^2} - \dfrac{z^2}{c^2} = 1$

The above method shows that this, too, is a central conicoid meeting the x axis at $(\pm a, 0, 0)$, the y axis at $(0, \pm b, 0)$, but not meeting the z axis.

A section by the plane $y = k$ is the hyperbola

$$\frac{x^2}{a^2} - \frac{z^2}{c^2} = 1 - \frac{k^2}{b^2},$$

which reduces to two lines in the plane for $k = \pm b$.

The section by the plane $z = k$ is the ellipse

$$\frac{x^2}{a^2} + \frac{y^2}{b^2} = 1 + \frac{k^2}{c^2},$$

an ellipse which has axes increasing as k increases. When $k = 0$ we get the real ellipse

$$\frac{x^2}{a^2} + \frac{y^2}{b^2} = 1.$$

Such a surface is called a *hyperboloid of one sheet*.

(c) The Equation $-\dfrac{x^2}{a^2} - \dfrac{y^2}{b^2} + \dfrac{z^2}{c^2} = 1$

This is a central conicoid meeting the z axis in real points $(0, 0, \pm c)$ but not meeting the other axes. The section by $z = 0$ is imaginary, since we obtain

$$\frac{x^2}{a^2} + \frac{y^2}{b^2} = -1,$$

and this ellipse does not give real values for x and y until the section is by the plane $z = k$, where $k \geqslant c$. We then get the real ellipse

$$\frac{x^2}{a^2} + \frac{y^2}{b^2} = \frac{k^2}{c^2} - 1,$$

a curve whose axes increase with k.

Sections of (c) by $x = k$, for example, give the hyperbola

$$\frac{z^2}{c^2} - \frac{y^2}{b^2} = 1 + \frac{k^2}{a^2}.$$

This surface is called a *hyperboloid of two sheets*, since unlike the above surface, it consists of two separated surfaces.

All the above surfaces are symmetrical with regard to any co-ordinate plane; thus if (x, y, z) is on one of the surfaces, so is $(x, -y, z)$ (since the only power of y is even), showing that the surface is symmetrical with respect to the x, z plane (or $y = 0$). Similarly, there is symmetry in opposite quadrants.

15.5. Surfaces of Revolution

Suppose $P(0, y, z)$ is any point on a curve $f(y, z) = 0$ in the yz plane. If this curve revolves round the z axis the distance of the point P from

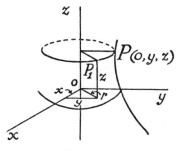

FIG. 59.

the z axis and its height above the xy plane remain constant. But during the rotation the first distance is no longer y but $\sqrt{(x^2 + y^2)}$. Since this magnitude satisfies the equation when used as y, we now have that throughout the rotation the co-ordinates of P_1 now (x, y, z) satisfy the equation

$$f(\sqrt{x^2 + y^2}, z) = 0.$$

Any surface whose equation can be put in this form is therefore a surface of revolution round the z axis. For example, the sphere

$$x^2 + y^2 + z^2 = a^2$$

can be written $\qquad [\sqrt{(x^2 + y^2)}]^2 + z^2 = a^2,$

and so may be considered, as is evident, generated by the rotation of the circle in the y, z plane

$$y^2 + z^2 = a^2$$

about the z axis.

Similarly, for any equation which can be written

$$F_1(\sqrt{y^2 + z^2}, x) = 0,$$

we have a surface of revolution round the x axis, and

$$F_2(\sqrt{z^2 + x^2}, y) = 0$$

a surface of revolution about the y axis.

The above example of a sphere can be written in either of these forms, and so, as is obvious, can be considered as generated by the revolution of the circle about the x or y axis.

EXERCISE 36

1. Find the axis and semi-vertical angle of the circular cones given by:
    (a) $x^2 + y^2 - z^2 = 0$; (b) $x^2 - 3y^2 + z^2 = 0$.

2. Find the equation to the right circular cone of semi-vertical angle α whose axis is the z axis.

3. Find the equation of the cone of revolution with vertex at the origin, axis the y axis and semi-vertical angle $30°$.

4. What surface is represented by:
    (a) $4x^2 + y^2 - 8x + 4y - 4 = 0$;
    (b) $xy + 2x - y - 6 = 0.$

5. It is given that

$$5x^2 + 8y^2 + 5z^2 + 4yz + 8zx - 4xy = 144$$

is a right circular cylinder, axis the line

$$x = 2y = -z.$$

Find the area of a section by the xOy plane, and hence the area of the section perpendicular to the axis.

6. Show that the cylinder with generators parallel to the z axis and passing through the intersection of the ellipsoid $3x^2 + 2y^2 + z^2 = 18$ and the plane $y - z = 3$ is right circular. Find the area of the section of the cylinder in the (x, y) plane and hence the area and also the centre of the section of the ellipsoid by the given plane. [L.U.]

7. Find the equation to the surface generated when the ellipse in the yz plane $y^2/a^2 + z^2/b^2 = 1$ rotates round the: (a) z axis; (b) y axis.

8. If water be projected with velocity V m/s from a point on the level ground, all possible paths [in two dimensions (x, z)] are included in the envelope

$$x^2 = -\frac{2V^2}{g}\left(z - \frac{V^2}{2g}\right).$$

Find the paraboloid of revolution obtained when this curve revolves round the vertical z axis. Hence find the area that can be wetted on a vertical wall whose equation may be taken as $x = a$.

15.6. The Tangent Plane to a Surface

Let the equation to a surface be

$$F(x, y, z) = 0,$$

so that for any small change in the values of the point (x, y, z) on the surface involving moving to a near point also on the surface, the theorem of Total Differential gives

$$\frac{\partial F}{\partial x}dx + \frac{\partial F}{\partial y}dy + \frac{\partial F}{\partial z}dz = 0. \quad . \quad . \quad . \quad (1)$$

If, however, the point (x, y, z) is joined to a near point $(x + dx, y + dy, z + dz)$ also on the surface the direction ratios of the join, which is ultimately a tangent line, are, being given by the difference of the co-ordinates,

$$dx, dy, dz \quad . \quad . \quad . \quad . \quad . \quad . \quad (2)$$

Since (1) and (2) are true for all such pairs of points on the surface we conclude from (1) that a line with direction ratios

$$\frac{\partial F}{\partial x}, \frac{\partial F}{\partial y}, \frac{\partial F}{\partial z}$$

is always perpendicular at the point (x, y, z) to any tangent line at the point. These direction ratios must then be those of the normal at the point.

The tangent plane at $(x_1 y_1 z_1)$ is therefore

$$(x - x_1)\frac{\partial F}{\partial x_1} + (y - y_1)\frac{\partial F}{\partial y_1} + (z - z_1)\frac{\partial F}{\partial z_1} = 0$$

where, for example, $\dfrac{\partial F}{\partial x_1}$ means that we have found $\dfrac{\partial F}{\partial x}$ and substituted the values $(x_1 y_1 z_1)$ in it.

The normal at the point is similarly

$$\frac{x - x_1}{\dfrac{\partial F}{\partial x_1}} = \frac{y - y_1}{\dfrac{\partial F}{\partial y_1}} = \frac{z - z_1}{\dfrac{\partial F}{\partial z_1}}.$$

Example 7. Find the equations of the tangent planes that can be drawn through the line

$$3x + 4y + 9z = 20, \qquad x = 3z$$

to touch the ellipsoid

$$x^2 + 4y^2 + 9z^2 = 10.$$

Find also the equations of the chord of contact of these two planes. [L.U.]

If
$$F(xyz) = x^2 + 4y^2 + 9z^2 - 10$$
$$\frac{\partial F}{\partial x} = 2x, \qquad \frac{\partial F}{\partial y} = 8y, \qquad \frac{\partial F}{\partial z} = 18z$$

therefore at a point $(x_1 y_1 z_1)$ on the surface the equation of the tangent plane is
$$(x - x_1)2x_1 + (y - y_1)8y_1 + (z - z_1)18z_1 = 0,$$
which, since
$$x_1{}^2 + 4y_1{}^2 + 9z_1{}^2 = 10,$$
reduces to
$$xx_1 + 4yy_1 + 9zz_1 = 10 \qquad \cdot \quad \cdot \quad \cdot \quad \cdot \quad \cdot \quad \cdot \quad (1)$$
Any plane through the line is
$$3x + 4y + 9z - 20 + k(x - 3z) = 0$$
or
$$(3 + k)x + 4y + (9 - 3k)z = 20 \qquad \cdot \quad \cdot \quad \cdot \quad \cdot \quad \cdot \quad (2)$$

If this is the tangent plane at $(x_1 y_1 z_1)$, since (1) and (2) are the same equation
$$\frac{x_1}{3 + k} = \frac{y_1}{1} = \frac{9z_1}{9 - 3k} = \frac{1}{2}$$
$$\therefore \quad x_1 = \tfrac{1}{2}(3 + k), \quad y_1 = \tfrac{1}{4}, \quad z_1 = \tfrac{1}{6}(3 - k) \quad \cdot \quad \cdot \quad \cdot \quad (3)$$

This point lies on the surface, therefore
$$\tfrac{1}{4}(3 + k)^2 + 4(\tfrac{1}{4})^2 + 9 \cdot \tfrac{1}{36}(3 - k)^2 = 10,$$
a quadratic equation whose roots are $k = \pm 3$. The two tangent planes are therefore
$$3x + 4y + 9z - 20 \pm 3(x - 3z) = 0$$
or
$$3x + 2y = 10 \quad \text{and} \quad 2y + 9z = 10.$$

From (3) their points of contact are respectively
$$(3, \tfrac{1}{2}, 0) \quad \text{and} \quad (0, \tfrac{1}{2}, 1).$$

The join of these points is
$$\frac{x - 3}{3} = \frac{y - \tfrac{1}{2}}{0} = \frac{z}{-1}.$$

EXERCISE 37

1. Find the equation of the tangent plane at the point $(1, 2, -2)$ to the sphere $x^2 + y^2 + z^2 + 2x - 6y + 1 = 0$.

2. Find the tangent plane and normal at the point $(1, -2, 2)$ to the sphere $x^2 + y^2 + z^2 = 9$.

3. Show that the tangent plane at the point (x_0, y_0, z_0) to the surface
$$ax^2 + by^2 + cz^2 = 1 \quad \text{is} \quad axx_0 + byy_0 + czz_0 = 1.$$

4. Show that $5x + 8y + 2z + 16 = 0$ is a tangent plane to the surface $x^2 + 2xz - 2y^2 + 16 = 0$, and find the point at which it touches the surface.

5. Show that the tangent plane at (x_0, y_0, z_0) to the ellipsoid
$$x^2/a^2 + y^2/b^2 + z^2/c^2 = 1 \quad \text{is} \quad \frac{xx_0}{a^2} + \frac{yy_0}{b^2} + \frac{zz_0}{c^2} = 1.$$

Find the volume included between this plane and the three co-ordinate planes.

6. Find the equations of the tangent planes to $3x^2 - 6y^2 + 2z^2 = 5$ which pass through the line of intersection of the planes $3x = 9y + z$, $6x - 3y + 3z = 5$.

[L.U.]

7. Obtain the condition that the plane $lx + my + nz = 1$ shall touch the ellipsoid
$$\frac{x^2}{a^2} + \frac{y^2}{b^2} + \frac{z^2}{c^2} = 1.$$

Prove that the plane $x + y + nz = 1$ will intersect the ellipsoid
$$3x^2 + 2y^2 + z^2 = 1$$
in real points only if $6n^2 \geqslant 1$.

[L.U.]

8. If the tangent plane at (X, Y, Z) on the ellipsoid $x^2/a^2 + y^2/b^2 + z^2/c^2 = 1$ meets the axes at L, M, N respectively, show that the area of the triangle LMN is $a^2b^2c^2/2pXYZ$, where p is the length of the perpendicular from the origin to the tangent plane. [L.U.]

9. Prove that the equation of the tangent plane at the point (α, β, γ) on the sphere

$$x^2 + y^2 + z^2 + 2gx + 2fy + 2hz + c = 0$$

is $(x - \alpha)(\alpha + g) + (y - \beta)(\beta + f) + (z - \gamma)(\gamma + h) = 0.$

Find the equations of the two planes through the line $x = -y + 3 = z$, which touch the sphere

$$x^2 + y^2 + z^2 - 4x + 2y = 1$$

and find the co-ordinates of their points of contact. [L.U.]

10. Find the equation of the sphere which has its centre at the point $(2, 3, -1)$ and touches the line

$$\frac{x - 13}{10} = \frac{y - 8}{3} = \frac{z + 7}{-8}.$$

Find also the equation of the tangent plane to the sphere which contains the above tangent line. [L.U.]

11. A straight line L passes through the origin and is a tangent to each of the spheres

$$x^2 + y^2 + z^2 + 2ax + p = 0, \qquad x^2 + y^2 + z^2 + 2by + q = 0.$$

Show that the angle ϕ between L and the z axis is given by

$$\sin^2 \phi = (p/a^2) + (q/b^2).$$

Also find the distance between the points of contact of L with the spheres. [L.U.]

12. Find the equations of the tangent plane and the normal at the point $(x_1 y_1 z_1)$ on the ellipsoid $ax^2 + by^2 + cz^2 = 1$.

At the points where the plane $x + y + z = 0$ meets the ellipsoid

$$x^2 + 2y^2 + 3z^2 = 6,$$

normals to the ellipsoid are drawn. Show that these normals meet the plane $z = 0$ on the ellipse $3x^2 + 9xy + 15y^2 = 2$. [L.U.]

13. Prove that the equation of the tangent plane to the surface $F(xyz) = 0$ at the point (α, β, γ) is

$$(x - \alpha) \frac{\partial F}{\partial \alpha} + (y - \beta) \frac{\partial F}{\partial \beta} + (z - \gamma) \frac{\partial F}{\partial \gamma} = 0.$$

Find the equation of the tangent plane at any point (α, β, γ) on the surface $x^2y - a^2z = 0$ and prove that the tangent planes at all points of the surface which lie in the plane $x = \alpha$ meet the plane $x = 0$ in parallel lines. [L.U.]

14. Obtain the equations to the tangent plane and the normal at the point $(a\alpha, a\beta, a\alpha\beta)$ on the surface $az = xy$.

Show that the tangent plane meets the surface in two straight lines and give their equations. [L.U.]

15. Show that the line joining the points $(a, h, 0)$ $(0, k, c)$ touches the ellipsoid $x^2/a^2 + y^2/b^2 + z^2/c^2 = 1$ if $2hk = b^2$. Show also that if in addition $h = b$, the point of contact is $(\tfrac{1}{3}a, \tfrac{2}{3}b, \tfrac{2}{3}c)$. [L.U.]

16. P is a point on the surface $xy = az$. Q is the reflection of P in the plane $z = 0$. Show that the tangent plane to the surface at P passes through the projection of Q on the axes of co-ordinates. [L.U.]

17. Find the condition that $lx + my + nz + p = 0$ should be a tangent plane to the sphere

$$x^2 + y^2 + z^2 + 2ux + 2vy + 2wz + d = 0.$$

Find the equations of the tangent planes to the sphere

$$x^2 + y^2 + z^2 - 2x - 4y + 2z - 219 = 0$$

which intersect in the line

$$3(x - 10) = -4(y - 14) = -6(z - 2).$$ [L.U.]

18. Prove that the normal at the point $(1, 2, 3)$ on the ellipsoid
$$3x^2 + 2y^2 + z^2 = 20$$
touches the sphere given by $17(x^2 + y^2 + z^2) = 38$. [L.U.]

19. Find the two tangent planes to the hyperboloid
$$x^2 + y^2 - 4z^2 = 4$$
which pass through the line $y - 2 = 0$, $x - 2z - 4 = 0$.

20. Find the equations of the tangent planes to the ellipsoid
$$x^2/4 + y^2/16 + z^2/9 = 1$$
at the points in which it is met by the line
$$x/2 = y/2 = z/3.$$
Find the equation of the projection on the xy plane of the section of the ellipsoid made by the plane through the centre parallel to the above tangent planes.
[L.U.]

21. Normals are drawn to the ellipsoid $x^2/3 + y^2/2 + z^2/1 = 1$ at the intersections with the ellipsoid $x^2/1 + y^2/2 + z^2/3 = 1$. Show that the locus of the points in which they meet the plane $z = 0$ is the ellipse $3x^2 + 2y^2 = 1$, $z = 0$.
[L.U.]

22. Show that the line $y = 2z - 15$, $x = 2z - 15$ is a normal to the surface $x^{2/3} + y^{2/3} + z^{2/3} = 6$ and find the point at which it is the normal.

23. Find the equations of the tangent plane and of the normal to the surface $z^3 = axy$ at the point $(at, a/t, a)$.
Show that the line of intersection of the tangent plane with the plane $z = 0$ is a tangent to the hyperbola $4xy = a^2$, $z = 0$. Show also that the normal meets the plane $z = 0$ in a point which lies on the hyperbola
$$9(x - 3y)(3x - y) + 64a^2 = 0 = z.$$
[L.U.]

24. Prove that the point $P(\theta^3 + 3\lambda\theta^2, \theta^2 + 2\lambda\theta, \theta + \lambda)$ lies on the surface
$$(yz - x)^2 = 4(y - z^2)(xz - y^2)$$
and show that the equation of the tangent plane at P is $x - 3y\theta + 3z\theta^2 - \theta^3 = 0$. Deduce that if θ remains constant but λ varies the locus of P is a line at every point of which the surface has the same tangent plane. [L.U.]

25. Find the equation of the tangent plane at the point $\{- 2a/(1 + \lambda),$ $- 2a/(1 - \lambda), a/2\}$ of the surface
$$\frac{1}{x} + \frac{1}{y} + \frac{1}{z} = \frac{1}{a}.$$
Show that the intersection of the tangent plane with the plane $z = 0$ touches the hyperbola given by
$$xy = a(x + y), \quad z = 0.$$
[L.U.]

CHAPTER 16

DOUBLE AND TRIPLE INTEGRALS

16.1. Double Integration

(a) The area shown in the diagram enclosed between $x = a$, b by $y = F(x)$, $y = f(x)$ would be found in the usual way by evaluating

$$\int_a^b (y_2 - y_1)dx$$

where $y_1 = f(x)$, $y_2 = F(x)$. There is another way which leads to important developments. Consider the element of area $dxdy$ whose

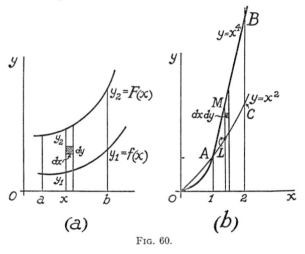

FIG. 60.

bottom left-hand corner is at the point (x, y). We may imagine the whole area divided into such elements, and if we can sum them, another method of obtaining the area is provided. A method of summing is as follows. Regarding the x and width dx as constant, sum for the vertical strip from y_1 to y_2. This can be written

$$dx\int_{y_1}^{y_2} dy$$

Having found the area of this strip, we now add up all such strips between $x = a$ and $x = b$ by the expression

$$\int_a^b dx\int_{y_1}^{y_2} dy$$

257

Evaluation of the right-hand integral gives

$$\int_a^b dx(y_2 - y_1)$$

and we can continue in the usual way. With this understanding that we will adopt such a method of evaluation, any area may be written

$$\iint dxdy$$

where, when required, we will later write it, for example, as above in the form

$$\int_a^b dx \int_{y_1}^{y_2} dy$$

and evaluate.

In the same way the co-ordinates of the centre of gravity of the area could be expressed as

$$\bar{x} = \frac{\iint xdxdy}{\iint dxdy}, \qquad \bar{y} = \frac{\iint ydxdy}{\iint dxdy}$$

for $xdxdy$ is the moment about the y axis of the element of area $dxdy$ supposed concentrated at the point (x, y).

Similarly, the second moment of the area about an axis through the origin perpendicular to the xy plane could be written

$$\iint (x^2 + y^2)dxdy,$$

since $x^2 + y^2$ is the square of the distance from the origin of the element of area.

Example 1. Find, giving a sketch of the field of integration,

$$\int_1^2 \int_{x^2}^{x^4} x^2ydydx. \qquad \text{[Fig. 60(b)]}$$

Since the limits of the right-hand integral involve x, we are required first to integrate the function with respect to y from $y = x^2$ the lower limit to $y = x^4$ the upper limit. We therefore draw the curves $y = x^2$ and $y = x^4$. Since x is maintained constant during this integration draw a line LM parallel to the y axis and distant x from it. If this meets $y = x^2$ in L and $y = x^4$ in M, then we have integrated from L to M at this stage of the process. The result of the above integration must now be integrated from $x = 1$ the lower limit of the left-hand integral to $x = 2$ the upper limit. We therefore draw the lines $x = 1$ and $x = 2$. The field of integration is therefore ABC and LM is a line within the range $x = 1$ to 2.

The value of the integral is

$$I = \int_1^2 x^2dx \int_{x^2}^{x^4} ydy = \int_1^2 x^2\left(\frac{1}{2}y^2\right)_{x^2}^{x^4} dx$$

$$= \frac{1}{2}\int_1^2 x^2(x^8 - x^4)dx = \underline{85 \cdot 9}.$$

Note that when an integral is given as

$$\iint f(x, y)dxdy$$

with limits, it is these that show the variable x or y for the first integration. Thus if the right-hand integral sign has limits involving x, then the first integration is with respect to y, and it would help to write the integral as

$$\int dx \int f(x, y)dy.$$

We integrate with respect to y and for y in the integrated function substitute the given limits involving x. A function of x only remains, to be integrated with respect to x. Similarly, if the right-hand limits of integration involve y only then the first integration is with respect to x. When all the limits of integration are constant the data must show which limits are for x or y but generally

$$\iint f(xy)dxdy$$

means $\int \left(\int f(x, y)dx \right) dy \quad$ or $\quad \int dy \int f(x, y)dx.$

EXERCISE 38

1. Show with a sketch that the same area is given by

$$\int_a^b ydx \quad \text{and} \quad \int_a^b dx \int_0^{f(x)} dy \quad \text{or} \quad \int_a^b \int_0^{f(x)} dydx$$

where $y = f(x)$ is a curve enclosing, with the x axis, an area between $x = a$ and b.

2. Find

(a) $\displaystyle\int_1^4 x\left(\int_1^2 dy \right) dx$ (b) $\displaystyle\int_0^2 ydy \int_0^1 xdx$ (c) $\displaystyle\int_0^a \left(\int_0^b (y^3 + x^2y)dx \right) dy$

3. Show that in each of the above examples the same answer is obtained when the order of integration is interchanged, *i.e.*,

$$\int_1^4 \left(\int_1^2 xdy \right) dx = \int_1^2 \left(\int_1^4 xdx \right) dy.$$

(This is always true when all the limits of integration are constant.)

4. Show that the area enclosed by the x axis and $y = x(2 - x)$ is given by the following integral and evaluate it.

$$\int_0^2 \int_0^{x(2 - x)} dydx$$

5. Find

(i) $\displaystyle\int_0^a dx \int_0^{\sqrt{a^2 - x^2}} dy;$ (ii) $\displaystyle\int_0^a dy \int_0^{\sqrt{a^2 - y^2}} x^3dx;$

(iii) $\displaystyle\int_0^1 dx \int_0^{x^2} x(x^2 + y^2)dy;$ (iv) $\displaystyle\int_0^1 dx \int_0^{x^2} \frac{dy}{\sqrt{x^2 - y^2}}.$

6. Give sketches showing the region over which each of the following double integrals is taken. Evaluate each and note that with each pair the region and the value is the same, the difference being that the first integral is evaluated with respect to y and then x, whilst in the second double integral the order has been reversed.

(i) $\displaystyle\int_0^a dx \int_x^a (x^2 + y^2)dy,$ $\displaystyle\int_0^a dy \int_0^y (x^2 + y^2)dx;$

(ii) $\displaystyle\int_0^a \int_0^{\sqrt{a^2-x^2}} xydydx,$ $\displaystyle\int_0^a \int_0^{\sqrt{a^2-y^2}} xydxdy;$

(iii) $\displaystyle\int_0^2 \int_{x^2}^{2x} (2x + 3y)dydx,$ $\displaystyle\int_0^4 \int_{\frac{1}{2}y}^{\sqrt{y}} (2x + 3y)dxdy;$

(iv) $\displaystyle\int_0^4 \int_0^{x(4-x)} dydx,$ $\displaystyle\int_0^4 \int_{2-\sqrt{4-y}}^{2+\sqrt{4-y}} dxdy;$

7. Show that the co-ordinates of the centroid of an area are given by

$$\bar{x} = \iint xdxdy \Big/ \iint dxdy, \quad \bar{y} = \iint ydxdy \Big/ \iint dxdy.$$

Where it is understood that the limits of integration will be so chosen that we integrate over the whole of the given area.

Find the centroid of the area enclosed by the curves $y^2 = 4ax$, $x^2 = 4ay$.

8. Show that the second moment of an area about an axis through the origin perpendicular to it is given by

$$\iint (x^2 + y^2)dxdy.$$

Find the second moment of the area of Question 7 about this axis and also about the x axis.

9. Find the volume formed when the above area revolves round the x axis.

10. The density of a semicircular lamina is proportional to the distance from the bounding diameter. If the radius is r, show that the C.G. is distant $\frac{3}{16}\pi r$ from the diameter.

11. Show that the area bounded by the isothermal curves $pv = a$, $pv = b$ and the adiabatic curves $pv^\gamma = c$, $pv^\gamma = d$ is

$$\frac{b-a}{\gamma-1}\log\left(\frac{d}{c}\right).$$

12. The electrical attraction at a point situated at a distance r from an infinite plane due to a suface density of electricity γ is given by

$$r\gamma \int_0^\infty \int_0^{2\pi} \frac{xd\theta dx}{(r^2 + x^2)^{3/2}}.$$

Find the value of this attraction.

13. A calculation of the coefficient of inductance of a circular solenoid requires the evaluation of

$$\frac{\mu n i}{\pi} \int_0^a \int_0^\pi \frac{rd\theta dr}{b - r\cos\theta}.$$

Show that this value is $\mu n i[b - (b^2 - a^2)^{\frac{1}{2}}]$.

14. By double integration find the area in the first quadrant enclosed by the curves $4x = 3y$, $x^2 + y^2 = 25$, and $3x^2 = 16y$.

15. Evaluate

$$\frac{P}{4\pi^2 c^2 N} \int_0^{\pi/2} \int_0^{2c\cos\theta} \left[r^2 \log\frac{r}{a} + \frac{1}{2}(a^2 - r^2) \right] r\,dr\,d\theta$$

to find the deflection at the centre of a circular plate due to a load P situated at a distance c from the centre and uniformly spread over a small circle of radius a.

16.2. Volumes

The box of height z standing on the area $dxdy$ as shown in the figure has a volume $z\,dxdy$. If we integrate this element of volume over such

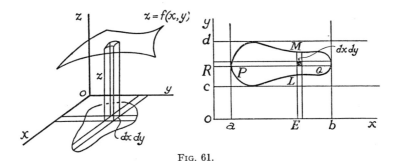

FIG. 61.

a range as to cover the curve shown in the x, y plane, we will have found the volume cut off on the cylinder standing on this curve by the surface $z = f(x, y)$. We may therefore write the volume in the form

$$V = \iint z\,dxdy$$

where z will be known in terms of x and y.

We may evaluate this integral in two ways. We may first integrate with respect to y from $y = EL$ to $y = EM$, finding the volume standing on the ground area of length LM and width dx. We then integrate this result from $x = a$ to b to cover the entire area and sum all such volumes. In this way

$$V = \int_a^b \int_{EL}^{EM} z\,dy\,dx.$$

We may, however, find it more convenient to integrate first with regard to x, finding the volume standing on the area of length PQ and width dy. We then integrate this result from $y = c$ to d to cover the entire range of y. In this case

$$V = \int_c^d \int_{RP}^{RQ} z\,dxdy.$$

Example 3. Find the volume of the paraboloid of revolution $x^2 + y^2 = 2z$ cut off by the plane $z = 2$.

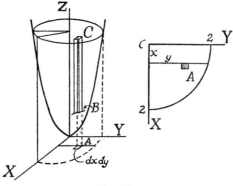

FIG. 62.

The length intercepted between the plane $z = 2$ and the surface is AC — AB, which gives

$$BC = 2 - \frac{x^2 + y^2}{2}$$

The element of volume on the base $dydx$ is then

$$\left(2 - \frac{x^2 + y^2}{2}\right)dydx$$

It is evident that we will add all such elements if we integrate over the first quadrant of the circle $x^2 + y^2 = 4$ and multiply this result by 4, since we require the volume in the four positive octants.

Integration first with respect to y requires, for a given x, y to move from 0 to $\sqrt{(4 - x^2)}$. We then integrate for x from $x = 0$ to 2.

$$\therefore \quad V = 4\int_0^2 \int_0^{\sqrt{(4 - x^2)}} \left(2 - \frac{x^2 + y^2}{2}\right)dydx$$

$$= 4\int_0^2 dx \cdot \left[2y - \tfrac{1}{2}x^2y - \tfrac{1}{6}y^3\right]_0^{\sqrt{(4 - x^2)}}$$

$$= 4\int_0^2 [2\sqrt{(4 - x^2)} - \tfrac{1}{2}x^2\sqrt{(4 - x^2)} - \tfrac{1}{6}(4 - x^2)^{3/2}]dx.$$

With $x = 2\sin\theta$,

$$V = 4\int_0^{\frac{1}{2}\pi} (4\cos\theta - 4\sin^2\theta\cos\theta - \tfrac{4}{3}\cos^3\theta)(2\cos\theta d\theta)$$

$$= 32\int_0^{\frac{\pi}{2}} (\cos^2\theta - \sin^2\theta\cos^2\theta - \tfrac{1}{3}\cos^4\theta)d\theta$$

$$= 32\left[\frac{1}{2}\cdot\frac{\pi}{2} - \frac{3 \cdot 1}{4 \cdot 2}\cdot\frac{\pi}{2} - \frac{1}{3}\frac{3 \cdot 1}{4 \cdot 2}\cdot\frac{\pi}{2}\right] = \underline{4\pi \text{ cubic units.}}$$

16.3. Polar Co-ordinates

In some cases it is convenient to use polar co-ordinates. We must therefore find the expression for the element of area to be used instead of $dydx$.

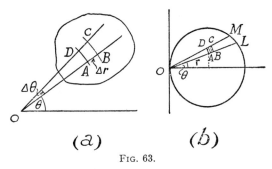

Fig. 63.

If OB, OC are two radii angle $\Delta\theta$ apart with AD an arc of a circle radius r and BC an arc of radius $r + \Delta r$ and A the point (r, θ), then ABCD the area enclosed (Fig. 63(a)) is

$$\tfrac{1}{2}(r + \Delta r)^2\Delta\theta - \tfrac{1}{2}r^2\Delta\theta$$
$$= r\Delta r\Delta\theta + \tfrac{1}{2}(\Delta r)^2\Delta\theta.$$

Retaining the term of lowest order only, and therefore ignoring the third-order term

$$\text{area ABCD} = r\Delta r\Delta\theta$$

In the limit the element of area is then

$$rdrd\theta$$

and we may regard ABCD as a rectangle of sides $AD = rd\theta$ and $AB = dr$.

Example 4. Find the position of the C.G. of a circular disc if its density at any point is $\mu \times$ (square of distance from 0), a point on the circumference (Fig. 63(b)).

If A is the point (r, θ) and ABCD the element of area we may consider it all concentrated at A so that the mass of $rdrd\theta$ is $rdrd\theta \times \mu r^2$ at the point (r, θ). From this

$$\Sigma m = \iint \mu r^2(rdrd\theta)$$

With the given point O as origin and the diameter through O as axis the equation to the circle is $R = d\cos\theta$, where d is the diameter. Integrating first from $r = 0$ to $r = R$ ($= d\cos\theta$) gives the mass of the sector OLM. We will then integrate with respect to θ from 0 to $\tfrac{1}{2}\pi$ and double to include the lower half of the circle.

$$\Sigma m = 2\int_0^{\frac{1}{2}\pi}\int_0^{d\cos\theta} \mu r^3drd\theta = \tfrac{3}{32}\pi\mu d^4.$$

Since the point A is distant $r\cos\theta$ from the y axis the moment of the element ABCD about the y axis is

$$\mu r^2 (rdrd\theta) \times r\cos\theta$$

$$\therefore \quad \Sigma mx = 2\int_0^{\frac{1}{2}\pi}\int_0^{d\cos\theta} \mu r^4 \cos\theta drd\theta = \tfrac{1}{16}\mu\pi d^5.$$

$$\bar{x} = \frac{\Sigma mx}{\Sigma m} = \frac{\tfrac{1}{16}\mu\pi d^5}{\tfrac{3}{32}\mu\pi d^4} = \tfrac{2}{3}d$$

The C.G. is clearly on the axis and $\bar{y} = 0$.

16.4. Cylindrical Co-ordinates

In some problems it is very convenient to use the mixture of Cartesian and polar co-ordinates called cylindrical co-ordinates.

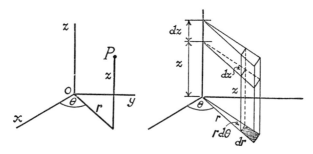

Fig. 64.

In this system the position of the point on the ground $(x, y$ plane) is fixed by polar co-ordinates and the element of area is $rdrd\theta$. The height of the point above the plane is given by z, so that the element of volume standing on the base $rdrd\theta$ is

$$zrdrd\theta$$

and the volume V is given by

$$\iint zrdrd\theta$$

taken over the base area. We therefore need the equations to the surfaces concerned in terms of r, θ and z only.

Example 5. Find the volume common to the sphere $x^2 + y^2 + z^2 = 16$ and the cylinder $x^2 + y^2 = 9$.

In cylindrical co-ordinates the surfaces are

$$r^2 + z^2 = 16, \qquad r^2 = 9$$

since $\qquad x = r\cos\theta, \qquad y = r\sin\theta.$

We will integrate over one-quarter of the circular area $r = 3$ and multiply the result by 8.

At any point (r, θ) inside the circle $r = 3$, the height is given by the distance up to the sphere and is $z = \sqrt{(16 - r^2)}$

$$\therefore \quad V = 8 \iint \sqrt{(16 - r^2)} r \, dr \, d\theta,$$

where r varies from 0 to 3 and then θ varies from 0 to $\frac{1}{2}\pi$. Therefore

$$V = 8 \int_0^{\frac{1}{2}\pi} \int_0^3 r\sqrt{(16 - r^2)} \, dr \, d\theta$$

$$= 8 \int_0^{\frac{1}{2}\pi} \left[-\tfrac{1}{3}(16 - r^2)^{3/2} \right]_0^3 d\theta$$

$$= \tfrac{1}{3}\pi(256 - 28\sqrt{7})$$

It is sometimes helpful to transform an integral from Cartesian into polar co-ordinates.

Example 6. Evaluate by transforming into polar co-ordinates

$$\int_0^\infty dx \int_0^\infty \frac{dy}{(x^2 + y^2 + a^2)^2}.$$

x and y vary from 0 to ∞, so that every point in the first quadrant is covered. The values for r and θ must be chosen so that the element of area $r \, d\theta \, dr$ does the same. This will occur if r varies from 0 to ∞ and θ from 0 to $\frac{1}{2}\pi$.

Since $x^2 + y^2 + a^2 = r^2 \cos^2 \theta + r^2 \sin^2 \theta + a^2 = r^2 + a^2$, the integral becomes

$$\int_0^{\frac{1}{2}\pi} \int_0^\infty \frac{r \, d\theta \, dr}{(r^2 + a^2)^2}.$$

Either order of integration will give the result easily. We will evaluate the integral as

$$\int_0^{\frac{1}{2}\pi} d\theta \int_0^\infty \frac{r \, dr}{(r^2 + a^2)^2} = \int_0^{\frac{1}{2}\pi} d\theta \left[-\frac{1}{2} \cdot \frac{1}{r^2 + a^2} \right]_0^\infty$$

$$= \frac{1}{2a^2} \int_0^{\frac{1}{2}\pi} d\theta = \frac{\pi}{4a^2}$$

EXERCISE 39

1. Show that the volume enclosed by the surface $xy = z$ and standing on the area enclosed by $x = 0$, $x = b$, $y = a$, $y = c$ is $\frac{1}{4}b^2(c^2 - a^2)$.

2. Find the volume bounded by the cylinder $x^2 + y^2 = 9$ and the planes $z = 3y$ and $z = y$.

3. A right cylinder has as base a semicircular area radius a and diameter AB. If a plane is drawn through AB making $45°$ with the base, show that a volume $\frac{2}{3}a^3$ is enclosed between it and the base.

4. A circular hole of radius b is made centrally through a sphere of radius a. Find the volume of the sphere remaining.

5. Find the volume in the first octant common to the cylinders $x^2 + y^2 = r^2$ and $x^2 + rz = r^2$.

6. Find the volume common to a sphere of radius a and a cylinder of radius $\frac{1}{2}a$ with one of its generators passing through the centre of the sphere.

7. Two equal right circular cylinders have axes intersecting at right angles. Show that the volume common to both is $16a^3/3$, where a is the radius of a cylinder.

[L.U.]

8. Find the area outside the circle $r = 2$ and inside the circle $r = 4\cos\theta$. Find also the M.I. of this area about an axis through O perpendicular to the area if the density at any point varies inversely as the distance from the pole.

9. Find the centroid of the uniform area bounded by the cardioid $r = a(1 + \cos\theta)$.

10. A right cylinder has as base one loop of the curve $r^2 = a^2\cos 2\theta$ (a figure eight). Find the volume of this cylinder cut off by the sphere centre O and radius a.

11. Find the area outside the cardioid $r = a(1 + \sin\theta)$ and inside the circle $r = 3a\sin\theta$.

12. Find the volume in the first octant enclosed by the three co-ordinate planes and the surface $x^2 + y^2 + z^2 = a^2$.

13. A square hole of side $2b$ is cut symmetrically and horizontally through a vertical cylinder of radius a, the sides of the hole being horizontal and vertical. Find the volume removed if $a > b$.

14. Transform to polars and hence show that the area of a loop of the curve

$$(x^2 + y^2)^3 = 16a^2x^2y^2$$

is $\frac{1}{2}\pi a^2$.

15. Transform to polars the equation

$$(x^2 + y^2)^2 = a^2(x^2 - y^2).$$

Hence show that the area of one loop is $\frac{1}{2}a^2$ and that its centre of gravity is $\pi a/4\sqrt{2}$ from the pole. Show also that the second moment of this loop about an axis through the pole perpendicular to the plane is $\pi a^4/16$.

16. Interpret

$$\int_0^a dx \int_{\sqrt{(ax - x^2)}}^{\sqrt{(a^2 - x^2)}} \frac{dy}{\sqrt{(a^2 - x^2 - y^2)}}$$

as an integral taken over an area, showing the area in a diagram. Evaluate the integral by transforming to polar co-ordinates.

[L.U.]

17. Prove that

(a) $\displaystyle\int_{-\infty}^{+\infty}\int_{-\infty}^{+\infty} \frac{dx\,dy}{(x^2 + y^2 + 1)^{3/2}} = 2\pi.$

(b) $\displaystyle\int_0^2 \int_{1-(2x-x^2)^{1/2}}^{1 + (2x - x^2)^{1/2}} \frac{dx\,dy}{(x^2 + y^2)^2} = \pi.$

18. The length of the side of a square plate ABCD is $2a$, and O is the mid point of AB. If the surface density at any point P on the plate is λOP^2, λ being a constant, express the mass of the plate as a double integral. Show that the distance of the centre of mass of the plate from AB is $1\cdot4a$ and find the moment of inertia of the plate about AB.

[L.U.]

19. By considering the pressure on an element of area $dx\,dy$ at the point (x, y) on a vertical plane area immersed in liquid of weight w per unit volume, show that if the axes be in the surface and vertical the co-ordinates of the centre of pressure are given by

$$\iint wx^2 dx\,dy \Big/ \iint wx\,dx\,dy, \quad \iint wxy\,dx\,dy \Big/ \iint wx\,dx\,dy$$

where the integrals are taken over the area.

20. Find $\iint (a - x)^2 dx dy$ taken over half the circle $x^2 + y^2 = a^2$. A horizontal boiler has a flat bottom, and its ends are plane and semicircular. If it is just filled with water, show that the depth of the centre of pressure of either end is $0.7 \times$ total depth very nearly. [L.U.]

21. A circular plate of radius a lies on a rough plane (coefficient of friction μ) inclined at angle α to the horizontal. The plate is of mass M, is pinned at a point on its circumference to the plane and lies with its diameter through the pin horizontal. Show that the frictional torque preventing the motion of an element $r d\theta dr$ at the point (r, θ) (where the pin is the pole and the diameter is the initial line) is

$$\mu \, \frac{Mg \cos \alpha}{\pi a^2} \, r^2 d\theta dr.$$

Hence show that the total torque is

$$\int_{-\pi/2}^{+\pi/2} \int_0^{2a \cos \theta} \frac{\mu \, Mg \cos \alpha}{\pi a^2} \, r^2 d\theta dr$$

and evaluate this integral.

22. A footstep bearing carries a total load W distributed over a circle of radius a; the coefficient of friction between the bearing surfaces is μ. When new the load is uniformly distributed over the circle, but when worn, the intensity varies inversely as the distance from the centre. Calculate the frictional torque in each case and show that for a worn bearing it is $\frac{3}{4}$ of that for a new one. [L.U.]

23. A heavy uniform plate of mass M in the form of the cardioid

$$r = a(1 + \cos \theta)$$

lies on a rough horizontal table (coefficient of friction μ), and is movable about a vertical axis through the pole of co-ordinates. Show that the couple which must be applied to the plate in order to make it move is $\frac{10}{9}\mu a Mg$.

24. A heavy uniform square plate lies on a rough plane (coefficient of friction μ) inclined at angle α to the horizontal. If the plate is pinned at one corner, show that it will rest with the diagonal through the pinned point in any possible position if

$$3 \tan \alpha < \mu[2 + \sqrt{2} \log (1 + \sqrt{2})].$$

25. A uniform solid sphere of radius a has a cylindrical hole radius b bored centrally through it. Prove that the square of the radius of gyration about the axis of the hole is $(2a^2 + 3b^2)/5$. For a diameter of the sphere perpendicular to the above, show that the result is $(4a^2 + b^2)/10$.

26. Show that the equation $(x^2 + y^2 + z^2 + c^2 - a^2)^2 = 4c^2(x^2 + y^2)$ represents an anchor ring. Sketch the surface showing how it lies with respect to the axes of co-ordinates. Prove that the co-ordinates of any point on its surface may be expressed in the form $x = (c + a \cos \theta) \cos \phi, y = (c + a \cos \theta) \sin \phi, z = a \sin \theta$, and evaluate the surface area by calculating

$$\iint a(c + a \cos \theta) d\theta d\phi. \qquad \text{[L.U.]}$$

27. A plate in the form of a quadrant of the ellipse $x^2/a^2 + y^2/b^2 = 1$ is of small but varying thickness, the thickness at any point being proportional to the product of its distances from the axes. Show that the co-ordinates of the centroid are $\frac{8a}{15}, \frac{8b}{15}$. [L.U.]

28. $x^2/a^2 + y^2/b^2 = 1$ gives the contour of the base of a right circular cylinder; the height is c. The cylinder is bevelled down so that the height z of any point

(x, y) of the base in the first quadrant is given by $z = c\left(1 - \dfrac{x}{a}\right)\left(1 - \dfrac{y}{b}\right)$. Express the volume left in this quadrant as a double integral and show that its value is $\tfrac{1}{4}abc\left(\pi - \tfrac{13}{6}\right)$.

[L.U.]

16.5. The Integral $\displaystyle\int_0^\infty e^{-x^2}dx$

In Statistics and Mathematical Physics the value of the above integral is needed. It may be evaluated by a simple " trick ". (When the limits are not a zero or an infinity (plus or minus) it cannot be evaluated exactly and tables must be consulted.)

Put $x = az$ where a is a constant so that the limits of z are also zero and infinity. If I denote the integral

$$I = \int_0^\infty e^{-a^2 z^2}a\,dz \qquad \left(= \int_0^\infty e^{-x^2}dx\right) \quad \cdot \quad \cdot \quad \cdot \quad (1)$$

Multiply both sides by e^{-a^2}, which on the right-hand side may be placed under the integral as a is a constant and the integration is with respect to z.

$$\therefore \quad Ie^{-a^2} = \int_0^\infty e^{-(1 + z^2)a^2}a\,dz.$$

We will now let a vary from 0 to ∞ and integrate both sides with respect to a between these limits.

$$I\int_0^\infty e^{-a^2}da = \int_0^\infty da\int_0^\infty e^{-(1 + z^2)a^2}a\,dz.$$

The integral on the left-hand side is I. On the right-hand side all the limits of integration are constant, and we may therefore interchange the order of integration (see Example 3, Exercise 38. We will assume this permissible, even when the constant upper limit is infinite.)

It follows that

$$I^2 = \int_0^\infty dz\int_0^\infty e^{-(1 + z^2)a^2}a\,da$$

$$= \int_0^\infty dz\left[\frac{e^{-(1 + z^2)a^2}}{-2(1 + z^2)}\right]_{a=0}^\infty$$

$$= \int_0^\infty \frac{dz}{2(1 + z^2)} = \left[\tfrac{1}{2}\tan^{-1}z\right]_0^\infty = \frac{\pi}{4}.$$

Since I is positive

$$I = \int_0^\infty e^{-x^2}dx = \tfrac{1}{2}\sqrt{\pi} \quad \cdot \quad \cdot \quad \cdot \quad \cdot \quad \cdot \quad (2)$$

The curve $y = e^{-x^2}$ is symmetrical with regard to the y axis, therefore considering the area between it and the x axis

$$\int_{-\infty}^{+\infty} e^{-x^2}dx = 2\int_0^{\infty} e^{-x^2}dx = \sqrt{\pi}.$$

Using the form of (1),

$$\int_0^{\infty} e^{-a^2z^2}a\,dz = \tfrac{1}{2}\sqrt{\pi}$$

Put $a = 1/\sigma\sqrt{2}$ where later σ will be defined as the standard deviation of the curve about the y axis.

$$\therefore \int_0^{\infty} e^{-z^2/2\sigma^2}\frac{1}{\sigma\sqrt{2}}\,dz = \frac{1}{2}\sqrt{\pi}$$

and

$$\frac{1}{\sigma}\sqrt{\frac{2}{\pi}}\int_0^{\infty} e^{-z^2/2\sigma^2}dz = 1 \quad . \quad . \quad . \quad . \quad (3)$$

Also

$$\int_{-\infty}^{+\infty} e^{-z^2/2\sigma^2}\frac{1}{\sigma\sqrt{2}}\,dz = \sqrt{\pi},$$

$$\therefore \frac{1}{\sigma\sqrt{2\pi}}\int_{-\infty}^{+\infty} e^{-z^2/2\sigma^2}dz = 1 \quad . \quad . \quad . \quad . \quad (4)$$

This last form is often used.

To save space e^{-x^2} is often written as $\exp(-x^2)$. The student will therefore meet with the expression

$$\frac{1}{\sigma\sqrt{2\pi}}\int_{-\infty}^{+\infty} \exp(-x^2/2\sigma^2)dx = 1 \quad . \quad . \quad . \quad (5)$$

where we have also used the more usual variable x instead of z, which was introduced to avoid confusion.

16.6. Triple Integrals

To find the mass of a body of constant density ρ we could begin with a volume $z\,dx\,dy$ whose mass is $\rho z\,dx\,dy$. A double integral then gives the total mass. If, however, the density varies we must begin with an elementary volume $dx\,dy\,dz$, a box of sides dx, dy, dz parallel to the axes (Fig. 65). These sides are so small that we may assume the density throughout the box to be the same as that at the corner nearest the origin whose co-ordinates are (x, y, z). If this density is the function of x, y, z represented by $\rho(xyz)$, then the mass of the element is

$$\rho(xyz)dx\,dy\,dz$$

and the total mass is the triple integral

$$\iiint \rho(xyz)dx\,dy\,dz$$

where the limits are chosen to cover the whole volume.

In evaluating the above integral we follow the order, as in double integrals, of the differentials and in the above as written would integrate first with respect to x then y and then z.

The following example shows how a problem that can be solved by using a double integral is simpler with a triple integral.

Example 7. The points P$(a, 0, 0)$, Q$(0, b, 0)$, R$(0, 0, c)$ are joined and with the origin O form a solid tetrahedron. Find the position of its centre of gravity.

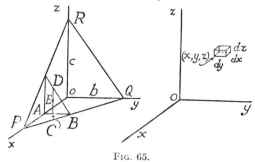

FIG. 65.

Let E be the point (x, y, z) the corner of an element of volume $dxdydz$. To find the total volume we will first integrate along CD, *i.e.*, from $z = 0$ to the value given by the equation to the plane PQR on which D lies, *i.e.*,

$$\frac{x}{a} + \frac{y}{b} + \frac{z}{c} = 1,$$

so that
$$z = c\left(1 - \frac{x}{a} - \frac{y}{b}\right).$$

We now integrate with regard to y from A to B, *i.e.*, since the line joining PQ is

$$\frac{x}{a} + \frac{y}{b} = 1$$

from $y = 0$ to $y = b(1 - x/a)$.

Next integrate for x from O to P, *i.e.*, 0 to a. This gives

$$V = \int_0^a dx \int_0^{b(1 - x/a)} dy \int_0^{c(1 - x/a - y/b)} dz$$

$$= \int_0^a dx \int_0^{b(1 - x/a)} c\left[1 - \frac{x}{a} - \frac{y}{b}\right] dy \quad \cdot \quad \cdot \quad \cdot \quad \cdot \quad \cdot \quad \cdot \quad (1)$$

$$= \int_0^a c\left[y - \frac{xy}{a} - \frac{y^2}{2b}\right]_0^{b(1 - x/a)} dx$$

$$= \int_0^a c\left[b\left(1 - \frac{x}{a}\right) - \frac{x}{a}b\left(1 - \frac{x}{a}\right) - \frac{1}{2b}b^2\left(1 - \frac{x}{a}\right)^2\right] dx.$$

$$= \tfrac{1}{6}abc$$

$\therefore$ $\Sigma m = \rho\tfrac{1}{6}abc$, where ρ is the constant density.

Similarly, taking moments for the element $\rho\,dxdydz$ about the x axis and integrating as above

$$\Sigma my = \rho \int_0^a \int_0^{b(1-x/a)} \int_0^{c(1-x/a-y/b)} y\,dz\,dy\,dx$$

$$= \rho \int_0^a \int_0^{b(1-x/a)} cy\left(1 - \frac{x}{a} - \frac{y}{b}\right) dy\,dx$$

$$= \rho \int_0^a c\left(\frac{y^2}{2} - \frac{xy^2}{2a} - \frac{y^3}{3b}\right)_0^{b(1-x/a)} dx$$

$$= \rho \int_0^a c\,\frac{b^2}{6}\left(1 - \frac{x}{a}\right)^3 dx = \rho\,\frac{ab^2c}{24},$$

$$\therefore \quad \bar{y} = \frac{\Sigma my}{\Sigma m} = \frac{b}{4}.$$

Similarly for $\bar{x}$ and $\bar{z}$ so that the centre of gravity is at the point $(\frac{1}{4}a, \frac{1}{4}b, \frac{1}{4}c)$.
The student should find CD $= z$ for use in the element of volume $z\,dxdy$ and solve the above problem as a double integral, noting that he begins at equation (1).

16.7. Spherical Polar Co-ordinates

Suppose the plane zOP to revolve through a given angle ϕ from the plane zOx. In this plane in its final position fix the position of the

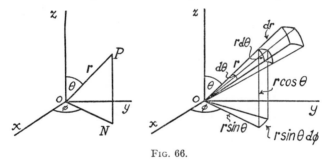

Fig. 66.

point P by means of polar co-ordinates (r, θ) with Oz as the initial line. These co-ordinates (r, θ, ϕ) are the *spherical polar co-ordinates* of P. Since $z = r \cos \theta$, and ON $= r \sin \theta$, it is clear that if P is also the point (x, y, z)

$$x = r \sin \theta \cos \phi, \qquad y = r \sin \theta \sin \phi, \qquad z = r \cos \theta.$$

If OP increases by dr and OP sweeps round in its plane so that θ increases by $d\theta$, the area generated is $r\,d\theta\,dr$. If now ON rotates so that ϕ increases by $d\phi$, then since N describes an arc of a circle $r \sin \theta\,d\phi$, the total element of volume generated by the element of area $r\,d\theta\,dr$ is $r\,d\theta\,dr \times r \sin \theta\,d\phi$ or

$$r^2 \sin \theta \, dr \, d\theta \, d\phi$$

A volume is therefore in spherical polars

$$\iiint r^2 \sin \theta \, dr \, d\theta \, d\phi$$

where the limits of integration are chosen so as to cover the volume concerned.

Example 8. Find the position of the centre of gravity of an octant of a sphere, cut off by the three perpendicular planes through its centre.

In spherical polars with the origin at the centre of the sphere its equation is $r = a$. With the usual axes taken as the lines of intersection of the perpendicular planes, the x co-ordinate of the element of volume $r^2 \sin \theta dr d\theta d\phi$ is $r \sin \theta \cos \phi$. Therefore if the density is ρ, taking moments about the y axis

$$\rho \tfrac{1}{6} \pi a^2 \bar{x} = \iiint r \sin \theta \cos \phi \, . \, \rho r^2 \sin \theta dr d\theta d\phi.$$

To cover the octant r must vary from 0 to a,

$$\begin{array}{cccc} \theta & ,, & ,, & 0 \text{ to } \tfrac{1}{2}\pi, \\ \phi & ,, & ,, & 0 \text{ to } \tfrac{1}{2}\pi, \end{array}$$

therefore the triple integral must be written

$$\rho \int_0^{\pi/2} \cos \phi d\phi \int_0^{\pi/2} \sin^2 \theta d\theta \int_0^a r^3 dr$$

$$= \rho \int_0^{\pi/2} \cos \phi d\phi \int_0^{\pi/2} \sin^2 \theta d\theta \, \frac{a^4}{4}.$$

$$= \frac{\rho a^4}{4} \int_0^{\pi/2} \cos \phi \, d\phi \, \frac{1}{2} \cdot \frac{\pi}{2} = \rho \, \frac{\pi a^4}{16},$$

from which

$$\bar{x} = \tfrac{3}{8}a.$$

Clearly $\bar{y}$ and $\bar{z}$ have the same value.

EXERCISE 40

1. Find:

(a) $\displaystyle\int_{-2}^{+2} dx \int_0^1 dy \int_1^2 (2x + 3y + 4z)dz$;

(b) $\displaystyle\int_0^2 \int_1^3 \int_1^2 xy^2 z dz dy dx$;

(c) $\displaystyle\int_0^a \int_0^b \int_0^c (x^2 + y^2)dz dy dx$;

(d) $\displaystyle\int_0^r \int_0^\alpha \int_0^\beta (x^2 + y^2 + z^2)dx dy dz$;

  where $\alpha = \sqrt{(r^2 - z^2)}$, $\beta = \sqrt{(r^2 - y^2 - z^2)}$;

(e) $\displaystyle\int_0^a \int_0^{\sqrt{(a^2 - y^2)}} \int_0^{\sqrt{(a^2 - x^2)}} x dz dx dy$;

(f) $\displaystyle\int_0^\pi \int_0^{\frac{1}{2}\pi} \int_0^r \rho^2 \sin\theta d\rho d\theta d\phi;$

(g) $\displaystyle\int_{-1}^{+1} dz \int_0^z dx \int_{x-z}^{x+z} (x + y + z)dy.$

2. Find $\displaystyle\int_0^a \int_0^a \int_0^{a-x} xyzdzdydx,$

and state the region through which the triple integral is found.

3. Show that

$$\iiint \frac{dxdydz}{(1 + x + y + z)^3}$$

integrated through the volume bounded by the co-ordinate planes and the plane $x + y + z = 1$ can be written

$$\int_0^1 \int_0^{1-r} \int_0^{1-x-y} \frac{dzdydx}{(1 + x + y + z)^3}$$

and evaluate this.

4. Find by triple integration

   (a) The volume in the first octant bounded by the co-ordinate planes and $x + 2y + 3z = 6.$

   (b) The volume enclosed by the cylinder $x^2 + y^2 = 9$ and the planes $z = 5 - x$ and $z = 0.$

   (c) The volume enclosed by the cylinders $y^2 = z$, $x^2 + y^2 = a^2$ and the plane $z = 0.$

5. Find the moment of inertia about the x axis of the solid, of uniform density ρ, formed by the cylinder $x^2 + y^2 = r^2$, and the planes $z = 0$, $z = a.$

6. The water face of a dam is vertical and has the form of a trapezium height h and length $2a$ at the top and a at the bottom. The thickness of the dam increases uniformly from zero at the top to a at the bottom. Show that the C.G. of the dam is $5a/16$ from the vertical face and find its distance below the top of the dam. [L.U.]

7. A spherical globe of radius a has a density at any point proportional to the depth below the tangent plane at the topmost point. Show that the mass centre is at a depth $6a/5$ and that the square of the radius of gyration of the globe about a horizontal axis through the mass centre is $9a^2/25.$ [L.U.]

8. A right circular cone has a height h and semi-vertical angle α. Find the total mass if the density at any point in the cone is k times its distance from the vertex.

9. The elastic energy of a volume of a material is $\tau^2V/2EI$, where τ N/m^2 is the stress and V the volume, E, I being constants. Find the elastic energy of a cylindrical volume radius R m, length l m, in which the stress τ varies as the distance from the axis, being zero at the axis and τ_0 at the outer surface.

10. Green's Theorem, an important theorem in Mathematical Physics, states that for certain functions

$$\iint (Pdydz + Qdzdx + Rdxdy)$$

$$= \iiint \left(\frac{\partial P}{\partial x} + \frac{\partial Q}{\partial y} + \frac{\partial R}{\partial z}\right) dxdydz,$$

where the first integral is taken over a *surface* and the second through the corresponding *volume*. Apply this to evaluate

$$\iint (x^3 dydz + x^2 ydzdx + x^2 zdxdy)$$

over the surface bounded by $z = 0$, $z = c$ and $x^2 + y^2 = a^2$.

16.8. Changing the Order of Integration

It will have been noticed that in some cases it is easier to integrate first with respect to one variable than the other. In other cases it is impossible to evaluate the integral as given, and in such cases we try a change in the given order of integration.

Example 9. Find

$$\int_0^{\sqrt{(\pi/2)}} \int_y^{\sqrt{(\pi/2)}} \sin (x^2) dxdy.$$

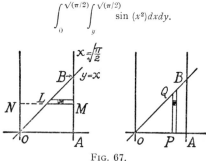

Fig. 67.

Since $\sin (x^2)$ has not an obvious integral we will change the order of integration so as to integrate with respect to y first. For this we must draw a diagram of the field of integration. It is given that when y is constant x moves from $x = y$ (the lower limit) to $x = \sqrt{(\frac{1}{2}\pi)}$ (the upper limit). We therefore draw these two lines. y then moves from $y = 0$ to $y = \sqrt{(\frac{1}{2}\pi)}$. The field of integration is then the triangle OAB as shown. (In the integral with the order changed we must take care that this area is covered by the new limits of integration chosen.) In the first integration

$$\int_y^{\sqrt{(\frac{1}{2}\pi)}} \sin (x^2) dx$$

we may assume that if $y(= ON)$ is a given value of y within the field of integration, then we have integrated along the strip LM parallel to the x axis. The second integration then sums the values for all strips between $y = 0 \ (= A)$ and $y = \sqrt{(\frac{1}{2}\pi)} \ (= B)$.

To change the order of integration we first integrate along a strip PQ parallel to the y axis where P is on $y = 0$ and Q on $y = x$. Having integrated along PQ, we must now choose the limits for x so that the field OAB is covered by the integration with respect to x. Clearly $x = 0$ to $x = \sqrt{(\frac{1}{2}\pi)}$ will ensure this. The integral with the order changed is therefore

$$\int_0^{\sqrt{(\frac{1}{2}\pi)}} dx \int_0^x \sin (x^2) dy$$

$$= \int_0^{\sqrt{(\frac{1}{2}\pi)}} x \sin (x^2) dx = \left[-\tfrac{1}{2} \cos (x^2) \right]_0^{\sqrt{(\frac{1}{2}\pi)}} = \underline{\tfrac{1}{2}}$$

EXERCISE 41

1. Indicate by a sketch the region over which the double integral is taken

$$\int_0^a dx \int_x^a (x^2 + y^2)dy.$$

Write down the integral with the order changed and hence, or directly, evaluate it. [L.U.]

2. Show from a consideration of the field of integration that the two integrals

$$\int_0^3 \int_{\frac{1}{2}y}^y y^2 dx dy + \int_3^6 \int_{\frac{1}{2}y}^3 y^2 dx dy$$

can be reduced to one. Hence evaluate it.

3. Show that a change in the order of integration of

$$\int_0^3 \int_{4x/3}^{\sqrt{25 - x^2}} f(xy)dydx$$

would require two integrals.

4. Show that if the field of integration is a rectangle of sides $x = a, b, y = c, d$, the integral of $f(x, y)$ over this field may be written

$$\int_a^b dx \int_c^d f(xy)dy \quad \text{or} \quad \int_c^d dy \int_a^b f(xy)dx,$$

so that in this case (of all four limits of integration constant) the limits are the same for both orders of integration.

5. Show with the aid of a sketch of the field of integration that

$$\int_0^{2a} dx \int_{x^2/4a}^{3a - x} f(x, y)dy = \int_0^a dy \int_0^{2\sqrt{ay}} f(x, y)dx + \int_a^{3a} dy \int_0^{3a - y} f(x, y)dx.$$

6. Change the order of integration and hence evaluate

$$\int_0^1 \int_{2y}^2 e^{x^2} dx dy.$$

7. Evaluate

$$\int_0^\infty dy \int_{ay/b}^\infty e^{-x^2/d^2} dx.$$

8. Evaluate

$$\int_0^1 \int_{\sqrt{y}}^1 \frac{dxdy}{\sqrt{(x^2 - y^2)}}.$$

9. Change the order of integration in the double integral

$$\int_0^a \left(\int_y^a \frac{x^2 dx}{\sqrt{(x^2 + y^2)}} \right) dy.$$

Hence or otherwise evaluate it. [L.U.]

10. Show by means of a diagram the area over which the double integral

$$\int_{a/2}^a dx \int^a \frac{xdy}{(x^2 + y^2)^{3/2}}$$

is taken. By inverting the order of integration or otherwise show that the value of the integral is

$$\log \left(\frac{2 + \sqrt{5}}{1 + \sqrt{2}}\right) - \frac{1}{\sqrt{2}} \log 2.$$ [L.U.]

11. Show that

$$\int_y^a \frac{dx}{\sqrt{(a - x)(x - y)}} = \pi.$$

By changing the order of integration find

$$\int_0^a \int_0^r \frac{dxdy}{(y + a)\sqrt{(a - x)(x - y)}}.$$ [L.U.]

12. Show that

$$\int_0^\infty e^{-ax} \cos bx dx = \frac{a}{a^2 + b^2}.$$

Integrate both sides of this relation with respect to b from o to b and change the order of integration. Hence prove

$$\int_0^\infty e^{-ax} \frac{\sin bx}{x} dx = \tan^{-1} \frac{b}{a}.$$

13. Change the order of integration and hence evaluate:

(a)
$$\int_0^a dy \int_0^y \frac{xdx}{\sqrt{[(a^2 - x^2)(a - y)(y - x)]}};$$

(b) and show

$$\int_0^2 dx \int_{\sqrt{(2x)}}^2 \frac{dy}{\sqrt{(1 + x^2 + y^2)} \log (1 + y^2)} = 1.$$ [L.U.]

14. Show in a diagram the region over which the integral

$$\int_0^1 dx \int_0^{\sqrt{(x - x^2)}} \frac{4xy}{x^2 + y^2} e^{- (x^2 + y^2)} dy$$

extends. Transform to polar co-ordinates and hence or otherwise show that the integral has the value e^{-1}. [L.U.]
[Show that the region is the upper half of the circle centre $(\frac{1}{2}, 0)$ radius $\frac{1}{2}$. In polars the integrand becomes $2 \sin 2\theta \, e^{-r^2}$ and the element of area is $rdrd\theta$. To cover the same region the limits for r are 0 to $\cos \theta$ and for θ, 0 to $\pi/2$.]

CHAPTER 17

THE MOTION OF A RIGID BODY

"... the same law takes place in a system of many bodies as in a single body. For the progressive motion whether of one single body or of a whole system of bodies is always to be estimated from the motion of the centre of gravity."

I. Newton, *Principia Philosophiae* (1686).

17.1. Linear Momentum

Consider a body composed of separate particles such as a mass m_r at the point (x_r, y_r), where $r = 1, 2, \ldots n$. If this particle is acted on by external forces with components (X_r, Y_r) then we have for the component accelerations

$$m_r\ddot{x}_r = X_r, \qquad m_r\ddot{y}_r = Y \qquad (r = 1, \ldots n).$$

Summing over all the particles,

$$\Sigma m_r\ddot{x}_r = \Sigma X_r, \qquad \Sigma m_r\ddot{y}_r = \Sigma Y_r, \quad . \quad . \quad . \quad . \quad (1)$$

[This is not strictly correct, for in addition to the external force X_r there will also be an internal force, X_r' say, due to the interactions between the particles. But since action and reaction are equal and opposite, the effect of these forces cancel each other, so that $\Sigma X_r' = 0$. We will therefore simplify the work by omitting these forces throughout.]

If in a given direction, the x direction suppose, the sum of the forces is zero so that $\Sigma X_r = 0$, then

$$\Sigma m_r\ddot{x}_r = 0.$$

Integrating : $\qquad \Sigma m_r\dot{x}_r = \text{constant}.$

This is the *principle of conservation of linear momentum*, which states that if no external forces act on a system of bodies in a certain direction, the total momentum of the system in that direction remains constant.

Example 1. A wedge of angle α lies on a smooth horizontal plane, and a mass m slides down its smooth inclined face. Find the acceleration of the particle down the wedge. The mass of the wedge is M.

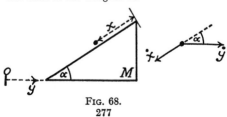

Fig. 68.

Suppose the wedge to have moved a distance y whilst the mass m has moved a distance x down the face. Then $\dot{y}$ is the velocity of the wedge whilst $\dot{x}$ is the velocity of the particle relative to the wedge. Its actual velocity is as shown in the accompanying figure.

On the system as a whole there is no horizontal force acting, and so the horizontal momentum, initially zero, remains so. This gives

$$M\dot{y} - m(\dot{x}\cos\alpha - \dot{y}) = 0$$

or
$$(M + m)\dot{y} = m\dot{x}\cos\alpha \quad \cdot \quad \cdot \quad \cdot \quad \cdot \quad \cdot \quad (1)$$

For the motion of m down the face

$$mg\sin\alpha = m(\ddot{x} - \ddot{y}\cos\alpha) \quad \cdot \quad \cdot \quad \cdot \quad \cdot \quad \cdot \quad (2)$$

If we differentiate (1) and substitute for $\ddot{y}$ in (2)

$$g\sin\alpha = \ddot{x} - \cos\alpha\left(\frac{m\ddot{x}\cos\alpha}{M + m}\right),$$

$$\ddot{x} = \frac{(M + m)g\sin\alpha}{M + m\sin^2\alpha}.$$

17.2. Angular Momentum

We had the equations

$$m\ddot{x} = X, \qquad m\ddot{y} = Y$$

for the particle at the point (x, y). Multiply each equation by the corresponding y or x respectively and subtract to obtain

$$m(x\ddot{y} - y\ddot{x}) = Yx - Xy.$$

Summation for all the particles gives

$$\Sigma m(x\ddot{y} - y\ddot{x}) = \Sigma(Yx - Xy) \quad \cdot \quad \cdot \quad \cdot \quad \cdot \quad (2)$$

But the moment about the origin of the force (X, Y) acting at (x, y) is $(Yx - Xy)$, so that $\Sigma(Yx - Xy)$ is the total moment of the external forces about the origin. Also, since the origin is fixed in space, the left-hand may be rewritten, as can be seen by differentiation, and we have

$$\frac{d}{dt}\Sigma m(x\dot{y} - y\dot{x}) = \Sigma(Yx - Xy) \quad \cdot \quad \cdot \quad \cdot \quad \cdot \quad (3)$$

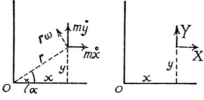

Fig. 69.

Now $\qquad \Sigma m(x\dot{y} - y\dot{x}) \quad$ or $\quad \Sigma[(m\dot{y})x - (m\dot{x})y] \quad \cdot \quad \cdot \quad (4)$

may clearly be called the moment of momentum (or angular momentum) about O so (3) above may be stated as: *the rate of change of angular momentum about any fixed axis is equal to the sum of the moments of the external forces about that axis.*

When a rigid body is rotating at the rate ω round a fixed axis the diagram shows that

$$\dot{y} = \omega r \cos \alpha = \omega x, \qquad \dot{x} = -\omega r \sin \alpha = -\omega y.$$

With these values

$$x\dot{y} - y\dot{x} = (x^2 + y^2)\omega = r^2\omega$$

and (3) becomes

$$\frac{d}{dt}\{\Sigma mr^2\omega\} = \Sigma(Yx - Xy).$$

Σmr^2 is constant with respect to time for a rigid body rotating round a fixed axis, and is Mk^2 the moment of inertia of the body about the axis. Putting θ for ω, we then have

$$Mk^2 \frac{d}{dt}(\dot{\theta}) = \Sigma(Yx - Xy)$$

or $$Mk^2\ddot{\theta} = C \qquad . \quad . \quad . \quad . \quad . \quad . \quad (5)$$

where C is the moment of the external forces about the axis. This has been used in Vol. I as the *Torque Equation.*

If there are no external forces acting or if the external forces have a total of zero moment about the fixed axis, then by (3)

$$\frac{d}{dt}\Sigma m(x\dot{y} - y\dot{x}) = 0$$

integrate $$\Sigma m(x\dot{y} - y\dot{x}) = \text{constant}$$

or, for rotation round a fixed axis,

$$Mk^2\omega = \text{constant}.$$

This is called the *principle of the conservation of angular momentum.*

Example 2. A hollow rod closed at each end is of mass M, length $2a$, and can rotate in a horizontal plane about a vertical axis at its centre. It contains a fine stream of powder of total mass m uniformly distributed along the length $2a$ of the hole. If the system is set rotating with an angular velocity ω find the angular velocity when the powder has accumulated equally at each end.   [L.U.]

The initial angular momentum about the axis of rotation is $(M + m)\frac{a^2}{3}\omega$.

If the final value of ω is Ω, then finally it is $\left(M\frac{a^2}{3} + 2\frac{m}{2}a^2\right)\Omega$.

These are equal, since the forces acting have no moment about the axis of rotation, therefore

$$\Omega = \frac{M + m}{M + 3m}\omega.$$

Example 3. A mass m lies held at rest on a smooth horizontal table and is attached to a string passing through a hole in it and carrying a mass M. m is projected perpendicular to the string with velocity v when it is distant a from the hole. Find its velocity along the string when a has decreased to $\frac{1}{2}a$.

Since the forces acting on m have no moment about the hole, its moment of momentum about the hole remains constant. If u is the velocity perpendicular to the string when m is $\frac{1}{2}a$ from the hole

$$(mu)\tfrac{1}{2}a = (mv)a$$

$$u = 2v.$$

M has descended a distance $\frac{1}{2}a$, and this work done has been transformed into kinetic energy. If V is the velocity along the string at this instant, equating the increase in kinetic energy to the work done,

$$Mg \cdot \tfrac{1}{2}a = \tfrac{1}{2}(M + m)V^2 + \tfrac{1}{2}mu^2 - \tfrac{1}{2}mv^2,$$

$$Mga = (M + m)V^2 + 3mv^2,$$

$$V^2 = (Mga - 3mv^2)/(M + m).$$

This result only has meaning if $Mga > 3mv^2$, which is the condition that when m is projected M falls instead of rising upwards.

EXERCISE 42

1. A smooth wedge of mass M and angle α rests on a smooth horizontal table. Another wedge of same mass and angle rests on it so that its upper surface is horizontal. A smooth particle of mass m is placed on this upper surface and the whole system allowed to move from rest under gravity. If the two wedges can move only in directions perpendicular to their edges, show that m descends vertically with an acceleration

$$\frac{2g(m + M) \sin^2 \alpha}{M + (2m + M) \sin^2 \alpha}.$$

2. A plank of mass M and length l is initially at rest along a line of greatest slope of a fixed smooth plane inclined at angle α to the horizon and a man of mass m starting at the upper end walks down the plank so that it remains at rest. Show that he reaches the other end in time

$$[2ml/(M + m)g \sin \alpha]^{\frac{1}{2}}.$$

3. A bead of mass M can slide on a smooth straight horizontal wire and a particle of mass m is attached to the bead by a light string of length l. The particle is held in contact with the wire with the string taut and is then let go. Prove that when the string is inclined to the wire at an angle θ the bead will have slipped a distance $ml(1 - \cos \theta)/(M + m)$ along the wire and that the angular velocity ω of the string will be given by the equation

$$(M + m \cos^2 \theta)l\omega^2 = 2(M + m)g \sin \theta.$$

4. An equilateral triangle ABC formed of wire of total mass M is resting vertex A downwards on smooth pegs so that the base BC is horizontal. There are rings of mass m_1, m_2, on the side BA, CA respectively. These rings are initially at B, C respectively, and begin to slide down their respective sides. Show that the velocity of the triangle at any instant when both rings are moving on it is equal to the difference of the speeds of the rings relative to the wire and that the acceleration of the triangle is, for $m_1 > m_2$

$$\sqrt{3}(m_1 - m_2)g/(4M + 3m_1 + 3m_2).$$

5. Two particles A and B, each of mass m, are attached to the ends of a light spiral spring AB, and the system is placed on a smooth horizontal table. A blow of impulse I is applied to A in the direction AB. Prove that the greatest compression is $I/\sqrt{(2mS)}$, where S is the force needed to extend the spring by unit length. Prove also that when the spring regains its natural length for the first time, it has moved forward a distance $\pi I/2\sqrt{(2mS)}$. [L.U.]

6. A circular turntable can rotate freely about a vertical axis through its centre, its moment of inertia about this axis being I. On this turntable a small toy engine of mass m runs on a circular track of radius r and with O as centre. Initially the system is at rest. Show that when the engine has a velocity v relative to the table, this latter has an angular velocity $mrv/(I + mr^2)$.

7. A uniform circular disc of mass M and radius a rotates freely about a fixed vertical axis through its centre perpendicular to its plane and carries a particle of mass $\frac{1}{4}$M which is free to move along a smooth radial groove. Initially the disc rotates with angular velocity ω and the particle is at rest at the centre. Prove that when the particle, after being slightly disturbed, has moved a distance r along the radius, the angular velocity of the disc is $4a^2\omega/(4a^2 + r^2)$.

[L.U.]

8. A thin straight tube of small bore is movable about its centre on a smooth horizontal table, and it contains a uniform thin rod of the same length and mass, whose centre of gravity is nearly at the middle point of the tube. Prove that if the system be set in motion with angular velocity ω the angular velocity of the tube as the rod leaves it is $\omega/7$.

9. A horizontal wheel with buckets on its circumference revolves about a frictionless axis. Water falls into the bruckets at a uniform rate of mass m per unit of time. Treating the buckets as small compared with the wheel, find the angular velocity of the wheel after time t, if Ω is its initial value and show that if I be the moment of inertia of the wheel and buckets about the vertical axis and r the radius of the circumference on which the buckets are placed, the angle turned through by the wheel in time t is

$$\frac{I\Omega}{mr^2} \log \left(1 + \frac{mr^2t}{I}\right).$$

10. A block of mass M at rest on a smooth horizontal table has a smooth-walled cylindrical hole of radius a with its axis horizontal, and a small bead of mass m is at rest in the hole, in the vertical plane through the centre of mass of the block. If the block is then suddenly given a velocity V along the table in a direction normal to the axis of the hole, show that the bead will just rise to the level of the axis if $V^2 = 2ga (M + m)/M$. Prove also that when the bead is next at its lowest level the velocity of the block is $(M - m)V/(M + m)$. [L.U.]

11. A man stands on a swing, and for the purposes of this question he may be regarded as a particle whose distance from the smooth horizontal axis of the swing is l when he crouches and $l - h$ when he stands. As the swing falls the man crouches, and as it rises he stands, the changeover being assumed instantaneous. If the swing falls through an angle α and then rises through an angle β, show that

$$\sin \tfrac{1}{2}\beta = \left(\frac{l}{l - h}\right)^{3/2} \sin \tfrac{1}{2}\alpha.$$

Explain how the man can continuously increase his angle of swing and how the extra energy is obtained. [L.U.]

12. A particle on a smooth table is attached by a string passing through a small hole in the table and carries an equal particle hanging vertically. The former particle is projected along the table at right angles to the string with velocity $\sqrt{(2gh)}$ when at a distance a from the hole. If r is the distance from the hole at time t, prove the results:

(1) $2 \left(\dfrac{dr}{dt}\right)^2 = 2gh \left(1 - \dfrac{a^2}{r^2}\right) + 2g(a - r).$

(2) The tension of the string is $\frac{1}{2}mg \left(1 + \dfrac{2a^2h}{r^3}\right)$, m being the mass of each particle.

13. Masses m, m_1 are attached to the ends of a light inextensible string AOB and rest on a smooth horizontal table. The string is in contact with a smooth fixed peg at O, and the portions OA($= a$) and OB($= b$) of the string are in a straight line. The mass m is now projected horizontally with velocity u perpendicular to OA. If the string remains in contact with the peg and all the motion takes place in a horizontal plane, prove that the mass m_1 reaches the peg with velocity

$$\frac{u}{a + b} \sqrt{\frac{mb(2a + b)}{m + m_1}}.$$

17.3. The Motion of the Centre of Gravity

If M is the mass of the whole system of particles and $(\bar{x}, \bar{y})$ the co-ordinates of the centre of gravity referred to fixed axes,

$$M\bar{x} = \Sigma mx, \qquad M\bar{y} = \Sigma my.$$

From these by differentiation, since the axes are fixed

$$M\dot{\bar{x}} = \Sigma m\dot{x}, \qquad M\dot{\bar{y}} = \Sigma m\dot{y},$$

showing that the linear momentum of the system is the same as that of a particle whose mass is the total mass and which moves with the velocity of the centre of gravity.

Again differentiating :

$$M\ddot{\bar{x}} = \Sigma m\ddot{x}, \qquad M\ddot{\bar{y}} = \Sigma m\ddot{y},$$

but $$\Sigma m\ddot{x} = \Sigma X, \qquad \Sigma m\ddot{y} = \Sigma Y,$$

so that $$M\ddot{\bar{x}} = \Sigma X, \qquad M\ddot{\bar{y}} = \Sigma Y \quad . \quad . \quad . \quad . \quad (6)$$

showing that the centre of gravity moves as if all the mass and all the forces were concentrated there.

We can also show that the motion of translation of the centre of gravity and of motion relative to the centre of gravity are quite independent.

17.4. Pressure on the Axis of Rotation

A body of mass M is rotating round a fixed axis at O. Its centre of gravity G is h distant from O, and its radius of gyration about the axis at G is k.

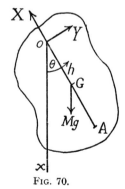

Fig. 70.

If when OG makes angle θ with the downward vertical, the reaction at the axis has components X, Y along and perpendicular to OG respectively, then for the motion of G

$$X - Mg \cos \theta = Mh\dot{\theta}^2,$$

since G is describing a circle with O as centre, and

$$Y - Mg \sin \theta = Mh\ddot{\theta}$$

since $h\ddot{\theta}$ is the (instantaneous) linear acceleration of G perpendicular to OG.

The Torque Equation gives

$$- Mgh \sin \theta = M(k^2 + h^2)\ddot{\theta}.$$

If we multiply each side by $2\dot{\theta}$ and integrate

$$2Mgh \cos \theta = M(k^2 + h^2)\dot{\theta}^2 + C,$$

an equation which can be written down at once by stating that the work done equals the gain in kinetic energy.

Suppose the body to consist of a rod OA of length $2h$ which falls from rest in the position with A vertically above O. For the position shown in the figure above

$$\tfrac{1}{2}M \left(4 \frac{h^2}{3}\right) \dot{\theta}^2 = Mgh(1 + \cos \theta),$$

since the work done by the descent of the centre of gravity has been converted into the kinetic energy of rotation of the rod about O. By differentiation

$$\ddot{\theta} = -\frac{3g}{4h} \sin \theta$$

Hence using the above values for X, Y

$$X = Mg \cos \theta + Mh \cdot \frac{3g}{2h} (1 + \cos \theta)$$

$$= \tfrac{1}{2}Mg(3 + 5 \cos \theta)$$

$$Y = Mg \sin \theta + Mh\ddot{\theta}$$

$$= Mg \sin \theta + Mh \left(-\frac{3g}{4h} \sin \theta\right)$$

$$= \tfrac{1}{4}Mg \sin \theta.$$

Example 4. A uniform cylinder of radius a rolls down an inclined plane of angle α. To find the acceleration and the condition for no slipping.

Let F be the frictional force and R the normal reaction on the cylinder of mass M. The centre of gravity moves as if all the forces were acting there. For its motion down the plane

$$Mg \sin \alpha - F = M\ddot{x} \quad . \quad . \quad . \quad . \quad . \quad . \quad \textbf{(1)}$$

If θ is the angle turned through by the cylinder in rolling a distance x

$$x = a\theta,$$

therefore
$$\dot{x} = a\dot{\theta}, \qquad \ddot{x} = a\ddot{\theta}.$$

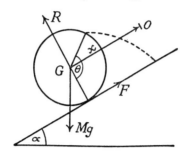

FIG. 71.

Taking moments about the centre of gravity as though it is a fixed point,

$$Fa = M \frac{a^2}{2} \ddot{\theta}$$

therefore
$$F = \tfrac{1}{2}M\ddot{x}.$$
Using this in (1)
$$Mg \sin \alpha - \tfrac{1}{2}M\ddot{x} = M\ddot{x}$$
$$\ddot{x} = \tfrac{2}{3}g \sin \alpha \quad \ldots \ldots \quad (2)$$
and
$$F = \tfrac{1}{2}M\ddot{x}$$
$$= \tfrac{1}{3}Mg \sin \alpha.$$
Since there is no motion perpendicular to the plane
$$R = Mg \cos \alpha.$$
For no slipping we must have
$$F \leqslant \mu R$$
$$\tfrac{1}{3}Mg \sin \alpha \leqslant \mu Mg \cos \alpha$$
$$\tan \alpha \leqslant 3\mu \quad \ldots \ldots \ldots \quad (3)$$

With $\mu = 0\cdot2$ this shows that the cylinder will not slip provided the angle of slope is not greater than $\tan^{-1} (0\cdot6) = 31°$.

If slipping does occur, the geometrical relation $x = a\theta$ no longer holds, but we have instead
$$F = \mu R$$
$$= \mu Mg \cos \alpha.$$
This and (1) give
$$\ddot{x} = g (\sin \alpha - \mu \cos \alpha). \quad \ldots \ldots \quad (4)$$
instead of the value from (2).

Example 5. A uniform rod is held in a vertical position with one end on a very rough horizontal table and released from rest. Find the motion.

If R is the normal reaction and F the frictional force when the rod has turned through angle θ, since the centre of gravity moves as if all the forces were applied there,

$$F = M\ddot{x} = M \frac{d^2}{dt^2} (a \sin \theta)$$
$$= Ma (\cos \theta \, \ddot{\theta} - \sin \theta \, \dot{\theta}^2),$$

$$R - Mg = M\ddot{y} = M \frac{d^2}{dt^2} (a \cos \theta)$$

$$= - Ma (\sin \theta \, \ddot{\theta} + \cos \theta \, \dot{\theta}^2).$$

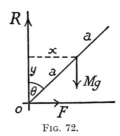

FIG. 72.

Moments about O:

$$Mga \sin \theta = M \frac{4}{3} a^2 \ddot{\theta}$$

or

$$\ddot{\theta} = \frac{3g}{4a} \sin \theta.$$

Multiply both sides by $2\dot{\theta}$ and integrate:

$$\dot{\theta}^2 = - \frac{3g}{2a} \cos \theta + C.$$

$\theta = 0$ when $\theta = 0$:

$$0 = - \frac{3g}{2a} + C,$$

therefore

$$\dot{\theta}^2 = \frac{3g}{2a} (1 - \cos \theta),$$

which could have been written down at once using the principle that the gain in kinetic energy equals the work done.

Using these values of $\dot{\theta}^2$ and $\ddot{\theta}$,

$$F = \tfrac{3}{4}Mg \sin \theta(3 \cos \theta - 2), \qquad R = \tfrac{1}{4}Mg(1 - 3 \cos \theta)^2.$$

These equations show that:

(1) F, the friction, changes its direction as θ passes through the value $\cos^{-1} (2/3)$.

(2) R, the reaction, vanishes but does not change in sign as θ passes through the value $\cos^{-1} (1/3)$. The end at O does not therefore leave the table.

(3) The ratio F/R becomes infinite when $\cos \theta = \tfrac{1}{3}$, so that in practice sliding will occur before this value is reached, sliding backwards or forwards according as the slipping occurs before or after the angle $\cos^{-1} (2/3)$ is reached (when F changes sign).

EXERCISE 43

1. A thin hoop rolls without slipping down an inclined plane of slope α. Find the acceleration of its centre.

2. Solve the above problem for a solid sphere.

3. A light thread is wound round a reel which may be regarded as a solid cylinder. The end of the thread is held and the reel falls vertically, keeping its axis horizontal. Find the acceleration of the reel and the tension in the thread.

4. A hollow cylinder whose inner radius is one-half the outer rolls down an inclined plane of slope α. Find the acceleration of the centre of the cylinder. If the slope is so great that slipping also occurs, find the distance travelled from rest in time t.

5. A thin uniform rod of length $2a$ attached to a smooth hinge at one end O falls from a horizontal position. Show that the horizontal strain on the hinge is greatest when the rod is inclined to the vertical at 45° and that the vertical strain is then 11/8 times the weight of the rod.

6. A uniform circular lamina of mass M can turn in a vertical plane about an axis at right angles to its plane through a point in its circumference. If it starts from rest from the position in which the diameter through this point is horizontal, prove that the horizontal and vertical components of the pressure on the axis, when this diameter makes an angle θ with the horizontal are

$$\text{Mg sin } 2\theta \quad \text{and} \quad \tfrac{1}{4} \text{ Mg } (4 - 3 \cos 2\theta).$$

7. A uniform rod of mass M free to turn about a fixed smooth pivot at one end is held horizontally and released. Prove that when in the subsequent motion the rod makes an angle θ with vertical, the pressure on the pivot is

$$\tfrac{1}{4} \text{ Mg } \sqrt{(1 + 99 \cos^2 \theta)}.$$

8. A uniform rod AB of length $2a$ and mass M is free to turn about a fixed point in AB distant $\tfrac{1}{2}a$ from A and has a particle of mass M attached to the end B. The rod is held in a horizontal position and then released. Find the angular velocity when AB is vertical, and prove that the pressure on the point of support is then $82Mg/17$. [L.U.]

9. A uniform disc of mass M, radius a and centre C oscillates under gravity about a fixed horizontal tangent through the point O on the circumference. If the angular velocity of the disc in its lowest position is ω, find the total pressure on the axis for a deflection θ from this position.

Find the value of ω if in a position of instantaneous rest, the resultant pressure is perpendicular to the line OC.

10. A thin uniform square plate ABCD of side $2a$ and mass M can turn freely about an axis through AB which is horizontal. If the plate be released from a horizontal position, show that when it makes an angle of 30° with the horizontal, its angular speed is $\tfrac{1}{2}\sqrt{(3g/a)}$, and find the magnitude of the reaction at the axis in this position.

17.5. The Kinetic Energy of a Plane Rigid Body

Suppose the centre of gravity G of the body has velocity components u, v as shown and the body is also rotating with angular

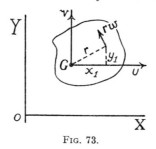

FIG. 73.

velocity ω. The velocity of any point P, an element of the body of mass m, is relative to G, $r\omega$ perpendicular to the join GP, and therefore its actual velocity components are if $(x_1 y_1)$ are the co-ordinates of P relative to G,

$$u - r\omega \sin \theta = u - \omega y_1$$
$$v + r\omega \cos \theta = v + \omega x_1$$

The kinetic energy of the particle is then

$$\tfrac{1}{2}m[(u - \omega y_1)^2 + (v + \omega x_1)^2]$$

and of the whole body, by summation

$$\tfrac{1}{2}(u^2 + v^2)\Sigma m + \tfrac{1}{2}m\Sigma\omega(x_1{}^2 + y_1{}^2) - u\omega\Sigma my_1 + v\omega\Sigma mx_1$$

Now $\Sigma m = M$ the mass of the whole body, $\Sigma my_1 = 0 = \Sigma mx_1$, since G is the centre of gravity of the body and $\Sigma m(x_1{}^2 + y_1{}^2) = Mk^2$, the moment of inertia of the body about an axis through G perpendicular to the plane of the body. We therefore have that the kinetic energy of the body is

$$\tfrac{1}{2}M(u^2 + v^2) + \tfrac{1}{2}Mk^2\omega^2,$$

which can be interpreted as the kinetic energy of the whole mass of the body moving with the velocity of the centre of gravity plus the kinetic energy relative to the centre of gravity.

Example 6. A rod of length $2a$ and mass M stands vertically on a smooth table and falls from rest. Find the angular velocity when it becomes horizontal and the reaction of the table at this instant.

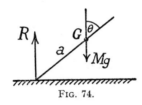

FIG. 74.

Since there is no horizontal force, the centre of gravity of the rod descends vertically. When the rod makes angle θ with the vertical the centre of gravity has a velocity of $\dfrac{d}{dt}(a \cos \theta)$ only. The angular velocity is $\dot\theta$, so that the kinetic energy is

$$\tfrac{1}{4}Ma^2 \sin^2 \theta\, \dot\theta^2 + \tfrac{1}{2}M\frac{a^2}{3}\dot\theta^2.$$

This arises from the work done by the vertical descent of the centre of gravity. Equating the two:

$$\tfrac{1}{2}Ma^2\dot\theta^2(\sin^2 \theta + \tfrac{1}{3}) = Mga(1 - \cos \theta) . \quad . \quad . \quad . \quad (1)$$

when $\theta = \tfrac{1}{2}\pi$, this gives

$$\dot\theta = \sqrt{(3g/2a)}.$$

If the reaction at the end is R, for the vertical motion

$$R - Mg = M\frac{d^2}{dt^2}(a \cos \theta)$$
$$= - M[a \sin \theta\, \ddot\theta + a \cos \theta\, \dot\theta^2].$$

We have a value for $\dot{\theta}^2$ in (1). If also we differentiate (1) with regard to time, first rewriting it as

$$\dot{\theta}^2 = \frac{2g}{a}\frac{1-\cos\theta}{\sin^2\theta + \frac{1}{3}},$$

$$2\dot{\theta}\ddot{\theta} = \frac{2g}{a}\frac{(\sin^2\theta + \frac{1}{3})\sin\theta\,\dot{\theta} - (1-\cos\theta)2\sin\theta\cos\theta\,\dot{\theta}}{(\sin^2\theta + \frac{1}{3})^2}$$

or

$$\ddot{\theta} = \frac{g\sin\theta}{a(\sin^2\theta + \frac{1}{3})^2}(\tfrac{4}{3} - 2\cos\theta + \cos^2\theta).$$

When $\theta = \frac{1}{2}\pi$

$$\ddot{\theta} = \frac{3g}{4a},$$

and for this value of θ

$$R = Mg - Ma\left(\frac{3g}{4a}\right)$$
$$= \tfrac{1}{4}Mg \text{ newtons.}$$

Example 7. A sphere rolls in the bottom of a hollow fixed cylinder of radius b. If the line joining the centres of the sphere and cylinder has an angular velocity ω when it is vertical, find the angular velocity when it makes an angle θ with the vertical. Deduce the time of a small oscillation of the sphere.

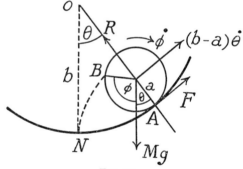

Fig. 75.

Suppose the point B on the sphere was originally at N. Since there is no slipping

$$\text{arc } AB = \text{arc } AN, \qquad a(\phi + \theta) = b\theta,$$

therefore

$$\phi = \frac{b-a}{a}\theta,$$

and

$$\dot{\phi} = \frac{b-a}{a}\dot{\theta} \qquad \cdots \cdots \quad (1)$$

where $\dot{\phi}$ is the angular velocity of the sphere about its centre. At the lowest point, the velocity of the sphere is $(b-a)\omega$, at the higher point it is $(b-a)\dot{\theta}$. Equating loss of kinetic energy to the work done,

$$Mg(b-a)(1-\cos\theta) = \tfrac{1}{2}M(b-a)^2[\omega^2 - \dot{\theta}^2] + \tfrac{1}{2}M\cdot\tfrac{2}{5}a^2\cdot\left(\frac{b-a}{a}\right)^2[\omega^2 - \dot{\theta}^2],$$

i.e.,

$$g(1-\cos\theta) = 7\frac{(b-a)}{10}[\omega^2 - \dot{\theta}^2],$$

an equation for the angular velocity of the line of centres, and hence $(b-a)\dot{\theta}$ the linear velocity of the sphere.

Differentiating this equation:

$$g \sin \theta = - 7 \frac{(b - a)}{5} \ddot{\theta}.$$

When θ is small, $\sin \theta \simeq \theta$, and this becomes

$$\ddot{\theta} = - \frac{5g}{7(b - a)} \theta,$$

showing that the motion is simple harmonic, and the time of a complete oscillation is given by

$$T = 2\pi \sqrt{\frac{7(b - a)}{5g}}.$$

If R is the normal reaction at A, and F the frictional force along the tangent at A, for the motion of the centre of gravity of the small sphere

$$R - Mg \cos \theta = M(b - a)\dot{\theta}^2,$$

$$Mg \sin \theta - F = - M(b - a)\ddot{\theta}.$$

Also, from moments about the centre of gravity

$$Fa = - M\tfrac{2}{5}a^2\ddot{\phi}$$

or by (1)

$$F = - M\tfrac{2}{5}(b - a)\ddot{\theta}.$$

Using this value for $\ddot{\theta}$ and the value for $\dot{\theta}^2$, R and F can be evaluated.

EXERCISE 44

1. A string passing over a rough pulley of mass M_1, radius r, radius of gyration k carries at its ends masses M_2, M_3 where $M_2 > M_3$. If no slipping occurs, show that the acceleration of the masses is $g(M_2 - M_3)/\left[M_2 + M_3 + \dfrac{M_1 k^2}{r^2}\right]$.

2. A truck of total mass M_1 has two axles carrying wheels of radius a; the moment of inertia of each axle and its wheels being $M_2 k^2$. If the truck rolls down an incline of angle α, show that its acceleration is

$$M_1 g \sin \alpha/(M_1 + 2M_2 k^2/a^2).$$

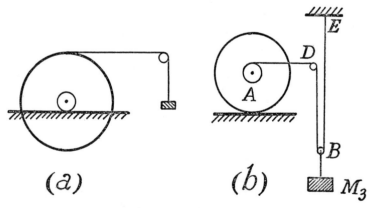

Fig. 76.

3. A disc of mass 5 kg and diameter 0·6 m has fastened to it on either side a concentric disc of mass 1 kg and diameter 0·3 m. If the combination rolls as shown on rough horizontal rails, moved by a horizontal string passing over a

fixed smooth peg and carrying a mass of 4 kg, show that the tension in the string is 13·7 N and the acceleration of the disc is 2·14 m/s² (Fig. 76(a)).

4. A reel A rolls on a horizontal table as shown (Fig. 76(b)). D is a small smooth fixed peg, and E a fixed point. The mass of A is M_1, its inner and outer radii a, b respectively, its radius of gyration k_1. A mass M_3 hangs by a smooth hook from B. Find the acceleration of M_3.

5. Two gear-wheels A and B, of radii a and b and moments of inertia I_1 and I_2, respectively, are mounted on parallel axes and run permanently in mesh. When in motion there are constant frictional torques P and Q on A and B respectively. A constant torque G is applied to A. Find:

    (a) the tangential force between the wheels;

    (b) the angular acceleration of A;

    (c) the number of revolutions made by B in acquiring from rest a speed of N revolutions per second.

Prove also that for motion to be possible, G must be greater than $P + Qa/b$.

[L.U.]

6. A rigid body is free to rotate about a fixed vertical axis, and at time t a horizontal line in the body makes an angle θ with a fixed horizontal line. The body has moment of inertia I, about the axis, and is acted on by a couple $a - \mu\theta$, which is in a horizontal plane and in the sense in which θ increases, a and μ being constants. Write down the equation of motion of the body and find its general solution. Show also that if

$$\theta = 0 \quad \text{and} \quad \frac{d\theta}{dt} = 0 \quad \text{at} \quad t = 0,$$

$$\theta = \frac{a}{\mu}\left\{1 - \cos\left(t\sqrt{\frac{\mu}{I}}\right)\right\}.$$

[L.U.]

7. A truck has a body of mass M and four wheels, each of mass m, radius r and radius of gyration k. It is driven by a torque T applied to the back axle. Find the acceleration and the frictional forces between each wheel and the ground.

[L.U.]

8. A uniform circular disc of mass M and radius a in a vertical plane can turn freely in its own plane about its centre. A particle of mass m is attached to a point of the rim of the disc. Show that ω_1 and ω_2 the angular velocities of the disc, when the particle is at the highest and lowest points of the rim satisfy the relation

$$a(M + 2m)(\omega_2{}^2 - \omega_1{}^2) = 8mg.$$

9. A pulley whose radius is r and moment of inertia about its axis I is mounted in a smooth horizontal bearing. A chain of weight W and length $3\pi r$ has one end fixed to the pulley at an extremity of a horizontal diameter, and then passes completely round the pulley, its other end hanging freely. If the system is allowed to move from rest, show that the angular velocity of the pulley after turning through half a revolution is

$$[Wrg(3\pi^2 - 4)/3\pi(I + Wr^2)]^{1/2}.$$

[L.U.]

10. Prove that the kinetic energy of a lamina of mass M moving in its own plane is $\frac{1}{2}M(V^2 + k^2\omega^2)$, where V is the velocity of the mass centre, ω its angular velocity and k its radius of gyration about the mass centre.

A uniform straight rod is placed with its ends in contact with a smooth vertical wall and a smooth horizontal plane, the vertical plane through the rod being perpendicular to the wall. If the rod is allowed to fall freely under gravity, show that it will lose contact with the wall when its centre has fallen through one-third of its original height above the plane.

[L.U.]

11. A uniform circular cylinder of mass M and radius a carries a particle of mass m fixed in its surface. It is placed on a rough horizontal table with the

particle in the highest position. If the system is slightly disturbed and the cylinder rolls without slipping, prove that when it has turned through an angle θ

$$a\dot{\theta}^2\{\tfrac{3}{2}M + 2m(1 + \cos\theta)\} = 2mg(1 - \cos\theta).$$ [L.U.]

12. Two equal uniform rods AB, AC, each of length $2a$ and of the same mass are smoothly hinged together at A and are placed in a vertical plane with the ends B, C on a smooth horizontal plane so that each rod makes the same acute angle α with the horizontal. If the system is released from rest, prove that when each rod makes an angle θ with the horizontal, the velocity of B is

$$\sin\theta[6ag\,(\sin\alpha - \sin\theta)]^{1/2}.$$ [L.U.]

13. A reel consists of two uniform discs, each of radius b and mass M, connected by a cylinder of radius a, also of mass M. It rests with its axis horizontal on a rough horizontal plane. A string having one end fixed to the cylinder is wound round its middle portion and leaves it horizontally in a direction perpendicular to the axis and below the axis. Show that a force P applied to the end of the string produces, if the plane is rough enough to prevent slip, an acceleration $2Pgb(b - a)/Mg(8b^2 + a^2)$ and find the least value of μ which prevents slip.

14. A thin plank of mass M is placed across two cylindrical rollers, each of mass m and radius a, and the system is allowed to move down a slope of angle α. Show that if no slipping occurs, the acceleration of the plank is

$$\frac{4(m + M)g\sin\alpha}{(3m + 4M)}.$$ [L.U.]

15. An engine of mass M has two pairs of equal wheels a forward pair each of radius a, and a rear pair each of radius b. The moments of inertia of the pairs with their axles about the axes of these axles are A for the forward pair and B for the rear pair. The engine is set in motion by a couple G applied to the forward axle. Prove that if none of the wheels slip, the friction that can be called into play between either of the forward wheels and the rail must exceed

$$\frac{M + B/b^2}{M + A/a^2 + B/b^2} \cdot \frac{G}{2a},$$

the rails being horizontal.   [L.U.]

16. A car of mass M moves in a circle of radius r with a velocity v. Its centre of gravity is at a height h above the road, and $2b$ is the distance between the wheels. Show that the inner vertical reaction and outer are, respectively,

$$\tfrac{1}{2}Mg - \frac{hM}{2b}\frac{v^2}{r}, \qquad \tfrac{1}{2}Mg + \frac{hM}{2b}\frac{v^2}{r},$$

and the speed at which it will tend to overturn is

$$v = \sqrt{\frac{bgr}{h}}.$$

17. A car is driven along a straight horizontal road by a torque applied to the axles. Its centre of gravity is at a height h above the road, c behind the front wheels and d in front of the rear wheels. When it is accelerating at a rate f, show that the vertical reaction on the front and rear wheels are, respectively,

$$\frac{Mg}{c + d}\left(d - \frac{fh}{g}\right), \qquad \frac{Mg}{c + d}\left(c + \frac{fh}{g}\right).$$

If the driving torque is applied to the rear wheels only and μ is the coefficient of friction, show that the maximum acceleration is

$$\frac{\mu cg}{c + d - \mu h},$$

but if the torque is applied to the front wheels only it is

$$\frac{\mu dg}{c + d + \mu h}.$$

18. A flywheel of moment of inertia I about its axis of rotation is acted upon by a couple whose moment at time t is

$$L_0 + L_1 \sin pt,$$

where L_0 and L_1 are constant, and is subjected to a resisting couple of moment $k\omega$, where ω is the angular velocity at any instant and k is constant. Find the value of ω at any time t, and show that as t becomes large ω will consist of a constant and a harmonic whose amplitude is

$$L_1/\sqrt{(p^2k^2 + I^2)}. \qquad \text{[L.U.]}$$

19. A solid cylinder of radius b rolls on the inside of a fixed hollow horizontal cylinder of radius a. Show that the periodic time of a small oscillation about the lowest point is

$$2\pi[3(a - b)/2g]^{1/2} \text{ sec.}$$

20. A uniform solid cylinder of radius r rolls without slipping on the inside of a rough hollow cylinder of radius $R (> r)$ which is fixed with its axis horizontal. If ω is the angular velocity of the cylinder in its lowest position and

$$3r^2\omega^2 > 11g(R - r),$$

show that the first cylinder will roll completely round the second. [L.U.]

21. A uniform bar of length a and mass M is freely pivoted at one end, and is let fall from a horizontal position. Determine the angular velocity when the rod has fallen through an angle θ.

Show that in this position, the tension in the rod at a distance x from the pivot is $\frac{1}{4}Mg \sin \theta(5 - 2x/a - 3x^2/a^2)$.

22. A uniform circular disc of mass $2m$ and radius a has a particle of mass m fixed to the circumference. The disc is projected with its plane vertical and the particle initially in its highest position, so as to roll without slipping on a horizontal rail. Prove that when the radius to the particle makes an angle θ with the upward vertical

$$a\dot{\theta}^2 = \{7a\Omega^2 + 2g(1 - \cos \theta)\}/(5 + 2 \cos \theta),$$

where Ω is the angular speed of projection.

Hence or otherwise find the vertical reaction on the rail when the disc has turned through one right angle. [L.U.]

23. A thin uniform rod AB of mass m and length $2a$ is held vertically with the end B on a smooth horizontal plane. B is attached to a point C in the plane by a light inextensible string. If the rod is released so that it falls in the vertical plane through BC and rotates towards C, show that the string becomes slack when $\theta = \cos^{-1}(2/3)$. Show also that immediately before the string becomes slack, A has a vertical acceleration equal to $3g/2$. [L.U.]

24. A wheel of moment of inertia I is fixed on a shaft of small radius r, and round the shaft is wound a length l of thin chain of mass m per unit length, one end being fixed to the shaft, while the other carries a mass equal to that of the whole chain. Initially the shaft is at rest with the attached mass level with the axis. If the mass be released, show that the length of chain unwound after t seconds is $2l \sinh^2 \lambda t$, where

$$\lambda = \frac{1}{2}\{mgr^2/(I + 2mlr^2)\}^{1/2}. \qquad \text{[L.U.]}$$

25. A uniform square plate of side l and mass M is free to turn in a vertical plane about a smooth horizontal axis through one corner. Show that the angular velocity ω, which must be given to it when hanging at rest in order that it may just turn through $180°$ is $(3\sqrt{2g/l})^{1/2}$. If a particle of mass kM is attached to the centre of the plate and it is set rotating, starting from its rest position with the same angular velocity ω, show that it will turn through an angle α given by

$$\cos \alpha = -\left(\frac{k + 2}{2k + 2}\right).$$

Show also that $\alpha > 120°$ whatever the value of k. [L.U.]

26. Two uniform rods AB, BC of equal length $2a$ and of equal mass m are freely hinged at B. C is free to slide on a fixed smooth vertical wall, and AB is free to turn in a vertical plane perpendicular to the wall about a smooth fixed horizontal axis through A at a distance $2a$ from the wall. Initially ABC is a straight line with C higher than A. The rods are then released and B falls. If the velocity of C is u when AC is horizontal, show that the velocity of the middle point of BC is then $\frac{1}{2}\sqrt{3}u$ and that

$$7u^2 = 36\sqrt{3}ga. \qquad\qquad \text{[L.U.]}$$

17.6. The Moment of Momentum of a Body Moving in Two Dimensions

We had in (4) of 17.2 the expression for the moment of momentum about the origin

$$\Sigma m(x\dot{y} - y\dot{x}).$$

If (x_1, y_1) are the co-ordinates of the point (x, y) relative to the centre of gravity $(\bar{x}, \bar{y})$ so that

$$x = x_1 + \bar{x} \qquad y = y_1 + \bar{y}$$

this may be written

$$\Sigma m\{(x_1 + \bar{x})(\dot{y}_1 + \dot{\bar{y}}) - (y_1 + \bar{y})(\dot{x}_1 + \dot{\bar{x}})\}.$$

Since $\qquad\qquad \Sigma mx_1 = 0 = \Sigma my_1$

so that $\qquad\qquad \Sigma m\dot{x}_1 = 0 = \Sigma m\dot{y}_1$

this becomes, with $\Sigma m = M$, the total mass,

$$M[\bar{x}\dot{\bar{y}} - \bar{y}\dot{\bar{x}}] + \Sigma m[x_1\dot{y}_1 - y_1\dot{x}_1].$$

The first factor is the moment of momentum about O of a particle of the mass M situated at G its centre of gravity and moving with it. The second factor is the moment of momentum of the body relative to G and, as above, can be changed to

$$Mk^2\dot{\theta}$$

so that the above can be written

$$M[\bar{x}\dot{\bar{y}} - \bar{y}\dot{\bar{x}}] + Mk^2\dot{\theta} \quad . \quad . \quad . \quad . \quad (1)$$

[If p is the length of the perpendicular from the origin upon the direction of the velocity of the centre of gravity, this may be written in a form which is sometimes more useful than the above

$$Mvp + Mk^2\dot{\theta}] \quad . \quad . \quad . \quad . \quad (2)$$

From (1) we may regard the whole momentum of a body as consisting of two parts, one, acting at the centre of gravity with components $M\ddot{x}$, $M\ddot{y}$, and another, a spin couple $Mk^2\dot{\theta}$, for the moment of these about any point O gives the moment of momentum of the body about O as in (1).

From this we can deduce the usual equations for $\ddot{x}$, $\ddot{y}$ and $\ddot{\theta}$, and also a new and important equation : The moment of the external forces about any axis is equal to the moment of the rate of change of momentum about this axis, i.e.,

$$M[\bar{x}\ddot{\bar{y}} - \bar{y}\ddot{\bar{x}}] + Mk^2\ddot{\theta} \quad . \quad . \quad . \quad . \quad (3)$$

Note: (*a*) The expressions (1), (2) above should be noted, as they are required later.

(*b*) In (3) moments are taken about the origin, but may be taken about any point using the three components $M\ddot{x}$, $M\ddot{y}$ and $Mk^2\ddot{\theta}$. We choose any convenient point, *e.g.*, the instantaneous point of contact of a sphere with the ground so that the forces there do not have a moment about the axis there or appear in the equation of moments. If, for example, we choose to take moments about the point through which the centre of gravity is instantaneously passing, the first factor in (3) is zero and the moment of the external forces about the centre of gravity is

$$Mk^2\ddot{\theta}$$

as previously obtained.

Example 8. A sphere rolls down a plane inclined at angle α to the horizon. The first diagram shows the external forces acting, and the second shows the

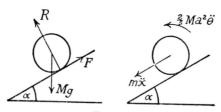

Fig. 77.

rates of change of momentum, the two being equivalent ($\ddot{y}$ is obviously zero).

In a general case note that $m\ddot{x}$, $m\ddot{y}$, $mk^2\ddot{\theta}$ must be taken as acting in the positive directions of x, y, θ, respectively.

Taking moments about the point of contact of the sphere and the plane

$$Mg \sin \alpha \,.\, a = M\ddot{x}a + M\tfrac{2}{3}a^2\ddot{\theta}$$

or since $x = a\theta$,

$$\ddot{x} = \tfrac{5}{7}g \sin \alpha$$

as previously obtained.

Example 9. The lock of a train door will engage only if the angular velocity at closing is greater than Ω. The door has a radius of gyration K about a

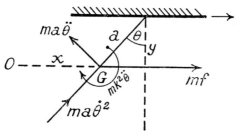

Fig. 78.

vertical axis at the hinges, and its centre of gravity is distant a from this axis. The door is at rest and open perpendicular to the side of the stationary train. If this begins to move with acceleration f, show that the door will not close unless

$$f > \tfrac{1}{2}\Omega^2 K^2/a.$$

If the door has turned through an angle θ in time t from the start of the train, the equivalent forces relative to the line of hinges are as shown, excluding the force mf. Since the train is subject to the acceleration, the total forces (or rates of change of momentum) acting on the door are as shown. Taking moments about the line of hinges, since the only external forces, apart from the weight, act at the line of hinges,

$$m(k^2 + a^2)\ddot{\theta} - maf \cos \theta = 0$$

or $$K^2\ddot{\theta} = af \cos \theta$$

This can be written

$$K^2\omega \frac{d\omega}{d\theta} = af \cos \theta$$

therefore $$K^2\omega^2 = 2af \sin \theta + C$$

$\theta = 0, \ \omega = 0$: $$0 = C$$

Finally $$\omega^2 = \frac{2af \sin \theta}{K^2}$$

$$= 2af/K^2 \quad \text{at} \quad \theta = \tfrac{1}{2}\pi.$$

The lock catches then if

$$\omega^2 > \Omega^2$$

or $$f > \tfrac{1}{2}K^2\Omega^2/a.$$

As an alternative method:

Let $x =$ displacement of the centre of gravity of the door in the direction of motion, $y =$ the distance of this from the side of the train,

$\therefore \quad x = \tfrac{1}{2}ft^2 - a \sin \theta,$ $\qquad\qquad y = a \cos \theta,$

$\dot{x} = ft - a \cos \theta \, \dot{\theta},$ $\qquad\qquad \dot{y} = -a \sin \theta \, . \, \dot{\theta},$

$\ddot{x} = f + a \sin \theta \, . \, \dot{\theta}^2 - a \cos \theta \, . \, \ddot{\theta},$ $\qquad \ddot{y} = -a \cos \theta \, . \, \dot{\theta}^2 - a \sin \theta \, . \, \ddot{\theta}.$

As above, the moment of rate of change of momentum about the line of hinges is zero, i.e.,

$$m\ddot{x} \times a \cos \theta + m\ddot{y} \times a \sin \theta - mk^2\ddot{\theta} = 0$$

which reduces to, as above,

$$af \cos \theta - K^2\ddot{\theta} = 0.$$

EXERCISE 45

1. The door of a stationary railway carriage stands open and perpendicular to the length of the train. The train starts off with acceleration f, and at the same time the door is given an angular velocity Ω in the direction towards the front of the train so as to shut the door. Show that, if the door can be regarded as a smoothly hinged uniform rectangular plate of width $2a$, then Ω must be at least of magnitude $\sqrt{(3f/2a)}$ in order to close the door.   [L.U.]

2. A uniform bar of length $2a$ has a small ring attached to one end and hangs by it free to slide along a fixed smooth horizontal wire. If the bar makes small oscillations under gravity, show that the period is $2\pi\sqrt{(a/3g)}$.   [L.U.]

3. A sphere of radius a, whose mass centre G is distant $\tfrac{1}{4}a$ from its centre C, is placed on a rough horizontal plane, and is held at rest with CG horizontal. If the sphere rolls when CG is released, prove that when CG makes an angle

ϕ with the horizontal, the horizontal and vertical components of the acceleration of G are respectively

$$a\{(1 - \tfrac{1}{2}\sin\phi)\ddot{\phi} - \tfrac{1}{2}\cos\phi \cdot \dot{\phi}^2\}$$

and

$$\tfrac{1}{2}a(-\cos\phi \cdot \ddot{\phi} + \sin\phi \cdot \dot{\phi}^2).$$

If the radius of gyration of the sphere about a horizontal axis through G at right angles to CG is $\tfrac{3}{5}a$, and the sphere rolls as soon as CG is released, prove that the coefficient of friction is not less than 32/73. Assuming that the motion is one of pure rolling, find the angular velocity of the sphere when CG is vertical.

[L.U.]

4. A wheel of mass m and radius a is eccentrically loaded so that its centre of mass G is at a distance b from its axis C. The wheel is constrained to roll with uniform velocity V on a rough horizontal track. Prove that the horizontal force P which must be applied at C to maintain this motion at the instant when CG has turned through an angle θ from the downward vertical is given by

$$P = \frac{mb}{a^2}(V^2 + ag)\sin\theta. \qquad [\text{L.U.}]$$

5. A uniform solid sphere of mass M and radius a has a particle of mass M embedded in its surface and rests on a perfectly rough horizontal plane with the particle in its highest position. If the sphere is slightly displaced, show that when the particle just touches the plane, the linear velocity of the sphere is $\sqrt{(20ga/7)}$ and that the normal thrust on the plane is $34Mg/7$. [L.U.]

6. A flywheel with an axle of radius r, total mass M and radius of gyration k, rolls with its axle on two parallel horizontal rails, rough enough to prevent slip. Motion is caused by a mass m attached to a light cord wrapped round the axle with w hanging freely. Show that motion is possible in which the hanging portion of the cord makes a steady angle α with the vertical given by

$$\sin\alpha/(1 - \sin\alpha)^2 = mr^2/M(r^2 + k^2)$$

and that the acceleration of the axis of the flywheel is then $g\tan\alpha$. [L.U.]

17.7. Impulsive Forces

It can be shown that if impulsive forces, whose components are P, Q and moment N about the centre of gravity of a body of mass M, act, then the changes in linear and angular momentum are given by

$$M(\dot{x} - u) = P$$
$$M(\dot{y} - v) = Q$$
$$Mk^2(\dot{\theta} - \omega) = N$$

where u, v, ω are the initial values and $\dot{x}$, $\dot{y}$, $\dot{\theta}$ the values after the impulsive action has occurred. Owing to the instantaneous nature

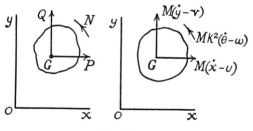

Fig. 79.

of the action in this case, we may use the equivalence shown in the diagram above to take moments about any point in the plane.

Example 10. A uniform square lamina is rotating in its own plane about one corner A, which is fixed, with angular velocity Ω, when suddenly that corner becomes free and an adjacent corner B becomes fixed. Prove that the angular speed is reduced to $\frac{1}{4}\Omega$ and that the impulse makes an angle $\tan^{-1}(\frac{5}{3})$ with AB.

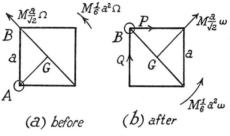

(a) before   (b) after

Fig. 80.

At any instant the momentum of the lamina consists of the two components $Ma\Omega/\sqrt{2}$ acting along GB and the spin couple $M\frac{1}{6}a^2\Omega$. After B has been fixed let the angular velocity be ω. Since the impulse acted at B, there will be no change in the moment of momentum about B. This gives

$$M\tfrac{1}{6}a^2\Omega = \left(M\frac{a}{\sqrt{2}}\omega\right)\frac{a}{\sqrt{2}} + M\tfrac{1}{6}a^2\omega,$$

$$\omega = \tfrac{1}{4}\Omega.$$

If P, Q are the components at B of the impulse as shown, equating them to the change in linear momentum in their respective directions,

$$P = \left(M\frac{a}{\sqrt{2}}\omega\right)\cos 45 - \left(-M\frac{a}{\sqrt{2}}\Omega\right)\cos 45$$

$$= M\frac{a}{2}(\omega + \Omega) = \tfrac{5}{8}aM\Omega,$$

$$Q = \left(M\frac{a}{\sqrt{2}}\omega\right)\cos 45 - \left(M\frac{a}{\sqrt{2}}\Omega\right)\cos 45$$

$$= M\frac{a}{2}(\omega - \Omega) = -\tfrac{3}{8}aM\Omega$$

Q therefore acts in the opposite direction to that shown, and the resultant impulse at B makes an angle $\tan^{-1}\left(\dfrac{P}{-Q}\right)$ with AB $= \tan^{-1}\left(\dfrac{5}{3}\right)$.

Example 11. Two equal rods AB, BC, each of length $2a$ and mass m, lie in line on a smooth table being smoothly jointed at B. An impulse P is applied at A perpendicular to the rod AB. Find the initial velocities of the rods.

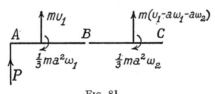

Fig. 81.

Let the velocity of the mass centre of AB be u, ω, the angular velocity of AB and ω_2 the angular velocity of BC relative to B in the direction shown. Then the velocity of the centre of mass of BC is $u_1 - a\omega_1 - a\omega_2$.

Taking moments about B for the rod BC to eliminate the impulsive reaction at B from the equation :

$$m(u_1 - a\omega_1 - a\omega_2)a - \tfrac{1}{3}ma^2\omega_2 = 0 . \qquad \ldots \quad (1)$$

For the rod AB, moments about B,

$$P.2a = mu_1a + \tfrac{1}{3}ma^2\omega_1 \qquad \ldots \quad (2)$$

For the system as a whole

$$P = mu_1 + m(u_1 - a\omega_1 - a\omega_2) \qquad \ldots \quad (3)$$

The solution of these equations is

$$u_1 = \frac{5P}{4m}, \qquad a\omega_1 = \frac{9P}{4m}, \qquad a\omega_2 = -\frac{3P}{4m}.$$

In some cases it would be simpler to assume that the actual velocity of the centre of BC is u_2, and the angular velocity of BC is ω_2 about this centre. The above equations are now :

$$mu_2a - \tfrac{1}{3}ma^2\omega_2 = 0;$$
$$P.2a = mu_1a + \tfrac{1}{3}ma^2\omega_1;$$
$$P = mu_1 + mu_2.$$

We need another equation, since there are four unknowns, and must express the geometrical fact that the velocity of B is the same regarded as a point on AB or BC, *i.e.*,

$$u_1 - a\omega_1 = u_2 + a\omega_2.$$

Example 12. A uniform sphere of mass M and radius a rolls with velocity v on a horizontal plane perpendicular to a block of height $\tfrac{1}{4}a$, which it strikes. Find the condition that the sphere surmounts the block.

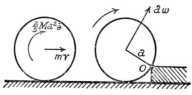

Fig. 82.

When the sphere strikes the block there is an impulsive reaction at O. Therefore there will be no change in the moment of momentum about O, an axis about which the impulse at O will have zero moment. Before the blow there was a linear momentum mv and an angular momentum $\tfrac{2}{5}mav$ (since $a\theta = v$), see 17.6. If the angular velocity immediately afterwards is ω, the angular momentum of the sphere rotating about O is $m(\tfrac{2}{5}a^2 + a^2)\omega$. We therefore have

$$\tfrac{7}{5}ma^2\omega = mv . \frac{a}{2} + \tfrac{2}{5}mav$$

$$\omega = \frac{9v}{14a}.$$

To surmount the block the kinetic energy of rotation round O must be sufficient to raise the centre of gravity through a height $\frac{1}{4}a$ at least. The condition required is then

$$\frac{1}{2}M\left(\frac{7}{5}a^2\right)\left(\frac{9v}{14a}\right)^2 > Mg\frac{a}{2}$$

or

$$v^2 > \frac{140}{81}ga.$$

Example 13. A uniform circular wire of radius a lies on a smooth horizontal table and is movable about a fixed point O on its circumference. An insect of mass equal to that of the wire starts from the other end of the diameter through O and crawls along the wire with a uniform velocity v relative to the wire. Show that at the end of time t the wire has turned through an angle

$$\frac{vt}{2a} - \frac{1}{\sqrt{3}}\tan^{-1}\left[\frac{1}{\sqrt{3}}\tan\frac{vt}{2a}\right].$$

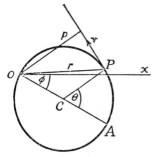

FIG. 83.

Suppose at time t the diameter OA has rotated through an angle ϕ from its initial position Ox. If angle ACP $= \theta$ where P is the position of the insect, arc AP $= a\theta = vt$ or $\theta = vt/a$.

The moment of momentum of the system, wire plus insect, about O is unchanged, and so remains at zero value throughout the motion. We will consider it as consisting of the wire plus insect moving with angular velocity $\dot\phi$ about O

$$m(a^2 + a^2)\dot\phi + mr^2\dot\phi$$

and the component due to the relative velocity v

$$- mvp,$$

where p is the perpendicular from O on to the tangent at P.

Since angle OPC $= \frac{1}{2}\theta$,

$$r = 2a\cos\tfrac{1}{2}\theta, \qquad p = r\cos\tfrac{1}{2}\theta = 2a\cos^2\left(\tfrac{1}{2}\theta\right).$$

For the zero value of the total moment of momentum about O

$$m2a^2\dot\phi + m \cdot 4a^2\cos^2\left(\tfrac{1}{2}\theta\right)\dot\phi - mv\{2a\cos^2\left(\tfrac{1}{2}\theta\right)\} = 0,$$

$$\frac{a}{v}\frac{d\phi}{dt} = \frac{\cos^2\left(\tfrac{1}{2}\theta\right)}{1 + 2\cos^2\left(\tfrac{1}{2}\theta\right)} = \frac{\cos^2\left(\dfrac{vt}{2a}\right)}{1 + 2\cos^2\left(\dfrac{vt}{2a}\right)},$$

Integrating between O and t $\left[T = \dfrac{vt}{2a}\right]$

$$\frac{a}{v}\phi = \frac{2a}{v}\int_0^{vt/2a} \frac{\cos^2 T}{1 + 2\cos^2 T}\,dT,$$

$$\phi = \int_0^{vt/2a} \left(1 - \frac{1}{1 + 2\cos^2 T}\right) dT$$

$$= \frac{vt}{2a} - \int_0^{vt/2a} \frac{\sec^2 T \, dT}{\sec^2 T + 2}$$

$$= \frac{vt}{2a} - \int_0^{vt/2a} \frac{d(\tan T)}{3 + \tan^2 T}$$

$$= \frac{vt}{2a} - \frac{1}{\sqrt{3}} \tan^{-1}\left(\frac{1}{\sqrt{3}} \tan T\right)_0^{vt/2a}$$

$$= \underline{\frac{vt}{2a} - \frac{1}{\sqrt{3}} \tan^{-1}\left(\frac{1}{\sqrt{3}} \tan \frac{vt}{2a}\right).}$$

17.8. Small Oscillations

Many such problems may be solved by the use of the *Torque Equation*. By this means the angular acceleration of the body is found when it is displaced say θ from its equilibrium position. Since the displacement is small, we may put θ for $\sin \theta$ and 1 for $\cos \theta$ in the equation, which then reduces to the standard type

$$\ddot{\theta} = -n^2\theta$$

so that the period of the simple harmonic motion may be written down.

Example 14. A uniform rigid rod AB of mass $2m$ and length a, is smoothly hinged to a fixed point at A and has a particle of mass m attached to B, which is connected by a light elastic string of modulus mg and natural length $a/2$ to a point C vertically below A. If $AC = 2a$, show that the period of small oscillations about the position of equilibrium with B vertically below A is $\pi(5a/3g)^{1/2}$.   [L.U.]

If the acute angle at A, angle $CAB = \theta$ and angle $ABC = \frac{1}{2}\pi + \alpha$, with T the tension in the string BC, moments about A gives

$$2mg\left(\frac{a}{2}\sin\theta\right) + mg \cdot a\sin\theta + T\cos\alpha \cdot a = -\left(2m \cdot \frac{a^2}{3} + m \cdot a^2\right)\ddot{\theta}.$$

But

$$BC^2 = 4a^2 + a^2 - 2 \cdot 2a \cdot a\cos\theta$$

and

$$BC = a\sqrt{(5 - 4\cos\theta)}.$$

Also by the Sine rule

$$\frac{\sin(\frac{1}{2}\pi + \alpha)}{\sin\theta} = \frac{2a}{a\sqrt{(5 - 4\cos\theta)}}$$

so that

$$\cos\alpha = \frac{2\sin\theta}{\sqrt{(5 - 4\cos\theta)}}.$$

With these values the above Torque equation becomes

$$2mg \, a\sin\theta + a \cdot \frac{mg}{\frac{1}{2}a}\left(a\sqrt{(5 - 4\cos\theta)} - \frac{1}{2}a\right)\frac{2\sin\theta}{\sqrt{(5 - 4\cos\theta)}} = -ma^2\frac{5}{3}\ddot{\theta}$$

or

$$6\sin\theta - \frac{2\sin\theta}{\sqrt{(5 - 4\cos\theta)}} = -\frac{5a}{3g}\ddot{\theta}$$

With θ for $\sin\theta$, 1 for $\cos\theta$, this reduces to

$$\ddot{\theta} = -\frac{12g}{5a}\theta$$

and the result follows.

Many such problems including the type dealt with above may be solved by expressing, generally, that the sum of the kinetic and potential energies is constant throughout the motion—if it is in a conservative system.

Example 15. Four equal uniform rods each of mass m and length $2a$ are smoothly jointed to form a rhombus ABCD which is hung from a fixed point at A. A and C are connected by a light elastic string of natural length a and modulus $2mg$. Show that the energy equation for that motion of the system in which C remains vertically below A is given by

$$a(1 + 3 \sin^2 \theta)\dot{\theta}^2 - 6g \cos \theta(1 - \cos \theta) = \text{constant}$$

where the angle BAD $= 2\theta$ and $\cos \theta > \frac{1}{4}$. Differentiate this equation to obtain the second order differential equation satisfied by θ and deduce that the length of the equivalent simple pendulum for small oscillations about the position $\theta = \frac{1}{3}\pi$ is $13a/18$.   [L.U.]

Since $\cos \theta > \frac{1}{4}$, $4a \cos \theta > a$ and the elastic string AC is stretched and therefore in tension throughout the motion.

The K.E. of AB $= \frac{1}{2}m(\frac{4}{3}a^2)\dot{\theta}^2$.

Since relative to A and AC the co-ordinates of the C.G. of BC are

$$(3a \cos \theta, a \sin \theta),$$

the K.E. of BC $= \frac{1}{2}m \left(\frac{a^2}{3}\right) \dot{\theta}^2 + \frac{1}{2}m[(a \cos \theta . \dot{\theta})^2 + (-3a \sin \theta . \dot{\theta})^2].$

The P.E. of AC $= \frac{1}{2} . \frac{2mg}{a} (4a \cos \theta - a)^2$.

The gravitational P.E. of the rods referred to a horizontal through A as the zero position is

$$-2mga \cos \theta - 2mg3a \cos \theta = -8mga \cos \theta$$

Expressing that for the four rods

$$\text{K.E.} + \text{P.E.} = \text{constant},$$

$$m\frac{4}{3}a^2\dot{\theta}^2 + m\frac{a^2}{3}\dot{\theta}^2 + ma^2[\cos^2 \theta + 9 \sin^2 \theta]\dot{\theta}^2 + mga(4 \cos \theta - 1)^2$$
$$-8mga \cos \theta = C.$$

This reduces to

$$a(1 + 3 \sin^2 \theta)\dot{\theta}^2 - 6g \cos \theta(1 - \cos \theta) = B.$$

Differentiate this with respect to time, divide by $\dot{\theta}$ and since for small oscillations $\dot{\theta}^2 \simeq 0$, we are left with

$$\ddot{\theta} = \frac{3g \sin \theta(2 \cos \theta - 1)}{a(1 + 3 \sin^2 \theta)}$$

From this we deduce, as could be done initially by statical methods, that the equilibrium position of the rhombus occurs when

$$2 \cos \theta - 1 = 0, \qquad \theta = \pi/3.$$

The small oscillations will then be about this equilibrium position, so put $\theta = \pi/3 + \alpha$ where it is known that α remains small. The equation above becomes

$$\ddot{\alpha} = \frac{3g}{a} . \frac{\sin \left(\frac{2\pi}{3} + 2\alpha\right) - \sin \left(\frac{\pi}{3} + \alpha\right)}{1 + 3 \sin^2 \left(\frac{\pi}{3} + \alpha\right)}$$

Expand and put $\cos 2\alpha = \cos \alpha = 1$, $\sin 2\alpha = 2\alpha$,

$$\ddot{\alpha} = \frac{3g}{a} . \frac{[\frac{1}{2}\sqrt{3} + (-\frac{1}{2})2\alpha] - [\frac{1}{2}\sqrt{3} + \frac{1}{2}\alpha]}{1 + 3[\frac{1}{2}\sqrt{3} + \frac{1}{2}\alpha]^2}$$

Ignore α^2 :

$$\ddot{a} = \frac{3g}{a} \cdot \frac{-\frac{3}{2}\alpha}{\frac{13}{4} + \frac{3\sqrt{3}}{2}\alpha} = -\frac{18g}{13a}\alpha\left(1 - \frac{6\sqrt{3}}{13}\alpha \cdots\right)$$

$$= -\frac{18g}{13a}\alpha$$

The length of the E.S.P. is therefore $13a/18$.

EXERCISE 46

1. A circular disc is rotating round its centre on a smooth horizontal table. If a point on the rim is suddenly fixed, find the new angular velocity in terms of the initial one, and the impulse at the fixing point.

2. A square lamina ABCD of side $2a$ lies on a smooth horizontal table. If the corner A is made to move along BA produced with velocity u, find the initial angular velocity of the lamina.

3. A uniform rod of mass m and length $2a$ is lying on a smooth horizontal table and is struck by a blow P perpendicular to its length at one end. Find the velocity of this end.

4. A disc of radius a lies on a smooth horizontal table when a point on the edge is made to move along the tangent at the point with velocity u. Prove that the disc begins to turn with angular velocity $2u/3a$.

5. A sphere of radius a and radius of gyration k about any axis through its centre rolls with linear velocity v on a horizontal plane, in a direction perpendicular to a fixed block of height h where $h < a$. Show that after the (inelastic) impact the sphere will surmount the block if, assuming no slipping,

$$(a^2 - ah + k^2)^2v^2 > 2gha^2(a^2 + k^2).$$

6. A body hangs vertically from a fixed axis O distant h from its centre of gravity G, the moment of inertia of the body about the fixed axis being $M(h^2 + k^2)$. Find the point of application of a horizontal blow on the line OG if there is to be no impulsive reaction at the axis O. Show that the distance from O is the length of the Simple Equivalent Pendulum for the body about O.

7. A circular target of radius 0·3 m is made of uniform thin sheet metal and has a mass of 10 kg. It is freely suspended from a horizontal tangential hinge, and while hanging at rest it is struck by a bullet of mass 50 g, which is moving at right angles to the plane of the target, the point of impact being half-way between the centre and the hinge. If the bullet coalesces with the target on impact and the latter swings back through an angle of 60°, show that the velocity of the bullet just before impact is approximately 770 m/s.

8. Two equal uniform thin rods AB, BC, each of mass m, length l, are freely jointed at B and are in line moving perpendicular to their length with velocity u. The mass centre of AB is suddenly brought to rest. Find the angular velocity of each rod immediately after impact and prove that four-sevenths of the original kinetic energy is lost by the impact. [L.U.]

9. A uniform toothed disc of mass m and radius $2a$ rotating with angular velocity ω about a thin fixed shaft through its centre is suddenly thrown into gear with another uniform disc of mass m and radius a, smoothly mounted on a parallel light thin shaft through its centre distant $3a$ from the fixed shaft about which it is free to revolve. Show that the angular velocity of the smaller disc after contact is $4\omega/5$ and that the velocity of its centre is $2a\omega/5$. [L.U.]

10. The moments of inertia of three gear-wheels A, B and C are in the ratio 4 : 2 : 1, and the radii of their pitch circles in the ratio 4 : 3 : 2. Initially B and C are in mesh and at rest, while A is rotating at n revs per minute. If A is suddenly brought into mesh with B, show that its speed is reduced to $9n/26$, and find the

ratio of the impulse between A and B to that between B and C. (The shafts on which the wheels are mounted remain fixed.)   [L.U.]

11. A thin uniform rod of mass m and length $2a$ falls freely with its length vertical. When the rod is moving with speed v the lower end strikes a smooth inelastic plane fixed at an angle of 30° to the horizontal. Prove that the magnitude of the impulsive reaction of the plane is $2\sqrt{3}mv/7$ and find the speed after impact of the end striking the plane.   [L.U.]

12. A uniform cube of mass M and length of edge a rests on a horizontal table with one edge freely hinged to the table. The upper opposite edge receives a horizontal blow I perpendicular to the edge and at its mid point. Show that if the cube is to topple over

$$I^2 \geqslant \tfrac{2}{3}M^2ag(\sqrt{2}-1)$$

13. A uniform solid cube of mass M and edge of length $2a$ rests on one face on a smooth horizontal table. It is given a horizontal impulse I at the mid point of one edge of its top face and perpendicular to that edge. Show that the impulsive reaction at the table is $3I/5$ and show also that the cube will overturn in the subsequent motion if $I^2 > 10M^2ga(\sqrt{2}-1)/3$.   [L.U.]

14. A uniform rod of length $2a$ and mass m is hinged at one end and lies on a horizontal table. It is struck a blow P at the other end perpendicular to its length. Find the initial impulse at the hinge and the initial angular velocity. If the coefficient of friction between the rod and table is μ, through what angle does the rod rotate before stopping.

15. A cubical block of edge $2a$ stands on a horizontal plane rough enough to prevent sliding. If the plane is suddenly given a horizontal velocity v parallel to two vertical faces of the block, prove it will upset if $v^2 > \tfrac{16}{3}ag(\sqrt{2}-1)$.
                    [L.U.]

16. A uniform inelastic sphere of radius a rolling without slipping along a horizontal plane with constant velocity v comes in contact with a step of height $\tfrac{2}{5}a$ perpendicular to its plane of motion. Assuming that the step is sufficiently rough to prevent slipping, prove that the sphere will surmount the step if $v^2 > \tfrac{420}{121}ag$.   [L.U.]

17. A uniform heavy circular cylinder is rolling along a horizontal plane with speed V when it meets a plane inclined to the horizontal at an angle $\alpha = \cos^{-1}(\tfrac{3}{4})$, the line of intersection of the two planes being parallel to the axis of the cylinder. Assuming the impact to be inelastic and no slipping to occur, find with what speed the cylinder will begin to roll up the plane and show that the magnitude of the impulse on the cylinder is $2MV/3$, where M is the mass of the cylinder.
                    [L.U.]

18. A uniform straight rod of mass m and length $2l$ standing upright on a table is slightly disturbed and allowed to fall, no slipping occurring during the motion. When it reaches an inclination of 60° to the vertical it strikes a small fixed peg at a distance $\tfrac{1}{2}l$ from the lower end, and begins to turn about it. Show that the angular velocity immediately after impact is $\dfrac{5}{7}\sqrt{\dfrac{3g}{l}}$, and find the impulse on the peg.   [L.U.]

19. A uniform rod AB of mass M and length $2a$ is free to move on a smooth horizontal table about a pivot at A. Initially the rod is at rest and a particle of mass m is attached by a light string to the end B, and is at rest at B. If the particle is projected with a velocity V along the table at right angles to AB, show that the angular velocity with which the rod begins to move is

$$3mV/2a(M+3m)$$

and find the impulse on the pivot and the impulsive tension in the string.
                    [L.U.]

20. A uniform rod of mass M and length $2a$ moving parallel to itself with

velocity v strikes a stationary particle of mass m which adheres to the rod, at a distance x from the centre. Show that the magnitude of the impulse is

$$Mma^2v/\{Ma^2 + m(a^2 + 3x^2)\},$$

and find the loss of kinetic energy. [L.U.]

21. Two equal uniform straight rods AB, BC, each of mass m and length a, are jointed together at B. When the rods rest on a smooth horizontal table with AB, BC in line, C is jerked into motion with a velocity v perpendicular to the lengths of the rods. If ABC remains a straight line after motion begins, prove that friction at the joint B must supply an impulsive couple of magnitude $\frac{1}{8}mva$. [L.U.]

22. A uniform solid sphere of mass M and radius a resting on a table is given a horizontal blow J in a vertical plane, containing the centre, at a height $3a/4$ above the table. Calculate the linear velocity of the centre and the angular velocity immediately after the blow and also the amount of kinetic energy which it generates. [L.U.]

23. A uniform rectangular block of length $2b$ and square section of side $2a$ stands on one of its square ends on a smooth horizontal floor. It receives a horizontal blow J at a height $b/2$ above the floor normal to and in the centre line of one face. Determine the initial motion of the block and show that if

$$J^2 > 8m^2g(4a^2 + b^2)\{\sqrt{(a^2 + b^2)} - b\}/3b^2$$

the block will topple over. [L.U.]

24. Equal uniform bars PQ, QR each of mass m, freely jointed at Q lie at rest on a smooth horizontal table inclined to each other at an obtuse angle $\pi - \alpha$. A horizontal blow of impulse I is applied at P in the direction perpendicular to PQ. Show that the velocity given to P has a component along PQ of magnitude

$$6I \sin \alpha \cos \alpha/m(16 + 9 \sin^2 \alpha).$$ [L.U.]

25. A small insect moves along a uniform bar of mass equal to itself and of length $2a$, the ends of which are constrained to remain on the circumference of a fixed circle of radius $2a/\sqrt{3}$. If the insect starts from the middle point of the bar and moves along the bar with relative velocity V, show that the bar in time t will turn through an angle

$$\frac{1}{\sqrt{3}} \tan^{-1}\left(\frac{Vt}{a}\right).$$

26. A horizontal turntable of mass M in the form of a uniform circular disc of radius a can rotate freely about a fixed vertical axis through the centre O. A man of mass m stands at P a point on the circumference of the disc, both being at rest. The man now walks on the disc and describes relative to the disc a circle on OP as diameter. Show that when the man has returned to P, the disc has turned through an angle

$$\pi\left[1 - \sqrt{\frac{M}{M + 2m}}\right].$$ [L.U.]

27. A uniform circular disc of mass M and radius a, can rotate freely with its plane horizontal about its centre which is fixed. A groove is cut in the disc along a radius and an insect of mass $\frac{1}{4}M$ is at rest at the end of the groove on the circumference of the disc. If the disc is set rotating with angular velocity Ω and at the same instant the insect begins to move along the groove with uniform velocity V relative to the disc, show that when the insect reaches the centre the disc has turned through an angle $\pi a\Omega/2V$. [L.U.]

28. A stationary uniform circular disc of radius a can rotate freely in a horizontal plane about its centre O, which is fixed. An insect of the same mass as the disc is also at rest on its circumference, and begins to walk on the disc, describing relative to it a circle on OP as diameter. Show that when the insect has described the first quadrant of this circle, the disc has turned through an angle

$$\pi(9 - 2\sqrt{3})/36.$$ [L.U.]

29. A taut wire in tension T when fastened to two points $2l$ apart on a smooth horizontal table has a mass m attached to its mid point. This mass is pulled aside perpendicular to the line of the wire through a small distance x and released. If the tension is constant at T obtain the equation

$$m\ddot{x} = -2Tx/\sqrt{(l^2 + x^2)}$$
$$= -2Tx/l \qquad (x \text{ small}).$$

Hence show that the periodic time of an oscillation is

$$2\pi \sqrt{\left(\frac{ml}{2T}\right)} \text{ seconds.}$$

30. A uniform circular disc of mass M and radius a can revolve in a vertical plane about a fixed horizontal axis at its centre. A particle of mass m is fastened to its circumference. Assume a displacement θ from the equilibrium position and : (i) use the Torque equation; (ii) use the energy equation to show that the periodic time of a small oscillation is

$$2\pi \sqrt{\frac{a(M + 2m)}{2mg}}$$

31. An endless light elastic string of unstretched length $2a$ passes round two small smooth pegs a distance a apart in a horizontal line. A heavy particle is attached to the string which in equilibrium forms an equilateral triangle. Show that if the particle is displaced vertically through a small distance and then released it will oscillate with a period

$$2\pi \left(\frac{2\sqrt{3}a}{7g}\right)^{\frac{1}{2}} \qquad\qquad \text{[L.U.]}$$

32. A particle of mass m is attached to the mid point of an unstretched elastic string of natural length a and modulus mg. The string with the mass attached is then stretched between two points in the same vertical line distant $2a$ apart. Show that in equilibrium the particle is $a/4$ below the centre point of the two fixed points.

If the particle is now slightly displaced from its equilibrium position in either a vertical or a horizontal direction show that in each case the ensuing motion is simple harmonic but that the period of a horizontal oscillation is $(15/7)^{\frac{1}{2}}$ times the period of a vertical oscillation. [L.U.]

33. A uniform rod of mass m and length l turns freely in a vertical plane about a fixed axis through one end O. A light elastic string of modulus mg and natural length $\frac{1}{4}l$ has one end attached to the mid point of the rod and the other to a fixed point A at distance $\frac{1}{2}l$ vertically above O. If θ is the angle made by the rod with the upward vertical OA, show that the equation of motion of the rod is

$$2l\ddot{\theta} = 3g(1 - 2\sin\tfrac{1}{2}\theta)\cos\tfrac{1}{2}\theta.$$

Show also that the period of small oscillations about the position of equilibrium $\theta = \frac{1}{3}\pi$ is the same as that of a simple pendulum of length $8l/9$. [L.U.]

34. A particle of mass m is clamped to the rim of a disc of mass M and radius a. The disc can roll on a rough horizontal rail with its plane vertical and initially the particle is vertically below the centre of the disc. Show by considerations of energy or otherwise that

$$\{3M + 4m(1 - \cos\theta)\}a\dot{\theta}^2 - 4mg\cos\theta$$

is constant where θ is the angle through which the disc has rolled in time t.

If the system is slightly disturbed from the position of stable equilibrium show that

$$T = 2\pi \left(\frac{3Ma}{2mg}\right)^{\frac{1}{2}}$$

for the periodic time of oscillation. [L.U.]

35. A uniform rod AB, of mass m and length $2a$, can turn freely about a horizontal axis at A. An elastic string of natural length a and modulus λ is attached to B and to a point C vertically over A, where $AC = 2a$. If $3\lambda > mg$ prove that there is a stable position of equilibrium given by

$$\sin\left(\frac{\alpha}{2}\right) = \frac{\lambda}{4\lambda - mg}$$

where α is the angle the rod makes with the upward vertical.

If the rod is slightly displaced from this position show that the period of a small oscillation is

$$4\pi\left\{\frac{(4\lambda - mg)am}{3(5\lambda - mg)(3\lambda - mg)}\right\}^{\frac{1}{2}}$$ [L.U.]

36. A wheel of mass M and centre O consists of a uniform circular disc of radius a in which have been cut four circular holes. The radius of each hole is $a/4$ and their centres, which are at a distance of $a/2$ from O, form the vertices of a square. Show that the moment of inertia of the wheel about an axis through O normal to the plane of the wheel is $55Ma^2/96$.

The wheel is free to rotate about this axis. A light elastic string of modulus $2Mg$ and natural length a joins P, a point on the circumference of the wheel to a fixed point A in the plane of the wheel. If $AO = 3a$, show that the length of the equivalent simple pendulum for small oscillations of the wheel about its stable position of equilibrium is $55a/288$. [L.U.]

37. A uniform solid sphere of radius a can roll in a vertical plane on the inside of a fixed hollow sphere of radius $4a$, the friction being sufficient to prevent sliding. Show that when the line of centres makes an angle θ with the downward vertical the angular velocity of the solid sphere is $3\dot{\theta}$.

Show that as long as the spheres remain in contact $21a\ddot{\theta} + 5g\sin\theta = 0$ and determine the period of small oscillations of the solid sphere about its position of equilibrium. [L.U.]

38. A uniform hemisphere of radius a lies with its curved surface on a horizontal plane. Prove that when slightly disturbed it will oscillate in the same period as a simple pendulum of length $26a/15$ if the plane is rough and $83a/120$ if it is smooth. [L.U.]

39. Each of two perfectly rough wires is bent to a circle of radius $10a$ and they are fixed with their planes perpendicular to the line joining their centres which is horizontal and the distance between the centres is $3a$. A homogeneous solid sphere of diameter $5a$ is placed on the wires and slightly disturbed from the position of stable equilibrium. Prove that the length of the equivalent simple pendulum is $13a$. [L.U.]

CHAPTER 18

THE METHOD OF LEAST SQUARES AND CORRELATION

" The object of statistical studies is to facilitate the discovery and expression of relationships between different groups of characters. . . . Is there a correlation between the hooked or aquiline nose and Jewish descent? The discovery that in fair samples of Jews only fourteen per cent. have the ' characteristic Jewish nose ' is an unambiguous answer."

Cohen and Nagel, *Logic and Scientific Method.*

18.1. The Method of Least Squares

The student will have met with the following type of problem : an experiment is performed, and the series of corresponding values given below obtained for P the effort and W the load lifted by P. It is required to find a relation between P and W known to be of the form $P = aW + b$.

W	.	.	14	42	84	112	newtons
P	.	.	5·1	13·3	26·0	35·3	newtons

The usual method is to plot the pairs of values on graph paper and, using one's judgment, draw a line lying evenly between the points. The equation to this line is then found, and represents the relation between the variables P and W. It will be found that when one of the values of W above is substituted in this equation the corresponding value of P is not obtained. It is assumed that the value of P obtained from the equation is the correct value, whilst that obtained by experiment is somewhat in error, due to all the usual causes that affect experimental results.

This method has a number of disadvantages. The line drawn between the points depends for its position entirely upon the judgment of the student, so that a class of thirty students will obtain thirty somewhat different equations between P and W. Also the method can only be applied to relations that are of, or can be changed into, the straight-line form.

A method has been invented which overcomes both these disadvantages. It provides a unique equation to the line—we all get the same answer—and it can be applied to a large variety of relations.

Let (x_r, y_r) be a pair of corresponding values in which x_r is correct (*e.g.*, we have put a 10 kg mass on the machine, and therefore know this exactly), but y_r is the reading subject to experimental errors. If the exact relation between each x and its y is

$$y = ax + b$$

then when we substitute x_r we get the correct corresponding value for y

$$y = ax_r + b \quad . \quad . \quad . \quad . \quad . \quad . \quad (1)$$

307

The experimental value is y_r, and

$$y_r - y \quad \text{or} \quad y_r - ax_r - b . \quad . \quad . \quad . \quad . \quad (2)$$

is the " error " in the experimental value. We could endeavour to make the sum of these errors zero as a method of finding a and b. For n pairs of values this gives

$$\Sigma(y_r - ax_r - b) = 0$$

or $\qquad\qquad \Sigma y_r - a\Sigma x_r - nb = 0 \quad . \quad . \quad . \quad . \quad (3)$

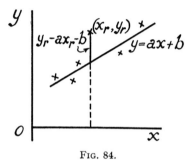

FIG. 84.

Division by n gives

$$\bar{y} - a\bar{x} - b = 0 \quad . \quad . \quad . \quad . \quad . \quad (4)$$

where $(\bar{x}, \bar{y})$ is the mean position of the points on the graph paper. This shows that the line passes through this mean position [since $(\bar{x}, \bar{y})$ satisfies the equation], but does not help further in finding a and b. *The Method of Least Squares* adopts the criterion that the sum of the squares of the errors given by (2) should be a minimum, *i.e.*,

$$\Sigma(y_r - ax_r - b)^2$$

is to be a minimum. For functions of one variable the derivative with respect to the variable equated to zero gives the equation from which the required value of the variable is obtained. With functions of more than one variable it can be shown (Chapter 5) that the partial derivative with respect to each variable in turn, equated to zero, will provide a system of simultaneous equations from which the unknowns may be found. Applying this we have for the partial derivative with respect to a equated to zero

$$\Sigma 2(y_r - ax_r - b)(-x_r) = 0$$

or $\qquad\qquad \Sigma x_r y_r - a\Sigma x_r^2 - b\Sigma x_r = 0 \quad . \quad . \quad . \quad . \quad (5)$

Similarly, for the variable b

$$\Sigma 2(y_r - ax_r - b)(-1) = 0$$

or $\qquad\qquad \Sigma y_r - a\Sigma x_r - nb = 0 \quad . \quad . \quad . \quad . \quad (6)$

The student will notice that this equation (6) is the same as (3), so that a line obtained by the method of least squares also passes through the mean position of the points.

The solution of (5) and (6) will give a and b the unknown values required for the equation of the line $y = ax + b$.

We will apply it to the above example, where $W(= x)$ is correctly known, but $P(= y)$ is subject to error.   We now have

x_r	y_r	$x_r y_r$	x_r^2
14	5·1	71·4	196
42	13·3	558·6	1764
84	26·0	2184·0	7056
112	35·3	3953·6	12 544
252	79·7	6767·6	21 560

Substitution of these values in the equations (5) and (6) gives
$$21560a + 252b - 6767·6 = 0$$
$$252a + 4b - 79·7 = 0$$

Solving these :
$$a = 0·307, \qquad b = 0·567.$$

The equation is therefore
$$y = 0·307x + 0·567,$$
whilst the usual graphical solution gave
$$y = 0·3x + 1·1.$$

Example 1.   It is known that the pressure p and the volume v of a gas are related by the equation $pv^\gamma = C$.   From the following data find a value for γ.

p	.	.	.	0·5	1·0	1·5	2·0	2·5	3·0
v	.	.	.	1·62	1·00	0·75	0·62	0·52	0·46

If the equation was unknown we could plot the pairs of values on graph paper and discover that the (p, v) curve was not a straight line, but that $\log p$ plotted with $\log v$ did appear to give a line. [Note that the method of least squares applied to the (p, v) data would give a line —the best line fitting the data— which is a disadvantage of the method.]   With $x = \log p$, $y = \log v$ we have the table of values :

(log p) x_r.	(log v) y_r.	$x_r y_r$.	x_r^2.
−0·3010	0·2095	−0·0631	0·0906
0	0	0	0
0·1761	−0·1249	−0·0220	0·0310
0·3010	−0·2076	−0·0625	0·0906
0·3979	−0·2840	−0·1130	0·1583
0·4771	−0·3372	−0·1609	0·2276
1·0511	−0·7442	−0·4215	0·5981

The equations for a and b in $y = ax + b$ are :
$$0·5981a + 1·0511b + 0·4215 = 0,$$
$$1·0511a + 6b + 0·7442 = 0,$$
from which
$$a = -0·7033, \qquad b = -0·0008.$$

We now have

$$y = - 0.7033x - 0.0008,$$

or

$$\log v = - 0.7033 \log p - 0.0008,$$

$$\log p + 1.42 \log v = - 0.0011,$$

$$\log (pv^{1.42}) = - 0.0011$$

$$pv^{1.42} = 10^{-0.0011} = 0.9971.$$

The value of γ is therefore 1.42.

18.2. A simple example which illustrates the application of the method to measurements in general is as follows.

Example 2. A, B, C are three points in order on a line. AB is measured as 2 mm, BC as 3 mm but AC as 5·1 mm. To find the best value of AB using all the data available, *i.e.*, the " corrected " value of AB.

Let x, y be the respective corrected values of AB, BC, then $x - 2$, $y - 3$, $x + y - 5.1$ are the respective errors in AB, BC and AC. To make the sum of the squares a minimum we need the values of $\frac{\partial S}{\partial x} = 0$, $\frac{\partial S}{\partial y} = 0$ from

$$S = (x - 2)^2 + (y - 3)^2 + (x + y - 5.1)^2.$$

These are:

$$4x + 2y - 14.2 = 0,$$

$$2x + 4y - 16.2 = 0$$

giving

$$x = 2.033, \qquad y = 3.033.$$

The best value we can use for AB is therefore 2·033 mm.

18.3. When Both Variables are Subject to Error

So far we have dealt with the case where it was clear that one of the variables (x_r, y_r) could be regarded as free from error, and this was chosen as the x co-ordinate. When both variables are subject to error we apply the method of least squares by making a minimum of the sum of the squares of the perpendicular lengths of the points (x_r, y_r) from the assumed line.

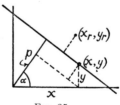

Fig. 85.

The figure shows that if the perpendicular from the origin on to the line is of length p and makes angle α with the positive x axis, the equation to the line may be written

$$x \cos \alpha + y \sin \alpha = p \quad . \quad . \quad . \quad . \quad (7)$$

Since $\sqrt{(\cos^2 \alpha + \sin^2 \alpha)} = 1$, the length of the perpendicular from (x_r, y_r) to the line is (Vol. I, p. 57)

$$\pm (x_r \cos \alpha + y_r \sin \alpha - p).$$

Using α and p as the two variables that determine the line, we obtain their values by finding the conditions uat S should be a minimum where

$$S = \Sigma(x_r \cos \alpha + y_r \sin \alpha - p)^2.$$

We have then

$$\frac{\partial S}{\partial p} = \Sigma 2(x_r \cos \alpha + y_r \sin \alpha - p)(-1) = 0$$

or

$$\Sigma(x_r \cos \alpha + y_r \sin \alpha - p) = 0.$$

For n pairs of values this is

$$\cos \alpha \, \Sigma x_r + \sin \alpha \Sigma y_r - np = 0 \quad \cdot \quad \cdot \quad \cdot \quad (8)$$

which may be written

$$\cos \alpha \, \frac{\Sigma x_r}{n} + \sin \alpha \, \frac{\Sigma y_r}{n} - p = 0,$$

showing that the mean position $(\bar{x}, \bar{y})$ of the points (x_r, y_r) lies on (7). For convenience we will transfer to this point $(\bar{x}, \bar{y})$ as origin by writing $x_r = \bar{x} + X_r, y_r = \bar{y} + Y_r$. Equation (8) now becomes

$$\cos \alpha(\Sigma X_r + n\bar{x}) + \sin \alpha(\Sigma Y_r + n\bar{y}) - np = 0.$$

But $\Sigma X_r = 0 = \Sigma Y_r$, since each is the sum of the deviations from its mean. The equation is now

$$\bar{x} \cos \alpha + \bar{y} \sin \alpha - p = 0 \quad \cdot \quad \cdot \quad \cdot \quad (9)$$

The partial derivative with respect to α gives

$$\frac{\partial S}{\partial \alpha} = \Sigma 2(x_r \cos \alpha + y_r \sin \alpha - p)(-x_r \sin \alpha + y_r \cos \alpha) = 0$$

or with the same transfer as above

$$\Sigma(X_r \cos \alpha + Y_r \sin \alpha)(X_r \sin \alpha - Y_r \cos \alpha + \bar{x} \sin \alpha - \bar{y} \cos \alpha) = 0.$$

This simplifies to

$$(\cos^2 \alpha - \sin^2 \alpha)\Sigma X_r Y_r = \sin \alpha \cos \alpha \Sigma(X_r^2 - Y_r^2)$$

or

$$\tan 2\alpha = \frac{2\Sigma X_r Y_r}{\Sigma(X_r^2 - Y_r^2)} \quad \cdot \quad \cdot \quad \cdot \quad (10)$$

We assumed for the equation to the line

$$x \cos \alpha + y \sin \alpha - p = 0,$$

and found (9)

$$\bar{x} \cos \alpha + \bar{y} \sin \alpha - p = 0.$$

By subtraction the line is

$$(x - \bar{x}) \cos \alpha + (y - \bar{y}) \sin \alpha = 0 \quad \cdot \quad \cdot \quad \cdot \quad (11)$$

where α is determined by (10).

The two values of α obtained from (10) differ by 90°, so that there are two possible lines, both passing through the mean position. A rough graph of the points will indicate which value is to be chosen for the best possible line, the line perpendicular to it being the worst possible line in the sense that the sum of squares of the deviations is the maximum for all lines passing through $(\bar{x}, \bar{y})$.

Example 3. In an experiment on the discharge of water over a weir the following results were obtained.

Quantity (Q)	.	.	.	2·34	4·23	6·38	9·12
Head (H)	.	.	.	4·80	7·21	9·61	16·20

If both values are subject to experimental errors, find a law of the form $Q = AH^n$ giving the best values for A and n.

The linear form of the law is

$$\log Q = \log A + n \log H$$

or

$$y = c + nx.$$

In tabular form the required calculations are:

					Sum.
$\log Q$ (y_r).	0·3692	0·6263	0·8048	0·9600	2·7603
$Y_r = y_r - \bar{y}$	−0·3209	−0·0638	0·1147	0·2699	
Y_r^2	0·1030	0·0041	0·0132	0·0728	0·1931
$\log H$ (x_r)	0·6812	0·8579	0·9827	1·2095	3·7313
$X_r = x_r - \bar{x}$	−0·2516	−0·0749	0·0499	0·2767	
X_r^2	0·0633	0·0056	0·0025	0·0766	0·1480
$X_r Y_r$	0·0807	0·0048	0·0057	0·0747	0·1659

$$\bar{y} = \frac{2\cdot7603}{4} = 0\cdot6901, \qquad \bar{x} = \frac{3\cdot7313}{4} = 0\cdot9328.$$

From equation (10) above

$$\tan 2\alpha = \frac{2(0\cdot1659)}{0\cdot1480 - 0\cdot1931} = -7\cdot3570$$

$$2\alpha = -82\cdot26° \quad \text{or} \quad 180° - 82\cdot26°$$

$$\alpha = -41\cdot13° \quad \text{or} \quad 48\cdot87°.$$

The equation to the line is

$$(x - 0\cdot9328) \cos \alpha + (y - 0\cdot6901) \sin \alpha = 0$$

or

$$y = -\cot \alpha(x - 0\cdot9328) + 0\cdot6901 \quad . \quad . \quad . \quad . \quad (12)$$

A rough graph of the points (x_r, y_r) shows that the line will make an acute angle with the positive x axis so that in equation (12) the coefficient of x must be positive. Choosing then the value $\alpha = -41\cdot13°$, the equation to the line is

$$y = \cot (41\cdot13)(x - 0\cdot9328) + 0\cdot6901$$

$$= 1\cdot1451x - 0\cdot3780$$

This is

$$\log Q = 1\cdot145 \log H - \log (2\cdot388)$$

or

$$\underline{Q = 0\cdot419 \; H^{1\cdot145}}$$

The student about to start on these exercises which require a lot of arithmetical work should remember Kepler. Despite a lifetime of trouble, including, as a minor incident, the defence of his mother against burning as a witch, he undertook monumental arithmetical calculations to verify his hypotheses, finally arriving at the laws which perpetuate his name. About the year 1600, towards the end of his life, he wrote (in Latin):

> Precious time and labour and expense wasted
> Oh sorrow, how much mounting work in vain.

But he carried on !

EXERCISE 47

1. Suppose x_1, x_2, . . . x_n are n values of a measurement. Assume that A is the correct value, and by applying the method of least squares to

$$\Sigma(x_r - A)^2$$

show that the arithmetic mean of the n values is the best value we can use.

2. A, B, C are three points in order on a line. AB is measured as 3·20 mm, BC as 5·70 mm, and AC as 8·95 mm. In this case let x equal the excess of AB over 3·20 mm, y the excess of BC over 5·70 mm, so that $x + y$ equals the excess of AC over 8·90 mm. Therefore $x + y = 0.5$. We now find x, y from the condition that

$$S = x^2 + y^2 + (x + y - 0.5)^2$$

is a minimum, and hence obtain the corrected values of AB and BC.

3. OA, OB, OC are the arms of angles vertex at O and taken in order. Angle AOB is measured as 5° 10′ 13·2″, angle BOC as 17° 4′ 3·7″ and angle AOC as 22° 14′ 17·1″. Find the " corrected " value of angles AOB, BOC using the method of Question 2.

4. Find the best values of x and y from the series of equations: $x + y - 3 = 0$, $2x - y = 0$, $x + 4y - 10 = 0$, $3x + 2y - 8 = 0$. (Find the x and y that makes $S = \Sigma e^2$ a minimum, where e represents one of the equations.)

5. Calculate the mean of the following values of X and Y.

X .	.	.	0·25	1·00	2·25	4·00	6·25
Y .	.	.	0·12	0·90	2·13	3·84	6·07

The corresponding values of X, Y satisfy approximately the equation

$$Y = mX + c.$$

By the method of least squares obtain the best values for m and c, if the error is in the Y values only. [L.U.]

6. To find the resistance r of a wire it was connected in a Wheatstone Bridge circuit and the following series of equations obtained where R is the resistance of the rest of the circuit:

$$0·1r + R = 0·224, \quad 0·2r + R = 0·430, \quad 0·3r + R = 0·615,$$
$$0·4r + R = 0·826, \quad 0·5r + R = 1·040, \quad 0·6r + R = 1·208,$$
$$0·7r + R = 1·412, \quad 0·8r + R = 1·603, \quad 0·9r + R = 1·798,$$
$$1·0r + R = 1·990.$$

Find the best value for r.

7. An experiment provides the pairs of values:

x .	.	.	12	15	21	26	33
y .	.	.	173·1	270·3	529·5	811·5	1307

show that a law of the form $y = ax^2 + b$ holds, and find the best values for a and b.

8. By plotting log y against log x show that x and y in the following series of values are connected by a law of the form $y = ax^n$. Find the best values for a and n.

x .	.	.	5·7	11·1	23·4	31·5	42·7
y .	.	.	7·71	23·95	85·06	141	236·9

9. A test on the conductive heating of mild-steel pipes at 32° C. gave the figures

Current i (amperes/unit area)	Power p (watts/unit area)
151	2·59
120·8	1·825
90·6	1·157
30·2	0·201

Obtain a law of the form $p = Ki^n$ finding the best values for K and n.

10. An experiment on the life of a cutting tool at different cutting speeds gave the values:

Speed, v m/min .	.	35	40	50	60
Life, T min	. .	61	26	7	2·9

Show that log T plotted against log v gives a straight-line law, and find it in the form $v = AT^n$.

11. The following table gives the observed values of y corresponding to the given values of x:

x .	.	0·0	0·2	0·4	0·6	0·8	1·0	1·2
y .	.	−1·85	−1·20	−0·55	0·15	0·80	1·35	2·00

Find by the method of least squares a relation $y = ax + b$ connecting x and y.
[L.U.]

12. Use the method of least squares to find the best solution of the four equations

$$2u - v = 3, \quad u + 3v = 702, \quad u + v = 301, \quad 3u + v = 497,$$

giving your answer to the nearest whole number. [L.U.]

13. It is expected that the following observed readings satisfy a relation of the form $y = \lambda x^2 + \mu$:

x .	.	.	1	2	3	4	5
y .	.	.	20·9	24·1	28·9	36·1	44·7

Calculate λ and μ by the method of least squares, giving values correct to one place of decimals. [L.U.]

14. To fit a parabola $y = a + bx + cx^2$ to data by the method of least squares we find the values of a, b, c that will make a minimum of

$$\Sigma(y - a - bx_r - cx_r^2)^2.$$

Show that the " normal " equations are

$$\Sigma(y_r - a - bx_r - cx_r^2) = 0$$
$$\Sigma x_r(y_r - a - bx_r - cx_r^2) = 0$$
$$\Sigma x_r^2(y_r - a - bx_r - cx_r^2) = 0.$$

Fit the above parabolic equation to the data:

x .	.	.	0·5	1·0	2·0	3·0	5·0
y .	.	.	3·1	6·0	11·2	14·8	20

15. The yield of a chemical process was measured at three temperatures, each with two concentrations of a particular reactant, as recorded below:

Temperature $t°$ C..	.	40	40	50	50	60	60
Concentration, x .	.	0·2	0·4	0·2	0·4	0·2	0·4
Yield, y .	.	38	42	41	46	46	49

Use the method of least squares to find the best values of the coefficients a, b, c in the equation

$$y = a + bt + cx,$$

and from your equation estimate the yield at 70° C with concentration 0·5.
[L.U.]

In the following exercises Nos. 16–19 it should be assumed that both variables are subject to error.

16. Find a law of the form $y = ax + b$ for

x .	.	.	2·1	3·4	5·6	7·9	8·3
y .	.	.	10·3	12·9	17·3	21·9	22·7

17. Find a law of the form $x^n y = A$ for

x .	.	.	37·36	31·34	26·43	14·04
y .	.	.	10·16	12·26	14·70	28·83

18. The following pairs of values were found for the potential difference V volts and the current A amperes in an electric arc:

V	.	.	.	50·3	47·9	46·8	45·1	43·6
A	.	.	.	1·96	2·98	3·96	5·96	7·97

Find a law of the form $V = a + \dfrac{b}{A}$.

19. If v is the velocity in m/s and t the rise in temperature in degrees Celsius in a split bearing using gas engine oil, an experiment gave the values

v	.	.	.	3·93	6·52	9·44	12·85
t	.	.	.	26·4	39·3	52·5	63·2

Find a law of the form $t = Cv^n$.

20. A set of helicoidal propellers of different pitch ratios (p/D) tested in a wind tunnel at constant thrust for different torque ratios (J) gave the following results:

p/D	0·3	0·5	0·7	1·0	1·25	1·5	1·8	2·2	2·5
J	0·3286	0·5464	0·7581	1·0770	1·3500	1·6200	1·9659	2·4358	2·8040

Find by least squares the best line fitting this data if J only is subject to error.

18.4. Correlation

We have by this time become aware of the extent to which Science is concerned with discovering the relationship between characteristics. We are concerned, for example, with investigating the extent to which being good at Mathematics is associated with being good at say Physics, or the extent to which the abrasion loss of a material and its hardness are associated. To find the hardness of a material is simple, but to find its abrasion loss is relatively difficult, so that a scientist requiring the abrasion loss of a given material and possessing a mathematical equation between them would find simply its hardness and obtain the abrasion loss to a sufficient degree of accuracy by substituting this value in the given equation.

We will consider only first degree, i.e., linear relationships represented by an equation $y = ax + b$; many others can be put into this form. Consider the relationship between ability at Mathematics and at Physics as shown by the marks in each subject obtained by a large number, n, of students.

If being good at Mathematics, i.e., marks above the average in Mathematics, is associated with being good at Physics, i.e., marks above the average in Physics, and marks below the average in Mathematics is associated with marks below the average in Physics, then a graph of the corresponding pairs of marks (x_r, y_r) will give a diagram such as (a) with all the points in quadrants (1) and (3) where the two lines drawn across the graph are drawn through the mean value of the marks $\bar{x}$ for Mathematics and $\bar{y}$ for Physics. For any point in the quadrant (1) $(x_r - \bar{x})$ and $(y_r - \bar{y})$ are both positive, whilst both are negative for points in (3). We thus find that

$$\sum_{r=1}^{n} (x_r - \bar{x})(y_r - \bar{y})$$

316

summed over all the pairs of values will consist of a sum of all positive values, and so be relatively large.

If being good at Mathematics is associated with being poor at Physics, *i.e.*, marks above the average at Mathematics is associated

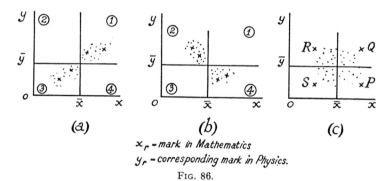

x_r = mark in Mathematics
y_r = corresponding mark in Physics.

FIG. 86.

with marks below the average in Physics, and vice versa, the graph will be shown by figure (*b*), with the points tending to be concentrated in quadrants (2) and (4). In this case

$$\Sigma(x_r - \bar{x})(y_r - \bar{y})$$

will consist of terms all negative, and so the sum will be large and negative.

Finally, suppose there is no connection between ability at Mathematics and at Physics. Then, in a sufficiently large group of students for every student whose marks are P [figure (*c*)] there will be others at Q, R and S, where Q, R, S, P are symmetrically placed about $(\bar{x}, \bar{y})$. In this case

$$\Sigma(x_r - \bar{x})(y_r - \bar{y})$$

will be exactly zero. In most cases it will consist of positive and negative terms, and the sum will be relatively small.

We therefore find that the *statistic*

$$\sum_{r=1}^{n} (x_r - \bar{x})(y_r - \bar{y}) \quad . \quad . \quad . \quad . \quad (13)$$

is a good measure of the degree of relation between the two variables, but for comparison purposes it must be standardised.

(*a*) In one case we may have 10 000 pairs of values, and in another only 100. We must then divide (13) by n the number of pairs of values over which it is summed.

(*b*) We might have a case of 2 pairs of numbers where the x's were 0 and 100, giving an average $\bar{x}$ of 50 and another case where the x's were 49 and 51, again giving an average $\bar{x}$ of 50. In the first case the statistic (13) will be much larger owing entirely to the larger dispersion or spread of the x's about the mean. This will affect the standard deviation of the x's equally, and so can be corrected for by dividing by σ_x the standard deviation of the x's.

(c) The same holds for the spread of the y's about their mean, and we must therefore divide by σ_y, the S.D. of the y's, to allow for this.

The standardised formula now is

$$r = \frac{\Sigma(x - \bar{x})(y - \bar{y})}{n\sigma_x\sigma_y}$$

an expression denoted by r and called the coefficient of correlation.

For purposes of calculation it is often convenient to rearrange this. Since

$$\Sigma(x - \bar{x})(y - \bar{y}) = \Sigma xy - \bar{y}\Sigma x - \bar{x}\Sigma y + \Sigma\bar{x}\bar{y}$$
$$= \Sigma xy - \bar{y} \cdot n\bar{x} - \bar{x} \cdot n\bar{y} + n\bar{x}\bar{y}$$
$$= \Sigma xy - n\bar{x}\bar{y},$$

$$\sigma_x^2 = \frac{1}{n}\Sigma(x - \bar{x})^2 = \frac{1}{n}[\Sigma x^2 - 2\bar{x}\Sigma x + \Sigma\bar{x}^2]$$

$$= \frac{1}{n}[\Sigma x^2 - n\bar{x}^2],$$

$$\sigma_y^2 = \frac{1}{n}\Sigma(y - \bar{y})^2 = \frac{1}{n}[\Sigma y^2 - n\bar{y}^2],$$

therefore another form for r is

$$r = \frac{\Sigma xy - n\bar{x}\bar{y}}{\sqrt{(\Sigma x^2 - n\bar{x}^2)(\Sigma y^2 - n\bar{y}^2)}}.$$

As an example consider the pairs of values obtained by the relation $y = 3x$. In such a case there is clearly perfect correlation with x and y decreasing or increasing together. If we take, say, the seven pairs of values :

										Sum
x .	.	.	1	2	3	4	5	6	7	28
y .	.	.	3	6	9	12	15	18	21	84
xy	.	.	3	12	27	48	75	108	147	420

we find : $\bar{x} = 4$, $\bar{y} = 12$

$$\Sigma xy = 420 \qquad \Sigma x^2 = 140 \qquad \Sigma y^2 = 1260$$

therefore
$$r = \frac{420 - 7 \times 4 \times 12}{\sqrt{(140 - 7 \times 16)(1260 - 7 \times 144)}}$$

$$= \frac{84}{\sqrt{(28 \times 252)}} = 1.$$

If $y = -3x$ so that again there is perfect correlation, but y decreases as x increases, it is clear from the above that r would be -1. We have therefore a coefficient which fulfils, so far, our requirements, by giving the numerical value of 1 for perfect correlation, and showing by its sign whether the variables increase and decrease together or vice versa.

18.5. r as a Measure of Common Causal Factors

The following example illustrates another aspect of the meaning of r, and provides a useful interpretation of it. Suppose we have arranged three sets, each of n numbers,

$$x_1, x_2, x_3, \ldots x_n,$$
$$y_1, y_2, y_3, \ldots y_n,$$
$$z_1, z_2, z_3, \ldots z_n,$$

such that, for convenience, all have the same average 0 and the same standard deviation σ, i.e.,

$$\Sigma x_r = \Sigma y_r = \Sigma z_r = 0$$

and

$$n\sigma^2 = \Sigma x_r{}^2 = \Sigma y_r{}^2 = \Sigma z_r{}^2.$$

Suppose also that there is no correlation between the three sets of numbers, so that, from the formula for r,

$$\Sigma x_r y_r = \Sigma x_r z_r = \Sigma y_r z_r = 0.$$

With this data we will form two new series of numbers, a series A_r, such that

$$A_r = x_r + y_r + z_r \qquad (r = 1, 2 \ldots n)$$

and a series

$$B_r = y_r + z_r \qquad (r = 1, 2 \ldots n)$$

so that in a sense we can say that the numbers A and B have two out of three separate causes in common.

For the coefficient of correlation between the A's and B's we have, since the average of each, being the average of its constituents, is zero,

$$r^2 = \frac{(\Sigma A_r B_r)^2}{\Sigma A_r{}^2 \Sigma B_r{}^2}$$

Now

$$\begin{aligned}
\Sigma A_r B_r &= \Sigma(x_r + y_r + z_r)(y_r + z_r) \\
&= \Sigma x_r y_r + \Sigma x_r z_r + \Sigma y_r{}^2 + \Sigma y_r z_r + \Sigma z_r y_r + \Sigma z_r{}^2 \\
&= 0 + 0 + n\sigma^2 + 0 + 0 + n\sigma^2 \\
&= 2n\sigma^2
\end{aligned}$$

Similarly,

$$\begin{aligned}
\Sigma A_r{}^2 &= \Sigma(x_r + y_r + z_r)^2 \\
&= 3n\sigma^2,
\end{aligned}$$

$$\Sigma B_r{}^2 = \Sigma(y_r + z_r)^2 = 2n\sigma^2.$$

Therefore

$$r^2 = \frac{(2n\sigma^2)^2}{3n\sigma^2 \times 2n\sigma^2} = \frac{2}{3},$$

so that the square of the correlation coefficient is a measure of the proportion of causes common to the two quantities A and B. This result can be generalised, and we would obtain $r^2 = n/N$ when A and B had n causes in common out of a total of N. We can now make the obvious deduction that $r = 0$ when there are no causal factors in common and $r = 1$ when both are determined by the same totality of causes.

18.6. Nonsense Correlation Coefficients

The amount of money (X_r) spent annually on cosmetics has been steadily increasing, and so has the amount of money (Y_r) spent annually on education in Nigeria. A correlation coefficient between these pairs of values (X_r, Y_r) would therefore give a value near to unity. It is clearly nonsense to assert that one type of expenditure causes the other, or that they have any causal factors in common. A correlation coefficient can never be used to establish a dependence, but only to measure a dependence known to exist. It can also be used to help choose between alternative relations each known to exist. Thus a student calculated the coefficient of correlation between the business profits each year and the number of profit-sharing schemes instituted in the same year and obtained a fairly high value for r. When, however, he calculated r for the business profits of each year and the number of profit-sharing schemes instituted the next year he obtained a much higher value, thus verifying that although a causal relation exists, it takes time—say a year—before a firm making profit can decide, institute a profit-sharing scheme and have its inception appear in the official records. In short, the profits of one year cause the profit-sharing scheme of the next year, and not of the same year.

18.7. Lines of Regression

In fitting a line by the method of least squares, we assumed that the x values were correct and only the y values subject to error. To find the unknowns a and b in $y = ax + b$ we obtained the equations :

$$a\Sigma x_r^2 + b\Sigma x_r - \Sigma x_r y_r = 0,$$
$$a\Sigma x_r + bn - \Sigma y_r = 0.$$

From these

$$a = \frac{n\Sigma x_r y_r - \Sigma x_r \Sigma y_r}{n\Sigma x_r^2 - (\Sigma x_r)^2}$$

$$= \frac{n\Sigma x_r y_r - (n\bar{x})(n\bar{y})}{n\Sigma x_r^2 - (n\bar{x})^2}$$

$$= \frac{\Sigma x_r y_r - n\bar{x}\bar{y}}{\Sigma x_r^2 - n\bar{x}^2}$$

(see 18.4)
$$= \frac{\Sigma(x_r - \bar{x})(y_r - \bar{y})}{n\sigma_x^2}.$$

If we denote by p the expression $\dfrac{1}{n}\Sigma(x_r - \bar{x})(y_r - \bar{y})$, which is the mean value of the product moment of the values about the means of x and y, then

$$a = \frac{p}{\sigma_x^2}.$$

Since the line $y = ax + b$ fitted by least squares passes through the mean $(\bar{x}, \bar{y})$, it may now be written

$$y - \bar{y} = \frac{p}{\sigma_x^2} (x - \bar{x}).$$

This gives the best y associated with a given x, and in this form the line is called the *line of regression* of y on x, and p/σ_x^2 is called the *coefficient of regression* of y on x.

In the same way we could regard y as correct and find a line giving the best x for any given y. This is clearly obtained by interchanging x and y in the above equation to obtain

$$x - \bar{x} = \frac{p}{\sigma_y^2} (y - \bar{y}).$$

This is called the *line of regression* of x on y and p/σ_y^2 the *coefficient of regression* of x on y. These two lines are in general quite distinct, only having the point $(\bar{x}, \bar{y})$ in common.

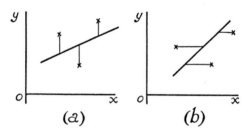

(a) Regression of y on x. Vertical deviations a minimum. $y - \bar{y} = r\frac{\sigma_y}{\sigma_x}(x - \bar{x})$

(b) Regression of x on y. Horizontal deviations a minimum. $x - \bar{x} = r\frac{\sigma_x}{\sigma_y}(y - \bar{y})$

Fig. 87.

The student should appreciate that if we had an exact linear relationship between the variables there would be only one line which could be written as $y = ax + b$ to give y from x or $x = y/a - b/a$ to give x from y, but here we have not such a relationship but only a correlation. Each line of regression is based on minimising the uncertainty estimate of one of the variables when the other is assumed to be given exactly.

It can be seen that

$$r = \frac{p}{\sigma_x \sigma_y},$$

where r is the coefficient of correlation already defined.

It could be shown from this basis that $r = 0$ for no relationship, otherwise always lies between ± 1, and attains one of these values for perfect relationship. Thus if there is a perfect relationship, the two lines of regression are the same line giving

$$\frac{p}{\sigma_x^2} = \frac{\sigma_y^2}{p}$$

or

$$\frac{p^2}{\sigma_x^2 \sigma_y^2} = 1$$

so that

$$r = \pm 1.$$

Example 4. In an experiment the following pairs of values were obtained.

| x . | . | . 58 | 86 | 148 | 166 | 188 | 202 | 210 |
| y . | . | . 0 | 4 | 18 | 29 | 51 | 73 | 90 |

Find the two lines of regression, the coefficient of correlation and the best possible line fitting the data, on the assumption that x and y are both subject to error.

In tabular form:

x_r	y_r	$x_r y_r$	x_r^2	y_r^2
58	0	0	3364	0
86	4	344	7396	16
148	18	2664	21,904	324
166	29	4814	27,556	841
188	51	9588	35,344	2601
202	73	14,746	40,804	5329
210	90	18,900	44,100	8100
1058	265	51 056	180 468	17 211

$$\bar{x} = \frac{1058}{7} = 151 \cdot 1 \qquad \bar{y} = \frac{265}{7} = 37 \cdot 86$$

$$p = \frac{1}{n}\Sigma(x_r - \bar{x})(y_r - \bar{y})$$

$$= \frac{1}{n}\Sigma x_r y_r - \bar{x}\bar{y}$$

$$= \tfrac{1}{7} \times 51\ 056 - 151 \cdot 1 \times 37 \cdot 86$$

$$= 7293 \cdot 7 - 5720 \cdot 6 = 1573 \cdot 1$$

$$\sigma_x^2 = \frac{1}{n}\Sigma x_r^2 - \bar{x}^2 = \tfrac{1}{7} \times 180\ 468 - (151 \cdot 1)^2$$

$$= 25\ 781 \cdot 1 - 22\ 831 \cdot 2 = 2949 \cdot 9$$

$$\sigma_y^2 = \frac{1}{n}\Sigma y_r^2 - \bar{y}^2 = \tfrac{1}{7} \times 17\ 211 - (37 \cdot 86)^2$$

$$= 2458 \cdot 7 - 1433 \cdot 4 = 1025 \cdot 3$$

The equation to the line of regression of y on x is

$$y - \bar{y} = \frac{p}{\sigma_x^2}(x - \bar{x})$$

which is

$$y - 37 \cdot 86 = \frac{1573 \cdot 1}{2949 \cdot 9}(x - 151 \cdot 1)$$

$$= 0 \cdot 5332(x - 151 \cdot 1)$$

$$\underline{y = 0 \cdot 5332x - 42 \cdot 7} \quad . \quad . \quad . \quad . \quad . \quad (a)$$

The line of regression of x on y is

$$x - \bar{x} = \frac{p}{\sigma_y^2}(y - \bar{y})$$

$$x - 151 \cdot 1 = \frac{1573 \cdot 1}{1025 \cdot 3}(y - 37 \cdot 86)$$

$$x = 1 \cdot 5342y + 93 \cdot 0$$

or

$$y = 0 \cdot 6518x - 60 \cdot 6 \qquad . \quad . \quad . \qquad . \quad (b)$$

The coefficient of correlation is given by

$$r^2 = \frac{p^2}{\sigma_x^2 \sigma_y^2} = 0 \cdot 5332 \times 1 \cdot 5342$$

therefore

$$r = 0 \cdot 90$$

The best possible line when both x and y are subject to error is by equation (11)

$$(x - \bar{x}) \cos \alpha + (y - \bar{y}) \sin \alpha = 0$$

where

$$\tan 2\alpha = \frac{2\Sigma(x_r - \bar{x})(y_r - \bar{y})}{\Sigma(x_r - \bar{x})^2 - \Sigma(y_r - \bar{y})^2} = \frac{2p}{\sigma_x^2 - \sigma_y^2}$$

$$= \frac{2 \times 1573 \cdot 1}{2949 \cdot 9 - 1025 \cdot 3} = \frac{3146 \cdot 2}{1924 \cdot 6} = 1 \cdot 6347$$

therefore

$$2\alpha = 58° \, 33' \quad \text{or} \quad 180 + 58° \, 33'$$

$$\alpha = 29° \, 17' \quad \text{or} \quad 90° + 29° \, 17'$$

A rough graph of the points (x_r, y_r) shows that the line has a positive gradient. We must then choose $\alpha = 90° + 29° \, 17'$, and the equation to the line is

$$y - 37 \cdot 86 = - \cot (119° \, 17')(x - 151 \cdot 1)$$

or

$$y = 0 \cdot 561x - 47 \quad . \quad . \quad . \quad . \quad . \quad . \quad . \quad . \quad (c)$$

EXERCISE 48

1. Find the correlation coefficient between x and y given:

x	.	.	.	5	12	14	16	18	21	22	23	25
y	.	.	.	11	16	15	20	17	19	25	24	21

State whether it appears to have significance. [L.U.]

2. Choose, say, 8 equal intervals such as $(0, 3)$, $(6, 9)$, $(8, 11)$ . . . In each interval choose 2 numbers (x, y). Find the coefficient of correlation between the 8 x's and the corresponding y's. [In such a case the (x_r, y_r) are almost linearly related and r is nearly 1.]

3. The marks obtained in mathematics and physics by twelve candidates in an examination were:

Mathematics	.	.	67	63	72	54	41	56	44	89	78	51	60	45
Physics	.	.	46	48	71	60	50	46	35	92	69	32	51	48

Find the correlation coefficient and comment on its significance. [L.U.]

4. The following pairs of associated values of two variables x and y were observed:

x	.	.	.	34	39	39	40	42	42	46	52	58	65
y	.	.	.	32	29	32	27	40	43	41	45	46	49

Calculate (i) the mean $\bar{x}$ and the standard deviation σ_x of the values of x; (ii) the mean and s.d. of the values of y; (iii) the correlation coefficient ρ given by

$$\rho = \frac{\Sigma(x - \bar{x})(y - \bar{y})}{n\sigma_x \sigma_y},$$

where n is the number of pairs of observations. [L.U.]

5. The product moment of x and y about the values $x = a$, $y = b$ is defined as

$$\mu_{11}(a,\ b) = \frac{1}{n}\Sigma(x_r - a)(y_r - b).$$

Show that (see 18.7)

$$p = \mu_{11}(a,\ b) - (\bar{x} - a)(\bar{y} - b).$$

For the following pairs of values $(x,\ y)$ find the product moment about $x = 0$, $y = 0$, and hence find p by the above formula.

x	.	.	. 1	2	3	4	5
y	.	.	. 3	7	9	10	11

6. Let x denote the order of merit of 6 students in one examination, and y their order of merit in another examination. If the corresponding orders are

x	.	.	. 1	2	3	4	5	6
y	.	.	. 2	4	1	3	6	5

show that
$$\bar{x} = \bar{y} = 3\cdot5,\qquad \sigma_x{}^2 = \sigma_y{}^2 = 2\cdot92,$$
$$p = 1\cdot92 \qquad\qquad r = 0\cdot66$$

In the case of n students show that:

$$\bar{x} = \bar{y} = (n + 1)/2;$$
$$\sigma_x{}^2 = \sigma_y{}^2 = (n^2 - 1)/12;$$
$$\frac{1}{n}\Sigma(x_r - y_r)^2 = \sigma_x{}^2 + \sigma_y{}^2 - 2p.$$

7. If $(X_t,\ Y_t)$, (where $t = 1,\ 2\ \ldots\ n$) denote deviations from the corresponding means of x, y, show that

$$r^2 = (\Sigma X_t Y_t)^2 / \Sigma X_t{}^2 \Sigma Y_t{}^2.$$

Now show that (where $s = 1,\ 2 \ldots n$ and $s \neq t$)

$$\Sigma X_t{}^2 \Sigma Y_t{}^2 - (\Sigma X_t Y_t)^2 = \Sigma(X_t Y_s - X_s Y_t)^2$$

Deduce that :

(a) in general $- 1 < r < 1$;

(b) $r^2 = 1$ only if all points $(X_t,\ Y_t)$ lie on a straight line through the origin so that correlation is perfect and the two regression lines coincide.

8. X_1 and X_2 are two variates whose correlation coefficient is r and whose variances are $\sigma_1{}^2$ and $\sigma_2{}^2$ respectively. Show that if a is a constant the variance of $X_1 + aX_2$ is

$$\sigma_1{}^2 + a^2\sigma_2{}^2 + 2ar\sigma_1\sigma_2.$$

If $Y_1 = X_1 + aX_2$ and $Y_2 = X_1 + bX_2$ determine values a and b which are independent of r and such that Y_1 and Y_2 are uncorrelated.

CHAPTER 19

MATHEMATICAL AND STATISTICAL PROBABILITY

". . . Time and chance happeneth to them all."

Ecclesiastes.

" There are many difficulties and troubles with which a factory management has to contend—dies which wear; bearings that get loose; stock which is undersize, oversize, or dirty; loose fixtures; careless, tired or untrained employees. For these reasons it would seem that there is no mathematical method which takes into account all these factors. However there is a kind of mathematics which is applicable in just such conditions and that is the mathematics of probability."

L. T. Radar, Putting Quality into Quantity.

" The phrase ' The Laws of Chance ' is self contradictory. If it is chance then there are no laws, if there are laws then it is not chance."

Lord Samuel, Essays in Physics.

" To-day in Physics chance has become the primary notion, mechanics an expression of its quantitative laws and the overwhelming evidence of causality with all its attributes in the realm of ordinary experience is satisfactorily explained by the statistical laws of large numbers."

Max Born, The Natural Philosophy of Cause and Chance.

19.1. Introduction

The words probable and probability are often used in everyday conversation in such statements as that it will probably rain tomorrow. In such statements there is often a large degree of wishful thinking combined with a certain amount of reliance upon the weather forecast, the present state of the weather, etc. We cannot be concerned with such statements and their significance, and deal only with *mathematical* and *statistical probability*.

19.2. Consider a pack of fifty-two playing-cards containing as usual four kings. As far as we know, there is no reason why any one card should tend to be drawn from the pack more frequently than any other, *i.e.*, the drawings—each called an event—are all *equally likely*. Suppose each event to consist of drawing a card, noting its kind, and returning it to the pack, which is well shuffled before the next event. In such a case the events are all mutually *exclusive*, each event being a repetition of the first and giving rise to a card which is or is not, for example, a king.

Suppose the event of a king is called a success, then on the average it is reasonable to expect that in every fifty-two events as above, four should be successes, and therefore the ratio of successes to total drawings or the *probability* of a success is taken as 4/52. Based on this reasoning, we have the following definition of *mathematical probability*.

Definition

Suppose an event can happen in a total of N ways all mutually exclusive and equally likely to happen. Amongst these there are n ways, each occurrence of which is regarded as a success. The probability of a success is then n/N, *i.e.*, the ratio of the number of ways favourable to the event to the total possible number of ways.

Example 1. If a die is tossed once, what is the probability of obtaining: (*a*) the number **3**; (*b*) an even number?

(*a*) Of the *six* ways in which the die can fall only *one* will show the figure 3. The probability is therefore 1/6.

(*b*) Amongst the six ways, three of them will show an even number. The probability is therefore $3/6 = \frac{1}{2}$.

The probability of a success is usually denoted by p, so that

$$p = n/N.$$

The probability of a failure is usually denoted by q, so that

$$q = (N - n)/N = 1 - p.$$

From this
$$p + q = 1.$$

By this definition a probability is always a positive fraction lying between 0 and 1. If an event cannot happen, *i.e.*, n, the number of successes, is zero, then the probability of its happening is $0/N = 0$, whilst if it is certain to happen each time, *i.e.*, $n = N$, then its probability is $N/N = 1$.

19.3. The Addition Theorem

If an event can happen in a number of different and mutually exclusive ways, the probability of its happening at all is the sum of the separate probabilities that each event happens.

This theorem may be regarded as self-evident and implicit in the definition of probability. We have already used it in Example 1(*b*), where the method used in effect adds the separate probabilities of obtaining the numbers 2, 4, 6. It should be noted that the equation obtained above,

$$p + q = 1,$$

is an illustration of this theorem, for since an event either happens or does not happen, the sum of the separate probabilities must amount to 1, *i.e.*, certainty.

The following proof is useful for general purposes: suppose we require the probability of success in one series of events where it is n_1/N_1 or in another mutually exclusive series of events where it is n_2/N_2. Then out of the total of $N_1 N_2$ events there are $n_1 N_2$ in which success is obtained only in the first series and $n_2 N_1$ only in the second series. The chance of success is therefore

$$\frac{n_1 N_2 + n_2 N_1}{N_1 N_2} = \frac{n_1}{N_1} + \frac{n_2}{N_2}$$

i.e.,
$$p = p_1 + p_2.$$

A similar proof holds when there are more than two mutually exclusive ways in which an event can occur.

Example 2. Find the chance of throwing 9 at least in a single throw of 2 dice.

For each way in which one die can fall, the other has 6 ways of falling so that there are $6 \times 6 = 36$ ways in which the two dice can fall.

Since 9 can occur in 4 ways, the chance of throwing a 9 is 4/36. The chance of throwing a 10 is similarly 3/36, 2/36 for the chance of throwing 11 and 1/36 for the chance of throwing 12. Therefore the chance of throwing at least 9 is

$$\frac{4}{36} + \frac{3}{36} + \frac{2}{36} + \frac{1}{36} = \frac{10}{36}.$$

19.4. The Multiplication Theorem

If a compound event is made up of a number of separate and independent sub-events, the probability of the compound event is the product of the separate probabilities of each sub-event.

Proof: Using the notation of the previous theorem, we see that there are $N_1 N_2$ combinations of all the different results, and out of these there are $n_1 n_2$ results which denote success in both events. The probability of success in both is therefore

$$p = \frac{n_1 n_2}{N_1 N_2} = p_1 p_2.$$

This proof can be clearly extended to more than two events, and for r events

$$p = p_1 p_2 p_3 \cdots p_r.$$

Example 3. What is the least number of dice that must be thrown so that it is more likely than not that at least one 6 is obtained?

The chance of not obtaining a 6 with a die is 5/6. By the above theorem the probability of the compound event of no sixes with n dice is $(5/6)^n$. Therefore $1 - (5/6)^n$ is the probability that at least one six is obtained. We must find n so that

$$1 - (\tfrac{5}{6})^n > \tfrac{1}{2},$$
$$(\tfrac{5}{6})^n < \tfrac{1}{2},$$
$$n \log (\tfrac{5}{6}) < \log \tfrac{1}{2},$$
$$- n \log (1 \cdot 2) < - \log 2,$$
$$n > \frac{\log 2}{\log (1 \cdot 2)} \simeq 3 \cdot 8.$$

Therefore at least 4 dice must be thrown.

19.5. It is useful to note the difference between the *Addition* theorem and the *Multiplication* theorem. In the first we are concerned with the probability of *either* one thing *or* another, *e.g.*, the probability of Xmas day falling either on a Monday or on a Tuesday. We have clearly increased the probability of the result—in fact, doubled it. In the second we are concerned with the probability of *both* one thing *and* another, *e.g.*, the probability of Xmas day being both wet and cold. Here we have clearly decreased the probability by requiring not only wet but also cold. In the first case we therefore increase the probabilities by adding fractions, whilst in the second case we decrease each by multiplying together fractions—each less than unity.

19.6. The student should note the following example, which illustrates *contingent* probabilities.

Example 4. A sample of 52 machined parts is known to contain 4 defectives. What is the chance of drawing out 2 that are defective.

The chance of drawing the first defective is 4/52. The chance with the second, *the first having been drawn*, is 3/51. The compound probability is therefore

$$\frac{4}{52} \times \frac{3}{51} = \frac{1}{221}.$$

Note : If instead of 52 the sample is much larger, say 10 000, the two probabilities above would be 0·0004 and 0·0003 very approximately. The difference is so small that we could, in choosing a small number r of defectives from a large sample, regard the probability of choosing each in turn as constant at p, the probability of choosing the first. The compound probability of choosing the r defectives is then p^r. This approximation is used in practical applications but not here.

Example 5.* A batch of 20 articles contains 3 defectives. A sample of 5 is chosen at random. What is the probability of obtaining 0, 1, 2 or 3 defectives.

We can choose a sample of 5 out of 20 in $_{20}C_5$ ways. For 0 defectives we must consider the number of ways in which 5 good articles can be chosen from the total of 17 good ones. This number is $_{17}C_5$. Therefore if p_0 denotes the probability of choosing no defectives

$$p_0 = \frac{_{17}C_5}{_{20}C_5} = \frac{17\,!}{5\,!\,12\,!} \times \frac{5\,!\,15\,!}{20\,!} = \frac{91}{228} \quad \cdot \quad \cdot \quad \cdot \quad \cdot \quad (1)$$

The 1 defective will be accompanied by 4 good articles. The number of ways, then, of choosing this 1 defective is

$$_3C_1 \times {_{17}C_4}.$$

We therefore find

$$p_1 = \frac{_3C_1 \times {_{17}C_4}}{_{20}C_5}$$

$$= \frac{3\,!}{1\,!\,2\,!} \times \frac{17\,!}{4\,!\,13\,!} \times \frac{5\,!\,15\,!}{20\,!} = \frac{35}{76} \quad \cdot \quad \cdot \quad \cdot \quad \cdot \quad (2)$$

The 2 defectives will be accompanied by 3 good articles. Therefore the number of ways of choosing them is

$$_3C_2 \times {_{17}C_3}$$

and the probability of doing this is

$$p_2 = \frac{_3C_2 \times {_{17}C_3}}{_{20}C_5}$$

$$= \frac{3\,!}{2\,!\,1\,!} \times \frac{17\,!}{3\,!\,14\,!} \times \frac{5\,!\,15\,!}{20\,!} = \frac{5}{38} \quad \cdot \quad \cdot \quad \cdot \quad \cdot \quad (3)$$

Finally, the chance of choosing 3 defectives is given by

$$p_3 = \frac{_3C_3 \times {_{17}C_2}}{_{20}C_5}$$

$$= 1 \times \frac{17\,!}{2\,!\,15\,!} \times \frac{5\,!\,15\,!}{20\,!} = \frac{1}{114} \quad \cdot \quad \cdot \quad \cdot \quad \cdot \quad (4)$$

---

* A brief revision of permutations and combinations will be found at the end of this chapter.

It will be noticed that one of these mutually exclusive events—the choosing of 0, 1, 2 or 3 defectives—is bound to occur, *i.e.*, the probability of its occurrence is 1. This is verified, since

$$p_0 + p_1 + p_2 + p_3 = \frac{91}{228} + \frac{35}{76} + \frac{5}{38} + \frac{1}{114} = 1.$$

19.7. The Statistical Definition of Probability

' Basic to all the arguments of the Treatise on Probability is Keynes' view that probability should be regarded as an indefinable concept. . . . The question remains an open one. In my own person I cannot resist some uneasiness in regard to the indefinability of probability and hanker after some form of the frequency theory.''

The Life of J. M. Keynes, by R. F. Harrod.

In practical affairs we are not often concerned with tossing dice or pennies, but more often concerned with the results of an investigation in which a given number of articles were tested. Thus we might have tested 20 rubber tyres, all produced, as far as we are aware, under identical conditions and found that 2 were defective, *i.e.*, did not pass certain standards of endurance, etc. laid down in advance. Since the test involves the destruction of each tyre, we cannot continue indefinitely and must take the probability that a tyre, made under these conditions, is defective as $2/20 = 0\cdot1$. It might well be that if we continued with the test a somewhat different result would be obtained, *e.g.*, in 40 tyres tested, 5 (or even 3) might be found defective. However, for want of more data the figure $p = 0\cdot1$ must be accepted and used. Considerations such as these lead to the following definition :

A large number, m, of trials are made under the same essential conditions. In n of these trials a certain event occurs. The limit of this ratio n/m as m is indefinitely increased is called the probability that the event happens.

In symbols $$p = \lim_{m \to \infty} \left(\frac{n}{m}\right).$$

It will be noticed that we assume this limit to exist. Also by $m \longrightarrow \infty$ we mean that the trials will be repeated as often as is considered necessary or as often as is possible under the circumstances.

As an illustration, consider that 10 electric switches are to be tested, and the lot contains 3 defective ones which happen to be placed at the end of the row so that if G stands for a switch that will pass the test and D for one that will not, the row of switches may be represented as

G G G G G G G D D D.

The first to be tested satisfies the test and therefore so far, if p is the probability of finding a defective switch, $n = 0$, $m = 1$ and

$$p = \frac{0}{1} = 0$$

After the second switch has been tested $n = 0$, $m = 2$ and

$$p = \frac{0}{2} = 0.$$

Proceeding in this way, we have the succession of numbers (each a probability)

$$0,\ 0,\ 0,\ 0,\ 0,\ 0,\ 0,\ \tfrac{1}{8},\ \tfrac{2}{9},\ \tfrac{3}{10}.$$

The last fraction is the only one that is correct as representing the probability of finding a defective switch. Had these switches been arranged in any other of the many possible orders a totally different series of probabilities would have been obtained, but once again the final result would be $p = 3/10$. It is in this sense that the definition is taken as providing a working method of obtaining a value for the probability.

Example 6. On the average one work-piece in ten is rejected (in a certain process). Find the probability that in a test sample of five pieces: (a) at least four are passed; (b) at least one is passed.

We must assume that $p = \tfrac{1}{10}$ is the probability of a work-piece being rejected and $\tfrac{9}{10}$ is the probability that it is passed.

(a) The probability that the first four tested are passed and the fifth is rejected is

$$\tfrac{9}{10} \times \tfrac{9}{10} \times \tfrac{9}{10} \times \tfrac{9}{10} \times \tfrac{1}{10}.$$

But any other order of these five fractions will give finally the probability that four pass and one fail and there are five different orders. The probability of passing 4 pieces is therefore

$$5(\tfrac{9}{10})^4(\tfrac{1}{10}) \quad . \qquad . \qquad . \qquad . \qquad . \qquad . \qquad (1)$$

The probability of passing 5 pieces is

$$(\tfrac{1}{10})^5.$$

The required probability is therefore

$$5(\tfrac{9}{10})^4(\tfrac{1}{10}) + (\tfrac{1}{10})^5 = \underline{0 \cdot 918\ 54}$$

(b) The probability that all fail is $(\tfrac{1}{10})^5$. Therefore the probability that not all fail, i.e., at least one passes is

$$1 - (\tfrac{1}{10})^5 = \underline{0 \cdot 999\ 99}$$

Note : The result in (1) above when stated generally is known as *Bernoulli's Theorem*. This states that if the probability of an event occurring is p, a constant, the probability it occurs exactly r times in n trials is

$$_nC_r p^r (1 - p)^{n-r}.$$

19.8. Expectation

If p is the probability of an event happening in a single trial, then pN is the *expectation* or expected number of occurrences of the event in N trials. For, by the definition

$$p = \frac{\text{number of occurrences}}{\text{total number of trials } (= N)},$$

so that the number of occurrences expected is pN.

In actual practice the number pN will not be obtained any more than 25 heads will be obtained when a penny is tossed 50 times. But the more often this experiment of 50 tosses is repeated, the nearer we expect the average of the results to approach the theoretical figure of 25. Similarly, if a large number of random samples, each of size N, is drawn from a product having a fraction p defective, then pN is the expected average number of defectives per sample. Further, if p is the chance of winning a prize of £P, then £pP is called the *expectation*. This is based on the following reasoning. In N attempts the prize is expected to be won pN times, and the prize money received is then £$(p$N)P. The competitor can therefore for equality's sake afford to pay an entrance fee of £$(p$P) for each of the N entries.

Example 7. A die is tossed repeatedly until a 6 appears when the game is ended. If the 6 appears at the nth throw the player receives n pence. What is his expectation?

The chance a 6 appears first throw is $\frac{1}{6}$, so that the expectation from this throw is $\frac{1}{6} \times 1$. The chance it appears at the second throw, not having appeared at the first throw is $\frac{5}{6} \times \frac{1}{6}$, and the expectation is then $\frac{5}{6} \times \frac{1}{6} \times 2$. The total expectation E is therefore

$$\text{E} = \tfrac{1}{6} + \tfrac{5}{6} \times \tfrac{1}{6} \times 2 + (\tfrac{5}{6})^2 \times \tfrac{1}{6} \times 3 + (\tfrac{5}{6})^3 \times \tfrac{1}{6} \times 4 \ldots + (\tfrac{5}{6})^{n-1} \times \tfrac{1}{6} \times n + \ldots$$

To sum this infinite series we write

$$6\text{E} = 1 + \tfrac{5}{6} \times 2 + (\tfrac{5}{6})^2 \times 3 + (\tfrac{5}{6})^3 \times 4 + \ldots$$

Multiply both sides by 5/6.

$$\tfrac{5}{6} \cdot 6\text{E} = \tfrac{5}{6} \times 1 + (\tfrac{5}{6})^2 \times 2 + (\tfrac{5}{6})^3 \times 3 + \ldots$$

Subtract:

$$6\text{E}(1 - \tfrac{5}{6}) = 1 + \tfrac{5}{6} + (\tfrac{5}{6})^2 + (\tfrac{5}{6})^3 + \ldots$$

$$= \frac{1}{1 - \tfrac{5}{6}} = 6$$

Finally, $$\underline{\text{E} = 6 \text{ pence.}}$$

19.9. Permutations and Combinations

For the student who has not access to the earlier volume of this series dealing with the theory of arrangements we will repeat the main results. Each of the arrangements which can be made by taking some or all of a number of things is called a *permutation*. Thus from the three letters a, b, c we can, taking 2 at a time, form the permutations

$$ab, \quad ba, \qquad ac, \quad ca, \qquad bc, \quad cb.$$

The number of such is denoted by $_3\text{P}_2$, so that $_3\text{P}_2 = 6$. Generally, $_n\text{P}_r$ denotes the number of permutations of n things taken r at a time, and

$$_n\text{P}_r = n(n-1)(n-2) \ldots (n-r+1)$$
$$= \frac{n!}{(n-r)!}.$$

When we are concerned with the members of the group and not with their arrangements amongst themselves, the result is called a *com-

bination. Thus from the above letters *a*, *b*, *c* we can form the combinations two at a time of

$$ab, \qquad ac, \qquad bc$$

so that $\qquad\qquad {}_3C_2 = 3.$

Generally ${}_nC_r$ denotes the number of combinations of different things taken *r* at a time and

$$
{}_nC_r = \frac{n(n-1)(n-2) \ldots (n-r+1)}{1 \cdot 2 \cdot 3 \ldots r}
$$

$$
= \frac{n!}{r!\,(n-r)!}.
$$

Note 1.   From the formulae above, we see that

$$
{}_nP_r = r!\ {}_nC_r.
$$

This follows from the definition, since each combination of *r* different things can be arranged to give *r* ! permutations of its *r* members.

Note 2. $\qquad\qquad {}_nC_{n-r} = \dfrac{n!}{(n-r)!\,(n-n-r)!}$

$$
= \frac{n!}{(n-r)!\,r!} = {}_nC_r.
$$

This also follows from the definition, for each time we pick out a group (or combination) of *r* things we leave a group of $n-r$ things. The number of different ways of choosing the *r* things must therefore be equal to the number of ways of choosing the $(n-r)$ things out of the given *n*.

This formula often simplifies the arithmetic, since

$$
{}_{20}C_{18} = \frac{20 \cdot 19 \cdot 18 \ldots \ldots 3}{1 \cdot 2 \cdot 3 \ldots \ldots 18},
$$

but more easily

$$
{}_{20}C_{18} = {}_{20}C_2 = \frac{20 \cdot 19}{1 \cdot 2} = 190
$$

We also have relations such as

$$
{}_nC_{r-1} + {}_nC_r = \frac{n!}{(r-1)!\,(n-r+1)!} + \frac{n!}{r!\,(n-r)!}
$$

$$
= \frac{n!}{(r-1)!\,(n-r+1)!} \left[\frac{1}{n-r+1} + \frac{1}{r} \right]
$$

$$
= \frac{(n+1)!}{r!\,(n+1-r)!} = {}_{n+1}C_r.
$$

Note 3.   By the formula, when $r = n$

$$
{}_nC_n = \frac{n!}{n!\,0!}.
$$

There is only one way of choosing n things from n, and the above formula must equal 1. We therefore interpret

$$0! \equiv 1.$$

Example 8. A committee of 6 is to be formed from a group of 7 engineers and 4 mathematicians. How many different committees can be formed if at least 2 mathematicians are always to be included.

All possible combinations are obtained by forming committees of 2 mathematicians and 4 engineers, 3 mathematicians and 3 engineers, and 4 mathematicians and 2 engineers. Therefore the required number of ways is

$$_4C_2 \times {}_7C_4 + {}_4C_3 \times {}_7C_3 + {}_4C_4 \times {}_7C_2,$$

$$= \frac{4!}{2!\,2!} \times \frac{7!}{4!\,3!} + \frac{4!}{3!\,1!} \times \frac{7!}{3!\,4!} + 1 \times \frac{7!}{2!\,5!}$$

$$= 210 + 140 + 21 = \underline{371}.$$

EXERCISE 49

1. Find the chance of throwing a number greater than 4 with an ordinary die.

2. Find the chance of throwing a 6 at least once in two throws of a die.

3. A card is drawn from each of 2 packs of 52 cards. Find the probability that at least one is a five of diamonds.

4. Statistics show that 2 per cent of fruit boats arrive with their cargoes ruined. If 2 boats are due to arrive, what is the chance that: (a) both cargoes are ruined; (b) only one is ruined; (c) neither is ruined.

5. The four letters *s e n t* are placed in a row at random. Find the probability that an English word is formed.

6. Find the chance of throwing exactly 15 in one throw with 3 dice.

7. Find the chance that 4 white balls are drawn from a bag when 4 are drawn at random from the contents of 10 white, 4 black and 3 red.

8. The chance that A can solve a problem is p_1. The chance that B can do so is p_2. Show that the chance that it is solved when both try is

$$1 - (1 - p_1)(1 - p_2).$$

Show that the same result is given by the sum of the three mutually exclusive probabilities: (a) A and B both succeed; (b) A succeeds but B fails; (c) A fails but B succeeds.

9. A bag contains 20 red and 15 white discs. Find the chance that if 5 are chosen at random, the batch will contain 3 red.

10. A hits a target 80 times out of 100 shots, whilst B does so 90 times out of 100 shots. What is the chance that the target is hit if each fires once?

11. 4 persons are chosen at random from a group of 3 men, 2 women and 4 children. Find the chance that the chosen group contains exactly 2 children.

12. Find the chance that if 2 cards are drawn in succession from a pack of 52 cards (without returning the first card), then both are aces.

13. A sample of 5 is to be drawn for test purposes from an order of 20 articles containing 3 defectives. Find the chance it is accepted if orders containing more than one defective are rejected.

14. A biased coin is tossed a large number of times, and gives a head three times out of five. What is the chance that in 4 throws, 2 heads and 2 tails are obtained.

15. A bag contains 10 white, 4 black and 3 red balls. A group of 9 is chosen at random. What is the chance that the group consists of 4 white, 3 black and 2 red balls?

16. A box of 100 batteries contains 5 that are defective. A customer purchases 6. What is the probability that he gets at least one defective battery?

17. A public-opinion poll establishes that 3 out of 5 people are in favour of a certain proposal. What is the chance that if 3 people are taken at random they will provide a majority of opinion against the proposal?

18. Out of 10 springs 3 are defective. Two are chosen at random for a test. What is the chance that both are: (a) defective; (b) not defective?

19. A large number of machined parts contain 10 per cent defective. The number is so large that the probability of drawing a defective part can be taken as $\frac{1}{10}$. If 3 are drawn, find the chance that: (a) all 3 are defective; (b) 2 only are defective; (c) 1 only is defective; (d) none are defective. What is the sum of these 4 probabilities and why?

20. 1000 new cars are inspected and classified according to the faults discovered under the headings: B = bodywork, C = chassis, E = engine, I = other parts. If experience provides the data B = 20, C = 25, E = 10, I = 80 find the probability that a car has the faults: (i) B or C; (ii) C and I; (iii) at least one of B, C, E and I; (iv) no such faults.

21. A batch of electric lamps contains 5 per cent defective. What is the probability that in a test sample of five are found: (i) no defective; (ii) four defective? What is a fair price for a lot of 1000 if a non-defective lamp costs £0·10?

22. A firm manufacturing lemon-juice extractors wishes to advertise that its squeezers have been approved by the *Good Housewife Magazine*. This magazine requires that a random lot of 10 squeezers be submitted to it; 2 will be chosen at random and tested, and if found satisfactory the certificate of approval will be given. If 5 of the 10 squeezers are defective, what is the chance that the certificate of approval is issued?

23. An insecticide is tried on flies. It is found that 80 per cent are killed on the first application, but that those which survive develop a resistance, so that the *proportion* killed in any subsequent application is only one-half that of the immediately preceding application. Find the probability that a fly will survive 5 applications of insecticide.

24. A person takes steps each of length *l*, and is just as likely to go forward as backwards. Prove that after *n* steps he will have gone forward a distance *pl* with a probability

$$(\tfrac{1}{2})^n {}_nC_{\frac{1}{2}(p+n)}.$$

25. The probability of an event occurring on any one occasion is *p*. Prove that the probability of it occurring on exactly *r* of *n* occasions is

$$_nC_r p^r (1-p)^{n-r}.$$

If on the average rain falls on 12 days in every 30, find the probability:

    (i) that the first three days of a given week will be fine and the remainder wet;

    (ii) that rain will fall on just three days in a given week. [L.U.]

26. (i) If 2 persons are selected at random from a group consisting of 16 men and 4 women, find the probabilities that they are: (a) both men; (b) one a man, the other a woman; (c) both women.

(ii) From a bag containing 15 red balls and 10 black balls, a ball is drawn at random 6 times, each ball selected being returned to the bag before the next draw is made. What is the probability that a red ball will appear exactly *r* times out of 6? Find the most probable value of *r* and find its probability.

[L.U.]

27. A and B throw a die, the winner to be the first to throw a 6. If A throws first, what are their respective chances of winning?

[A's chance in his first throw is $\frac{1}{6}$. In his second it is $(\frac{5}{6})^2(\frac{1}{6})$, since both A and B must have failed once before A can have a second throw. In his third throw A's chance is $(\frac{5}{6})^4(\frac{1}{6})$ and so on.]

28. A and B take turns to throw 2 dice, the first to throw 9 winning. Show that their chances are as 9 : 8.

29. Three men in succession toss a coin, the winner to be the first to throw a head. Show that their chances are $\frac{4}{7}$, $\frac{2}{7}$, $\frac{1}{7}$ respectively.

30. Two dice are tossed. If 7 appears, the player receives £1, otherwise nothing. What is his expectation of gain ?

31. A penny is tossed repeatedly. If a head appears for the first time at the nth throw the player receives n pence and a new game begins. What is the player's expectation of gain per game ?

32. A penny is tossed until a head appears. If it appears at the first throw the banker pays £1, if at the second throw £x, if at the third throw £x^2 and so on. How much should the player pay the banker for a fair game ? Show that as $x \to 2$ this sum becomes infinite.

[This is known as the St. Petersburg Paradox.]

33. In the above problem suppose the banker to possess £100 and this or the sum due, whichever is less, is paid when a head first appears. Show that in this case £4·26 is the value of the player's expectation, *i.e.*, the amount be should pay the banker as an entrance fee.

34. The chance of being hit by a piece of shrapnel was, in certain circumstances, $\frac{1}{100}$. Show that if there are 3 pieces flying at random, the chance of escaping a hit is approximately $1 - e^{-3/100}$.

35. If $y = \phi(x)$ is a continuous curve such that $\phi(x)dx$ gives the probability that a variate (*i.e.*, a value of the data being measured) falls within the range $x - \frac{1}{2}dx$ to $x + \frac{1}{2}dx$, then $\phi(x)$ is called the probability density function. If x varies from a to b, then

$$\int_a^b \phi(x)dx = 1.$$

Show that for a distribution so defined

$$\bar{x} = \int_a^b x\phi(x)dx$$

$$\sigma^2 = \int_a^b (x - \bar{x})^2\phi(x)dx$$

$$= \int_a^b x^2\phi(x)dx - \bar{x}^2.$$

36. Show that for a frequency distribution given by $y = ae^{-ax}$ where x varies from 0 to ∞ the S.D. is $1/a$.

37. Two distributions have $(\sigma_1, \bar{x}_1, N_1)$, $(\sigma_2, \bar{x}_2, N_2)$ for their S.D., mean and number of variates. If they are combined, show that for the whole

$$(N_1 + N_2)\sigma^2 = N_1\sigma_1^2 + N_2\sigma_2^2 + \frac{N_1 N_2}{N_1 + N_2}(\bar{x}_2 - \bar{x}_1)^2.$$

38. The mean of 50 readings of a variable was 7·43, and their standard deviation (or root-mean-square deviation from this mean) was 0·28. The following ten additional readings become available : 6·80, 7·81, 7·58, 7·70, 8·05, 6·98, 7·78, 7·85, 7·21, 7·40.

If these are included with the original 50 readings find : (i) the mean ; (ii) the standard deviation of the whole set of 60 readings. [L.U.]

CHAPTER 20

PARTIAL DIFFERENTIAL EQUATIONS

" It is no paradox to say that in our most theoretical moods we may be nearest to our most practical applications."

A. N. Whitehead.

20.1. Introduction

Consider a thin bar of metal heated at one end. The temperature θ at any point distant x from that end may vary with the time. In this case for a given x we would be concerned with $\partial\theta/\partial t$. Or at a given time t, the temperature may vary from point to point along the bar, so that for a given t we are concerned with $\partial\theta/\partial x$. Generally we are concerned with both methods of variation, and we will obtain the partial differential equation for one dimensional heat flow

$$\frac{\partial^2\theta}{\partial x^2} = \frac{1}{K}\frac{\partial\theta}{\partial t}$$

By a solution of this for a particular problem is meant an expression for θ in terms of x and t free from all partial derivatives and satisfying the given conditions (or *boundary values*) that specify the problem.

Often the general theory is difficult, and we must be content with using our knowledge of the physical conditions in order to arrive at particular solutions which satisfy the needs of the problem. It can be shown in the cases dealt with that such solutions are unique, *i.e.*, are the only solutions.

20.2. Formation of Partial Differential Equations

(a) If
$$z = ax + by \qquad\qquad\qquad . \quad . \quad . \quad . \quad . \quad (1)$$
$$\frac{\partial z}{\partial x} = a, \quad \frac{\partial z}{\partial y} = b.$$

a, b can be eliminated to give

$$z = x\frac{\partial z}{\partial x} + y\frac{\partial z}{\partial y},$$

which is a first-order partial differential equation having (1) as a solution. Note that we have been able to eliminate two arbitrary constants a, b by means of a first-order equation.

(b) If
$$z = f(x - ct) + F(x + ct) \qquad . \quad . \quad . \quad . \quad (2)$$
where f, F are two arbitrary functions of the variables shown,

$$\frac{\partial z}{\partial x} = f'(x - ct) + F'(x + ct)$$

$$\frac{\partial^2 z}{\partial x^2} = f''(x - ct) + F''(x + ct) \qquad . \quad . \quad . \quad (3)$$

Similarly,
$$\frac{\partial z}{\partial t} = -cf'(x - ct) + cF'(x + ct)$$

$$\frac{\partial^2 z}{\partial t^2} = c^2 f''(x - ct) + c^2 F''(x + ct) \qquad . \quad . \quad . \quad (4)$$

335

From (3) and (4)

$$\frac{\partial^2 z}{\partial t^2} = c^2 \frac{\partial^2 z}{\partial x^2} \qquad \cdots \qquad (5)$$

a second-order equation resulting from the elimination of two arbitrary functions.

The example in (a) shows that the usual relation between the order of the equation and the number of constants eliminated need not hold here, as it does in ordinary differential equations. This does not concern us too much. The more important example is (b), where two completely general functions f and F have been eliminated, so that we must expect a solution of a partial differential equation to contain such general forms. Evidently the problem will be to choose the form that suits a given problem.

20.3. Some Simple Solutions

(a)
$$\frac{\partial^2 z}{\partial x^2} = 0.$$

Since the partial differential coefficient with respect to x of any function of y is zero, the first integral of this is

$$\frac{\partial z}{\partial x} = f(y).$$

From this
$$z = xf(y) + F(y),$$

where f, F are any two arbitrary functions.

(b)
$$\frac{\partial^2 z}{\partial x \partial y} = x^2 y.$$

We have
$$\frac{\partial z}{\partial y} = \tfrac{1}{3}x^3 y + f(y),$$

$$z = \tfrac{1}{6}x^3 y^2 + \int f(y)dy + F(x)$$

$$= \tfrac{1}{6}x^3 y^2 + G(y) + F(x),$$

where G is the integral of f, and hence an arbitrary function like f. F is also any arbitrary function.

(c) Solve $\dfrac{\partial^2 z}{\partial x \partial y} = 0$, given that $z(0, y) = y$, $\quad z(x, 0) = \sin x$.

These two conditions are known as boundary values. The first means that z, being a function of x and y, can be written $z(x, y)$ and for the values $(0, y)$ the function has the value y. The second boundary value means that when $y = 0$, for the value x the value of the function is $\sin x$.

From
$$\frac{\partial^2 z}{\partial x \partial y} = 0,$$

$$\frac{\partial z}{\partial y} = f(y),$$

$$z = \int f(y)dy + F(x)$$

$$= G(y) + F(x),$$

but $\quad z(0, y) = y, \qquad \therefore \quad y = G(y) + F(0)$

$z(x, 0) = \sin x, \quad \therefore \quad \sin x = G(0) + F(x).$

These are clearly satisfied by taking $F(x) = \sin x$ and $G(y) = y$, giving as the solution

$$z = y + \sin x.$$

EXERCISE 50

Eliminate the arbitrary constants from:
1. $z = ax^2 + by^2$.   2. $z = Ae^{mx} + Be^{-my}$.
3. $z = (x - a)^2 + (y - b)^2$.

4. Prove that the equation $\dfrac{\partial^2 z}{\partial x^2} + \dfrac{\partial^2 z}{\partial y^2} = 0$ is satisfied by $z = \tan^{-1}\left(\dfrac{y}{x}\right)$.

Eliminate the arbitrary functions from:
5. $z = f(x + 2y)$.   6. $z = f(y/x)$.
7. $z = f(x) + g(y)$.   8. $z = xf(x + y)$.
9. $z = f(x^2 + y^2)$.   10. $z = F(x + iy) + f(x - iy), (i \equiv \sqrt{-1})$.

11. Obtain a solution to the equation $\dfrac{\partial^2 z}{\partial x^2} = 0$, given that

$$z(0, y) = y^3, \qquad z(2, 0) = 2.$$

12. Solve the equation $\dfrac{\partial^2 z}{\partial x \partial y} = 2x$, given that when

$$z(0, y) = 0, \qquad \frac{\partial}{\partial x}\{z(x, 0)\} = x^2.$$

13. Show that $e^{-n^2 t} \sin nx$ is a solution of the equation $\partial u/\partial t = \partial^2 u/\partial x^2$. Hence if $A_1, A_2 \ldots A_N$ are constants, show that

$$u = \sum_{1}^{N} A_n e^{-n^2 t} \sin nx$$

is a solution with the value zero at $x = 0$ and $x = \pi$ for all values of t.

14. Show that $y = f(x + ct) + F(x - ct)$, is a solution of

$$c^2 \frac{\partial^2 y}{\partial x^2} = \frac{\partial^2 y}{\partial t^2}.$$

Taking both $f(x)$ and $F(x)$ to be of the form $A \sin ax$, show that the solution of the partial differential equation becomes

$$y = 2A \sin (ax) \cos (act).$$

Interpret this as a wave motion with y always zero at the points $x = 0$, $\pm \pi/a$, $\pm 2\pi/a$, etc., so that it is a " stationary " wave of variable amplitude $2A \cos (act)$.

15. Integrate:

$$(a) \quad \frac{\partial^2 z}{\partial y^2} = \frac{x}{y}; \qquad (b) \quad \frac{\partial^2 z}{\partial x^2} = e^{x+y}.$$

20.4. The Exponential Method of Trial Solutions

We found the exponential method of trial solutions very useful with ordinary linear differential equations with constant coefficients. It is equally useful here in giving particular solutions to important types of equations.

Example 1.   Find a solution to the equation

$$\frac{\partial^2 z}{\partial x^2} = \frac{\partial^2 z}{\partial t^2},$$

given that $z = 0$ when $x = +\infty$ and when $t = +\infty$.

It will be seen later that this equation represents wave motion, and the boundary values may be interpreted as showing, for example, that an initial displacement was given at $x = 0$ to start the motion, so that we know from physical considerations that when $t = +\infty$ the motion will have died away, i.e., $z = 0$, and also the effect at an infinite distance along the x axis must be zero, so that again $z = 0$ when $x = +\infty$.

Put
$$z = e^{ax+bt}$$

Substitution in the equation gives

$$a^2 e^{ax+bt} = b^2 e^{ax+bt},$$

or
$$a^2 = b^2$$

$$a = \pm b.$$

Both a and b must be negative, since the exponential is to be zero for infinite values of x and t. Therefore ignore the value $a = -b$ and put

$$a = b = -p^2,$$

where p is a real number and $-p^2$ obviously negative. A possible solution satisfying the conditions, then, is

$$z = e^{-p^2(x+t)}$$

and having for brevity's sake omitted the usual constant, we will now write this as

$$z = Ae^{-p^2(x+t)},$$

noting that so far p is not fixed, apart from being a real number.

Example 2. Find a solution of the equation

$$h^2 \frac{\partial^2 \theta}{\partial x^2} = \frac{\partial \theta}{\partial t}$$

which is zero when $t = +\infty$.

This, as stated, is one form of the equation governing the conduction of heat. The boundary value could be taken to mean that a bar was heated at time $t = 0$ and left to cool so that the temperature would be zero after an infinite time, where the temperature θ means the excess of the temperature above that of the air.

(a) Trying
$$\theta = e^{ax+bt}$$

gives
$$h^2 a^2 = b$$

so that
$$\theta = e^{ax + h^2 a^2 t}$$

is a possible solution for all values of a. Since θ must be zero when t is infinite, the coefficient of t must be negative. To obtain this put $a = ip$, $(i \equiv \sqrt{-1})$, so that $a^2 = -p^2$. The solution is now

$$\theta = e^{ipx - h^2 p^2 t}$$

$$= e^{-h^2 p^2 t}[\cos px + i \sin px].$$

This shows, as can be verified by trial, that

$$e^{-h^2 p^2 t} \cos px \quad \text{or} \quad e^{-h^2 p^2 t} \sin px,$$

alone will satisfy the equation. We take then as the solution

$$\theta = e^{-h^2 p^2 t}[A \cos px + B \sin px],$$

where A, B are arbitrary constants and p is any real number.

(b) Noting that $\theta = 0$ when $t = +\infty$, we could begin by trying

$$\theta = e^{-mt}X$$

where m is positive and X is a function of x only. Substitution of this in the differential equation gives

$$h^2 e^{-mt} X'' = -m e^{-mt} X$$

or since $e^{-mt} \neq 0$

$$X'' + \frac{m}{h^2} X = 0.$$

The solution of this is

$$X = A \cos \frac{\sqrt{m}}{h} x + B \sin \frac{\sqrt{m}}{h} x$$

and

$$\theta = e^{-mt} \left\{ A \cos \frac{\sqrt{m}}{h} x + B \sin \frac{\sqrt{m}}{h} x \right\}.$$

This solution is the same as that above, and was obtained a little more easily by starting with an appropriate trial solution.

Example 3. Given that $\theta = 0$ when $y = +\infty$ and also when $x = 0$, find a solution of

$$\frac{\partial^2 \theta}{\partial x^2} + \frac{\partial^2 \theta}{\partial y^2} = 0.$$

As usual let

$$\theta = e^{px + qy}.$$

Substitution gives

$$p^2 + q^2 = 0.$$

Since $\theta = 0$ when $y = +\infty$, q must be negative and real, so let

$$q = -n,$$

where n is a real positive number. [The student, if he so wished could use n^2 to emphasize that $-n^2$ is negative.]

We now have

$$p^2 = -n^2$$
$$p = \pm in$$

therefore

$$e^{inx - ny}, \qquad e^{-inx - ny},$$

are both possible solutions. We may take then as a solution

$$\theta = e^{-ny}[Ae^{inx} + Be^{-inx}]$$
$$= e^{-ny}[L \cos nx + M \sin nx].$$

But $\theta = 0$ when $x = 0$, therefore $L = 0$. The solution is now

$$\theta = Me^{-ny} \sin nx,$$

where n is not fixed, except that it is a real positive number. We may then take as the solution the infinite series

$$\theta = M_1 e^{-n_1 y} \sin n_1 x + M_2 e^{-n_2 y} \sin n_2 x + \ldots$$
$$= \sum_{r=1}^{\infty} M_r e^{-n_r y} \sin n_r x.$$

It appears, then, that we may impose upon this problem further boundary values to determine the values of n_r, M_r. This will be done later M_r being found by means, usually, of a Fourier series.

20.5. Heat Conduction in One Dimension

Consider a bar of constant cross-sectional area a, whose sides are insulated so that the heat flow is along the bar and one dimensional.

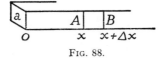

FIG. 88.

If θ is the temperature at the face A where $OA = x$ then $\theta + \dfrac{\partial \theta}{\partial x} \Delta x$

is the temperature at the face B a distance $x + \Delta x$ from O, for $\dfrac{\partial \theta}{\partial x}$ is the rate of increase of temperature with distance, so that $\frac{\partial \theta}{\partial x} \Delta x$ is the total increase in the small distance Δx. The average temperature of the small element of volume AB is then

$$\frac{1}{2}\left[\theta + \theta + \frac{\partial \theta}{\partial x} \Delta x\right] = \theta + \frac{1}{2}\frac{\partial \theta}{\partial x} \Delta x.$$

If s is the specific heat and ρ the density of the material, the quantity of heat in this volume is

$$s\rho a \Delta x \left[\theta + \frac{1}{2}\frac{\partial \theta}{\partial x} \Delta x\right].$$

The rate of change of this with regard to time is

$$s\rho a \Delta x \frac{\partial}{\partial t}\left[\theta + \frac{1}{2}\frac{\partial \theta}{\partial x} \Delta x\right]$$

$$= s\rho a \Delta x \left[\frac{\partial \theta}{\partial t} + \frac{1}{2} \Delta x \frac{\partial^2 \theta}{\partial t \partial x}\right] \quad \text{. . . . (1)}$$

But this rate of change is also due to the difference between the heat flowing in at the face A and that flowing out at the face B. By the law governing this flow, the rate of flow across an area is proportional to the area and the temperature gradient, *i.e.*, the rate of change of temperature with respect to the distance measured perpendicular to the area. In this case we take the heat flowing in over the face at A as

$$- Ka \frac{\partial \theta}{\partial x},$$

where $\partial \theta / \partial x$ is the temperature gradient with regard to distance and K is the constant of proportionality, called here the thermal conductivity. Since heat flows from a higher to a lower temperature $\partial \theta / \partial x$ is negative and $- \partial \theta / \partial x$ positive, so that K is a positive constant as required in its other uses. The heat flowing out over the face at B is

$$- Ka \frac{\partial}{\partial x}\left[\theta + \frac{\partial \theta}{\partial x} \Delta x\right].$$

The gain in heat of the element of volume is

$$\left\{- Ka \frac{\partial \theta}{\partial x}\right\} - \left\{- Ka \left(\frac{\partial \theta}{\partial x} + \frac{\partial^2 \theta}{\partial x^2} \Delta x\right)\right\} = Ka \frac{\partial^2 \theta}{\partial x^2} \Delta x.$$

Equating this to (1)

$$Ka \frac{\partial^2 \theta}{\partial x^2} \Delta x = s\rho a \Delta x \left[\frac{\partial \theta}{\partial t} + \frac{1}{2} \Delta x \frac{\partial^2 \theta}{\partial t \partial x}\right],$$

$$\therefore \quad \frac{K}{s\rho} \frac{\partial^2 \theta}{\partial x^2} = \frac{\partial \theta}{\partial t} + \frac{1}{2} \Delta x \frac{\partial^2 \theta}{\partial t \partial x}.$$

In the limit, as $\Delta x \to 0$, this gives

$$\frac{\partial^2 \theta}{\partial x^2} = \frac{1}{h^2} \frac{\partial \theta}{\partial t} \qquad \left[h^2 = \frac{K}{\rho s} \right]$$

the required equation of one dimensional heat flow.

20.6. Solution by Separation of Variables

Particular solutions of

$$\frac{\partial^2 \theta}{\partial x^2} = \frac{1}{h^2} \frac{\partial \theta}{\partial t}$$

have been found by assuming (20.4, Example 2)

$$\theta = e^{ax + bt}$$

Writing this as

$$\theta = e^{ax} \cdot e^{bt}$$

we note that it is the product of two functions, one of x only and the other of t only.   An equivalent method is then to assume

$$\theta = XT$$

where X is a function of x only and T of t only.   Since

$$\frac{\partial^2}{\partial x^2}(XT) = X''T, \qquad \frac{\partial}{\partial t}(XT) = XT',$$

substitution of this form of solution in the differential equation gives

$$X''T = \frac{1}{h^2} XT'$$

or, separating the variables

$$\frac{X''}{X} = \frac{1}{h^2} \frac{T'}{T}.$$

If X''/X contains x and T'/h^2T contains t, then as x, t vary independently, each expression can be made to assume an infinite number of values which have no connection with those due to the other expression.   The only possible conclusion is that each expression is a constant independent of x or t.   We have then

$$\frac{X''}{X} = \frac{1}{h^2} \frac{T'}{T} = \lambda \text{ (a constant)}$$

or

$$X'' = \lambda X, \quad T' = \lambda h^2 T.$$

There are now three forms of solution :

for $\lambda > 0 \qquad \theta = (Ae^{x\sqrt{\lambda}} + Be^{-x\sqrt{\lambda}})e^{\lambda h^2 t}$

$\lambda < 0 \qquad \theta = [C \cos (x\sqrt{-\lambda}) + D \sin (x\sqrt{-\lambda})]e^{\lambda h^2 t}$

$\lambda = 0 \qquad \theta = Ex + F.$

We will continue by means of a specific example.

Example 4.   A bar of length l has its sides insulated so that the heat flow is along the bar only.   It is at a uniform temperature θ_0, and the ends of the bar are suddenly changed to $0°$ and maintained at this temperature.   Find the temperature at any subsequent time t at a distance x from one end.

We must solve

$$\frac{\partial^2 \theta}{\partial x^2} = \frac{1}{h^2} \frac{\partial \theta}{\partial t}$$

subject to the boundary conditions

$$\theta(0, t) = 0 \qquad \theta(l, t) = 0$$

for all $t > 0$, since the ends are maintained at $0°$ from the beginning taken as $t = 0$, onwards.

Also
$$\theta(x, 0) = \theta_0 \text{ for } 0 < x < l.$$

For a given x, i.e., at a given point the temperature clearly decreases with the time t, so that λ in the previous set of solutions must be negative. The solution then for this problem is, with $\lambda = -p^2$, where p is a positive number

$$\theta = [C \cos px + D \sin px] e^{-p^2 h^2 t}.$$

Since $\theta(0, t) = 0$,

$$0 = Ce^{-p^2 h^2 t}, \qquad \therefore \quad C = 0,$$

and
$$\theta = D \sin px \cdot e^{-p^2 h^2 t}$$

Choosing the other zero value boundary condition $\theta(l, t) = 0$,

$$0 = D \sin pl \cdot e^{-p^2 h^2 t},$$

therefore $\sin pl = 0$, for $D = 0$ would give $\theta = 0$ for all t which is obviously incorrect.

From $\qquad \sin pl = 0$
(since $p, l > 0$) $\qquad pl = n\pi$, where n is any positive integer.

The solution is now

$$\theta = D \sin\left(\frac{n\pi x}{l}\right) e^{-n^2 \pi^2 h^2 t / l^2}$$

and more generally

$$\theta = \sum_{n=1}^{\infty} A_n e^{-n^2 \pi^2 h^2 t / l^2} \sin\left(\frac{n\pi x}{l}\right).$$

We still have the final condition to satisfy, $\theta(x, 0) = \theta_0$. Using this, we obtain from the above

$$\theta_0 = \Sigma A_n \sin\left(\frac{n\pi x}{l}\right) \quad \text{for} \quad 0 < x < l.$$

This is easily satisfied by finding a half-range Fourier sine series for θ_0 in the interval $0 < x < l$. With the usual formula (11.11)

$$A_n = \frac{2}{l} \int_0^l \theta_0 \sin\left(\frac{n\pi x}{l}\right) dx$$

$$= \frac{2\theta_0}{n\pi} (1 - \cos n\pi) \qquad \begin{aligned} &= 0 \qquad \text{for } n \text{ even} \\ &= \frac{4\theta_0}{n\pi} \qquad n \text{ odd.} \end{aligned}$$

The final form of the solution is then

$$\theta = \frac{4\theta_0}{\pi} \left[e^{-\pi^2 h^2 t / l^2} \sin\left(\frac{\pi x}{l}\right) + \frac{1}{3} e^{-9\pi^2 h^2 t / l^2} \sin\left(\frac{3\pi x}{l}\right) + \cdots \right]$$

$$= \frac{4\theta_0}{\pi} \sum_{n=1}^{\infty} \frac{1}{2n-1} e^{-(2n-1)^2 \pi^2 h^2 t / l^2} \sin\left(\frac{(2n-1)\pi x}{l}\right),$$

giving the temperature at any time t at a point distant x from the end chosen as origin. It could be shown that this solution is unique, i.e., is the only solution.

[It is not very difficult to evaluate a sufficient number of terms in the above series to obtain a result to any required degree of accuracy. Certain books on heat transmission contain the necessary tables.]

20.7. The Transverse Vibration of a String

We will suppose that the string was in equilibrium along the x axis and received a displacement such that the part AB of length Δx

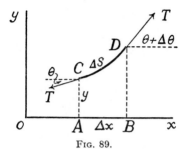

FIG. 89.

moved to CD. The string is supposed to be tightly stretched so that T is large and the effect of gravity can be ignored. Also the string has undergone very little extension, so that T the tension is constant throughout the string and CD ($= \Delta s$), the length of the displaced portion AB ($= \Delta x$), can be taken equal to Δx. Since

$$\frac{\partial s}{\partial x} = \sqrt{\left\{1 + \left(\frac{\partial y}{\partial x}\right)^2\right\}} \simeq 1 \quad \ldots \quad (1)$$

this is equivalent to assuming that the gradient at any point on the displaced string is so small that $(\partial y/\partial x)^2$ can be neglected.

The tension at C has a component parallel to the y axis of

$$\text{T} \sin \theta = \text{T} \frac{\partial y}{\partial s} = \text{T} \frac{\partial y}{\partial x} \frac{\partial x}{\partial s} = \text{T} \frac{\partial y}{\partial x} \quad \text{by (1).}$$

Therefore, the similar force at D is

$$\text{T} \frac{\partial y}{\partial x} + \frac{\partial}{\partial x}\left(\text{T} \frac{\partial y}{\partial x}\right) \Delta x$$

$$= \text{T} \frac{\partial y}{\partial x} + \text{T} \frac{\partial^2 y}{\partial x^2} \Delta x.$$

If ρ is the density of the string, for the motion of the length Δs perpendicular to the x axis

$$\left(\text{T} \frac{\partial y}{\partial x} + \text{T} \frac{\partial^2 y}{\partial x^2}\Delta x\right) - \text{T} \frac{\partial y}{\partial x} = \rho \Delta s \frac{\partial^2 y}{\partial t^2},$$

$$\therefore \quad \text{T} \frac{\partial^2 y}{\partial x^2} = \rho \frac{\partial s}{\partial x} \frac{\partial^2 y}{\partial t^2}$$

by (1) $$= \rho \frac{\partial^2 y}{\partial t^2}.$$

Finally, if $c^2 = \text{T}/\rho$

$$\frac{\partial^2 y}{\partial t^2} = c^2 \frac{\partial^2 y}{\partial x^2},$$

giving the partial differential equation governing the displacement y of any point on the string in terms of the distance x from one end and the time.

20.8. Solution of the Wave Equation

It has been seen (Exercise 57, No. 14) that

$$y = f(x - ct) + F(x + ct),$$

where f and F have any form, will satisfy the above equation. Before considering the method of solution, the special form

$$y = f(x - ct)$$

will be considered.

The curve $y = f(x)$ becomes $y = f(x - c)$ when the origin is moved a distance c to the left. We will as usual regard the axes as fixed so that $y = f(x - ct)$ is a curve moving to the right with velocity c for when $t = 0$, $1/c$, $2/c$, . . . we obtain successively for a given ordinate at x, $y = f(x)$, $f(x - 1)$, $f(x - 2)$ . . . so that the ordinate at $x - 1$ has moved to x and that at $x - 2$ to $x - 1$ . . . in time $1/c$. Similarly,

$$y = F(x + ct)$$

represents the curve $y = F(x)$ moving to the left with velocity c and

$$y = f(x - ct) + F(x + ct)$$

represents any wave form travelling left and/or right with a constant velocity c.

To solve the differential equation we can use the exponential method of trying

$$y = e^{px + qt}$$

or the method of separation of variables

$$y = XT.$$

If we know that the solution will be a vibration, a quicker method is to use

$$y = Xe^{i\omega t}$$

where X is a function of x only. Substitution of this gives

$$e^{i\omega t}c^2X'' + \omega^2 Xe^{i\omega t} = 0$$

or

$$X'' + \left(\frac{\omega}{c}\right)^2 X = 0$$

therefore

$$X = A\cos\left(\frac{\omega}{c}x\right) + B\sin\left(\frac{\omega}{c}x\right)$$

and

$$y = e^{i\omega t}\left[A\cos\left(\frac{\omega x}{c}\right) + B\sin\left(\frac{\omega x}{c}\right)\right],$$

showing that there are four types of solutions satisfying the equation

$$\cos\left(\frac{\omega x}{c}\right)\cos\omega t, \qquad \cos\left(\frac{\omega x}{c}\right)\sin\omega t,$$

$$\sin\left(\frac{\omega x}{c}\right)\cos\omega t, \qquad \sin\left(\frac{\omega x}{c}\right)\sin\omega t$$

and the sum of these each multiplied by a constant will be the general solution.

Example 5. A string is stretched and fastened to two points l apart. Motion is started by displacing the string into the form $y = y_0 \sin(\pi x/l)$ from which it is released at time $t = 0$. Find the displacement of any point x at time t.

The boundary conditions here are

$$y(0, t) = 0, \qquad y(l, t) = 0$$

$$y(x, 0) = y_0 \sin(\pi x/l)$$

$$\left(\frac{\partial y}{\partial t}\right)_{t=0} = 0 \qquad \text{for all } x \text{ in } 0 < x < l.$$

Proceeding as above

$$y = e^{i\omega t}\left[A\cos\frac{\omega x}{c} + B\sin\frac{\omega x}{c}\right].$$

Since $y(0, t) = 0$,

$$0 = e^{i\omega t}A, \qquad \therefore \quad A = 0.$$

Now using

$$y = e^{i\omega t}B\sin\left(\frac{\omega x}{c}\right),$$

which is equivalent to

$$y = L\sin\left(\frac{\omega x}{c}\right)\cos\omega t + M\sin\left(\frac{\omega x}{c}\right)\sin\omega t,$$

$$\frac{\partial y}{\partial t} = -L\omega\sin\left(\frac{\omega x}{c}\right)\sin\omega t + M\omega\sin\left(\frac{\omega x}{c}\right)\cos\omega t,$$

and since $\left(\dfrac{\partial y}{\partial t}\right)_{t=0} = 0$

$$0 = M\omega\sin\left(\frac{\omega x}{c}\right), \qquad \therefore \quad M = 0.$$

We now have

$$y = L\sin\left(\frac{\omega x}{c}\right)\cos\omega t \quad . \quad . \quad . \quad . \quad . \quad . \quad (1)$$

$y(l, t) = 0$:

$$0 = L\sin\left(\frac{\omega l}{c}\right)\cos\omega t.$$

We cannot have $L = 0$, for this gives $y = 0$ throughout the motion which is incorrect, therefore since $\cos\omega t \neq 0$ always

$$\sin\left(\frac{\omega l}{c}\right) = 0$$

$$\frac{\omega l}{c} = n\pi, \qquad (n = 1, 2, 3 \ldots)$$

and

$$\omega = \frac{n\pi c}{l},$$

giving the possible periods of vibrations.

From (1) a particular solution is

$$y = L_n \sin\frac{n\pi x}{l}\cos\frac{n\pi ct}{l} \qquad (n = 1, 2, 3 \ldots)$$

and the sum of such a series for all these values of n is also a solution. But the last condition is

$$y(x, 0) = y_0 \sin (\pi x/l).$$

Applying this,

$$y_0 \sin \frac{\pi x}{l} = \sum_{n=1}^{\infty} L_n \sin \frac{n\pi x}{l}, \quad \ldots \ldots \quad (2)$$

which is satisfied by taking the first term only of the right-hand side and $L_1 = y_0$, $L_n = 0$ $(n = 2, 3 \ldots)$. The solution required is then

$$y = y_0 \sin \frac{\pi x}{l} \cos \frac{\pi ct}{l}.$$

Suppose the problem was as above, but the initial shape was $y = y_0 \sin^3 (\pi x/l)$. We proceed as above, but instead of (2) would have

$$y_0 \sin^3 \left(\frac{\pi x}{l}\right) = \Sigma L_n \sin \frac{n\pi x}{l}.$$

This is

$$\frac{y_0}{4}\left[3 \sin \frac{\pi x}{l} - \sin \frac{3\pi x}{l}\right] = \Sigma L_n \sin \frac{n\pi x}{l},$$

which is satisfied with $L_1 = 3y_0/4$, $L_2 = 0$

$$L_3 = - y_0/4, \qquad L_n = 0 \ (n = 4, 5 \ldots).$$

The solution then is

$$y = \frac{y_0}{4}\left[3 \sin \frac{\pi x}{l} \cos \frac{c\pi t}{l} - \sin \frac{3\pi x}{l} \cos \frac{3c\pi t}{l}\right].$$

In much the same way we can obtain the equation

$$\frac{\partial^2 u}{\partial t^2} = c^2 \frac{\partial^2 u}{\partial x^2} \qquad \left[c^2 = \frac{E}{\rho}\right]$$

for the *longitudinal waves* of tension or compression that, for example, move along a bar hit at one end. u is the displacement of a given mark on the bar distant x from a fixed origin; E is Young's modulus for the material of the bar and ρ its density. By the above, it follows that these waves have a velocity $c = \sqrt{(E/\rho)}$.

20.9. Transverse Vibrations of a Uniform Beam

A beam was originally along the x axis, and at time t an element of length Δx moved a distance y perpendicular to the x axis. We will consider only this transverse vibration.

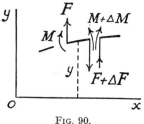

Fig. 90.

If M is the bending moment,

$$M = EI \frac{d^2y}{dx^2} \quad . \quad . \quad . \quad . \quad . \quad . \quad (1)$$

If F is the shear, taking moments about the centre of gravity for the equilibrium of the length Δx

$$F \cdot \tfrac{1}{2}\Delta x + M + (F + \Delta F)\tfrac{1}{2}\Delta x = M + \Delta M$$

as $\Delta x \longrightarrow 0$ this reduces to

$$. \quad F = \frac{dM}{dx} \quad . \quad . \quad . \quad . \quad . \quad . \quad (2)$$

If w is the weight per unit length, for the motion of the element perpendicular to the x axis (ignoring the small effect, if any, due to gravity),

$$\Delta F = -\frac{w}{g} \Delta x \frac{d^2y}{dt^2}$$

or

$$\frac{dF}{dx} = -\frac{w}{g} \frac{d^2y}{dt^2},$$

i.e., by (2)

$$\frac{d^2M}{dx^2} = -\frac{w}{g} \frac{d^2y}{dt^2},$$

and by (1)

$$EI \frac{d^4y}{dx^4} = -\frac{w}{g} \frac{d^2y}{dt^2} \quad . \quad . \quad . \quad . \quad . \quad (3)$$

or more appropriately, using partial derivative notation, since y is a function of x and t,

$$EI \frac{\partial^4 y}{\partial x^4} + \frac{w}{g} \frac{\partial^2 y}{\partial t^2} = 0 . \quad . \quad . \quad . \quad . \quad (4)$$

This may be solved by any of the previous methods. Since we are concerned with vibrations and their possible periods, assume

$$y = X e^{i\omega t} \quad . \quad . \quad . \quad . \quad . \quad . \quad (5)$$

where X is a function of x only and ω is the unknown period of the vibration.

Substitution in the equation gives

$$\frac{d^4X}{dx^4} - k^4X = 0 \qquad \left(k^4 \equiv \frac{\omega^2 w}{EIg}\right)$$

or

$$(D^4 - k^4)X = 0.$$

The solution of this is (8.2)

$$X = A \cos kx + B \sin kx + C \cosh kx + E \sinh kx \quad . \quad (6)$$

Since

$$M = EI \frac{\partial^2 y}{\partial x^2} \quad \text{and} \quad F = \frac{\partial M}{\partial x},$$

the end conditions will provide the data needed for further evaluation of a given problem. Thus, for example, at a free end M and F, *i.e.*, $\frac{\partial^2 y}{\partial x^2}$ and $\frac{\partial^3 y}{\partial x^3}$ are both zero, whilst at a freely hinged end y and M, *i.e.*, y and $\frac{\partial^2 y}{\partial x^2}$ are both zero.

Example 6. Find the natural frequencies of transverse vibrations for a uniform beam of length l freely hinged at its ends.

The line joining the ends is taken as the x axis and one end as the origin. Solving the equation (4) above, we get as in (6)

$$X = A \cos kx + B \sin kx + C \cosh kx + E \sinh kx.$$

Since $y = Xe^{i\omega t}$ and $y = 0$, $\frac{\partial^2 y}{\partial x^2} = 0$ at each end therefore $X = 0 = \frac{d^2X}{dx^2}$ at each end.

At $x = 0$ this gives
$$A + C = 0,$$
$$- A + C = 0,$$
so that
$$A = C = 0.$$

At $x = l$ from

$$X = B \sin kx + E \sinh kx,$$

$X = 0:$ $\qquad\qquad 0 = B \sin kl + E \sinh kl,$

$\frac{d^2X}{dx^2} = 0:$ $\qquad\qquad 0 = - B \sin kl + E \sinh kl.$

Addition gives

$$0 = E \sinh kl$$

or $E = 0$ since $kl \neq 0$.

Finally, $\qquad\qquad 0 = B \sin kl.$

If $B = 0$, then $X = 0$ and $y = 0$ always, giving the undeflected position. Otherwise

$$\sin kl = 0$$
$$kl = n\pi \qquad (n = 1, 2, \ldots)$$

From this

$$\frac{n^4\pi^4}{l^4} = k^4 = \frac{\omega^2 w}{EIg}$$

and

$$\omega^2 = \frac{n^4\pi^4 EIg}{wl^4} \qquad (n = 1, 2, \ldots),$$

so that the natural frequencies $\omega/2\pi$ are

$$\frac{n^2\pi}{2l^2} \left(\frac{EIg}{w}\right)^{\frac{1}{2}} \qquad (n = 1, 2, \ldots).$$

20.10. Uniform Transmission Line Equations

Consider a transmission line of two long parallel wires (the lead and return) with total resistance R, inductance L, capacity C and leakance

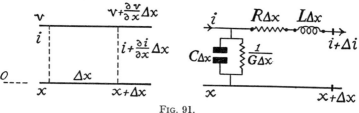

FIG. 91.

G per unit length. In the case of a single (telegraph) wire these constants would refer to unit length of the lead only.

At a distance x from the sending end let i be the current in the line,

v the voltage drop across the wires so that $i + \frac{\partial i}{\partial x} \Delta x$, $v + \frac{\partial v}{\partial x} \Delta x$ are the corresponding values a distance Δx further on. The drop in voltage is $-\frac{\partial v}{\partial x} \Delta x$ due to a resistance $R\Delta x$ and inductance $L\Delta x$. Therefore

$$R\Delta x . i + L\Delta x \frac{\partial i}{\partial t} = - \frac{\partial v}{\partial x} \Delta x$$

or

$$R i + L \frac{\partial i}{\partial t} = - \frac{\partial v}{\partial x}. \quad . \quad . \quad . \quad . \quad (1)$$

The loss in current $-\frac{\partial i}{\partial x} \Delta x$ is due to the capacity and the leakance. Due to the former increasing its charge by $C\Delta x \, \partial v/\partial t$ and the latter causing a leakage of $G\Delta x v$,

$$C\Delta x \frac{\partial v}{\partial t} + G\Delta x . v = - \frac{\partial i}{\partial x} \Delta x$$

or

$$C \frac{\partial v}{\partial t} + G v = - \frac{\partial i}{\partial x} \quad . \quad . \quad . \quad . \quad (2)$$

[If the line between x and $x + \Delta x$ is replaced by the equivalent four-terminal network as shown, the usual circuit relations would give these immediately.]

(1) and (2) are a pair of simultaneous linear partial differential equations of the first order in v and i. In a given case they must be solved with initial values of i and v given as functions of x and t, with given boundary values such as a given voltage or current at an end.

Writing (1) and (2) as

$$\left(R + L \frac{\partial}{\partial t} \right) i = - \frac{\partial v}{\partial x}, \qquad \left(C \frac{\partial}{\partial t} + G \right) v = - \frac{\partial i}{\partial x}$$

we can eliminate i as follows

$$\left(L \frac{\partial}{\partial t} + R \right) \left(C \frac{\partial}{\partial t} + G \right) v = - \left(L \frac{\partial}{\partial t} + R \right) \frac{\partial i}{\partial x}$$

$$= - \frac{\partial}{\partial x} \left(L \frac{\partial}{\partial t} + R \right) i$$

$$= - \frac{\partial}{\partial x} \left(- \frac{\partial v}{\partial x} \right)$$

or

$$CL \frac{\partial^2 v}{\partial t^2} + (CR + LG) \frac{\partial v}{\partial t} + RGv = \frac{\partial^2 v}{\partial x^2} \quad . \quad . \quad . \quad (3)$$

The elimination of v gives the same form of equation with i instead of v

$$CL \frac{\partial^2 i}{\partial t^2} + (CR + LG) \frac{\partial i}{\partial t} + RGi = \frac{\partial^2 i}{\partial x^2} \quad . \quad . \quad . \quad (4)$$

Some Special Cases

(a) The complete solution of (3) and (4) is complicated and, usually, special cases only are considered. Thus in (3) we expect the solution to represent wave motion of decreasing amplitude because of damping. Therefore assume

$$v = Ae^{-\alpha x} \sin p \left(t - \frac{x}{u} \right) \qquad \cdots \cdots \quad (5)$$

which represents a sine wave form of decreasing amplitude moving with velocity *u*, the wave having a frequency $p/2\pi$. $\left[\text{Compare } \sin p \left(t - \frac{x}{u} \right) \text{ with } f(x - ct) \right.$ of 20.8.$\Big]$

From this assumption:

$$\frac{\partial v}{\partial t} = Ae^{-\alpha x} p \cos p \left(t - \frac{x}{u} \right);$$

$$\frac{\partial^2 v}{\partial t^2} = Ae^{-\alpha x} (-p^2) \sin p \left(t - \frac{x}{u} \right);$$

$$\frac{\partial v}{\partial x} = Ae^{-\alpha x} \left\{ -\alpha \sin p \left(t - \frac{x}{u} \right) - \frac{p}{u} \cos p \left(t - \frac{x}{u} \right) \right\};$$

$$\frac{\partial^2 v}{\partial x^2} = Ae^{-\alpha x} \left\{ \left(\alpha^2 - \frac{p^2}{u^2} \right) \sin p \left(t - \frac{x}{u} \right) + \frac{2\alpha p}{u} \cos p \left(t - \frac{x}{u} \right) \right\}.$$

Substituting these values in (3) above and equating the coefficients of the sine and cosine terms,

$$RG - p^2 CL = \alpha^2 - \frac{p^2}{u^2}. \qquad \cdots \cdots \quad (6)$$

$$p(CR + LG) = \frac{2\alpha p}{u}$$

or

$$\alpha = \tfrac{1}{2} u(CR + LG) \qquad \cdots \cdots \quad (7)$$

By (6)

$$\frac{1}{u^2} = CL + \frac{\alpha^2 - RG}{p^2} \qquad \cdots \cdots \quad (8)$$

so that the velocity *u* depends in general on the frequency. For very high frequencies the second term on the right-hand side is very small and

$$u \simeq 1/\sqrt{LC}.$$

(b) The Distortionless Line

Since by (8) the waves of different frequencies travel at different rates, there will be distortion in the wave form. If, however,

$$\alpha^2 = RG \quad \text{or} \quad \alpha = \sqrt{RG}, \quad \text{then by (8)} \quad u = 1/\sqrt{CL}$$

exactly for all frequencies *p*. By (7) using these values for α and *u*

$$2\sqrt{(CLRG)} = CR + LG$$

or

$$(\sqrt{CR} - \sqrt{LG})^2 = 0,$$

$$\therefore \quad CR = LG \qquad \cdots \cdots \quad (9)$$

With this condition between the constants of the line, waves of all frequencies travel with the same velocity $1/\sqrt{CL}$, have the same constant α governing the rate of decrease of amplitude (α = $\sqrt{RG}$ is called the attenuation constant) and there is no distortion of the wave form.

(c) The Leaky Telegraph Wire

Suppose R and G are large in comparison with L and C. *i.e.* assume L = C = 0 in the equations (1) to (4) above.

(3) becomes

$$RGv = \frac{\partial^2 v}{\partial x^2},$$

the solution to which is, as usual,

$$v = Ae^{\sqrt{RG}x} + Be^{-\sqrt{RG}x}. \quad \bullet \quad \bullet \quad \bullet \quad \bullet \quad \bullet \quad (9)$$

Since (1) becomes

$$Ri = -\frac{\partial v}{\partial x}$$

$$= -\frac{\partial}{\partial x}[Ae^{\sqrt{RG}x} + Be^{-\sqrt{RG}x}],$$

therefore

$$i = \sqrt{\frac{G}{R}}[-Ae^{\sqrt{RG}x} + Be^{-\sqrt{RG}x}] \quad \bullet \quad \bullet \quad \bullet \quad (10)$$

(d) The Telegraph Equation

In certain cases, such as with underground and submarine cables, we may assume that the inductance and leakance are zero, *i.e.*, $L = G = 0$. Equations (1) and (2) then become

$$\frac{\partial v}{\partial x} = -Ri, \qquad \frac{\partial i}{\partial x} = -C\frac{\partial v}{\partial t} \quad \bullet \quad \bullet \quad \bullet \quad \bullet \quad \bullet \quad (11)$$

which give

$$\frac{\partial v}{\partial t} = \frac{1}{CR}\frac{\partial^2 v}{\partial x^2}, \qquad \frac{\partial i}{\partial t} = \frac{1}{CR}\frac{\partial^2 i}{\partial x^2} \quad \bullet \quad \bullet \quad \bullet \quad \bullet \quad (12)$$

equations resembling that governing the conduction of heat (and used by Kelvin in 1857 in connection with the first Atlantic cable).

We will solve one problem of practical importance. Suppose the sending end of the line is raised to a potential E_0, the far end being earthed through the receiving apparatus, whose resistance can be assumed zero. When the steady state has been reached, the voltage along the line will be given by

$$v = E_0(1 - x/l),$$

showing a steady fall from E_0 to 0 in the length l of the line.

Suppose, after this steady state has been established, that at time $t = 0$ the sending end is earthed. We must solve the voltage equation in (12) subject to

$$v(x, 0) = E_0(1 - x/l),$$
$$v(0, t) = 0 = v(l, t).$$

If we try the solution

$$v = Ae^{mx + nt}$$

substitution in (12) for v gives

$$n = m^2/CR$$

n must be negative or v would increase with t. Put $m = pj$ where p is real, then

$$n = -p^2/CR$$

and

$$v = Ae^{jpx - p^2t/CR}$$

$$= Ae^{-p^2t/CR}[\cos px + j\sin px],$$

showing that the cosine and sine terms separately are solutions, and hence we take

$$v = e^{-p^2t/CR}[L\cos px + M\sin px]$$

$v(0, t) = 0:$ $0 = L$

giving $v = Me^{-p^2t/CR}\sin px,$

$v(l, t) = 0:$ $0 = \sin pl,$

$$\therefore \quad pl = n\pi \qquad (n = 1, 2, \ldots),$$
$$p = n\pi/l \qquad (n = 1, 2, \ldots).$$

We now have

$$v = \sum_{n=1}^{\infty} M_n e^{-n^2 \pi^2 t / CRl^2} \sin \frac{n\pi x}{l}.$$

Using the final condition $v(x, 0) = E_0 (1 - x/l)$,

$$E_0 (1 - x/l) = \sum_{n=1}^{\infty} M_n \sin \frac{n\pi x}{l} \quad \text{for} \quad 0 < x < l.$$

Hence

$$M_n = \frac{2}{l} \int_0^l E_0 (1 - x/l) \sin \frac{n\pi x}{l} \, dx$$

$$= \frac{2E_0}{n\pi},$$

so that

$$v = \frac{2E_0}{\pi} \sum_{n=1}^{\infty} \frac{1}{n} e^{-n^2 \pi^2 t / CRl^2} \sin \frac{n\pi x}{l} \quad \cdot \quad \cdot \quad \cdot \quad \cdot \quad (13)$$

From this we deduce that if a line is dead with its receiving end earthed as above and the voltage E_0 is imposed at time $t = 0$ on the receiving end, then the voltage V at time t at the point x is given by, using v from (13),

$$V = E_0 (1 - x/l) - v.$$

Using

$$i = -\frac{1}{R} \frac{\partial v}{\partial x},$$

$$i = \frac{E_0}{Rl} \left[1 + 2 \sum_{n=1}^{\infty} e^{-n^2 \pi^2 t / CRl^2} \cos \frac{n\pi x}{l} \right]$$

from which the current at the sending end $(x = 0)$ or the receiving end $(x = l)$ can be obtained.

20.11. The Electromagnetic Equations

Finally, we will obtain Maxwell's electromagnetic equations using SI units.

(a) Ampere proposed the law, based on experimental work, that every electric current has a magnetic field associated with it, such that the line integral of the magnetic force taken round any closed curve is proportional to the current passing through the closed curve. In the SI system of units the constant of proportionality required in this law is taken as unity.

Before applying this law we must note that the line integral of any

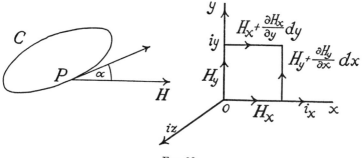

Fig. 92.

vector quantity H along a given curve C is obtained by multiplying the resolute of H along the tangent to the curve at any chosen point P by an element of length ds of the curve and summing this product over the total path or length of the curve. In Fig. 92 (left-hand diagram) this line integral is

$$\lim_{ds \to 0} \Sigma \text{ H} \cos \alpha \; ds = \int_{POP} \text{H} \cos \alpha \; ds$$

Now consider a point O in a conducting medium and three mutually perpendicular axes at O. Suppose that the components of H along these axes are H_x, H_y, H_z and the current densities i_x, i_y, i_z respectively. The line integral of H along the small rectangle shown of sides dx, dy and which is chosen to lie in the plane xy is (compare page 200)

$$H_x dx + \left(H_y + \frac{\partial H_y}{\partial x}dx\right) dy - \left(H_x + \frac{\partial H_x}{\partial y}dy\right) dx - H_y dy$$
$$= \left(\frac{\partial H_y}{\partial x} - \frac{\partial H_x}{\partial y}\right) dx dy,$$

The current through the rectangle is $i_z \, dx dy$, since we may assume that i_z is constant over this small area. Ampere's law now gives

$$\frac{\partial H_y}{\partial x} - \frac{\partial H_x}{\partial y} = i_z . \quad \ldots \ldots \quad (1)$$

Equations for i_x and i_y may be obtained in a similar manner or by a cyclical interchange of x, y and z. [Note that the conventional direction for the line integral is defined by the right-hand screw rule.]

(b) By applying a similar method to the components (i_x, i_y, i_z) as used for Laplace's equation (12.6), but considering a box of sides dx, dy, dz surrounding the point O, we will find the expression

$$\left(\frac{\partial i_x}{\partial x} + \frac{\partial i_y}{\partial y} + \frac{\partial i_z}{\partial z}\right) dx dy dz$$

for the total difference between the current entering and leaving this box. If there is no positive or negative accumulation of current, then

$$-\frac{\partial i_x}{\partial x} + \frac{\partial i_y}{\partial y} + \frac{\partial i_z}{\partial z} = 0.$$

More generally, if ρ is the density and $\rho dx dy dz$ the total charge enclosed, then the change is

$$-\frac{\partial}{\partial t}(\rho dx dy dz) = -\frac{\partial \rho}{\partial t} dx dy dz$$

giving, on equating the two expressions,

$$\frac{\partial i_x}{\partial x} + \frac{\partial i_y}{\partial y} + \frac{\partial i_z}{\partial z} = -\frac{\partial \rho}{\partial t} \quad \ldots \ldots \quad (2)$$

The minus sign is inserted because we have taken the outward-flowing current as positive so that a positive increment in this must lead to a negative increment in the charge within the box.

(c) Faraday proposed the law, based on his experimental work, that in a single loop circuit the e.m.f. induced in the circuit is proportional to the rate of change of flux through the circuit. Since e.m.f. may be taken as the line integral of field intensity, we have by considering an elemental loop as before

$$\left(\frac{\partial E_z}{\partial y} - \frac{\partial E_y}{\partial z}\right) dy\,dz = -\frac{\partial}{\partial t}(B_x dy\,dz)$$

where the negative sign must be taken to account for the fact that the direction of the e.m.f. is such as to drive currents which will set up a flux opposing the change in flux. This equation reduces to

$$\frac{\partial E_z}{\partial y} - \frac{\partial E_y}{\partial z} = -\frac{\partial B_x}{\partial t} \quad \cdots \cdots \quad (3)$$

Two similar equations may be obtained for the other components of B.

Since B may be defined by

$$B = \mu_0 \mu\, H$$

where μ is the relative permeability of the medium and μ_0 a constant introduced as a result of the system of units chosen, (3) becomes

$$\frac{\partial E_z}{\partial y} - \frac{\partial E_y}{\partial z} = -\mu_0 \mu \frac{\partial H_x}{\partial t} \quad \cdots \cdots \quad (4)$$

[μ_0 is usually called the permeability of free space and has a value $4\pi \times 10^{-7}$.]

(d) Finally, we return to the idea of an elemental box of sides dx, dy, dz and apply Gauss' Theorem. This states that the surface integral of the electric displacement over a closed surface equals the charge enclosed by the surface. The surface integral is taken by resolving the vector displacement at any chosen point normal to the surface there, multiplying by an element of area at the point and summing this product over the whole surface. For the box, this gives

$$\frac{\partial D_x}{\partial x} + \frac{\partial D_y}{\partial y} + \frac{\partial D_z}{\partial z} = \rho \quad \cdots \cdots \quad (5)$$

D is defined in terms of E by the equation

$$D = \kappa \kappa_0 E$$

where κ is the dielectric constant of the medium and κ_0 is a constant introduced by the choice of units. κ_0 is sometimes termed the dielectric constant of free space and has the value $10^{-9}/36\pi$.

Equation (5) may now be rewritten in terms of E in the form

$$\frac{\partial E_x}{\partial x} + \frac{\partial E_y}{\partial y} + \frac{\partial E_z}{\partial z} = \frac{\rho}{\kappa \kappa_0} \quad \cdots \cdots \quad (6)$$

Similar reasoning applied to the magnetic field gives

$$\frac{\partial H_x}{\partial x} + \frac{\partial H_y}{\partial y} + \frac{\partial H_z}{\partial z} = 0 \quad \cdots \cdots \quad (7)$$

since the total pole strength within any closed surface is zero.

20.12. Maxwell's Equations

The above were the electromagnetic equations before Maxwell's hypothesis of the *displacement current*.

Difficulties arise even in such a simple case as that of a current flowing when the plates of a condenser are joined—owing to the discontinuity at the plates in the usual *conduction* current. Maxwell invented a *displacement* current of magnitude

$$\frac{\partial D}{\partial t}$$

where D is the electric displacement between the plates. With the usual notation

$$D = \sigma = Q/A$$

so that for the total displacement current over the area A, in the case of the discharging condenser, above,

$$A \frac{\partial D}{\partial t} = A \cdot \frac{1}{A} \frac{\partial Q}{\partial t} = -i.$$

Thus the total displacement current equals the conduction current and, being in the opposite direction, provides us with a continuous circuit of current. The gap between the plates of the condenser now ceases to cause any (theoretical) difficulty.

The total current must now be taken as

$$i + \frac{\partial D}{\partial t},$$

the second term being zero when the field is not varying with time.

The equation (1) obtained in (*a*) above now becomes

$$\frac{\partial H_y}{\partial x} - \frac{\partial H_x}{\partial y} = i_z + \frac{\partial D_z}{\partial t} \quad \ldots \ldots \quad \text{(M.)}$$

whilst the equation of continuity of the conduction current in (*b*) becomes

$$\Sigma \frac{\partial}{\partial x} \left(i_x + \frac{\partial D_x}{\partial t} \right) = 0 \quad \text{or} \quad \Sigma \frac{\partial i_x}{\partial x} + \frac{\partial}{\partial t} \Sigma \frac{\partial D_x}{\partial x} = 0,$$

which gives, finally (since $\Sigma \partial D_x / \partial x = \rho$, as in (5)),

$$\Sigma \frac{\partial i_x}{\partial x} + \frac{\partial \rho}{\partial t} = 0$$

20.13. The Solution for Free Space

If we put $\kappa = \mu = 1$ and $\rho = i = 0$, the above equations reduce to the *free space* equations :

from (M.) $\quad \dfrac{\partial H_z}{\partial y} - \dfrac{\partial H_y}{\partial z} = \kappa_0 \dfrac{\partial E_x}{\partial t}.$ $\qquad$ from (3) $\quad \dfrac{\partial E_z}{\partial y} - \dfrac{\partial E_y}{\partial z} = -\mu_0 \dfrac{\partial H_x}{\partial t}$

$\qquad\qquad \dfrac{\partial H_x}{\partial z} - \dfrac{\partial H_z}{\partial x} = \kappa_0 \dfrac{\partial E_y}{\partial t} \qquad\qquad\qquad\qquad \dfrac{\partial E_x}{\partial z} - \dfrac{\partial E_z}{\partial x} = -\mu_0 \dfrac{\partial H_y}{\partial t}$

$\qquad\qquad \dfrac{\partial H_y}{\partial x} - \dfrac{\partial H_x}{\partial y} = \kappa_0 \dfrac{\partial E_z}{\partial t} \qquad\qquad\qquad\qquad \dfrac{\partial E_y}{\partial x} - \dfrac{\partial E_x}{\partial y} = -\mu_0 \dfrac{\partial H_z}{\partial t}$

from (6) $\quad \dfrac{\partial E_x}{\partial x} + \dfrac{\partial E_y}{\partial y} + \dfrac{\partial E_z}{\partial z} = 0.$ $\qquad$ from (7) $\quad \dfrac{\partial H_x}{\partial x} + \dfrac{\partial H_y}{\partial y} + \dfrac{\partial H_z}{\partial z} = 0.$

The first six equations contain both H and E, and we will investigate what happens when E is eliminated from the left-hand set.

Differentiating the first left-hand equation with respect to t gives

$$\kappa_0 \frac{\partial^2 E_x}{\partial t^2} = \frac{\partial^2 H_z}{\partial y \partial t} - \frac{\partial^2 H_y}{\partial z \partial t}$$

$$= -\frac{1}{\mu_0} \frac{\partial}{\partial y} \left(\frac{\partial E_y}{\partial x} - \frac{\partial E_x}{\partial y} \right) + \frac{1}{\mu_0} \frac{\partial}{\partial z} \left(\frac{\partial E_x}{\partial z} - \frac{\partial E_z}{\partial x} \right)$$

by using the second and third of the right-hand equations. This simplifies to

$$\kappa_0 \frac{\partial^2 E_x}{\partial t^2} = -\frac{1}{\mu_0} \left[\frac{\partial^2 E_y}{\partial y \partial x} - \frac{\partial^2 E_x}{\partial y^2} - \frac{\partial^2 E_x}{\partial z^2} + \frac{\partial^2 E_z}{\partial z \partial x} \right].$$

From the fourth left-hand equation

$$\frac{\partial E_z}{\partial x} = -\frac{\partial E_y}{\partial y} - \frac{\partial E}{\partial z},$$

which gives

$$\frac{\partial^2 E_z}{\partial x^2} = -\frac{\partial^2 E_y}{\partial x \partial y} - \frac{\partial^2 E_z}{\partial x \partial z},$$

and substitution in the above gives finally

$$\frac{1}{\mu_0 \kappa_0}\left[\frac{\partial^2 E_z}{\partial x^2} + \frac{\partial^2 E_z}{\partial y^2} + \frac{\partial^2 E_z}{\partial z^2}\right] = \frac{\partial^2 E_z}{\partial t^2} \quad \text{or} \quad \nabla^2 E_z = \frac{1}{\mu_0 \kappa_0}\frac{\partial^2 E_z}{\partial t^2}.$$

(A similar equation can be obtained for each component of E and H.)   This equation is the three-dimensional form of the equation obtained in 20.7, and represents a wave (or two oppositely directed waves) travelling with velocity $1/\sqrt{(\mu_0 \kappa_0)}$.   This shows that any condition of magnetic or electric intensity will travel outwards with velocity $1/\sqrt{(\mu_0 \kappa_0)}$—the velocity of light—in free space, and led Maxwell to assert that light waves were electromagnetic waves consisting of an electric plus magnetic intensity both travelling at this speed $1/\sqrt{(\mu_0 \kappa_0)}$.

As a further illustration we will consider the following.

20.14. Plane Waves in Free Space

Consider a plane wave, which for convenience is taken as perpendicular to the x axis.   As t varies, both E and H vary, but each will have the same value anywhere on the plane, and each is a function of x and t only.

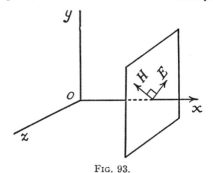

Fig. 93.

The equations of 20.13 reduce to

$$0 = \kappa_0 \frac{\partial E_z}{\partial t} \qquad\qquad 0 = -\mu_0 \frac{\partial H_z}{\partial t}$$

$$-\frac{\partial H_z}{\partial x} = \kappa_0 \frac{\partial E_y}{\partial t} \qquad\qquad -\frac{\partial E_z}{\partial x} = -\mu_0 \frac{\partial H_y}{\partial t}$$

$$\frac{\partial H_y}{\partial x} = \kappa_0 \frac{\partial E_z}{\partial t} \qquad\qquad \frac{\partial E_y}{\partial x} = -\mu_0 \frac{\partial H_z}{\partial t}$$

$$\frac{\partial E_z}{\partial x} = 0 \qquad\qquad \frac{\partial H_z}{\partial x} = 0$$

From the first and last pair E_z and H_z are both constant, but this constant is zero, since no magnetic dipoles or electric charges are present in the field.   The electric and magnetic intensities are therefore entirely parallel to the yz plane. We may therefore assume that

$$E_y = g(x - ct), \qquad E_z = h(x - ct), \qquad c = \frac{1}{\sqrt{(\mu_0 \kappa_0)}}$$

where g and h are arbitrary functions, for this form gives a wave progressing in the x direction with velocity c (see 20.8). Substitution gives :

$$-\frac{\partial H_z}{\partial x} = \kappa_0\{-cg'(x-ct)\}, \qquad -h'(x-ct) = -\mu_0\frac{\partial H_y}{\partial t};$$

$$\frac{\partial H_y}{\partial x} = \kappa_0\{-ch'(x-ct)\}, \qquad g'(x-ct) = -\mu_0\frac{\partial H_z}{\partial t}.$$

Therefore, the constants of integration being zero,

$$H_z = \sqrt{\frac{\kappa_0}{\mu_0}}\,g(x-ct) = \sqrt{\frac{\kappa_0}{\mu_0}}\,E_y,$$

$$H_y = -\sqrt{\frac{\kappa_0}{\mu_0}}\,h(x-ct) = -\sqrt{\frac{\kappa_0}{\mu_0}}\,E_z.$$

and since $H_x = 0 = E_x$, using the formula (14.7) from solid geometry for the angle between H and E

$$|H||E|\cos\theta = H_xE_x + H_yE_y + H_zE_z$$

$$= 0 + H_yH_z\sqrt{\frac{\mu_0}{\kappa_0}} + H_z\left(-\sqrt{\frac{\mu_0}{\kappa_0}}\,H_y\right)$$

$$= 0.$$

H and E are therefore perpendicular, forming with the x axis a set of mutually perpendicular directions as shown in the figure.

EXERCISE 51

1. Find a solution of $\dfrac{\partial z}{\partial x} + b\dfrac{\partial z}{\partial y} = 0$, given that z is never infinite and that $\dfrac{\partial z}{\partial x} = 0$ when $x = y = 0$.

2. Find a solution of $\dfrac{\partial^2 y}{\partial x^2} + a^2\dfrac{\partial^2 y}{\partial t^2} = 0$ such that $y = 0$ when $t = \infty$ and $y = 0$ when $x = 0$.

3. Find a solution of $\dfrac{\partial^2 z}{\partial x^2} = \dfrac{1}{a^2}\dfrac{\partial^2 z}{\partial y^2}$, given that z is never infinite for real values of x or y and that $z = 0$ when x or $y = 0$.

4. Find a solution of $\dfrac{\partial^2 y}{\partial x^2} - a^2\dfrac{\partial^2 y}{\partial t^2} = 0$ if $y = 0$ when $x = +\infty$ and when $t = +\infty$.

5. Find a solution of $\dfrac{\partial^2 V}{\partial x^2} = \dfrac{\partial V}{\partial t}$ given that $V = 0$ when $t = +\infty$ and when $x = 0$ or l.

6. Find a solution of $\dfrac{\partial^2 y}{\partial t^2} = a^2\dfrac{\partial^2 y}{\partial x^2}$ which is zero for $x = 0$ and reduces to $A\cos(pt + \alpha)$ when $x = c$. [This will solve the problem of a string tied at each end, of which a given point $x = c$ is made to move with the periodic vibration $A\cos(pt + \alpha)$.]

7. Find a solution of $\dfrac{\partial^2 u}{\partial t^2} = a^2\dfrac{\partial^2 u}{\partial x^2}$ in the form $f(x)\sin nt$ such that $\dfrac{\partial u}{\partial t} = k$, a constant, when $x = 0$ and $t = 0$; $\dfrac{\partial u}{\partial x} = 0$ when $x = 0$ for all t.

8. Find a solution of $\dfrac{\partial^2 z}{\partial x^2} + \dfrac{\partial^2 z}{\partial y^2} = 0$ such that when $y = n\pi/a$ (n being any integer), $z = 0$ and $\dfrac{\partial z}{\partial y} = e^{2ax}$.

9. Find a solution of $\dfrac{\partial^2 u}{\partial x^2} + \dfrac{\partial^2 u}{\partial y^2} = 0$, given that: (a) $u = \sin x$ when $y = 0$; (b) $u = 0$ when $x = \pm\pi$; (c) $u \to 0$ as $y \to \infty$.

10. Find a solution of $\dfrac{\partial^2 y}{\partial t^2} = a^2 \dfrac{\partial^2 y}{\partial x^2}$, such that: (a) $y = 0$ when $x = 0$; (b) $y = 0$ when $x = l$; (c) $y = 0$ when $t = 0$; (d) $\dfrac{\partial y}{\partial t} = A \sin\left(\dfrac{\pi x}{l}\right)$ when $t = 0$.

11. (a) Find a solution of the heat-transmission equation $\dfrac{\partial^2 \theta}{\partial x^2} = \dfrac{1}{\kappa}\dfrac{\partial \theta}{\partial t}$ applicable to the case of a long thin bar with a periodic change of temperature $\theta = \theta_0 \sin nt$ at the end $x = 0$.

[Physical considerations will show that the temperature in the solid must be periodic too. Assume

$$\theta = X e^{int},$$

where X is a function of x only. Substitution gives

$$\frac{d^2 X}{dx^2} - \frac{in}{\kappa} X = 0,$$

the general solution of which is, since $\sqrt{i} = \pm (1 + i)/\sqrt{2}$,

$$X = A e^{(1 + i)x(n/2\kappa)^{1/2}} + B e^{-(1 + i)x(n/2\kappa)^{1/2}},$$

and since X must be finite as $x \to \infty$, A = 0. Taking the imaginary part of the solution,

$$\theta = \theta_0 e^{-x(n/2\kappa)^{1/2}} \sin\left\{nt - x\left(\frac{n}{2\kappa}\right)^{1/2}\right\}$$

is the required solution for the surface temperature $\theta = \theta_0 \sin nt$.]

12. (a) If Y cos mx is a solution of $\dfrac{\partial z}{\partial y} = a^2 \dfrac{\partial^2 z}{\partial x^2}$, where Y is a function of y only, find a form for Y.

(b) Use exponential functions to determine particular solutions of the equation

$$\frac{\partial^2 u}{\partial x^2} + \frac{\partial^2 u}{\partial y^2} = 0,$$

such that $u = 0$ when $y = 0$ and $\dfrac{\partial u}{\partial x} = 0$ when $x = 0$ and 1.

13. If $V = x^n f\left(\dfrac{y}{x}\right)$, where f is any function, show that

$$x\frac{\partial V}{\partial x} + y\frac{\partial V}{\partial y} = nV.$$

If $\dfrac{\partial^2 V}{\partial x \partial y} = 0$ and u denotes $\dfrac{y}{x}$, show that

$$(n - 1)\frac{df}{du} = u\frac{d^2 f}{du^2}$$

and hence show that $V = ax^n + by^n$ where a and b are constants.   [L.U.]

14. If $x = r \cos \theta$, $y = r \sin \theta$, transform $\dfrac{\partial^2 V}{\partial x^2} + \dfrac{\partial^2 V}{\partial y^2} = 0$ to the form

$$\frac{\partial^2 V}{\partial r^2} + \frac{1}{r}\frac{\partial V}{\partial r} + \frac{1}{r^2}\frac{\partial^2 V}{\partial \theta^2} = 0.$$

Show that if $V = r^n f(\theta)$ is a solution of the equation, then

$$f(\theta) = A \cos n\theta + B \sin n\theta,$$

where A and B are arbitrary constants.   [L.U.]

15. Transform the equation $\dfrac{\partial^2 z}{\partial y^2} = a^2 \dfrac{\partial^2 z}{\partial x^2}$ by means of the substitution

$$x + ay = X, \qquad x - ay = Y.$$

$$\left[\text{Show that}\qquad \frac{\partial}{\partial x} = \frac{\partial}{\partial X} + \frac{\partial}{\partial Y}\right.$$

$$\frac{\partial}{\partial y} = a\left(\frac{\partial}{\partial X} - \frac{\partial}{\partial Y}\right)$$

and
$$a^2\frac{\partial^2 z}{\partial x^2} - \frac{\partial^2 z}{\partial y^2} = 4a^2\frac{\partial^2 z}{\partial X \partial Y}\right]$$

Hence show that the general solution is
$$z = f(x + ay) + F(x - ay),$$

where f, F are any two arbitrary functions.

16. If $x = \alpha + \beta$, $y = \alpha - \beta$ where (x, y) (α, β) are pairs of independent variables and if V is a function of x and y, and hence of α and β, show that

$$\frac{\partial^2 V}{\partial x^2} - \frac{\partial^2 V}{\partial y^2} = \frac{\partial^2 V}{\partial \alpha \partial \beta}.$$

Hence find a particular solution of

$$\frac{\partial^2 V}{\partial x^2} - \frac{\partial^2 V}{\partial y^2} = x^2 + y^2. \qquad \text{[L.U.]}$$

17. If the equation $x^2\dfrac{\partial^2 u}{\partial x^2} + x\dfrac{\partial u}{\partial x} + \dfrac{\partial^2 u}{\partial y^2} = 0$ has a solution of the form $u = XY$ where X, Y are respectively functions of x and y only, find the differential equations satisfied by X and Y and solve them when Y involves real trigonometrical functions only.

If $\dfrac{\partial u}{\partial x} = -\cos 2y$ when $x = a$ and u tends to zero as x tends to infinity, find u.

[L.U.]

18. Show that a solution of the equation $\dfrac{\partial^2 u}{\partial x^2} + \dfrac{\partial^2 u}{\partial y^2} = 0$ which will satisfy the conditions $u = 0$ when $x = 0$ and $x = \pi$, $u = 0$ when $y = \infty$, and $u = 1$ when $y = 0$ is

$$u = \frac{4}{\pi}(e^{-y}\sin x + \tfrac{1}{3}e^{-3y}\sin 3x + \tfrac{1}{5}e^{-5y}\sin 5x + \ldots).$$

19. Express $y = x\left(1 - \dfrac{x}{l}\right)$ as a sine series in the range $x = 0$ to l. Hence find a solution of $\dfrac{\partial^2 y}{\partial x^2} + \dfrac{\partial^2 y}{\partial t^2} = 0$ such that $y = 0$ when $t = \infty$ and when $x = 0$. Also when $t = 0$, $y = x(1 - x/l)$ for all values of x from $x = 0$ to l.

20. Find a solution of $\dfrac{\partial^2 y}{\partial t^2} = c^2\dfrac{\partial^2 y}{\partial x^2}$ such that y involves x trigonometrically; also:

(i) $y = 0$ when $x = 0$ or l for all t;

(ii) $\dfrac{\partial y}{\partial t} = 0$ when $t = 0$ for all x;

(iii) $y = \dfrac{2ax}{l}$ from $x = 0$ to $x = \tfrac{1}{2}l$ and $y = \dfrac{2a}{l}(l - x)$ from $x = \tfrac{1}{2}l$ to $x = l$ when $t = 0$.

21. A tightly stretched string with fixed end points $x = 0$ and l is initially at rest in its equilibrium position. If it is set vibrating by giving each point a velocity

$$\left(\frac{\partial y}{\partial t}\right)_{t=0} = 3(lx - x^2).$$

show that

$$y(x, t) = \frac{24l^3}{c\pi^4}\sum_1^\infty \frac{1}{(2n - 1)^4}\sin\frac{(2n - 1)\pi x}{l}\sin\frac{(2n - 1)c\pi t}{l}.$$

360

22. A rod of length l has its ends A and B kept at 0°C and 100°C respectively until steady-state conditions prevail. If the temperature of B is then reduced suddenly to 0°C and kept at this temperature while that of A is maintained constant, find the temperature $\theta(x, t)$ at a distance x from A at time t.

$$\left[\text{Solve } \frac{\partial \theta}{\partial t} = h^2 \frac{\partial^2 \theta}{\partial x^2} \text{ subject to } \theta(x, 0) = 100x/l. \quad \theta(0, t) = 0 = \theta(l, t). \right]$$

23. A homogeneous slab of conducting material is bounded by the planes $x = 0$, $x = \pi$, which are kept at zero temperature. The temperature $u(x, t)$ satisfies the equation

$$\frac{\partial u}{\partial t} = k \frac{\partial^2 u}{\partial x^2}.$$

Find solutions of this equation which satisfy the conditions on the faces of the slab and which do not become infinite as the time t increases indefinitely. If initially when $t = 0$ the slab is raised to temperature

$$u(x, 0) = \tfrac{1}{4}\pi Tx \qquad (0 \leqslant x \leqslant \tfrac{1}{2}\pi)$$
$$= \tfrac{1}{4}\pi T(\pi - x) \qquad (\tfrac{1}{2}\pi \leqslant x \leqslant \pi),$$

where T is a constant, show that the temperature at any subsequent time is

$$T(e^{-kt} \sin x - \tfrac{1}{9}e^{-kt} \sin 3x + \tfrac{1}{25}e^{-25kt} \sin 5x \ldots).$$

24. Obtain the equation

$$\frac{\partial^2 y}{\partial t^2} + \frac{Ek^2}{\rho} \frac{\partial^4 y}{\partial x^4} = 0$$

for the transverse vibrations of a uniform rod and verify that

$$y = (A \sin nx + B \cos nx + C \operatorname{sh} nx + D \operatorname{ch} nx) \cos (pt + \alpha)$$

is a solution provided $n^4 = \rho p^2/Ek^2$. Show that if the rod is of length l and mounted in short bearings at the ends, it has a simple vibration given by

$$y = A \sin \frac{\pi x}{l} \cos \left(\frac{\pi^2 t}{l^2} \sqrt{\frac{Ek^2}{\rho}} \right).$$

25. In the above problem if the rod is clamped at each end, show that it can vibrate periodically with a frequency f given by

$$f = 2\frac{\theta^2 k}{\pi l^2} \sqrt{\frac{E}{\rho}},$$

where θ is any root of the equation $\tan \theta = \pm \tanh \theta$.

26. In the above problem if the rod, a cantilever (clamped at one end and free at the other), is observed to vibrate with a minimum frequency f_0, show that

$$E = \frac{4\pi^2 \rho l^4 f_0^2}{k^2 \theta_0^4},$$

where θ_0 is the smallest positive root of the equation

$$\cosh \theta + \sec \theta = 0.$$

Show that $\tfrac{1}{2}\pi < \theta_0 < \pi$.

27. Obtain the equation

$$\frac{\partial^2 \theta}{\partial x^2} = \frac{1}{c^2} \frac{\partial^2 \theta}{\partial t^2}$$

for the free torsional oscillation of a uniform circular shaft.

A uniform circular shaft is clamped at one end and carries at the other end a thin flywheel whose axial moment of inertia is equal to that of the whole shaft. The lowest frequency of free torsional oscillations of this system is f, but if the flywheel is removed it becomes f_0. Show that

$$\frac{\pi f}{2f_0} = \cot \frac{\pi f}{2f_0}.$$

28. A vertical column of fluid of mass M is contained in a tube of constant cross-section and supports at the top a piston of mass m, which is free to move vertically. Obtain the equation

$$\frac{\partial^2 y}{\partial x^2} = \frac{1}{c^2}\frac{\partial^2 y}{\partial t^2}$$

for the longitudinal vibrations of the column and, assuming the density of the fluid to be independent of x, show that a normal mode of vibration has frequency $2\theta f_0/\pi$, where f_0 is the fundamental frequency when the piston is absent and θ is a root of the equation $\cot \theta = \frac{m}{M}\theta$.

29. The equation of the transverse vibrations of a membrane can be put in the form

$$\frac{\partial^2 w}{\partial r^2} + \frac{1}{r}\frac{\partial w}{\partial r} + \frac{1}{r^2}\frac{\partial^2 w}{\partial \theta^2} = \frac{1}{c^2}\frac{\partial^2 w}{\partial t^2}.$$

Show that this equation has a solution of the form

$$w = f(x) \cos(n\theta + \alpha) \cos(pt + \epsilon),$$

where $x = pr/c$ and α, ϵ are constants, provided that $f(x)$ satisfies the equation

$$x^2\frac{d^2 f}{dx^2} + x\frac{df}{dx} + (x^2 - n^2)f = 0.$$

30. A straight elastic rod oscillates along its length by stretching and compressing. It is of weight w per unit length and area of cross-section A. If T is the tension at a distance x from a fixed point in the line of the rod and u the extension, show that

$$\frac{w}{g}\frac{\partial^2 u}{\partial t^2} = \frac{\partial T}{\partial x}, \qquad EA\frac{\partial u}{\partial x} = T.$$

Deduce that in longitudinal motion waves travel along the rod with a velocity $(gEA/w)^{1/2}$ where E is Young's modulus for the rod.

31. If $y = e^{nx + j(wt + a)}$ satisfies

$$\frac{\partial^4 y}{\partial x^4} = k\frac{\partial^2 y}{\partial t^2},$$

where n, w, α, k are real constants, prove that $n = \pm(1 + j)m$ or $\pm(1 - j)m$ where $4m^4 = kw^2$. Deduce that $e^{mx}\sin(mx + wt + \alpha)$ and $e^{-mx}\sin(mx + wt + \alpha)$ also satisfy the given equation. [L.U.]

32. A voltage $V(t)$ is applied at time $t = 0$ to the end $x = 0$ of the line lying $0 \leqslant x \leqslant +\infty$, for which $R = G = 0$ (*i.e.*, a lossless line) in which the initial charge and current are zero. Show that the voltage at the point x is zero up to time x/c and is then $V(t - x/c)$ for $t > x/c$, *i.e.*, it is the voltage applied at $x = 0$ delayed by a time interval x/c.

For a distortionless line $(R/L = G/C)$ under the same conditions show that the voltage at x is zero up to time x/c and is then $e^{-Rx/Lc} V(t - x/c)$ for $t > x/c$, *i.e.*, has the same form as the applied voltage but is attenuated by the factor $\exp(-Rx/Lc)$.

33. An electric cable has resistance R per unit length, while the resistance of unit length of the insulation is S. If v is the voltage and i the current at a distance x from one end, show that: (i) $dv/dx + Ri = 0$; (ii) $d^2v/dx^2 = Rv/S$. If the cable of length L is supplied at one end at a voltage V and is insulated at the other, find the voltage drop on the cable and show that the current entering is

$$\{V/\sqrt{(RS)}\} \tanh\{L\sqrt{(R/S)}\}. \qquad \text{[L.U.]}$$

34. In the case of the leaky telegraph wire, using equations (9) and (10) show that if the voltage at the sending end $x = 0$ is E_0 and

(a) the line is very long, *i.e.*, $v = 0$, $i = 0$ when $x = \infty$, then

$$v = E_0 e^{-ax}, \qquad i = E_0\sqrt{\frac{G}{R}} \cdot e^{-ax}. \qquad (\alpha^2 = RG).$$

(b) the line is of length l and the farther end is earthed

$$v = E_0 \frac{\sinh \alpha(l - x)}{\sinh \alpha l}, \qquad i = E_0 \sqrt{\frac{G}{R}} \cdot \frac{\cosh \alpha(l - x)}{\sinh \alpha l}.$$

(c) the line is of length l and the farther end is insulated ($i = 0$ when $x = l$)

$$v = E_0 \frac{\cosh \alpha(l - x)}{\cosh \alpha l}, \qquad i = E_0 \sqrt{\frac{G}{R}} \cdot \frac{\sinh \alpha(l - x)}{\cosh \alpha l}.$$

35. The potential v at a distance x along a certain cable satisfies the equation

$$\frac{\partial^2 v}{\partial x^2} = rc \frac{\partial v}{\partial t},$$

where r and c are respectively the resistance and capacitance per unit length and are constant. After the cable has been charged to a uniform potential V both ends are short-circuited at the instant $t = 0$. Show that for $t > 0$, the total electrostatic energy in the cable is

$$\frac{4}{\pi^2} CV \sum_{n=0}^{\infty} \frac{1}{(2n + 1)^2} \exp\left(- \frac{\pi^2}{RC} (2n + 1)^2 t\right),$$

where R and C are respectively the total resistance and capacitance of the cable. (The electrostatic energy per unit length is $\frac{1}{2}cv^2$.)

ANSWERS

Exercise 1, *page* 4.

1. 0.   2. 1.   3. 1.   4. 0.   5. $\frac{1}{2}$.   6. $\frac{1}{2}$.

7. 1.   8. $-\frac{1}{10}$.   9. $\dfrac{Wx}{48EI}$ $(3l^2 - 4x^2)$.   10. $\dfrac{t}{2n} \sin nt$.

12. $\sec \theta = 1 + \frac{1}{2}\theta^2 + \frac{5}{24}\theta^4 + \dots \quad |\theta| < \frac{1}{2}\pi$.   13. (*a*) $4f'(c)$; (*b*) $f''(x)$.

14. $\frac{1}{2}a^2$.   15. $\frac{7}{360}$; $-\frac{1}{45}$.   18. $-\frac{1}{2}c(nt \cos nt - \sin nt)$.

23. $2x_1 - x_1 \log (1 - x_1{}^2) - \log \{(1 + x_1)/(1 - x_1)\}$.

Exercise 2, *page* 10.

1. $\rho = \dfrac{2}{15\sqrt{5}x^{\frac{1}{2}}} (1 + 125x^3)^{3/2}$; $\frac{9}{40}$.

2. $\dfrac{dy}{dx} = \cot \theta$, $\dfrac{d^2y}{dx^2} = -\dfrac{1}{a} \operatorname{cosec}^3 \theta$; $\rho = a$.   3. $\dfrac{2}{a^{\frac{1}{2}}} (a + x)^{3/2}$.

5. $\dfrac{(a^2 \sin^2 \theta + b^2 \cos^2 \theta)^{3/2}}{ab}$; $\left(\dfrac{x}{b}\right)^{2/3} + \left(\dfrac{y}{a}\right)^{2/3} = \lambda^{2/3}$.   7. $a \operatorname{ch}^2 \left(\dfrac{x}{a}\right)$.

8. $27ay^2 = 4(x - 2a)^3$.   9. $4a$.

10. $p^2y + x - 2cp = 0$; $py - p^3x + c(p^4 - 1) = 0$.   11. $\frac{125}{448}$; $(\frac{149}{448}, \frac{95}{112})$

12. $(x + a)^2 + \left(y - \dfrac{19a}{6}\right)^2 = \left(\dfrac{a}{6}\right)^2$.   13. 9/4.

17. $y - tx = 1 - t^3$.   21. $y^4 - 2axy^2 + 4a^2y^4 + 8a^4 = 0$.

22. (*a*) $tx + y = at^3 + 2at$; (*b*) $x = at^2 - l(1 + t^2)^{-\frac{1}{2}}$, $y = 2at + lt(1 + t^2)^{-\frac{1}{2}}$.

Exercise 3, *page* 14.

5. $x\{(\log x)^3 - 3 (\log x)^2 + 6 \log x - 6\}$.   9. $33.6 - 40\sqrt{5}/3 = 3.8$.

10. $13/15 - \frac{1}{4}\pi = 0.0813$.   11. (i) $\pi^5 - 20\pi^3 + 120\pi$; (ii) $\frac{1}{2}\pi^3 - 12\pi + 24$.

12. $\dfrac{a^7}{35} (432\sqrt{3} - 746)$.

Exercise 4, *page* 18.

1. $\frac{1}{4}\pi$.   2. $\frac{1}{2}\pi$.   3. 4/3.   4. $3\pi/16$.   5. 0.   6. $5\pi/16$.

7. (*a*) 4/3; (*b*) $3\pi/8$.   8. (*a*) $1/a^2$; (*b*) $\pi/4a$.   9. $3\pi/16$.   10. $5\pi^2/16$.

11. $21\pi^2/8$.   12. $(\frac{1}{4}\pi, 3/8)$.   13. (*a*) $4/3\pi$; (*b*) 3/8.

14. (*a*) $5\pi/2$; (*b*) $21\pi/16$.   15. 1.69.

16. $\frac{1}{3}(2 - \sqrt{2})$; $\frac{1}{8}(3 \log \tan \dfrac{3\pi}{8} - \sqrt{2})$.

17. $\frac{1}{2} \sin 4x + \sin 2x + x + C$; $\frac{1}{2} \sin 4x + 2 \sin 2x + 3x + D$.

18. 8/315.   19. (i) $3\pi/256$; (ii) 1/6.   20. 2/35.

21. (i) $5\pi/256$; (ii) 1/3.   22. (*a*) 4/35; (*b*) $5\pi/128$.

23. (*a*) $2a^7/35$; (*b*) $256a^{15}/45\,045$.   24. (*a*) $3\pi/256$; (*b*) $\pi/16$.

25. $\pi a^6/32$.   26. $\frac{1}{2}\pi a^3$.   27. $5\pi a^4/8$.

28. 2π.   29. $5\pi/4$.

30. (1) $\frac{3}{32}\pi a^2$; (2) $\bar{x} = \bar{y} = \dfrac{256a}{315\pi}$; (3) $\frac{16}{105}\pi a^3$; (4) $\frac{3}{2}a$; (5) $\frac{6}{5}\pi a^2$.

Exercise 5, page 22.

1. 32/35. 2. 0. 3. $35\pi/128$. 4. 0.

5. $\pi l(l^2 + \frac{3}{2}m^2)$. 6. 32/315. 7. 0. 8. 0.

9. (a) $3\pi/8a$; (b) 0. 12. $4\pi^2 r d(d^2 + \frac{3}{2}r^2)$.

Exercise 6, page 29.

The constant of integration has been omitted.

1. $\frac{1}{4} \tan^{-1}(x + \frac{1}{2})$. 2. $\mathrm{ch}^{-1}\left(\dfrac{x-2}{5}\right)$.

3. $\frac{1}{2}(x-2)\sqrt{(x^2 - 4x - 21)} - \frac{25}{2}\,\mathrm{ch}^{-1}\left(\dfrac{x-2}{5}\right)$. 4. $\frac{1}{2}\sin^{-1}\frac{1}{4}(2x - 1)$.

5. $\frac{1}{6}(3x - 5)\sqrt{(9x^2 - 30x - 119)} - 24\,\mathrm{ch}^{-1}\left(\dfrac{3x-5}{12}\right)$.

6. $\sin^{-1}(2x - 3)$. 7. $-\sin^{-1}\left(\dfrac{1+x}{2x}\right)$. 8. $-\mathrm{cosec}^{-1}(x+1)$.

9. $\frac{1}{2}(x+1)\sqrt{(x^2 + 2x + 6)} + \frac{5}{2}\,\mathrm{sh}^{-1}\left(\dfrac{x+1}{\sqrt{5}}\right)$. 10. $\frac{1}{2}\,\mathrm{ch}^{-1}(x^2)$.

11. $\frac{1}{4}x^2\sqrt{(x^4 - 1)} - \frac{1}{4}\,\mathrm{ch}^{-1}x^2$. 12. $x - \tan^{-1}x$.

13. $\log|x - 2| - \tan^{-1}\frac{1}{2}(x+1)$. 14. $-\dfrac{1}{2(x+1)} + \frac{1}{4}\log\dfrac{(x+1)^2}{x^2 + 1}$.

15. $-\frac{1}{2}e^{-x^2}$. 16. $-\sin\dfrac{1}{x}$. 17. $\frac{1}{2}\tan^{-1}(x^2)$.

18. $\frac{1}{2}\tan^{-1}(\frac{1}{2}\tan\theta)$. 19. $\dfrac{1}{(1 + \cos x)}$. 20. $\frac{1}{3}(1 + x^2)^{3/2}$.

21. $\dfrac{1}{\sqrt{3}}\log\dfrac{\sqrt{3} + \tan\frac{1}{2}\theta}{\sqrt{3} - \tan\frac{1}{2}\theta}$. 22. $\dfrac{2}{\sqrt{3}}\tan^{-1}\left(\dfrac{2e^x + 1}{\sqrt{3}}\right)$.

23. $2\sqrt{x} - 2\log|1 + \sqrt{x}|$. 24. $\dfrac{1}{3a}[(x + a)^{3/2} + (x - a)^{3/2}]$.

25. $\frac{1}{4}x^2 - \frac{1}{4}x + \frac{3}{8}\log(2x - 1)$. 26. $\sqrt{(x^2 + x - 2)} - \frac{3}{2}\,\mathrm{ch}^{-1}\left(\dfrac{2x+1}{3}\right)$.

27. $-\frac{1}{3}\log 2 - \log\frac{3}{4}$. 28. $\dfrac{13\pi}{8}$. 29. $4 + \dfrac{2}{\sqrt{3}}\log(2 + \sqrt{3})$.

30. $\pi/3\sqrt{3}$. 31. $\dfrac{a^2}{4}(2 + \pi)$. 32. $\frac{1}{2}\pi(a - b)$.

33. $\frac{1}{4}\pi - 2/3$. 34. $\frac{1}{8}(\pi + 2)$. 35. $\dfrac{2}{\sqrt{5}}\log\left(\dfrac{\sqrt{5}+1}{\sqrt{5}-1}\right)$.

36. $\pi/8$. 37. $\frac{1}{3}$. 38. A = 1, B = -2, C = 3; $3\pi/2 + \log 2$

39. $\dfrac{1}{41}\left(\dfrac{\pi}{16} + \dfrac{1}{5}\log 3\right)$. 40. $2\pi/3\sqrt{3}$.

42. $n^{p+3}\left[\dfrac{1}{p+1} - \dfrac{2}{p+2} + \dfrac{1}{p+3}\right]$.

43. (i) $\frac{1}{4}\pi - 1$; (ii) $4\log 2 - \log 3$; (iii) $\frac{1}{2}(\log 3 - 1)$.

44. (i) $80/9\sqrt{3}$; (ii) $\frac{1}{4}\pi$. 45. $3a \sin \alpha/4\alpha$. 46. $(24e^{\pi/2} - 41)/85$.

47. (i) $\frac{1}{2}\log(x^2 + 1) - \log|x + 1|$; (ii) $\frac{8}{3}\log 2 - 7/9$; (iii) $\frac{1}{192}(3\pi + 4)$.

48. (i) $\frac{1}{3}\tan^3\theta - \tan\theta + \theta$; (ii) $\frac{1}{4}\theta - \frac{5}{6}\tan^{-1}\left(\dfrac{1}{3}\tan\dfrac{\theta}{2}\right)$;

 (iii) $\frac{1}{2}\log(a^2 + x^2) + \frac{1}{2}\cdot\dfrac{a^2}{a^2 + x^2}$; (iv) $1\cdot 79$.

49. (i) 2; (ii) $\dfrac{1}{4\sqrt{3}}\log\left(\dfrac{\sqrt{6}+2}{\sqrt{6}-2}\right) = \dfrac{1}{2\sqrt{3}}\log(\sqrt{3} + \sqrt{2})$.

50. (i) $\tan^{-1}(x + 2)$; (ii) $\frac{1}{2}\log\left|\dfrac{x+1}{x+3}\right|$; (iii) $-x^2\cos x + 2x\sin x + 2\cos x$;

 (iv) $\frac{1}{2}\tan^{-1}(\frac{1}{2}\tan\frac{1}{2}\theta)$.

52. (a) $-\mathrm{ch}^{-1}(\cos\theta/\cos\alpha)$; (b) $\sin^{-1}(\sin\theta/\sin\alpha)$; (c) $0\cdot 914$; (d) $1\cdot 148$.

53. (i) $\sin^{-1}x + (1 - x^2)^{\frac{1}{2}}$; (ii) $2\tan^{-1}(e^n)$; (iii) $0\cdot 209$.

54. (a) πa^2; (b) $\frac{7}{4}\pi a^4$.

55. (i) $-(4 + 3x - x^2)^{\frac{1}{2}} + \frac{1}{2}\sin^{-1}\left(\dfrac{2x - 3}{5}\right)$; (ii) $-\dfrac{1}{\sqrt{6}}\cosh^{-1}\left(\dfrac{x + 11}{5x - 5}\right)$;

(iii) $\frac{2}{5}x - \frac{3}{5}\log |3 + 5\cos x| - \frac{3}{10}\log\left|\dfrac{2 + \tan\frac{1}{2}x}{2 - \tan\frac{1}{2}x}\right|$.

57. (i) $\dfrac{3\pi}{128}$; (ii) $\dfrac{\pi}{2\sqrt{(1 + a^2)(1 + b^2)}}$; (iii) $\dfrac{\pi}{6}$.   **58.** $\dfrac{\pi(2 + k^2)}{2(1 - k^2)^{5/2}}$.

59. $\dfrac{\pi^2}{16}$.   **60.** area $= \dfrac{\pi}{6}\left[\dfrac{1}{(1 - a)^2} + \dfrac{4}{1 + a^2} + \dfrac{1}{(1 + a)^2}\right]$.

61. $3\cdot 21$ mm

62. (i) $\frac{1}{2}\log |x^2 + 2x - 4| - \dfrac{1}{2\sqrt{5}}\log\left|\dfrac{x + 1 - \sqrt{5}}{x + 1 + \sqrt{5}}\right|$;

(ii) $\frac{1}{6}x^3 - \frac{1}{4}x^2\sin 2x - \frac{1}{4}x\cos 2x + \frac{1}{8}\sin 2x$.

63. (i) $8\frac{1}{4} + \log 4$; (ii) $\pi\sqrt{3}/18$.   **64.** (b) $x + \log\left\{\dfrac{(1 + e^x)}{(1 + 2e^x)^2}\right\}$.

Exercise 7, page 42.

1. (i) $147 + 416(x - 2) + 504(x - 2)^2 + 328(x - 2)^3 + 121(x - 2)^4 +$
$24(x - 2)^5 + 2(x - 2)^6$;

(ii) $\dfrac{147}{(x - 2)^4} + \dfrac{416}{(x - 2)^3} + \dfrac{504}{(x - 2)^2} + \dfrac{328}{(x - 2)} + 121 + 24(x - 2) +$
$2(x - 2)^2$.

2. $-\dfrac{2}{(x - 1)^2} - \dfrac{11}{x - 1} + 12\log(x - 1) + 5x + \frac{1}{2}(x - 1)^2 + C$.

3. $-\left(x - \dfrac{\pi}{2}\right) + \dfrac{1}{3!}\left(x - \dfrac{\pi}{2}\right)^3 - \dfrac{1}{5!}\left(x - \dfrac{\pi}{2}\right)^5 + \dfrac{1}{7!}\left(x - \dfrac{\pi}{2}\right)^7 - \cdots$;
$- 0\cdot 0175$.

4. (a) $(x - 1) - \frac{1}{2}(x - 1)^2 + \frac{1}{3}(x - 1)^3 \cdots$;

(b) $e^2\left[1 + (x - 2) + \dfrac{1}{2!}(x - 2)^2 + \dfrac{1}{3!}(x - 3)^3 + \cdots\right]$.

5. $4\cdot 19$.   **8.** $y_n = 2^{n/2}e^x\cos\left(x + \dfrac{n\pi}{4}\right)$; $y = \sum\limits_{n=0}^{\infty}\dfrac{x^n}{n!}2^{n/2}\cos\dfrac{n\pi}{4}$.

9. $y = \Sigma\dfrac{x^n}{n!}2^{n/2}\sin\dfrac{n\pi}{4}$.   **10.** (a) $\frac{10}{3}$; (b) $1\cdot 0392$.

11. $\pm\sqrt{0\cdot 0616} = \pm 0\cdot 25$.   **13.** $x + \dfrac{2}{4!}x^4 + \dfrac{10}{7!}x^7 + \dfrac{80}{10!}x^{10} + \cdots$

14. (a) $2x^{\frac{1}{2}} + \frac{1}{2}(\frac{2}{3}x^{3/2}) + \dfrac{1\cdot 3}{2\cdot 4}(\frac{2}{5}x^{5/2}) + \dfrac{1\cdot 3\cdot 5}{2\cdot 4\cdot 6}(\frac{2}{7}x^{7/2}) + \cdots + \cdots$
$\dfrac{1\cdot 3\cdot 5\cdot 7\cdots (2n - 1)}{2\cdot 4\cdot 6\cdot 8\cdots 2n}\left(\dfrac{2}{2n + 1}x^{\frac{2n + 1}{2}}\right) + \cdots$;

(b) $0\cdot 2003$.

15. $0\cdot 984\ 808$; $2\cdot 14 \times 10^{-11}$.

17. $1 + ax + \dfrac{a^2}{2!}x^2 + a\dfrac{(1^2 + a^2)}{3!}x^3 + \cdots$;

x^{2r}: $a^2(a^2 + 2^2)(a^2 + 4^2)\cdots (a^2 + (2r - 2)^2)/(2r)!$;
x^{2r+1}: $a(a^2 + 1^2)(a^2 + 3^2)\cdots (a^2 + (2r - 1)^2)/(2r + 1)!$

18. $x - \dfrac{2}{3!}x^3 - \dfrac{4\cdot 2^2}{5!}x^5 - \dfrac{6\cdot 4^2\cdot 2^2}{7!}x^7 \cdots$

20. $y = (\log a)^2 + \dfrac{2\log a}{a}x + \dfrac{2}{a^2}\dfrac{x^2}{2!} - \dfrac{2\log a}{a}\cdot\dfrac{1^2}{a^2}\dfrac{x^3}{3!} - \dfrac{2}{a^2}\cdot\dfrac{2^2}{a^2}\dfrac{x^4}{4!}\cdots$;

x^{2p}: $(-1)^{p-1}\dfrac{2}{a^2}\cdot\dfrac{2^2}{a^2}\cdot\dfrac{4^2}{a^2}\cdot\dfrac{6^2}{a^2}\cdots\dfrac{(2p - 2)^2}{a^2}\cdot\dfrac{1}{(2p)!}$;

x^{2p-1}: $(-1)^{p-1}\dfrac{2\log a}{a}\cdot\dfrac{1^2}{a^2}\cdot\dfrac{3^2}{a^2}\cdots\dfrac{(2p - 3)^2}{a^2}\cdot\dfrac{1}{(2p - 1)!}$

22. $y = ax + \dfrac{1}{2!}\, 2x^2 + \dfrac{1}{3!}\, ax^3 + \dfrac{1}{4!}\, 2^2 \cdot 2x^4 + \dfrac{1}{5!}\, 3^2 \cdot ax^5 + \ldots;$

$\qquad x^{2p} : \dfrac{1}{(2p)!}\, \{2 \cdot 2^2 \cdot 4^2 \cdot 6^2 \ldots (2p-2)^2\};$

$\qquad x^{2p+1} : \dfrac{1}{(2p+1)!}\, \{a \cdot 3^2 \cdot 5^2 \ldots (2p-1)^2\}.$

23. $1 - \dfrac{x^2}{2^2} + \dfrac{x^4}{2^2 \cdot 4^2} - \dfrac{x^6}{2^2 \cdot 4^2 \cdot 6^2} + \ldots$

24. 3·83.

25. 1·06 s.

26. 284π; (2·5, 0).

27. 29·1 min.

28. 70·2 km/h.

29. 39 333 units of area.

Exercise 8, page 47.

1. (a) 2; (b) -2; (c) 0; (d) 1.

3. (a) $\frac{13}{5}$, $\frac{7}{5}$; (b) $\frac{3}{2}$, $-\frac{5}{2}$; (c) $\frac{5}{7}$, $-\frac{5}{7}$.

4. (a) $\frac{23}{10}$, $\frac{1}{10}$.

5. $\dfrac{x}{6} = \dfrac{y}{-29} = \dfrac{z}{13}.$

6. (a) $x = \pm 1$; (b) $x = -\frac{1}{3}$, -1.

Exercise 9, page 51.

1. (a) 319; (b) 4; (c) 0.

2. (a) 0, ± 2; (b) 0, $\frac{1}{2}$.

3. $(1, 2, -1)$.

4. $\frac{1}{203}(865, 79, -311)$.

5. $\frac{1}{41}(22, 18, 1)$.

6. $(\frac{16}{5}, \frac{9}{5}, 0)$.

7. $\frac{1}{24}(95, -85, -65)$.

8. 1, 1, 2, -1.

9. (a) 1st and 3rd are parallel planes;
(b) only two equations since $(1) + (2) = (3)$.

12. $i_1 = 2\cdot26$; $i_2 = 0\cdot960$; $i_3 = 0\cdot411$.

13. $i_1 = 3\cdot46$; $i_2 = 1\cdot17$; $i_3 = 1\cdot50$.

15. (a) equation is " stable "; (b) -1, -2, $(-1 \pm j\sqrt{3})/2$.

Exercise 10, page 59.

1. (a) $(a-b)(b-c)(c-a)$; (b) 2·5, 3, -4.

2. (a) Value is unchanged when sign of j is changed. Value $= -6$; (b) -2.

3. $\lambda = -1$, 1, 12; $\lambda = -1$: $x = -1/11$, $y = -15/11$;

$\qquad\qquad\qquad \lambda = 1$: $\quad x = -5$, $\quad y = 1$;

$\qquad\qquad\qquad \lambda = 12$: $\quad x = 1/2$, $\quad y = 1$.

4. -3, $\pm\sqrt{3}$.

5. (a) 4; (b) $\begin{vmatrix} 1 & 1 & 1 \\ a & b & c \\ a^2 & b^2 & c^2 \end{vmatrix} = 0$, $a = b$, or $b = c$, or $c = a$.

6. $x + y - 5 = 0$; $3(x^2 + y^2) - 11x - 5y - 2 = 0$.

8. $ab + ac + ad + cd + bc + bd = 0$.

11. $\omega^4 - \omega^2 \left(\dfrac{p_1 + p_3}{m_1} + \dfrac{p_2 + p_3}{m_2} \right) + \dfrac{p_1 p_2 + p_2 p_3 + p_1 p_3}{m_1 m_2} = 0.$

12. $\omega^2 = 6(\sqrt{3} \pm 1)$.

Exercise 11, page 63.

1. $\begin{vmatrix} 7 & 19 \\ 11 & 22 \end{vmatrix}.$

2. $\begin{vmatrix} 9 & 8 \\ 16 & 17 \end{vmatrix}.$

3. $\begin{vmatrix} 9 & 20 & 29 \\ 12 & 6 & 1 \\ 9 & 26 & 38 \end{vmatrix}.$

4. $\begin{vmatrix} a & h & g \\ h & b & f \\ g & f & c \end{vmatrix}.$

5. $\begin{vmatrix} g & h & a \\ f & b & h \\ c & f & g \end{vmatrix}.$

Exercise 12, *page* **66**.

1. $x = 4$ mm, $\theta = 60°$.   2. $x = -3$, $y = 0$.   3. Each part $= \frac{1}{3}a$.

4. $8abc/3\sqrt{3}$.   7. $y = 6\cdot72 + 0\cdot0092x^2$.

8. $4\cdot42$ ohms.   10. $5i/24,\ 5i/24,\ 14i/24$.

Exercise 13, *page* **74**.

2. $\dfrac{a}{a^2 + b^2}$; (i) $\dfrac{a^2 - b^2}{(b^2 + a^2)^2}$; (ii) $\dfrac{2ab}{(a^2 + b^2)^2}$.

3. $\dfrac{1}{a - b} \log \dfrac{a}{b}$; $\dfrac{1}{(a - b)^2} \log \dfrac{a}{b} - \dfrac{1}{a(a - b)}$.

4. $\dfrac{\pi}{2ab(a + b)}$; $\dfrac{\pi(2a + b)}{4a^3b(a + b)^2}$.   5. $\log a$.

6. (a) $\dfrac{\pi a}{(a^2 - b^2)^{3/2}}$; (b) $\dfrac{-\pi b}{(a^2 - b^2)^{3/2}}$; (c) $\dfrac{\pi(aa' - bb')}{(a^2 - b^2)^{3/2}}$.   8. $\frac{1}{12}$.

9. $R = \dfrac{3wl}{8}$; $\dfrac{w^2l^5}{640\ EI}$.   11. (a), (b), (c), (d) $= 1$.

12. (a) $-3\cdot5$; (b) $\frac{1}{2} - 3\pi$; (c) $-0\cdot7$.   13. 0.

15. (a) $12\frac{1}{3}$; (b) $10\frac{1}{2}$; (c) $\frac{1}{2}$; (d) $e^3 - e + 2 - 3\log 3$.

17. (i) $-\dfrac{x_1}{m} \cos mx_1 + \dfrac{1}{m^2} \sin mx_1 + m^2x_1 \sin x_1 + m^2 \cos x_1 - m^2$;

    (ii) $\frac{1}{2}y_1^2 \cos x_1$.   18. $\frac{1}{4}a^2(\pi + 4) - \pi a^3$.

Exercise 14, *page* **82**.

1. $50\cdot522$ m; $3\cdot141$ m.   2. $l(\sqrt{2} - 1)$.

3. $T_0 = 16$ N; $T = 34$ N; 320 Nm.   4. 6 m.

5. $30°$; $60°$; $2 + \sqrt{3} + 1/\sqrt{3}$.   8. 111 m.   12. $6\cdot3$ m.

15. $l(\sqrt{8} - \sqrt{5})$.

Exercise 15, *page* **85**.

1. $\dfrac{100}{\sqrt{3}} \log_e (2 + \sqrt{3})$; $50/\sqrt{3}$.   2. $W\sqrt{2}$.

4. $\frac{1}{8}l(3 + \sqrt{3}) \log (3 + 2\sqrt{3}) + \frac{1}{2}\sqrt{3}l$.

Exercise 16, *page* **92**.

1. $y = -x^{-1} - \frac{3}{2}x^{-2} + \frac{7}{3}x^{-3} + C$.   2. $y = -4 \log \cos x + C$.

3. $\frac{1}{2} \tan^{-1} (y/2) = \frac{1}{3} \tan^{-1} (x/3) + C$.   4. $\log (y^2 + 4) = x + C$.

5. $(1 + y^2) = K(1 + x^2)$.   6. $e^x(x - 1) - (1 - y^2)^{\frac{1}{2}} = C$.

7. $(1 + x^2)^{\frac{1}{2}}(1 - y) = C$.

8. $\log (xy) + \dfrac{1}{x} - \dfrac{1}{y} = C$.   10. $4\cdot2$.

12. $r = \left(\dfrac{T}{f\pi}\right)^{\frac{1}{2}} e^{-wx/2f}$.   13. $\dfrac{1000}{3} d^{3/2}$ sec.

14. $5 - 9e^{-t/50}$ kg.; 5 kg.   15. $8\cdot33$ kg.

17. $y = \frac{1}{4}x^3 + \dfrac{C}{x}$.   18. $y = 1 + Ce^{-\sin x}$.   19. $y \sin x = \frac{1}{3}x^3 + C$.

20. $y = C\sqrt{x} + \dfrac{2}{x}$.

21. $i = \dfrac{E_0L(R \sin \omega t - L\omega \cos \omega t)}{R^2 + L^2\omega^2} + \dfrac{E_0L^2\omega e^{-Rt/L}}{R^2 + L^2\omega^2}$.

22. $i = \dfrac{E_0\omega C}{1 + \omega^2R^2C^2} (\cos \omega t + \omega RC \sin \omega t) - \dfrac{E_0\omega Ce^{-t/RC}}{1 + \omega^2R^2C^2}$.

23. $P = 2(1 - e^{-t/10}) - 0.15te^{-t/10}$; 1.66 kg.

26. $e^x + x + y + C = 0$. 27. $x^2 \log y + C = 0$.

28. $y \tan^{-1} x = C$. 29. $\alpha = -2$, $\beta = 0$; $x^2 y^2 + y = cx$.

30. $\log (abx + b^2 y + a) = bx + D$. 31. $\sin \left(\dfrac{y}{x}\right) = Ax$.

32. $\frac{1}{2} \log (x^2 + y^2) = x + C$. 33. $y = \sin (C \pm x)$.

34. $y = A \sin (C \pm x)$. 35. $y = x^2 + C \log x + D$.

36. $3x^2 + y^2 = Cx$. 37. $x^2 - y^2 = C$.

38. $y^2 + 2x^2 = C$. 39. $x^2 + y^2 = 2 \log Ax$.

Exercise 17, *page* 108.

1. $14e^{3x}$.

2. $e^{2x}(6 - 2x^2)$.

3. $- e^{4x}\{236 \sin 2x + 318 \cos 2x\}$.

4. $\dfrac{e^{ax}}{a^2 + b^2} (a \cos bx + b \sin bx)$.

5. $\dfrac{e^{px}}{p^2 + q^2} (p \sin qx - q \cos qx)$.

6. $\frac{1}{4}e^{4x}\{x^3 - \frac{3}{4}x^2 + \frac{3}{8}x - \frac{3}{32}\}$.

7. $\frac{1}{25}(3 \sin 3x - 4 \cos 3x)$. 8. $-\frac{1}{2}\cos x$. 9. $-\frac{1}{6}x \cos 3x$.

10. $-\dfrac{e^{2x}}{145}(-12 \sin 3x + \cos 3x)$. 11. $x^2 + x - 1$.

12. $\dfrac{1}{6}x^3 + \dfrac{7}{12}x^2 + \dfrac{29}{36}x + \dfrac{139}{216}$. 13. e^{-2x}.

14. $\dfrac{1}{17\,711}(144 \sinh 5x - 55 \cosh 5x)$. 15. $65 + 12x + x^2$.

16. $\frac{1}{12}x^4 - \frac{1}{3}x^3 + 2x^2$. 17. $\frac{1}{4}x^2 + \frac{3}{2}x + \frac{7}{4} + xe^x$.

18. $-\dfrac{x}{2}\cos x$. 19. $\frac{1}{5}e^{2x}\{x^2 - \frac{8}{5}x + \frac{22}{25}\}$.

20. $-\dfrac{e^{3x}}{377}\{16 \cos 4x + 11 \sin 4x\}$.

21. $y = Ae^{(1 + \sqrt{2})x} + Be^{(1 - \sqrt{2})x} + \cos x - \sin x$.

22. $y = A + Be^{-\frac{1}{2}x} - \frac{1}{34}(4 \cos 2x - \sin 2x)$.

23. $y = Ae^{2x} + Be^{-2x} - \frac{1}{2}(\frac{2}{5}\cos x + x \sin x)$.

24. $y = e^t\{A \cos t + B \sin t\} + \dfrac{e^{-t}}{65}\{\sin 2t + 8 \cos 2t\}$.

25. $y = e^{-x}\{A \cos \sqrt{2}x + B \sin \sqrt{2}x\} + \frac{1}{3}\{x^2 + \frac{2}{3}x - \frac{10}{9}\}$.

26. $y = (A + Bx)e^{2x} + \frac{1}{169}(12 \cos 3x - 5 \sin 3x)$.

27. $y = \frac{1}{5}e^x + \frac{2}{5}e^{\frac{1}{2}x}\{2 \cos x + \sin x\}$. 28. $x = \frac{7}{2}e^t - \frac{3}{2}e^{-\frac{1}{3}t} - 2t + 5$.

29. $s = e^{-2t}(2 - \cos t)$.

30. $y = \frac{1}{26}(3 \cos 3t - 2 \sin 3t) + \dfrac{e^t}{52}(9 \sin 2t - 6 \cos 2t)$.

31. (a) $\frac{1}{8}(x^2 \cos x - 3x \sin x)$; (b) $-\frac{1}{32}x^2 \sin 2x$.

32. $\frac{3}{4}(x^2 - x + \frac{3}{8}) + \frac{1}{2}\cos 4x$; $y = -\dfrac{e^{-4x}}{32}(25 + 76x) + \frac{3}{4}(x^2 - x + \frac{3}{8}) + \frac{1}{2}\cos 4x$.

33. (a) $y = A + B \sin 2x + C \cos 2x - 2 \cos x$;
   (b) $y = Ae^x + Be^{-x} + C \cos x + D \sin x - \frac{1}{5}\cos x \cosh x$;
   (c) $y = (1 + ax + x^4)e^{-ax}$.

34. (a) $y = e^{-x}(A \cos 2x + B \sin 2x) + \frac{1}{5}(x^2 + \frac{1}{2}x + \frac{13}{25})$;
   (b) $y = A \cos 4x + B \sin 4x + C \cos 2x + D \sin 2x + \dfrac{x}{48}\sin 2x - \dfrac{x}{96}\sin 4x$;
   (c) $y = A \cos 2x + B \sin 2x + Ce^{2x} + De^{-2x} - \frac{1}{20}e^{-2x}\cos 2x$.

35. $y = e^{-\frac{1}{2}t}[\cos 3t - \frac{1}{4}\sin 3t - \frac{1}{4}t \cos 3t]$.

37. (i) $y = Ae^{-2x} + Be^{-x} - \frac{1}{10}e^{-x}\{\cos 2x + 2 \sin 2x\}$; (ii) $y = e^{-ax}[A + Bx + \frac{1}{12}x^4]$.

38. $y = \cos 2x + \sin 2x - \cos x.$

39. $y = \dfrac{A}{x^2} + Bx^3 - \tfrac{1}{6}\log x + \tfrac{1}{36}.$

40. $y = Ax^4 + \dfrac{B}{x} - \tfrac{1}{6}x^2.$

41. $y = Ax^3 + \dfrac{B}{x} - \tfrac{1}{3}x^2(\log x + \tfrac{2}{3}).$

42. $y = x(1 - \log x) + \tfrac{1}{4}x(\log x)^3.$

43. $y = x(A + B \log x) + x^3.$

44. $y = \dfrac{A}{x^3} + \dfrac{B}{x} + \tfrac{1}{3}\{\log x - \tfrac{4}{3}\} + \dfrac{1}{4x}\{(\log x)^2 - \log x\}.$

45. $\theta = \dfrac{\theta_1 \log \dfrac{r}{r_2} + \theta_2 \log \dfrac{r_1}{r}}{\log \dfrac{r_1}{r_2}}.$

46. $\theta = \dfrac{r_2\theta_2 - r_1\theta_1}{r_2 - r_1} + \dfrac{r_1 r_2(\theta_1 - \theta_2)}{r(r_2 - r_1)}.$

47. $n = -2.$ $\dfrac{d^2z}{dx^2} + 4\dfrac{dz}{dx} = \cos x.$ $x^2 y = A + Be^{-4x} + \tfrac{1}{17}(4\sin x - \cos x).$

48. $m = 1;\ y = x\left(A + Be^{-\tfrac{1}{2x}}\right).$

49. $y = A\cos x^2 + B\sin x^2.$

Exercise 18, page 120.

1. $y = \dfrac{wEI}{P^2}\{\cos mx + \tan \tfrac{1}{2}ml \sin mx\} + \dfrac{w}{2P}\left\{x^2 - lx - \dfrac{2EI}{P}\right\}\ (m^2EI = P);$

$\dfrac{wEI}{P^2}\left(\sec \dfrac{ml}{2} - 1\right) - \dfrac{wl^2}{8P}.$ **3.** $y = \dfrac{wl^2}{8(P - Q)} \sin\left(\dfrac{\pi x}{l}\right).$

5. $y = \dfrac{W}{2P}\sqrt{\dfrac{EI}{P}} \cdot \sec\left(\dfrac{ml}{2}\right) \cdot \sin mx - \dfrac{W}{2P}x;\ \dfrac{W}{2}\sqrt{\dfrac{EI}{P}} \tan\left(\tfrac{1}{2}l\sqrt{\dfrac{P}{EI}}\right).$

9. $EIy_2 = W\{(l - x)\sin\theta + (a - y)\cos\theta\}.$ **10.** $7pl^4/240EI;\ 5pl^2/24.$

12. $x = \dfrac{v}{n\sqrt{(1 - h^2)}}\, e^{-hnt} \sin[nt\sqrt{(1 - h^2)}].$ **14.** $T = \tfrac{1}{5}\pi$ seconds; $\dfrac{3\sqrt{5}}{200}$ m.

15. $q > p^2;\ T = 2\pi/\sqrt{(q - p^2)};\ x = \dfrac{200}{139\,683}\{119\sin 9t - 180\cos 9t\}.$

16. (i) $\ddot{z} + 1{\cdot}5\,\dot{z} + \dfrac{10g}{3} = \dfrac{10g}{9}\sin 4t$

(z measured from position of static equilibrium);

(ii) $z = 0{\cdot}578\sin 4t - 0{\cdot}208\cos 4t$

17. For motion $c\sin(nt + \beta).$

$x = A\cos t\sqrt{\dfrac{\lambda}{ma}} + B\sin t\sqrt{\dfrac{\lambda}{ma}} + \dfrac{gam}{\lambda} + \dfrac{cn^2}{\dfrac{\lambda}{ma} - n^2}\sin(nt + \beta);$

18. $V = 100 - 100e^{-10t}\{\cos 500t + \tfrac{1}{50}\sin 500t\}.$

20. $i = \tfrac{4}{5}e^{-200t} - e^{-100t} + \tfrac{1}{5}\cos 100t + \tfrac{3}{5}\sin 100t.$

23. (i) $L\dfrac{d^2q}{dt^2} + R\dfrac{dq}{dt} + \dfrac{1}{C}q = E\cos pt;$

(ii) $q = (A + Bt)e^{-t/\sqrt{LC}} + \dfrac{EC}{(1 + LCp^2)^2}\{(1 - LCp^2)\cos pt - 2p\sqrt{LC}\sin pt\},$

where $R^2 = 4L/C;$

(iii) $C = 0{\cdot}0001;$ (iv) $0{\cdot}01$ amp.

26. $L = C(R^2 + \omega^2 L^2).$

27. $\sqrt{\dfrac{(R_1 + R_2)^2 + \omega^2 L_1^2}{R_1^2 + \omega^2 L_1^2}} \cdot E_0 \sin(\omega t + \alpha - \beta),$

where $\tan\alpha = \dfrac{\omega L_1}{R_1 + R_2},\quad \tan\beta = \dfrac{\omega L_1}{R_1}.$

29. $\sqrt{\left(\dfrac{R^2 + \omega^2 L^2}{9R^2 + 4\omega^2 L^2}\right)} \cdot E\cos(\omega t + \phi),$ where $\tan\phi = \dfrac{\omega LR}{3R^2 + 2\omega^2 L^2}.$

30. $L\dfrac{d^2i}{dt^2} + R\dfrac{di}{dt} + \dfrac{i}{C} = \omega E\cos\omega t;$

$i = e^{-Rt/2L}\{A\cos\alpha t + B\sin\alpha t\}\ (= \text{transient})$

$\qquad + \dfrac{\omega EC}{(1 - \omega^2 CL)^2 + \omega^2 C^2 R^2}\{(1 - \omega^2 CL)\cos\omega t + \omega CR\sin\omega t\},$

where $\alpha^2 = \dfrac{4L}{C} - R^2 > 0.$

Exercise 19, *page* 131.

1. $y = e^{\frac{1}{2}x}\left(\text{L}\cos\frac{\sqrt{3}}{2}x + \text{M}\sin\frac{\sqrt{3}}{2}x\right) + 7;$

$z = 5 + \frac{1}{2}e^{\frac{1}{2}x}\left[\left(\frac{3}{2}\text{L} - \frac{\sqrt{3}}{2}\text{M}\right)\cos\frac{\sqrt{3}}{2}x + \left(\frac{3}{2}\text{M} + \frac{\sqrt{3}}{2}\text{L}\right)\sin\frac{\sqrt{3}}{2}x\right].$

2. $y = \frac{1}{2}x^2 - 2x + \text{A}$; $z = 1$.   3. $x = \text{A}e^{\frac{1}{2}t} - 4$; $y = \text{A}e^{\frac{1}{2}t}$.

4. $x = e^{2t}(\text{A} + \text{B} + \text{B}t)$; $y = (\text{A} + \text{B}t)e^{2t}$.

5. $x = \frac{1}{3}\sin t + \cos 2t + \frac{7}{3}\sin 2t$; $y = -\frac{1}{3}\cos t - \sin 2t + \frac{7}{3}\cos 2t$.

6. $i_1 = 1\cdot9e^{-0\cdot573t} + 0\cdot10e^{-3\cdot926t}$; $i_2 = 3\cdot07e^{-0\cdot573t} - 0\cdot07e^{-3\cdot926t}$.

7. $x = \frac{1}{2}(e^{-t} + e^{-3t})$; $y = -\frac{1}{2}(e^{-t} - e^{-3t})$.

8. (i) $x = \frac{1}{3}\text{V}(\sin t + 2t)$, $y = \frac{2}{3}\text{V}(t - \sin t)$; (ii) $t = \pi$; (iii) $x = \frac{1}{3}\text{V}$, $y = \frac{4}{3}\text{V}$.

9. $x = \frac{3}{5} + \frac{1}{6}e^t - \frac{3}{4}e^{-t} - \frac{1}{60}e^{-5t}$; $y = \frac{2}{5} + \frac{1}{3}e^t - \frac{3}{4}e^{-t} + \frac{1}{60}e^{-5t}$.

10. $x = e^t + e^{-t}$; $y = \sin t + e^{-t} - e^t$.

11. $x = 2a\cos\dfrac{t}{\sqrt{2}} + 2\sqrt{2}b\sin\dfrac{t}{\sqrt{2}}$; $y = -a\cos\dfrac{t}{\sqrt{2}} - b\sqrt{2}\sin\dfrac{t}{\sqrt{2}}$.

12. $\theta = \text{L} + \text{M}e^{px}$, $\text{T} = \text{L} + m\text{M}e^{px}$, where $p = k(1 - 1/m)$; $\theta = 47\cdot6$, $\text{T} = 44\cdot7$.

13. $x = \dfrac{\text{E}}{\text{R}}\left(\frac{2}{3} - \frac{1}{2}e^{-\text{R}t/\text{L}} - \frac{1}{6}e^{-3\text{R}t/\text{L}}\right)$; $y = \dfrac{\text{E}}{\text{R}}\left(\frac{1}{3} - \frac{1}{2}e^{-\text{R}t/\text{L}} + \frac{1}{6}e^{-3\text{R}t/\text{L}}\right)$.

14. $x = \dfrac{a}{n^2}(nt - \sin nt)$, $y = \dfrac{a}{n^2}(1 - \cos nt)$.

15. (i) $x = \text{A}e^{2t} + \text{B}e^t + \text{C}e^{-t}$, $y = \frac{1}{3}(5\text{A}e^{2t} + 4\text{B}e^t - 4\text{C}e^{-t})$;
(ii) $x = 2(e^{2t} - e^t)$, $y = 5e^{2t} - 4e^t$.

16. (i) $x = \text{A}e^z + \text{B}e^{2z} + \text{C}e^{3z} + \frac{1}{6}(z + \frac{11}{6})$,
$y = -9\text{A}e^z - 12\text{B}e^{2z} - 17\text{C}e^{3z} - \frac{8}{6}(z + \frac{11}{6})$;
(ii) $x = \frac{1}{6}(z + \frac{11}{6})$, $y = -\frac{8}{6}(z + \frac{11}{6})$, locus: $8x + y = 0$.

17. (a) $u = -\frac{2}{3} - e^{-x}(\text{A} + \frac{1}{4} - \frac{1}{2}x) + \text{B}e^{3x}$; $v = \frac{1}{3} + (\text{A} - \frac{1}{2}x)e^{-x} + \text{B}e^{3x}$.
(b) $13x = 18a(e^{-3t/7} - e^{t/2})$, $13y = a(4e^{-3t/7} + 9e^{t/2})$.

19. $\omega = 3$ or 2 rad./sec.   20. $\text{T} = 2\pi\sqrt{\dfrac{\text{J}_1\text{J}_2}{\text{C}(\text{J}_1 + \text{J}_2)}}$.

21. $\text{A}(\text{I}_1p^2 - \text{C}_1 - \text{C}_2) + \text{B}\text{C}_2 + \text{F}_0 = 0$; $\text{A}\text{C}_2 + \text{B}(\text{I}_2p^2 - \text{C}_2) = 0$.

22. $\text{A} = \dfrac{\text{E}}{\text{L}p(\text{S}^2 + \text{N}^2p^2)}\{-2\text{N}p\text{S} + j(\text{N}^2p^2 - \text{S}^2)\}$, $\text{B} = \dfrac{\text{EM}(\text{S} - jp\text{N})}{\text{L}(\text{S}^2 + p^2\text{N}^2)}$.

23. $a\left\{\dfrac{\lambda^2 + k^2p^2}{(\lambda - mp^2)^2 + k^2p^2}\right\}^{\frac{1}{2}}$.

28. $y = c\left[1 + \dfrac{x^2}{2} + \dfrac{x^4}{2^2\cdot2!} + \cdots \dfrac{x^{2n}}{2^n\cdot n!} + \cdots\right]$
$+ \left[-1 + x + \dfrac{x^3}{1\cdot3} + \cdots \dfrac{x^{2n+1}}{1\cdot3\cdots(2n+1)} + \cdots\right].$

29. $y = \text{A ch }\sqrt{x} + \text{B sh }\sqrt{x}$.   30. $y = a_1x + a_0\left[1 - \dfrac{x^2}{2!} - \dfrac{x^4}{4!} - \dfrac{3x^6}{6!}\cdots\right]$.

31. $c = 0, -\dfrac{3}{2}$; $a_0\left[1 + \dfrac{3}{5}x + \dfrac{3}{7}x^2 + \dfrac{1}{3}x^3 + \dfrac{3}{7}x^4 + \cdots\right]$.

32. $a\left[1 - \dfrac{x}{8} + \dfrac{x^2}{8\cdot21}\cdots\right] + b\left[\dfrac{1}{5}x - \dfrac{x^2}{5\cdot14} + \cdots\right]$.

33. $a_0\left[1 - \dfrac{1}{2}x^2 + \dfrac{1}{2\cdot4}x^4 - \dfrac{1}{2\cdot4\cdot6}x^6\cdots\right] + a_1\left[x - \dfrac{1}{3}x^3 + \dfrac{1}{3\cdot5}x^5\cdots\right]$.

34. $a_0\left[1 + \dfrac{x^2}{2!} - \dfrac{x^3}{3!} + \dfrac{x^4}{4!} - \dfrac{4x^5}{5!}\cdots\right] + a_1\left[x + \dfrac{x^3}{3!} - \dfrac{2x^4}{4!} + \dfrac{x^5}{5!}\cdots\right]$.

Exercise 20, *page* 144.

12. $\text{C} = 67/5$ $\text{D} = -32/5$, $\text{A} = -67/5$, $\text{B} = 72/5$.

13. (i) $\dfrac{1}{p + a}$; (ii) $\dfrac{1}{p^2}$, $f(t) = e^{-t} - e^{-2t}$.   14. $\frac{1}{2} - e^{-t} + \frac{1}{2}e^{-2t}$.

15. $\dfrac{1}{a^2}(1 - \cos at)$.

16. $\dfrac{1}{a^2}\left(t - \dfrac{1}{a}\sin at\right)$.

17. $\dfrac{1}{b^2}(e^{-bt} - 1 + bt)$.

18. $\dfrac{1}{a - b}(ae^{-at} - be^{-bt})$.

19. te^{-at}.

20. $e^{-at}(1 - at)$.

21. $\dfrac{1}{2a^3}(\sin at - at\cos at)$.

22. $\dfrac{t}{2a}\sin at$.

23. $\dfrac{1}{2a}(\sin at + at\cos at)$.

24. $t\cos at$.

25. $\frac{1}{4}t - \frac{1}{3}\sin t + \frac{1}{24}\sin 2t$.

26. $e^{-at} + e^{\frac{1}{2}at}\left(\sqrt{3}\sin\dfrac{\sqrt{3}}{2}at - \cos\dfrac{\sqrt{3}}{2}at\right)$.

27. $\dfrac{1}{4a^3}(e^{at} - e^{-at} - 2\sin at)$.

30. $q = ECe^{-t/RC}$.

31. $q = EC(1 - e^{-t/RC})$.

32. $T = T_0e^{\mu\theta}$.

33. $i = 20\sin t + 10\cos t - 10e^{-\frac{1}{2}t}$.

34. $x = \frac{1}{9}(7e^4 + 2e^{-5t})$.

35. $x = \frac{1}{10}(11e^{5t} + 9e^{-5t})$.

36. $r = \frac{1}{2}(9e^{-\theta} - 5e^{-3\theta})$.

37. $x = \frac{1}{2}\sin 2t$.

38. $x = 2e^{-\frac{1}{2}t}\cos\dfrac{\sqrt{3}}{2}t$.

39. $x = 2e^{-3t}(1 + 5t)$.

40. $x = \cos 3t + \frac{17}{27}\sin 3t + \frac{1}{9}t$.

41. $x = \cos 2t + \frac{3}{2}\sin 2t + \frac{5}{4}t\sin 2t$.

42. $x = \frac{4}{25}e^{2t} - \frac{4}{25}e^{-3t}(1 + 5t)$.

43. $x = \frac{5}{36}e^{-3t}(6t + 7) + \frac{1}{36}e^{3t}$.

44. $x = \frac{1}{2} - \frac{1}{3}e^{-t} - \frac{3}{10}e^{-6t/11}$; $y = \frac{1}{5}e^{-t} - \frac{1}{5}e^{-6t/11}$.

45. $x = \dfrac{11}{37}\cos t + \dfrac{8}{37}\sin t - \dfrac{e^{-t}}{111}\left\{33\,\mathrm{ch}\,\dfrac{t}{\sqrt{3}} - 17\sqrt{3}\,\mathrm{sh}\,\dfrac{t}{\sqrt{3}}\right\}$;

$y = -\dfrac{6}{37}\cos t - \dfrac{1}{37}\sin t + \dfrac{e^{-t}}{37}\left\{6\,\mathrm{ch}\,\dfrac{t}{\sqrt{3}} - \dfrac{16\sqrt{3}}{3}\,\mathrm{sh}\,\dfrac{t}{\sqrt{3}}\right\}$.

46. $x = \dfrac{3}{2}e^t - \dfrac{143}{350}e^{-5t} - \dfrac{2}{5}t - \dfrac{13}{25} + \dfrac{3}{7}e^{2t}$;

$y = \dfrac{3}{2}e^t + \dfrac{143}{350}e^{-5t} - \dfrac{3}{5}t - \dfrac{12}{25} + \dfrac{4}{7}e^{2t}$.

47. $x = 4\,\mathrm{ch}\,2t + 2\,\mathrm{sh}\,2t$; $y = 6\,\mathrm{sh}\,2t - e^{-t}$.

48. $x = (A + Bt)e^t + (E + Ft)e^{-t}$;
    $y = \frac{1}{2}(B - A - Bt)e^t - \frac{1}{2}(E + F + Ft)e^{-t}$.

49. $i_1 = 4\sin t + 3\cos t - \frac{1}{2}e^{-\frac{1}{2}t} - \frac{5}{2}e^{-t}$;
    $i_2 = \sin t + 2\cos t + \frac{1}{2}e^{-\frac{1}{2}t} - \frac{5}{2}e^{-t}$.

50. $x = 3\sin t - 2\cos t + e^{-2t}$; $y = -\frac{7}{2}\sin t + \frac{9}{2}\cos t - \frac{1}{2}e^{-3t}$.

51. $RCL\ddot{q} + L\dot{q} + Rq = ERC$.

52. $\theta = \alpha\cos\left(t\sqrt{\dfrac{3g}{10a}}\right) + \dfrac{3\alpha}{4}\cos\left(t\sqrt{\dfrac{6g}{a}}\right)$;

$\phi = \dfrac{5\alpha}{4}\cos\left(t\sqrt{\dfrac{3g}{10a}}\right) - \dfrac{\alpha}{4}\cos\left(t\sqrt{\dfrac{6g}{a}}\right)$.

Exercise 21, page 158.

1. $y = \dfrac{wx^2}{24EI}(l - x)^2$.

2. $y = \dfrac{x^2}{120EI}(x^3 + 2l^3 - 3xl^2)$.

8. $x = \dfrac{umc}{He}\sin\left(\dfrac{He}{mc}t\right)$; $y = -\dfrac{umc}{He}\left\{1 - \cos\left(\dfrac{He}{mc}t\right)\right\}$.

19. 1st term = instantaneous exchange of charge between the condensers;
    2nd term = subsequent discharge of condensers.

Exercise 23, page 177.

3. $\dfrac{\pi}{4} - \dfrac{2}{\pi}\left(\cos x + \dfrac{1}{3^2}\cos 3x + \ldots\right) + \sin x - \frac{1}{2}\sin 2x + \frac{1}{3}\sin 3x - \ldots$

4. $\dfrac{2}{\pi}[\sin x + \frac{1}{2}\sin 2x + \frac{1}{3}\sin 3x + \ldots]$.

5 $\dfrac{4a}{\pi}\left[\sin x + \tfrac{1}{3}\sin 3x + \tfrac{1}{5}\sin 5x + \ldots\right].$

6. $\dfrac{4}{\pi}\left[\dfrac{\cos x}{1^2} + \dfrac{\cos 3x}{3^2} + \ldots\right].$

7. $\dfrac{E}{\pi} - \dfrac{2E}{\pi}\left[\dfrac{\cos 2x}{3} + \dfrac{\cos 4x}{15} + \ldots \dfrac{\cos 2nx}{4n^2-1} + \ldots\right] + \tfrac{1}{2}\sin x.$

8. $\dfrac{4E}{\pi}\left[\dfrac{1}{2} - \dfrac{\cos x}{3} - \dfrac{\cos 2x}{15} - \ldots - \dfrac{\cos nx}{4n^2-1} - \ldots\right].$

11. $\dfrac{\sin a\pi}{a\pi}\left[1 + 2a^2 \sum\limits_{1}^{\infty} \dfrac{(-1)^n}{a^2-n^2}\cos nx\right].$

12. $\dfrac{2}{\pi}\Sigma \dfrac{n[1-(-1)^n e^\pi]}{n^2+1}\sin nx;\quad \dfrac{e^\pi-1}{\pi} - \dfrac{2}{\pi}\Sigma \dfrac{1-(-1)^n e^\pi}{n^2+1}\cos nx.$

17. $2\sqrt{2/3}\pi.$

20. $\dfrac{12}{\pi^3}\left(\sin \pi x + \dfrac{1}{2^3}\sin 2\pi x + \dfrac{1}{3^3}\sin 3\pi x + \ldots \dfrac{1}{n^3}\sin n\pi x + \ldots\right).$

21. $\dfrac{2}{\pi}(\sin x - \tfrac{2}{3}\sin 3x + \tfrac{1}{5}\sin 5x + \tfrac{1}{7}\sin 7x \ldots).$

Exercise 24, page 182.

1. $\dfrac{1}{2} + \dfrac{2}{\pi}\left(\sin \dfrac{\pi x}{2} + \dfrac{1}{3}\sin \dfrac{3\pi x}{2} + \dfrac{1}{5}\sin \dfrac{5\pi x}{2} + \ldots\right).$

2. $-1 - \dfrac{4}{\pi}\sum\limits_{1}^{\infty}\dfrac{(-1)^n}{n}\sin n\pi x.$ 3. $-\dfrac{2}{\pi}\Sigma\dfrac{1}{n}\sin 2n\pi x.$

4. (a) $\dfrac{18}{\pi^3}\left[\left(\dfrac{\pi^2}{1} - \dfrac{4}{1^3}\right)\sin \dfrac{\pi x}{3} - \dfrac{\pi^2}{2}\sin \dfrac{2\pi x}{3} + \left(\dfrac{\pi^2}{3} - \dfrac{4}{3^3}\right)\sin \dfrac{3\pi x}{3} - \dfrac{\pi^2}{4}\sin \dfrac{4\pi x}{3}\right.$
$$\left. - \left(\dfrac{\pi^2}{5} - \dfrac{4}{5^3}\right)\sin \dfrac{5\pi x}{3} \ldots\right];$$

    (b) $3 + \dfrac{36}{\pi^2}\sum\limits_{1}^{\infty}\dfrac{(-1)^n}{n^2}\cos \dfrac{n\pi x}{3}.$

5. $\dfrac{8}{\pi^2}\left[\sin \dfrac{\pi x}{l} - \dfrac{1}{3^2}\sin \dfrac{3\pi x}{l} + \dfrac{1}{5^2}\sin \dfrac{5\pi x}{l} \ldots\right].$

Exercise 25, page 186.

1. $y = 0{\cdot}02 + 0{\cdot}64 \cos x - 0{\cdot}88 \cos 2x + 0{\cdot}65 \cos 3x$
          $+ 1{\cdot}76 \sin x - 0{\cdot}46 \sin 2x$

2. $y = 11{\cdot}11 + 9{\cdot}39 \sin \theta + 0{\cdot}98 \sin 2\theta$
          $+ 3{\cdot}4 \cos \theta - 0{\cdot}16 \cos 2\theta.$

3. $x = 1{\cdot}63 - 0{\cdot}04 \cos \theta + 1{\cdot}40 \sin \theta - 0{\cdot}04 \cos 2\theta - 0{\cdot}19 \sin 2\theta.$

4. $x = 6{\cdot}27 - 1{\cdot}99 \cos x - 0{\cdot}67 \cos 2x - 0{\cdot}21 \cos 3x$
          $+ 0{\cdot}71 \sin x - 0{\cdot}69 \sin 2x - 0{\cdot}31 \sin 3x$

Exercise 26, page 195.

1. $n[\tan (x + b) - \tan x]$ metres.   3. $59° 59{\cdot}7'.$

7. (a) $(x^2 + y^2)^{1/2}.$ 11. (a) $x = y = z = \tfrac{1}{2}.$

12. (a) $y - x/t^2(= 0)$;   (b) $[3x^2(1 - 2t^3) + 3y^2 t(2 - t^3)]/(1 + t^3)^2$;
     (c) $yz(e^{2t}\cos t + 2e^{2t}\sin t) + zx(e^{-t}\sec t \tan t - e^{-t}\sec t)$
                             $+ xy(-\operatorname{cosec}^2 t)\ [= e^t]$

13. $kr^2 (\sin \theta \cos \theta + \theta \cos 2\theta).$

14. $(y + z) + (x + z)2t + (y + x)\cos t.$

15. $-3x/y,\ 2x/z.$ 16. (a) $3x(x - y)$; (b) $2(x^4 - y^4)/x^3 y^4.$

17. $2r - t,\ -2s + t,\ -r + s.$ 18. $x(4 - 9vx)/6u(v - u).$

19. $(v - x^2)/u(v - u)$; $(u - x^2)/v(u - v)$; $(1 - vy)/u(v - u)$;
    $(1 - uy)/v(u - v).$

21. $\dfrac{150\sqrt{2}}{16}$ lux/s.

22. (a) $y(y_1{}^3 - a^2x_1) + x(x_1{}^3 - a^2y_1) = 2a^2x_1y_1;$
$y(x_1{}^3 - a^2y_1) - x(y_1{}^3 - a^2x_1) = y_1(x_1{}^3 - a^2y_1) - x_1(y_1{}^3 - a^2x_1).$

32. Each expression is the Jacobian of the transformation from an area on the (T, ϕ) to an area on the (P, V) diagram, or vice versa. Since each is 1, the two areas are equal, *i.e.*, the Carnot cycle has the same area on each diagram.

34. (i) $\dfrac{\partial u}{\partial x} = \cos x \operatorname{ch} y - \sin x \operatorname{sh} y/(2x + 2y + 1);$

$\dfrac{\partial u}{\partial z} = 3(x + y) \sin x \operatorname{sh} y/z(1 + 2x + 2y).$

37. $x^2 + y^2 = 1.$ 38. $x^4 + y^4 = a^4.$

39. $y + tx = 2at + at^3;$ $4(x - 2a)^3 = 27 ay^2.$

40. $(y - x - c)^2 = 4xc.$ 41. $x^2 = -\dfrac{2V^2}{g}\left(y - \dfrac{V^2}{2g}\right).$

Exercise 27, page 203.

3. (ii) Decreasing at $32/21$ m/s. 4. (i) $2u.$ 9. $a = -3.$

10. $\dfrac{\partial V}{\partial t} = -\dfrac{z}{2t}f'(z);$ $\dfrac{\partial^2 V}{\partial x^2} = \dfrac{1}{4t}f''(z).$ 11. $\dfrac{B}{x^2y^2} + C.$

13. (iii) $a^2 + b^2 + c^2 = 0.$

Exercise 28, page 210.

1. $-\frac{1}{25}(18 - j);$ $0\cdot721, 176°\ 49'.$ 2. $(u - 1)^2 + v^2 = 9.$

3. $0 + j3.$ 4. $3 + j, 1 \pm j4.$ 5. $-6 + j9.$ 7. (a) $(7 - j)\sqrt{2}$

8. (a) $2/5, 143°\ 8';$ (b) $\frac{1}{2}(1 + j\sqrt{3}).$

11. (i) R.P. $= \frac{7}{10}$, I.P. $= \frac{9}{10};$ (ii) C, $(1 - i2)$, D. $(3 + i2);$ $0 + i.$

13. (i) $1e^{j5\pi/6};$ (ii) $u = [2(x + 3) + y]/[(x + 3)^2 + y^2],$
$v = [(x + 3) - 2y]/[(x + 3)^2 + y^2].$

15. (a) $\sqrt{2}, \pi/4;$ $\sqrt{2}, 3\pi/4;$ $\sqrt{2}, -\pi/4.$

16. $[b\omega^2 \cos \omega t - \omega(c - a\omega^2) \sin \omega t]/[(c - a\omega^2)^2 + b^2\omega^2].$ 17. $\frac{1}{32}(\frac{10}{3}\pi - \frac{9}{2}\sqrt{3}).$

18. (a) $\sin x \operatorname{ch} y - j \cos x \operatorname{sh} y;$
(b) $\frac{1}{2}(1 + \cos 2x \operatorname{ch} 2y) - j\frac{1}{2} \sin 2x \operatorname{sh} 2y;$
(c) $\operatorname{ch} x \cos y + j \operatorname{sh} x \sin y;$
(d) $(2 \sin x \operatorname{ch} y - j2 \cos x \operatorname{sh} y)/(\operatorname{ch} 2y - \cos 2x);$
(e) $\exp (\sin x \operatorname{ch} y)\{\cos (\cos x \operatorname{sh} y) + j \sin (\cos x \operatorname{sh} y)\}$

19. $2(1 + \cos 2x \operatorname{ch} 2y)/(\cos 2x + \operatorname{ch} 2y).$

22. $5 \exp (j(0\cdot927));$ $1\cdot61 + j(0\cdot927).$

23. $x = a(3 + \cos \theta)(7 + 3 \cos \theta)/2(5 + 3 \cos \theta);$
$y = 3a \sin \theta(1 + \cos \theta)/2(5 + 3 \cos \theta).$

24. $u = -\dfrac{\pi}{3};$ $v = \log 3.$

Exercise 29, page 222.

3. (i) Circles $r = a;$ (ii) lines $\theta = b.$

6. (i) Parabolas $y^2 = 4a^2(a^2 - x);$ (ii) Parabolas $y^2 = 4b^2(x + b^2).$

7. Circle $| w - \frac{17}{3} | = \frac{1}{3}.$

8. (i) Circle: $u^2 + v^2 = e^{2a};$ (ii) line: $u = v \cot b.$

13. $z - ae^{j\frac{n\pi}{3}},$ where $n = 0$ to 5 to give A to F.

14. $(3R, -4R);$ radius $= 5R.$ 15. $r^2 = 2 \cos 2\theta.$

16. (i) $\dfrac{u^2}{4a^2} + \dfrac{v^2}{4b^2} = 1.$ **17.** $x = \dfrac{2v}{u^2 + v^2},\ y = \dfrac{2u}{u^2 + v^2}.$

Parts of the curves in order $u^2 + (v - 1)^2 = 1$, $u + v = 0$, $u^2 + (v - \frac{1}{2})^2 = \frac{1}{4}$, $u - v = 0$, beginning at $(1, 1)$, along the upper half of the first circle to $(- 1, 1)$, down the line $u + v = 0$ to the second circle, along the upper half to the point $(\frac{1}{2}, \frac{1}{2})$ and up the line $u - v = 0$ to the point $(1, 1)$.

19. (a) $0 \cdot 797 - i(0 \cdot 368)$; (b) $\dfrac{1}{r} + r,\ \dfrac{1}{r} - r,\ x^2 - y^2 = 2.$

21. (a) $\exp\{j(2n + 1)\pi/4\}$ where $n = 0$ to 3; (b) $4a$.   **22.** $2(b^2 + a^2)$.

25. w moves $(1, 0)$ to $(2, 0)$ along u axis then along $v^2 = 8 - 4u$ to $(1 + i2)$ and down to $(1, 0)$ along $u = 1.$

26. $\alpha = \frac{1}{2} \log \left(\dfrac{a + b}{a - b} \right).$

27. $u = \{(x - 1) \cos \alpha + y \sin \alpha\} \left\{ 1 + \dfrac{1}{(x - 1)^2 + y^2} \right\}$;

$v = \{- (x - 1) \sin \alpha + y \cos \alpha\} \left\{ 1 - \dfrac{1}{(x - 1)^2 + y^2} \right\}.$

28. $r = 1$; clockwise.   **31.** $e^x,\ y.$

32. $\tan^{-1} \left(\dfrac{a^2 - b^2}{2ab} \right)$, where $\alpha = a + ib.$

33. Given equation where λ is given by
$(z_3 - z_1)(\bar{z}_3 - \bar{z}_2) + \lambda(\bar{z}_3 - \bar{z}_1)(z_3 - z_2) = 0.$

Exercise 30, *page* 227.

1 to 4. Planes.   5. Sphere.

6. Circular cylinder, x axis as axis.   7. Two planes.

8. Cylinder, horizontal cross-section a parabola.

9. Cylinder, horizontal cross-section an ellipse.

10. Cylinder, horizontal cross-section an hyperbola.

11. Line given as intersection of two planes.

12. Circle as intersection of a circular cylinder and a plane.

13. Circle as intersection of sphere and plane.

14. Curve as intersection of parabolic cylinder and plane.

15. Hyperbola as intersection of hyperbolic cylinder and plane.

16. (i) $y = b$; (ii) $x = a$; (iii) $z = c.$

17. (i) $x = a,\ y = b$; (ii) $x = a,\ z = c$; (iii) $y = \dfrac{b}{a} x,\ z = c.$

Exercise 31, *page* 230.

1. $\sqrt{54}.$ 2. $2x - 8y - 4z + 3 = 0.$

3. $3(x^2 + y^2 + z^2) - 8x + 26y - 8z + 39 = 0.$

5. $48° \ 11',\ 131° \ 49',\ 70° \ 28'.$ 7. 7.

10. $(4, \frac{5}{3}, \frac{2}{3}).$ 12. $(\frac{16}{3}, \frac{4}{3}, 6).$ 13. $(\frac{23}{7}, 0, \frac{22}{7}).$

14. $\left(\dfrac{10 + n}{1 + n}, \dfrac{3 + 2n}{1 + n}, \dfrac{7 + n}{1 + n} \right)$, where $n = \pm \sqrt{\frac{93}{11}}.$

Exercise 32, *page* 233.

1. $7, 3\frac{1}{2}, - 2\frac{1}{3}.$ 2. $15, 10, 7\frac{1}{2}$; $187\frac{1}{2}.$

3. (i) $\dfrac{x}{1} + \dfrac{y}{2} - \dfrac{z}{3} = 1$; (ii) $\dfrac{x}{2a} - \dfrac{y}{3a} + \dfrac{z}{a} = 1$; (iii) $x + y + z = 3\sqrt{3}$;
(iv) $x + 2y - 3z = 4\sqrt{14}.$

ANSWERS

4. $x - 2y - z = 0.$ **5.** $x + y = 1.$ **6.** $x + 2y + 3z = 2.$
7. $x + y + z = 3.$ **8.** $3x + y = 0.$ **9.** (i) $(1, 0, 0)$; (ii) $(1, 2, 3).$
10. $-2x + z = 6.$ **11.** (a) $11/\sqrt{29}$; (b) $13/\sqrt{29}$. **12.** $1/\sqrt{3}.$
13. $x + 4y + 7z = 66.$ **15.** $x + y + z = 6$; $\sqrt{3}.$
16. (i) $x + y + \frac{1}{2}z = 10,\ x - 2z = 0$; (ii) $x - 2y - 2z = 1$; $1/3.$
17. 2. **18.** 2.

Exercise 33, page 238.

1. $\cos^{-1}(1/\sqrt{87})$; $\sin^{-1}(7/\sqrt{1102}).$ **2.** $90°.$
3. $\cos^{-1}(3/\sqrt{87}).$ **6.** $(2\frac{1}{2}, 1\frac{5}{6}, -3\frac{1}{8}).$
7. $\dfrac{x - 21/2}{1} = \dfrac{y}{-2} = \dfrac{z - 2}{1}.$ **8.** $\dfrac{x}{-3} = \dfrac{y}{2} = \dfrac{z}{4}.$
9. $\dfrac{x - 1}{1} = \dfrac{y}{3} = \dfrac{z + 2}{-3}$; $\sin^{-1}(14/\sqrt{266}).$ **10.** $(-3, 2, 1).$
12. $3x + 4y + 12z = 78$; $ON = 6$; $(18/13, 24/13, 72/13).$
13. $(1, 2, -3)/\sqrt{14}$; $\sqrt{(27/14)}.$ **14.** $13x - 11y - 5z + 20 = 0.$
15. $(1, 1, -2)/\sqrt{6}$; $x + y - 2z + 5 = 0.$
16. $12x - 4y + 3z = 0$; $\dfrac{x}{-9} = \dfrac{y}{3} = \dfrac{z}{40}.$ **19.** $(2, 1, -3).$
24. $a = b = 1$; $\dfrac{x - \frac{1}{2}}{1} = \dfrac{y + \frac{1}{2}}{1} = \dfrac{z}{2}.$ **25.** $(1, 2, 1)/\sqrt{6}$; $(2, -\frac{1}{2}, 5).$
26. $x + y - z = 0$; $60°.$ **27.** $(1, 1, 6)/\sqrt{38}.$
28. $\dfrac{x - 1}{5} = \dfrac{y - 3}{3} = \dfrac{z}{-4}$; $(-\frac{3}{2}, \frac{3}{2}, 2).$

Exercise 34, page 242.

1. $a/\sqrt{2}.$ **2.** $\frac{3}{7}\sqrt{101}.$ **3.** $2a.$ **4.** $5/\sqrt{6}.$
5. $6/7.$ **6.** $M(d^2 + \frac{1}{3}a^2 \sin^2\theta).$
7. (i) $r_1 = -\dfrac{327}{2009}, r_2 = \dfrac{-62}{287}$; (ii) $(3\cdot68, -0\cdot98, 0\cdot51), (2\cdot57, -0\cdot92, 1\cdot14)$;
   (iii) $d = 1\cdot27$; (iv) $\dfrac{x - 2\cdot57}{1\cdot11} = \dfrac{y + 0\cdot92}{-0\cdot06} = \dfrac{z - 1\cdot14}{-0\cdot63}.$
8. $7/\sqrt{6}.$ **10.** $x - 2y + z = 0$; $\sqrt{\frac{24}{29}}.$
11. $\dfrac{x - 2}{-1} = \dfrac{y + 1}{1} = \dfrac{z + 1}{3}.$ **12.** $5/\sqrt{(1286)}.$
13. $3\sqrt{30}$; $(-2, -5, 1)/\sqrt{30}.$ **14.** $x - 5y + 3z + 1 = 0$; $4/\sqrt{35}.$
15. $90°$; $23x - 13y + 32z - 93 = 0.$ **16.** $-3x + 2z - 1 = 0$; $\frac{1}{4}.$
17. $6(x^2 + y^2 + z^2) - 6xy + 6xz + 6yz = 9a^2$; $\pi\sqrt{3}a^2.$
18. $\dfrac{x}{2} = \dfrac{y}{-1} = \dfrac{z}{-4}$; $\left(\dfrac{2}{3}, -\dfrac{1}{3}, \dfrac{-4}{3}\right).$
19. $9x + 11y - 6z = 13$; $2\cdot195$ sq. units; $(117, 143, -78)/238.$
20. $2x + 3y + 6z = 38$; 7.
21. $\dfrac{x - \dfrac{2ab^2}{b^2 + c^2}}{0} = \dfrac{y}{c} = \dfrac{z}{b}$; $\dfrac{2bc}{\sqrt{(b^2 + c^2)}}.$

Exercise 35, page 245.

1. $x^2 + y^2 + z^2 - 4x + 4y - 2z = 0.$
2. $x^2 + y^2 + z^2 - 2x - 4y - 6z - 7 = 0.$
3. (a) $(2, -4, -5)$; $\sqrt{45}$; (b) $(\frac{1}{2}, 1, -\frac{5}{4})$; $\frac{1}{4}\sqrt{37}.$

4. $x^2 + y^2 + z^2 - 5x - 9y - 8z + 33 = 0$.

6. $x^2 + y^2 + z^2 - 3\frac{1}{4}x - 3y - 4\frac{1}{2}z + 6\frac{1}{4} = 0$.

7. $r + 1$, $2r + 1$, $r - 1$, where $r = \frac{1}{6}(-2 \pm \sqrt{82})$.

8. $1 \pm \sqrt{\frac{5}{7}}$, $2 \pm 2\sqrt{\frac{5}{7}}$, $\pm 3\sqrt{\frac{5}{7}}$.

9. (i) $6x + 3y + 2z = 38$; (ii) $2x - 3y - 6z + 7 = 0$.

12. $\frac{1}{4}\sqrt{29}$; $(-19, -10, 5)/9$.　　　　13. $\frac{400}{9}$ cu. units.　　15. $56\pi/3$ sq. units.

16. $(x + 1)^2 + (y + 2)^2 + (z + 3)^2 = 25$;
$(x - \frac{2}{5})^2 + (y - \frac{4}{5})^2 + (z - \frac{6}{5})^2 = \frac{121}{25}$.

17. $x^2 + y^2 + z^2 - 6x - 4y - 2z + 5 = 0$;
$3(x^2 + y^2 + z^2) - 22x - 8y - 2z + 19 = 0$.

19. Centre $(1, -1, -3)$; radius $= 4$.　　20. Centre $(1, 1, 1)$; radius $= 3$.

22. $x^2 + y^2 + z^2 + 20x + 20y + 18z + 72 = 0$; $x^2 + y^2 + z^2 - 2z - 8 = 0$.

23. $\dfrac{x}{-2} = \dfrac{y}{-2} = \dfrac{z}{3}$; $\cos^{-1}(8/9)$; $(-2, 23, 14)/9$.

Exercise 36, *page* 252.

1. (a) z axis, $45°$; (b) y axis, $60°$.　　2. $x^2 + y^2 = z^2 \tan^2 \alpha$.

3. $3(x^2 + z^2) = y^2$.

4. (a) Cylinder, axis parallel to z axis and through the point $(1, -2, 0)$.　Cross-section by $z = 0$ an ellipse axes $\sqrt{3}$, $\sqrt{12}$.
　　(b) Cylinder, axis parallel to z axis and through the point $(1, -2, 0)$.　Cross-section by $z = 0$ an hyperbola.

5. 24π; 16π.　　　　6. 4π; $4\pi\sqrt{2}$; $(0, 1, -2)$.

7. (i) $\dfrac{1}{a^2}(x^2 + y^2) + \dfrac{z^2}{b^2} = 1$; (ii) $\dfrac{1}{b^2}(x^2 + z^2) + \dfrac{y^2}{a^2} = 1$.

8. $x^2 + y^2 = -\dfrac{2V^2}{g}\left(z - \dfrac{V^2}{2g}\right)$; $\dfrac{2g}{3V^2}\left(\dfrac{V^4}{g^2} - a^2\right)^{3/2}$.

Exercise 37, *page* 254.

1. $2x - y - 2z - 4 = 0$.　　　　2. $x - 2y + 2z = 9$; $\dfrac{x}{1} = \dfrac{y}{-2} = \dfrac{z}{2}$.

4. $(2, -4, 3)$.　　　　5. $\dfrac{a^2b^2c^2}{6x_0y_0z_0}$.

6. $9x - 12y + 2z - 5 = 0$.; $3x + 6y + 4z - 5 = 0$.

9. $x + 2y + z = 6$, $(3, 1, 1)$; $x - y - 2z + 3 = 0$, $(1, 0, 2)$.

10. $x^2 + y^2 + z^2 - 4x - 6y + 2z + 5 = 0$; $x + 2y + 2z - 15 = 0$.

11. $\sqrt{q} \sim \sqrt{p}$.　　12. $axx_1 + byy_1 + czz_1 = 1$; $\dfrac{x - x_1}{ax_1} = \dfrac{y - y_1}{by_1} = \dfrac{z - z_1}{cz_1}$.

13. $2\alpha\beta x + \alpha^2 y - a^2 z = 3\alpha^2\beta - a^2\gamma$.

14. $\beta x + \alpha y - z = a\alpha\beta$; $\dfrac{x - a\alpha}{\beta} = \dfrac{y - a\beta}{\alpha} = \dfrac{z - a\alpha\beta}{-1}$.
　　Lines given by : $y - a\beta = 0$ and tangent plane; $x - a\alpha = 0$ and tangent plane.

17. $(p - lu - mv - nw)^2 = (l^2 + m^2 + n^2)(u^2 + v^2 + w^2 - d)$.
　　$3x + 4y - 86 = 0$,　　　　$2x + 2y + z - 50 = 0$.

19. $y = 2$, $x - y - 2z = 2$.

20. $\frac{1}{4}x + \frac{1}{12}y + \frac{2}{6}z \pm 1 = 0$; $32x^2 + 8xy + 5y^2 = 64$.

22. $(1, 1, 8)$.

23. $x + t^2y - 3tz + at = 0$; $t(x - at) = \dfrac{1}{t}\left(y - \dfrac{a}{t}\right) = -\frac{1}{3}(z - a)$.

25. $x(1 + \lambda)^2 + y(1 - \lambda)^2 + 16z = 4a$.

Exercise 38, *page* 259.

2. (a) $7\frac{1}{2}$; (b) 1; (c) $\frac{1}{12}ba^2(3a^2 + 2b^2)$.   4. 4/3.

5. (i) $\frac{1}{4}\pi a^2$; (ii) $\frac{2}{15}a^5$; (iii) 5/24; (iv) $\frac{1}{2}\pi - 1$.

6. (i) $\frac{1}{3}a^4$; (ii) $\frac{1}{8}a^4$; (iii) $\frac{136}{15}$; (iv) $\frac{32}{3}$.

7. $(\frac{2}{5}a, \frac{2}{5}a)$.   8. $\frac{288}{35}Aa^2$; $\frac{144}{35}Aa^2$ $(A = \frac{16}{3}a^2)$.

9. $\frac{96}{5}\pi a^3$.   12. $2\pi\gamma$.

14. $2 + \frac{25}{2}[\sin^{-1}\frac{4}{5} - \sin^{-1}\frac{3}{5}]$.   15. $\dfrac{P}{32\pi N}\left[6c^2\log\dfrac{c}{a} + 2a^2 - c^2\right]$.

Exercise 39, *page* 265.

2. 36.   4. $\frac{4}{3}\pi(a^2 - b^2)^{3/2}$.   5. $\frac{3}{16}\pi r^3$.   6. $\frac{2}{9}a^3(3\pi - 4)$.

8. $\frac{2}{3}(2\pi + 3\sqrt{3})$; $8M(9\sqrt{3} - \pi)/3(3\sqrt{3} + 2\pi)$.   9. $(\frac{5}{6}a, 0)$.

10. $a^3(3\pi + 20 - 16\sqrt{2})/18$.   11. πa^2 12. $\frac{1}{6}\pi a^3$.

13. $4b^2\sqrt{(a^2 - b^2)} + 4a^2b\sin^{-1}\left(\dfrac{b}{a}\right)$.   16. a.

18. $2\lambda\displaystyle\int_0^a dx\int_0^{2a}(x^2 + y^2)dy$; $\frac{164}{75}Ma^2$.   20. $\dfrac{a^4}{24}(15\pi - 32)$.

21. $32\mu a\, Mg\cos\alpha/9\pi$.   22. $\frac{3}{4}\mu aW$; $\frac{1}{4}\mu aW$.   26. $4\pi^2 ac$.

Exercise 40, *page* 272.

1. (a) 30; (b) 26; (c) $\frac{1}{3}abc(a^2 + b^2)$; (d) $\frac{1}{10}\pi r^5$; (e) $\frac{1}{4}a^4$; (f) $\frac{1}{3}\pi r^3$; (g) 0.

2. $\frac{1}{48}a^6$; Vol. bounded by $x = 0$, $x = a$; $y = 0$, $y = a$; $z = 0$, $z = a - x$.

3. $\frac{1}{2}(\log 2 - \frac{5}{8})$.   4. (a) 6; (b) 45π; (c) $\frac{1}{4}\pi a^4$.

5. $\frac{1}{12}\rho\pi a r^2(3r^2 + 4a^2)$.   6. $\frac{5}{8}h$.

8. $\frac{1}{6}\pi k h^4(\sec^3\alpha - 1)$.   9. $\pi\tau_0 LR^2/4EI$ joules.

10. $\frac{5}{4}\pi a^4 c$.

Exercise 41, *page* 275.

1. $\frac{1}{3}a^4$.   2. $\frac{180}{4}$.   6. $\frac{1}{4}(e^4 - 1)$.   7. $\frac{1}{3}ab$.

8. $\frac{1}{2}\pi - 1$.   9. $\frac{1}{3}a^3\log(1 + \sqrt{2})$.   11. $\pi\log 2$.

13. (a) πa.

Exercise 43, *page* 285.

1. $\frac{1}{2}g\sin\alpha$.   2. $\frac{5}{7}g\sin\alpha$.   3. $\frac{2}{3}g$; $\frac{1}{3}Mg$.

4. $\frac{8}{13}g\sin\alpha$; $\frac{1}{2}gt^2(\sin\alpha - \mu\cos\alpha)$.   8. $\sqrt{(24g/17a)}$.

9. x (along CO) $= Ma\omega^2 - \dfrac{Mg}{5}(8 - 13\cos\theta)$; $y = \frac{1}{3}Mg\sin\theta$; $\omega = \sqrt{(8g/5)}$.

10. $\frac{1}{3}Mg\sqrt{103}$.

Exercise 44, *page* 289.

4. $\dfrac{M_3(a + b)^2 g}{M_3(a + b)^2 + 4M_1(b^2 + k_1^2)}$.

5. (a) $[I_1bQ + I_2a(G - P)]/[I_1b^2 + I_2a^2]$;

   (b) $[b^2(G - P) - abQ]/[I_1b^2 + I_2a^2]$;

   (c) $\pi N^2[I_1b^2 + I_2a^2]/a[b(G - P) - aQ]$.

6. $\ddot{\theta} + \dfrac{\mu}{I}\theta = \dfrac{a}{I}$; $\theta = A\cos\left(t\sqrt{\dfrac{\mu}{I}}\right) + B\sin\left(t\sqrt{\dfrac{\mu}{I}}\right) + \dfrac{a}{\mu}$.

7. $\ddot{x} = T/r[(M + 4m) + 4mk^2/r^2]$;

(a rear wheel) $F_1 = \dfrac{T}{2r} \left[\dfrac{r^2(M + 4m) + 2mk^2}{r^2(M + 4m) + 4mk^2} \right]$ directed forwards.

(a front wheel) $F_2 = \dfrac{T}{r} \left[\dfrac{mk^2}{r^2(M + 4m) + 4mk^2} \right]$ directed backwards.

13. $P(2b^2 + 6ab + a^2)/3Mg(8b^2 + a^2)$.

18. $\omega = \left\{ \dfrac{L_1 I p}{I^2 p^2 + k^2} - \dfrac{L_0}{k} \right\} e^{-kt/I} + \dfrac{L_0}{k} - \dfrac{L_1}{I^2 p^2 + k^2} \{I p \cos pt - k \sin pt\}$.

21. $\sqrt{(3g \sin \theta/a)}$. 22. $\dfrac{m}{25} \{68g - 7a\,\Omega^2\}$.

Exercise 45, page 295.

3. $\dfrac{8}{5} \sqrt{\dfrac{g}{a}}$.

Exercise 46, page 302.

1. $\omega_1 = \tfrac{1}{3}\omega$; $\tfrac{1}{3}mr\omega$. 2. $\dfrac{3u}{8a}$. 3. $4P/m$.

8. $\omega = \omega_1 = \dfrac{6u}{7l}$. 10. $17 : 9$. 11. $\tfrac{8}{7}v$.

14. $\tfrac{1}{4}P$; $\dfrac{3P}{2am}$; $\dfrac{3P^2}{2\mu gam^2}$. 17. $\tfrac{5}{6}V$.

18. $\tfrac{1}{4}m\sqrt{3gl}$. 19. $\dfrac{mMV}{2(M + 3m)}$; $\dfrac{mMV}{M + 3m}$.

20. $Mma^2v^2/2[a^2(M + m) + 3mx^2]$. 22. $\dfrac{J}{M}$; $\dfrac{5J}{8aM}$; $\dfrac{37J^2}{64M}$.

23. For vel. of C.G. v (horiz.) $= \dfrac{J}{m}$; u (vert.) $= a\omega$, where $\omega = 3Jb/2m(4a^2 + b^2)$.

37. $2\pi \left(\dfrac{21a}{5g} \right)^{\frac{1}{2}}$.

Exercise 47, page 313.

2. $AB = 3\tfrac{11}{36}$; $BC = 5\tfrac{13}{16}$. 3. $5° \, 10' \, 13\cdot267''$; $17° \, 4' \, 3\cdot767''$.

4. $x = 283/249$, $y = 552/249$.

5. $\bar{X} = 2\cdot750$, $\bar{Y} = 2\cdot612$; $m = 0\cdot9884$, $c = -0\cdot1062$. 6. $1\cdot94$.

7. $a = 1\cdot20$, $b = 0\cdot33$. 8. $a = 0\cdot40$, $n = 1\cdot70$.

9. $K = 0\cdot000\,896$, $n = 1\cdot589$. 10. $v = 70\cdot2T^{-0\cdot160}$.

11. $y = 3\cdot214x - 1\cdot829$. 12. $u = 100$, $v = 200$.

13. $y = 1\cdot0x^2 + 20\cdot0$. 14. $y = 0\cdot157 + 6\cdot341x - 0\cdot474x^2$.

15. $y = 18\cdot32 + 0\cdot38t + 22x$; 56. 16. $y = 2x + 6\cdot1$.

17. $x^{1\cdot0653}$. $y = 480\cdot84$. 18. $V = 42 + 17\cdot1/A$.

19. $t = 12\cdot27v^{0\cdot625}$. 20. $J = 1\cdot117\dfrac{p}{D} - 0\cdot0264$.

Exercise 48, page 322.

1. $0\cdot89$; yes (but too little data).

3. $r = 0\cdot81$; should be significant, but twelve pairs is too little data.

4. $\bar{x} = 45\cdot7$, $\sigma_x = 9\cdot22$; $\bar{y} = 38\cdot4$, $\sigma_y = 7\cdot38$; $\rho = 0\cdot83$. 5. $p = 3\cdot8$.

8. $a = \sigma_1/\sigma_2$; $b = -\sigma_1/\sigma_2$.

Exercise 49, page 332.

1. $1/3$. 2. $11/36$. 3. $103/2704$.

4. (i) $0\cdot0004$; (ii) $0\cdot0392$; (iii) $0\cdot9604$. 5. $1/6$.

6. $5/108$. 7. $3/34$. 9. $0\cdot369$. 10. $0\cdot98$.

11. 10/21. **12.** 1/221. **13.** 49/57. **14.** 216/625.

15. 0·1037. **16.** 0·271. **17.** 44/125. **18.** (a) 1/15; (b) 7/15.

19. (a) 0·001; (b) 0·027; (c) 0·243; (d) 0·729; 1.

20. (i) 0·045; (ii) 0·002; (iii) 0·135; (iv) 0·865.

21. (i) 0·774; (ii) $10^{-5} \times 2·969$, £95. **22.** $\dfrac{2}{9}$. **23.** 0·082 08.

25. (i) $\left(\dfrac{3}{5}\right)^3 \left(\dfrac{2}{5}\right)^4$; (ii) $_7C_3 \left(\dfrac{2}{5}\right)^3 \left(\dfrac{3}{5}\right)^4$.

26. (i) (a) $\dfrac{12}{19}$, (b) $\dfrac{32}{95}$, (c) $\dfrac{3}{95}$; (ii) (a) $_6C_r \left(\dfrac{3}{5}\right)^r \left(\dfrac{2}{5}\right)^{6-r}$, (b) 4, 0·311.

27. $A = \frac{6}{11}$, $B = \frac{5}{11}$. **30.** £$\frac{1}{6}$. **31.** 2 pence.

32. £$\dfrac{1}{2-x}$. **38.** (i) 7·44; (ii) 0·30.

Exercise 50, page 337.

1. $2z = x\dfrac{\partial z}{\partial x} + y\dfrac{\partial z}{\partial y}$. **2.** $mz = \dfrac{\partial z}{\partial x} - \dfrac{\partial z}{\partial y}$.

3. $4z = \left(\dfrac{\partial z}{\partial x}\right)^2 + \left(\dfrac{\partial z}{\partial y}\right)^2$. **5.** $2\dfrac{\partial z}{\partial x} = \dfrac{\partial z}{\partial y}$.

6. $x\dfrac{\partial z}{\partial x} + y\dfrac{\partial z}{\partial y} = 0$. **7.** $\dfrac{\partial^2 z}{\partial x\partial y} = 0$.

8. $x\dfrac{\partial z}{\partial y} = x\dfrac{\partial z}{\partial x} - z$. **9.** $y\dfrac{\partial z}{\partial x} = x\dfrac{\partial z}{\partial y}$.

10. $\dfrac{\partial^2 z}{\partial x^2} + \dfrac{\partial^2 z}{\partial y^2} = 0$. **11.** $z = x\cos y + y^3$.

12. $z = x^2 y + \frac{1}{3}x^3$.

15. (a) $z = xy(\log y - 1) + yf(x) + g(x)$; (b) $z = e^{x+y} + xf(y) + F(y)$.

Exercise 51, page 357.

1. $z = A\cos p(y - bx)$. **2.** $y = Ae^{-nt}\sin(anx)$.

3. $z = L\sin(px)\sin(pay)$. **4.** $y = Ae^{-m(ax+t)}$.

5. $V = Ae^{-p^2 t}\sin px$, $(lp = \pi, 2\pi, \ldots)$.

6. $y\sin\left(\dfrac{pc}{a}\right) = A\cos(pt + \alpha)\sin\left(\dfrac{px}{a}\right)$.

7. $u = \dfrac{k}{n}\cos\left(\dfrac{nx}{a}\right)\sin(nt)$. **8.** $z = \dfrac{1}{2a}e^{2ax}\sin 2ay$.

9. $u = e^{-y}\sin x$. **10.** $y = \dfrac{Al}{\pi a}\sin\left(\dfrac{\pi x}{l}\right)\sin\left(\dfrac{\pi at}{l}\right)$.

12. (a) $Ae^{-a^2 m^2 y}$; (b) $u = A_n\cos(n\pi x)\,\text{sh}\,(n\pi y)$, $(n = 0, 1, \ldots)$.

16. $V = \frac{1}{12}(x^4 - y^4)$.

17. $Y = A\cos ky + B\sin ky$, $X = Cx^k + Dx^{-k}$; $u = a^3\cos 2y/2x^2$.

19. $y = \dfrac{8l}{\pi^3}\left[\dfrac{1}{1^3}e^{-pt}\sin px + \dfrac{1}{3^3}e^{-3pt}\sin 3px + \ldots\right]$, $(p = \pi/l)$.

20. $y = \dfrac{8a}{\pi}\left[\sin\left(\dfrac{\pi x}{l}\right)\cos\left(\dfrac{\pi ct}{l}\right) - \dfrac{1}{3^2}\sin\left(\dfrac{3\pi x}{l}\right)\cos\left(\dfrac{3\pi ct}{l}\right) + \ldots\right]$.

22. $\theta = \dfrac{200}{\pi}\sum_1^\infty \dfrac{(-1)^{n+1}}{n}\sin\left(\dfrac{n\pi x}{l}\right)\exp(-\kappa n^2\pi^2 t/l^2)$.

33. $V\left[1 - \text{sech}\left(L\sqrt{\dfrac{R}{S}}\right)\right]$.

INDEX

(The numbers refer to pages)

381

382